D1450347

ELEMENTS OF CARTOGRAPHY

ARTHUR ROBINSON
RANDALL SALE
JOEL MORRISON

DEPARTMENT OF GEOGRAPHY
UNIVERSITY OF WISCONSIN
MADISON, WISCONSIN

JOHN WILEY & SONS   NEW YORK   SANTA BARBARA   CHICHESTER   BRISBANE   TORONTO

Fourth Edition
ELEMENTS OF CARTOGRAPHY

Library of Congress Cataloging in Publication Data:

Robinson, Arthur Howard, 1915—
    Elements of cartography.

    Includes bibliographies and index.
    1. Cartography. I. Sale, Randall D., joint author. II. Morrison, Joel L., joint author.
III. Title.
GA105.3.R6 1978   526   78-1670
ISBN 0-471-01781-7

Printed in the United States of America

10 9 8 7 6 5 4

All areas of learning are dynamic, but in some the alterations are more profound than in others. It may be that the greater the technical component of a field, the more rapid will be the rate of its change. At any rate, cartography is in the midst of a revolution. The modifications in the field are dramatic, and they occur in all aspects of cartographic activity, from remote sensing at the compilation stage to color proofing in the map construction stage, to an increased use of feedback from map users. As in other editions, we have tried not to be lured too far from basics by the excitement of innovation, but have tried to integrate modern developments with the fundamentals of this age-old field.

The fourth edition is markedly different for several reasons. A new author has been added; two colors, instead of one, are used throughout; the International System of Units (metric) has been given precedence over the U.S. Customary whenever possible; a four-color insert has been added; but most important, the text has been largely rewritten to take account of the changes in various aspects of the field and the pedagogical coverage they require.

Several rather basic modifications may be noted. The two chapters on map projections and coordinate systems in the third edition have been shortened and more sharply focussed on the basic question about projections that now confronts a cartographer: the choice of a suitable system. Minimal directions for constructing some of the more common projections are placed in an appendix, leaving to the computer, the expert, and appropriate treatises the detailed analysis of this fundamental aspect of the field. The intricacies of remote sensing and the way these new methods can supply vast new quantities of data for mapping are dealt with as part of the basic air photography—photogrammetric material. The significance of measurement systems and the ordering of geographical data is treated in greater depth than before. The inevitable generalization in cartography is elevated to a chapter by itself, and the treatment of symbolization has been enlarged to reflect the basic significance of this communicative aspect of cartography.

The application of the computer and automated methods is included wherever appropriate throughout the book, reflecting the ubiquity of this relatively recent innovation. In addition, a separate chapter on computer-assisted cartography, with a glossary of terms has been added to tie together the variety of existing procedures which all too often seem unrelated to a neophyte.

A systematic approach to graphic design problems and a completely new approach to the analysis of color provides greater depth in these areas. The materials on typography, map reproduction and map construction once again have been brought up to date.

A table of conversions and useful numbers, an expanded table of trigonometric functions, and directions for calculating class intervals have been added to the appendices in this new edition.

It takes a great deal of help to produce a book like this. We were fortunate that Onno Brouwer was able to undertake the extensive primary responsibility to change over to two colors a majority of illustrations, many of which were originally prepared by Professor James J. Flannery of the University of Wisconsin-Milwaukee. He also prepared the dozens of new ones required. In this, Michael Czechanski was also very helpful. Three authors makes for much more than three times as much coordination, maintenance of consistency in preparing a variety of rough drafts and final manuscript and in keeping track of things among authors and between them and the publisher. Mrs. June Bennett did it all and was indispensible.

We owe much to our students, professional colleagues and institutions. During the past nine years many students, who must unfortunately go unnamed, made a variety of proposals for clarification. Professor George Jenks of the University of Kansas at Lawrence read the entire manuscript and made numerous important comments and suggestions. Our colleague, Phillip Muehrcke, helped in many ways. Professor Judy Olson of Boston University contributed especially helpful comments. All of us have been associated with the operation of the University of Wisconsin Cartographic Laboratory, a large versatile map production organization, and have benefited in many ways from the experience. As must always be the case, however, the ultimate responsibility is ours.

Authorship by university professors takes time away from ordinary family activities. Our wives were most understanding and helpful, not only by their patience and encouragement, but also in the tedious checking of the manuscript, tables and proofs.

ARTHUR H. ROBINSON

RANDALL D. SALE

JOEL L. MORRISON

Madison, WI

PREFACE

CONTENTS

ELEMENTS OF CARTOGRAPHY

All creatures presumably have some kind of awareness of their surroundings, and there is abundant evidence that some kind of mental imagery is a normal activity of the brain in at least the higher animals. How this takes place is still a mystery. This imagery ranges from purely pictorial recall such as a remembered face to artificial constructs such as might occur in a fanciful dream. An indispensable component of imagery is the spatial arrangement of objects. Some of this has to do with the geographical space of concern to humans, and their imagery of the spatial organization of observable things seems to be as normal as breathing. It can be simple and elementary, as when one is concerned with basic topological relations, such as inside or outside, near or far, in front or behind, or it can be quite sophisticated when objects are positioned in a conceptualized geometric space. The lower animals, primitive people, and young children probably construct only simple elemental spatial images; educated adults obviously are capable of highly complex constructions.

These mental spatial images are maps in that things are positioned in relative location, but such maps are, of course, unique to each individual. If one person were to describe a mental map to someone else, we may assume that the description would evoke a more or less similar image if all conditions were favorable. It would obviously be much easier, however, to substitute a drawing of the mental map for a word description in order to take advantage of the more efficient method of visual communication. This drawn representation of geographical space is what we commonly and loosely mean by the noun "map."

THE NATURE OF MAPPING

People must have assistance in observing and studying the great variety of phenomena that concerns them. Some things are very tiny, and we must use complex electronic and optical means (e.g., a microscope) to enlarge them so as to understand their configuration and structural relationships. In contrast, geographical things are so extensive that we must somehow reduce them to bring them into view. Cartography is a technique fundamentally concerned with reducing the spatial characteristics of a large area—a portion or all of the earth, or another celestial body—and putting it in a form that makes it observable. Just as

CHAPTER 1

INTRODUCTION TO CARTOGRAPHY AND MAPPING

spoken and written language allows people to communicate beyond the restriction of having to point to everything, a map allows us to extend the normal range of vision, so to speak, and makes it possible for us to see the broader spatial relations that exist over large areas.

A drawn geographical map is much more than a mere reduction. If it is well made, it is a carefully designed instrument for recording, calculating, displaying, analyzing and, in general, understanding the interrelation of things in their spatial relationship. Nevertheless, its most fundamental function is to bring things into view.

Maps range in size from the tiny portrayals that appear on some postage stamps to the enormous murallike wall maps used by civilian and military security groups to keep track of events and forces. They all have one thing in common: to add to the geographical understanding of the viewer. All beings live in a temporal and spatial environment in which everything is related to everything in one way or another. Since classical Greek times, curiosity about the geographical environment or milieu has steadily grown, and ways to represent it in a meaningful way have become more and more specialized. Today there are many different kinds of mapmaking, and the objectives and methods involved seem very different. It is important to realize, however, that all maps have the same basic objective of serving as an interpretation of the geographical milieu; therefore, however dissimilar the maps may seem, the cartographic methods involved are fundamentally alike.

Cartography: A Communication System

In order to transmit and record useful information, humans have developed several methods and the skills necessary to employ them. One method is the use of written language and the ability to deal with what we call literacy. Another is the employment of spoken language, which is called articulacy. Still another has to do with the use of numbers, and the corresponding skill has been called numeracy. Equally important is communicating through graphic devices; the skill of doing so is called graphicacy.*

Graphicacy consists of a variety of techniques ranging from photographs to drawn pictures, graphs, and diagrams. All graphics have one thing in common that distinguishes graphicacy from the other communication skills: the employment of the two dimensions of space to represent concepts and ideas. Spatial relations can be symbolized with words or numbers, but that is quite inefficient as recognized by the maxim "A picture is worth a thousand words." The map employs the two dimensions of geographical space in a systematic fashion, and this puts the skills of mapmaking and map reading in the graphicacy category.

To employ any communication system efficiently, its basic attributes must be understood. Just as we study the use of language, we must study how the map functions and, although the several methods of communication are quite different, they all have a great deal in common.

A typical communications network in its simplest form consists of a source, a channel that conveys the message, and a recipient who receives the message. Practically, the basic system will include an encoder between the source and the signal and a decoder between the signal and the recipient; such a system may be diagrammed as in Fig. 1.1. In everyday terms, using spoken

* W. G. V. Balchin, "Graphicacy," The American Cartographer, 3 (1), 33–38 (1976).

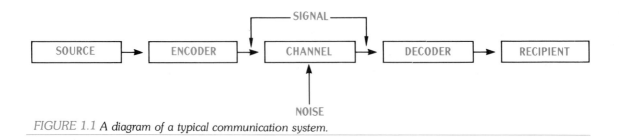

FIGURE 1.1 A diagram of a typical communication system.

language as an illustration, a speaking person may be the source, the voice mechanism will be the encoder, the sound waves generated will be the channel, and the ear-brain capabilities of the listener will be the encoder, transforming the sound waves back into thoughts. A distracting but apparently inevitable part of every communication system is an element generally called noise. Noise is any unwanted attribute in the signal-channel portion of the system that interferes with the efficiency of the transmission, such as static in a radio and snow and distortions on a TV screen.

In the cartographic communication system the real world is the source, the encoding is the symbolism of the map, and the signal is the two-dimensional graphic pattern created by the symbols; the signal consists of the light rays transmitted through the channel space to the decoder, which is the eye-mind mechanism of the recipient (Fig. 1.2). Noise is anything in the signal or channel that interferes with the transmission, such as distracting graphic patterns on the map or poor lighting, which lowers its visibility. Although the various components of the cartographic communication system, such as the encoding, may be quite different from a corresponding element in another system, the fundamental characteristics of all communication systems are similar.

In order to employ any system to the best advantage, the attributes of each component must be understood. Unless a map is prepared so that it is comprehensible to the person for whom it is intended, it will not perform its function. Consequently, a cartographer ideally will know as much about the various mental processes of the intended recipients as he or she will know about the techniques of cartographic symbolism.

The Scope of Cartography

In times past the terms cartography and cartographer referred only to the making of maps.

Even so, the terms were quite inclusive, often encompassing the survey operations involved in acquiring the data to be presented as well as the techniques of actually preparing the map. Since the mid-twentieth century, the scope of the field has been greatly enlarged to include the study of maps as documents. *The Multilingual Dictionary of Technical Terms in Cartography**[*]* defines cartography as follows.

The art, science and technology of making maps, together with their study as scientific documents and works of art. In this context maps may be regarded as including all types of maps, plans, charts, and sections, three-dimensional models and globes representing the Earth or any celestial body at any scale.

This concept of the field encompasses several elements that logically follow from cartography being a subdivision of the communication skill, graphicacy.

In the broad sense cartography now includes any activity in which the preparation and use of maps is a matter of basic interest. This includes teaching the skills of map use, studying the history of cartography, maintaining map collections and the associated cataloging and bibliographic activities, and designing and constructing maps, charts, plans, and atlases. Although each area involves highly specialized activities and often requires particular training, they all deal with maps, and it is the special character of the map as the central unique intellectual object that unites those who work with them. All maps are reduced (scale) rep-

---

* *International Cartographic Association, Commission II, E. Meynen, Chairman, Franz Steiner Verlag GMBH, Wiesbaden, 1973.*

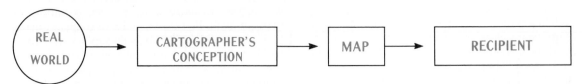

FIGURE 1.2 *The cartographic communication system.*

resentations of the earth or another celestial body, prepared according to a geometric plan, with generalized symbolic representations of reality. Although two maps may be very different, they will have more in common with each other than with any other form of nongraphic communication.

Maps Are Essential

The rapidly growing population of the earth and the increasing complexity of modern life, with its attendant pressures and contentions for available resources, has made necessary detailed studies of the physical and social environment, ranging from population to pollution and from food production to energy resources. The geographer, preeminently, as well as the planner, historian, economist, agriculturalist, geologist, and others working in the basic sciences and engineering, long ago found the map to be an indispensable aid.

A large map of a small region, depicting its land forms, drainage, vegetation, settlement patterns, roads, geology, or a host of other detailed distributions, makes available the knowledge of the relationships necessary to plan and carry on many works intelligently. The ecological complexities of the environment require maps for their study. The building of a road, a house, a flood-control system, or almost any other constructive endeavor requires prior mapping. Smaller maps of larger areas showing things such as flood plain hazards, soil erosion, land use, population character, climates, income, and so on, are indispensable to understanding the problems and potentialities of an area. Maps of the whole earth indicate generalizations and relationships of broad earth patterns with which we may intelligently consider the course of past, present, and future events.

The Definition of Map

Mental maps and their drawn counterparts can range from something highly structured and precise to something very general and impressionistic. Because of the wide variation in the character of these spatial representations, it is not easy to define the term "map" in a way that will make its meaning clear in all contexts. On the other hand, by tradition in the field of cartography, the word "map" has acquired some common constraints that are significant. One is that the representation is dimensionally systematic in that there

is a definable mathematical relationship among the objects shown: things such as distance, directions, or areas are dealt with according to an orderly plan. This relationship between reality and the representation is usually referred to as "scale." How this can be done (e.g., with the curved surface of the earth) will be considered later.

A second limitation is that a map is usually made on a flat surface. A flat medium is convenient for a variety of reasons, not the least of which being that it is easy to handle and to print on. One important exception to this is that the representation on a globe is also called a map, even though it is curved.

A third restriction is that a map can show only a selection of geographical phenomena that have been somehow generalized (simplified, classified, etc.). There is a considerable difference between the selecting done for a drawn map and that of an ordinary air photograph, which simply displays those items that affect the emulsion of a film. This difference is becoming more and more hazy because of modern developments in remote sensing as well as in methods of rectifying the geometry of imagery.

Recently it has become common to convert spatial phenomena to digital form and store the data on tapes or discs. These data can then be manipulated by a computer to supply answers to questions that formerly required a drawn map, such as what percentage of an area lies above a given elevation. This stored geographical information is referred to as a data bank; an interesting question is whether such a geographic data bank is itself a form of map, since it can be used as a map can. The terms "numerical map" and "digital map" have been suggested as appropriate names for this kind of map.

The Cartographic Method

Whatever the kind of map, cartography can be succinctly described as the art, science, and technology of making maps of the earth or other celestial bodies. Because there is such a variety of maps, it might be supposed that the operations in their preparation would be sufficiently varied as to make any generalization not very useful. On the other hand, a map is a unique form of communication; cartography has developed a distinct body of theory and practice that includes a series of processes that are peculiarly cartographic and common to the making of all maps.

All maps are reductions. This means that the first decision of the cartographer must have to do with the dimensional relationship between reality and the map; this is called "scale." The choice of scale is of primary importance, because it sets a limit on the information that can be included in the map and on the degree of reality with which it can be delineated. In addition, matters such as the expected use of the map affect the decision about scale (e.g., is the map to be used in a book, folded in an automobile, handled in the cockpit of an airplane or the chart room of a ship?), as do other factors such as the economical reproduction size and relation with a series of similar maps.

By their very nature all maps are the presentation of spatial relationships. Therefore, another of the cartographer's distinctive tasks is to employ a transformation of the spherical surface that changes it into a plane. Such a radical transformation introduces some unavoidable changes in directions, distances, areas, and shapes from the way they are on the spherical surface. A system of transformation from the spherical to the plane surface is called a map projection, and the choice of a projection must be made for every map. Furthermore, it is often convenient to employ on maps referencing systems called plane coordinate grids; these also involve the problem of map projection. Some cartographers specialize in the field of map projections. The derivation and manipulation of surface transformations requires a particular mathematical competence that is not within the abilities of most cartographers, but all cartographers must be familiar with the characteristics of transformations to employ them appropriately.

Every map, because it is a complex form of communication made for some general or specific purpose, must be carefully planned. Since every map is a reduction and has given objectives, a third major task of cartographers is to generalize. They must simplify where necessary, and plan to make either more or less graphically prominent (according to purpose) the categories of data that they intend to include.

Generalization is one of the most difficult of the cartographer's tasks. If we are to represent a coastline at a reduced scale, we must know the characteristics of that coastline or at least of its type. Similarly, if we are to generalize a river, we must know whether it is a dry-land or a humid-land stream, something of its meanderings, volume, and other important factors that might make it distinctive from or in character with others to be shown. Statistical data must be summarized and manipulated so that the map fairly and effectively conveys the essential character of the distribution. The cartographer's portrayal of the important and the subordination or elimination of the nonessential factors of the map data require that the mapmaker be well acquainted with the subject matter of the maps to be made.

A fourth major task of the cartographer is to design the graphic characteristics of the map. The map must be legible, the symbolism or notation must be suited to the objective of the map, and the whole must be fitted together to make an efficient graphic display. The preparation of a graphic communication, like that of any communication, must be planned carefully so that the receiver will be able to obtain efficiently the information being conveyed. This phase of the cartographic method involves a variety of operations: deciding on the methods of portrayal, choosing lettering sizes and styles, specifying widths of line, selecting colors and shadings, arranging various elements within the map, creating a legend, and the like. Cartographic design is complex but, like writing with words, communication with the graphic symbolism of maps can be learned.

The fifth major aspect of the cartographic method is the actual construction or drawing of the map and its reproduction. Formerly, one simply "drew" a map with pen and ink and then "sent it out" for duplication by a printer. During the past several decades, a major revolution has occurred in this phase of cartography. In some instances we still follow the old practice, but in modern cartography this is becoming more and more rare. The development of scribing, peel coat materials, film manipulation, proofing methods, and other processes has drawn the two aspects of map construction and map reproduction together so closely that it is impossible to think of them separately anymore. The cartographer must be familiar with all the processes involved in order to design his or her map efficiently and have it constructed and reproduced properly and economically.

Cartographic Techniques

The term "drafting" brings to mind pen, ink, T-squares, and the like; but today's cartographer is likely to be working much of the time with scribing instruments, films, opaque, stick-up lettering, and so on. Furthermore, with the rapid development of modern construction methods and automated techniques in cartography, the cartographer is likely to be operating machines such as digitizers, which record on punch cards or magnetic tape the x-y coordinates of the lines and positions on a map, as well as working with special films and photographic equipment.

It is important to understand the relation between the technical operations involved in producing a map and the conceptual aspects of its planning and design. When planning and designing, a cartographer is rather like an author or designer who combines the attributes of a planner, architect and engineer, while actual construction of the map calls for the attributes of a skilled craftsman. Many cartographers, who are themselves talented craftsmen, feel that skill in some particular technical aspect is indispensable to the cartographer. Such skills are desirable and useful, but they are by no means indispensable. A lack of a particular technical skill need deter no one from entering the field of cartography and, particularly, it should not deter any physical and social scientists from learning the principles of graphic expression. Modern cartography is a complex of technical and conceptual operations that are sufficiently interdependent that it is desirable that all who are involved have a broad understanding of all aspects of the field.

Science and Art in Cartography

Cartography has been described often as a meeting place of science and art. There seems no question that as we learn more and more about communication that more of the principles and precepts of cartography are being based on understanding and less on individual aesthetic intuition.

The cartographic process involves many intellectual activities that are scientifically based. We must apply logic in our approach to projections, generalizations, line characterizations, and so on; in this respect cartographers are kinds of practical scientists, much like engineers. They must study the characteristics of their building materials and know the ways and means of fitting them together so that the end product will convey the correct intellectual meaning to the reader.

Of equal, and sometimes greater, importance are the visual relationships inherent in this form of expression. People, or map readers, think and react in certain ways to visual stimuli. With knowledge of the principles and laws governing these reactions, cartographers can design the product to fit these perceptual patterns. The selection of lettering sizes and styles, circle sizes for representing statistics, and the colors and tones to be used in representing gradations of amount are examples of questions to be answered by reference to the principles of visual perception as applied to symbols. They constitute problems of visual logic.

Cartographers are scientific in other ways. One of the largest categories of information with which they work is that contained in other maps; since they can hardly have first-hand familiarity with all places in the world, they must be able to evaluate their source materials. They must have good geographical backgrounds and must be familiar with the state of topographic mapping and its geodetic foundations. To supplement map information, they must be able to evaluate and rectify census data, air photographs, documentary materials, and many other kinds of sources.

Cartography is not an experimental science in the sense that chemistry or physics is, nor is it searching for truth in the manner of the social sciences. Nevertheless, it employs the scientific method in the form of reason and logic in constructing its products. Its principles are derived through the analysis of scientific data. It has its foundations in the sciences of geodesy, geography, and psychology. In that it is based on sound principles and seeks to accomplish its ends through intellectual and visual logic, it is scientific in nature.

Before the last century the question of whether cartography was an art never arose, for it very definitely was. This is evident when we view the products of the earlier cartographers, who embellished their maps with all sorts of imaginative things such as fancy scrollwork, ornate lettering, and intricate compass roses. Special coloring methods and ingredients were carefully guarded secrets. Even as late as the middle of the nineteenth century the coloring of maps in one of Germany's greatest map houses was done by the society ladies of the town. Throughout the history of cartography, great emphasis has been laid on fine pen and brush skill; the aim has been to make something good to look at, and perhaps good enough to hang on a wall as a decoration. Now the concern is less for aesthetic appeal than it is for the efficiency with which the communicative objective is attained.

Maps today are strongly functional in that they are designed, like a bridge or a house, for a purpose. Their primary purpose is to convey information or to "get across" a geographical concept or relationship; it is not to serve as an adornment for a wall. On the other hand, one of the cartographer's concerns may be to keep from producing an ugly map; in this respect the cartographer is definitely an artist, albeit in a somewhat negative sense. Cartography is certainly a creative art in the way that careful, creative literary expression is an art.

The mapmaker is essentially a faithful recorder of given facts, and geographical integrity cannot be compromised to any great degree. Nevertheless, the range of creativity through scale, generalization, and graphic manipulation available to the cartographer is comparatively great. Just as some realistically painted things "are deadly mechanical records, so some faithful maps are alive while others leave us untouched."[*]

THE CLASSES OF MAPS

Because there are so many different kinds of maps, it is useful to classify them. As with many groups of diverse objects, no one approach is perfectly satisfactory. In order to provide a better basis for the appreciation of the similarities and differences among maps and cartographers, we will look at maps from three points of view: (1) their scale, (2) their communication objective, and (3) their subject matter and function.

Scale

All maps are representations of reality. The dimensions of reality must necessarily be reduced to

[*]*Rudolf Arnheim, "The Perception of Maps,"* The American Cartographer, 3 (1), 5–10 (1976).

the proportion that will accomplish the objectives and serve the function of the map. The proportion or ratio between the map dimensions and those of reality is called the map scale, and the various ways map scale can be stated and portrayed are treated in Chapter 3. Here it is only necessary to point out that the ratio between the size of a map and the size of the area it represents can range from very small to very large. When a small sheet is used to show a large area (e.g., a map of the United States, or even the world on a sheet the size of this page), that map is described as being a *small-scale* map. If a map the size of this page showed only a small part of reality (e.g., less than 1 km^2), it would be described as a *large-scale* map.

The terms large and small when combined with scale are related to the relative sizes at which objects are represented, not to the amount of reduction involved. Accordingly, when comparatively little reduction is involved and things such as buildings and dimensions are shown with considerable magnitude the map is termed a large-scale map. Such would be the case, for example, if everything on a map were reduced to one five-thousandth of its true size, in which case a road 20 m (65 ft) wide actually could be shown in correct proportion by a symbol 4 mm (0.16 in) wide. When great reduction has been employed, as for a small-scale map, most linear objects and other small features on the earth cannot be shown at a size proportional to the amount of reduction, but must be greatly magnified and symbolized to be seen at all. One may conclude that reality must be portrayed selectively and with considerable simplification on small-scale maps. On the other hand, although selection is often involved in large-scale maps, such maps can portray many aspects of reality in the actual proportion of the amount of reduction employed.

There are other differences between large- and small-scale maps, such as the amount of various kinds of distortion that is likely, within the map sheet, but the major difference is in the degree to which reality can be faithfully delineated in proportion to the reduction. When reality must be dealt with in simplified fashion, the process is called *cartographic generalization*. It is a complex subject and will be treated explicitly later. It is important now to remember that large-scale maps are usually not much generalized, but small-scale maps must always be.

There is no general consensus on the quantitative limits of the terms small, medium, and large scale, and there is no reason why there should be, since the terms are hardly more than relative. Most cartographers would argue, however, that a map with a reduction ratio of 1 to 50,000 or less (e.g., 25,000) would be a large-scale map. Maps involving ratios of reductions of 1 to 500,000 or more (e.g., 1,000,000) would probably be called small-scale maps.

Communication Objective

As we have just seen, the range from large scale to small scale is a continuum with no clear divisions separating the classes of map scale. Similarly, if we try to divide maps into classes based on their communication objectives, we find a great difference between the extremes, but the transition along the range from one class to the other is a gradual instead of an abrupt change.

The various geographical phenomena of the real world that may be represented on maps are almost infinite, and all we can say is that anything that can be observed (measured or identified) can be mapped in its spatial position. When we contemplate the diversity of the geographical environment, we can do so in two very different ways, which will be reflected in the kinds of maps we make. The two classes of maps are called *general* maps and *thematic* maps.

General Maps. General maps are those in which the objective is to portray the *spatial association* of a selection of diverse geographical phenomena. Things such as roads, settlements, boundaries, water courses, elevations, coastlines, and bodies of water are typically chosen to be portrayed on general maps. General maps that are also large-scale maps are usually called topographic maps (Fig. 1.3). They are issued in series of individual sheets and are very carefully made, usually with photogrammetric methods, by national or other public agencies. Maps of much larger scale are required for site location and other engineering purposes, and they employ only ground survey methods. Great attention is paid to their accuracy in terms of positional relationships

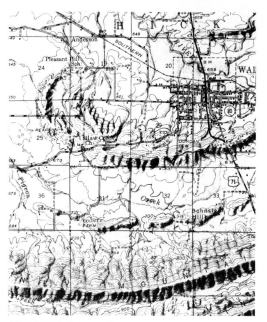

FIGURE 1.3 *A small part of a modern contoured topographic map. The landform is emphasized by colored shading. This monochrome reproduction, reduced from the original scale, cannot do justice to the excellence of the printed map in colors. (Waldron Quadrangle, Arkansas, 1:62,500, U.S. Geological Survey.)*

of the items mapped. In many cases they have the validity of legal documents and are the basis for boundary determination, assessing taxes, transfers of ownership, and other such functions that require great precision. In the United States and other countries, official map accuracy standards have been established for such general, large-scale maps. The map accuracy standards apply only to the metrical horizontal and vertical qualities of the maps, not to nonmetrical aspects such as labeling blunders, incompleteness, or being out of date.

The National Map Accuracy Standards in the United States may be summarized as follows. They are stated in English units.

1. **Horizontal accuracy.** For maps published at scales larger than 1:20,000 (approximately 3 in. or more to 1 mi) not more than 10% of well-defined tested points shall be in error by more than $1/30$ in. For maps published at scales

smaller than 1:20,000 (less than 3 in. to 1 mi), the error limit shall be ¹/₅₀ in.

2. **Vertical accuracy.** For contour maps at all publication scales not more than 10% of the elevations tested shall be in error by more than one-half the contour interval.

These standards, adopted in the 1940s, are applicable only to a limited range of large-scale maps; clearly, they cannot reasonably apply to very large-scale maps, such as a site survey map of 1 in. to 100 ft, in which the above horizontal accuracy standard would allow distance errors of more than 3 ft. Considerable discussion has been going on for years regarding the desirability of adopting a statistical approach in which the limits would be stated as an allowable standard error (or root mean square error).[*] The standard error, d, is defined as

$$d = \frac{\sqrt{\Sigma e^2}}{n}$$

in which e is error and n is the number of observa-

[*] *"Symposium on the National Map Accuracy Standards,"* Surveying and Mapping, 20 (4), 427–457 (1960), and Morris M. Thompson and George H. Rosenfield, *"On Map Accuracy Specifications,"* Surveying and Mapping, 31 (1), 57–64 (1971).

FIGURE 1.4 *A portion of a small-scale general map. (Base map of Wisconsin, 1:500,000, U.S. Geological Survey.)*

tions and they would be stated in linear terms for various map scales.

Small-scale general maps are typified by the maps of states, countries, and continents that are included in atlases (Fig. 1.4). These maps portray the association of phenomena similar to those shown on the large-scale general maps but, because they are small scale and the symbolization and representation must be much generalized, they cannot attain the standard of positional precision striven for on the large-scale maps.

Thematic Maps. Thematic maps are quite different from general maps. General maps attempt to portray the positional relationships of a variety of different geographical phenomena on one map; thematic maps concentrate on the spatial variations of a single phenomenon or the relationship between phenomena. In thematic maps the communication objective is to portray the *structure* of a distribution, that is, the character of the whole as consisting of the interrelation of the parts. There is no limit to the subject matter of thematic maps. They are typified by maps of average annual precipitation or temperatures, populations, atmospheric pressure, average annual income, grain production, and the land form (Fig. 1.5).

Just because a map deals largely with a single

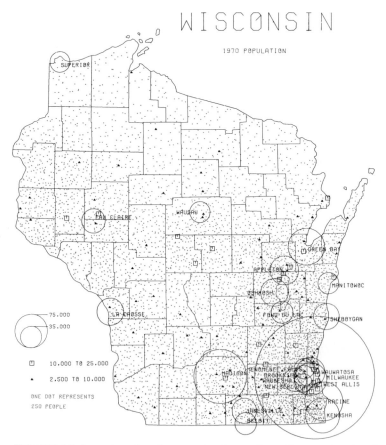

FIGURE 1.5 *A much reduced, computer-drawn thematic map showing the geographical structure of Wisconsin's 1970 population. Rural population is shown by dots and urban populations by distinctive symbols and graduated circles. (University of Wisconsin Cartographic Laboratory. From Proceedings, ACSM Fall Convention, 1972.)*

class of phenomenon does not necessarily mean that it is a thematic map. Maps showing the diversity of soils, bedrock geology, or population density can be properly classified as general maps if the primary objectives are simply to show the locations of types of soils, rock, or population density at particular places. On the other hand, maps made from the same data may employ methods of symbolization that focus attention on the structure of the distribution, and they would then be properly called thematic maps.

Thematic maps are commonly small-scale maps largely because many geographical distributions occur over considerable areas, and to portray their essential structure requires great reduction. Furthermore, it is easier for the map user to focus on the structural relationships of the distribution if the variations may be perceived without much shifting of the eyes or turning of the head. Nevertheless, this tends to be relative; when the

area of interest is a city, for example, maps intended to show the structure of individual phenomena may be of relatively large scale.

Accuracy in thematic mapping is less a matter of concern for the precision of individual map positions at small scales than it is for the truthfulness of the portrayal of the basic structural character of the distribution. For example, statistical data for individual locations are often biased by local influences, such as incorrect estimates by respondents, a weather station being in a valley, or unusual employment conditions because of a local business failure. Such aberrations may be quite atypical and to include them in a portrayal of the overall character of a geographical distribution may seriously bias the conceptions of a map user who, not knowing the data problems, could not make allowances for them.

Thematic maps often also require considerable data processing in order that the mapped in-

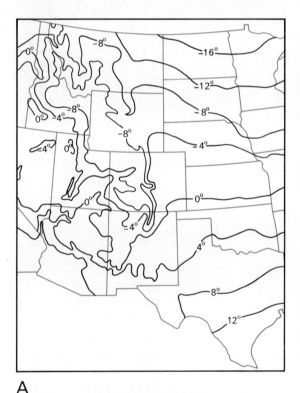

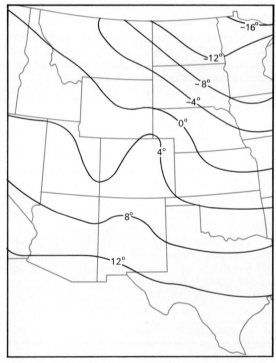

A   B

FIGURE 1.6 (A) Average January surface temperatures (celsius). (After Atlas of American Agriculture.) (B) Average January sea-level equivalent temperatures. (After Handbuch der Klimatologie.)

formation will reveal the structural characteristics intended by the cartographer. For example, suppose that a cartographer is interested in portraying the systematic effects of maritime and continental air masses on temperatures in the central United States in January and wished to do so with isotherms. If average January surface temperatures at reporting stations in the Midwest-Rocky Mountain states are mapped, as in Fig. 1. 6A, the map will not be very revealing, because the great variations in altitude naturally affect the temperatures. Figure 1.6A could be classed as being partly general and partly thematic. It is partly general in that it does show the locations of average January temperatures; it is partly thematic because the symbolization (isotherms) shows the structural relationships better than if only the temperature numbers had been plotted at the locations of the reporting stations.

When each temperature is processed by calculating its sea-level equivalent [surface temperature plus (temperature lapse rate × elevation)] and these values are mapped with isotherms, as in Fig. 1.6B, the general equatorward deflection of the isotherms is clearly evident. Figure 1.6B is definitely a thematic map because it portrays the basic structure of, in this case, a rather complex geographical relationship.

Subject Matter and Function

The variety of geographical phenomena and the myriad uses to which maps may be put combine to cause an enormous variety. Although they may all be classed as large scale or small scale and placed somewhere in the continuum from general to thematic, it is useful also to group maps on the basis of their subject matter and function. Several important categories may be recognized.

It is probable that among the earliest "permanent" maps were drawings to accompany the official register or list of property owners and their landholdings; these drawings were called cadastres. These cadastral maps showed the geographical relationships among the various parcels. They are common today, and they record property boundaries in much the way they did several thousand years ago (Fig. 1.7). A principal use of cadastral maps is to provide a basis on which to assess taxes—which may account in part for the fact that such maps have always been with us.

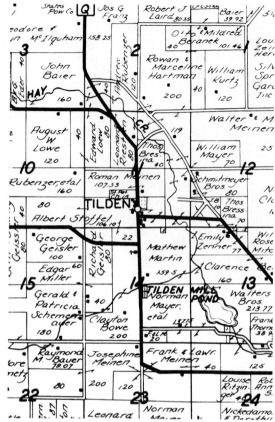

FIGURE 1.7 A section of a cadastral map showing ownership of land. The bold numbers are section numbers and the heavy lines are paved roads. (From the Chippewa County, Wisconsin Plat Book, 1966, courtesy Rockford Map Publishers.)

Closely allied to cadastral maps, but more general in nature, is a category of large-scale general maps called plans. These are detailed maps showing buildings, roadways, boundary lines visible on the ground, and administrative boundaries (Fig. 1.8). Plans of urban areas are likely to be very large scale. In countries that have a well-integrated mapping program, such as that of the Ordnance Survey of the United Kingdom, the very large-scale plans form the basis for the topographic map series.

A large group of maps designed to aid the navigator is given the general name of chart. There are various kinds of charts, such as those

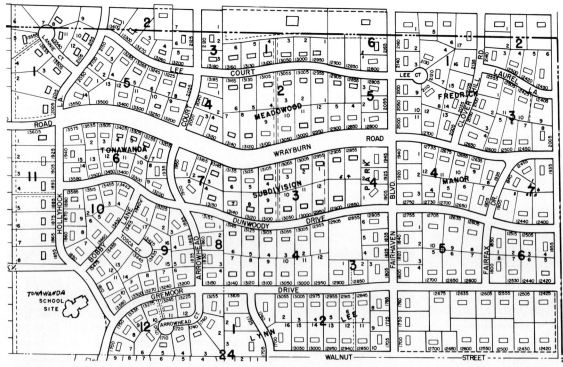

FIGURE 1.8 *A portion of a large-scale plan.* (*Village of Elm Grove, Wisconsin.*)

used by the mariner at sea, called nautical charts, and those used by the navigator of aircraft, called aeronautical charts. Because they are used for plotting positions and establishing bearings and courses, charts must be prepared to a high level of accuracy and must be frequently revised so as to be as nearly up to date as practicable. Charts range in scale from large-scale, such as air approach and mariner's harbor charts, to small-scale planning charts. Another class of maps, closely allied to charts in function, is the road map. Travelers on land ordinarily do not need to plot positions and calculate courses as do navigators on the sea and in the air, but they do need guidance as to which roads go where, how distant places are, and what the qualities of the various roads are. Road maps are very old but, with the advent of the automobile, road maps and road atlases became common. For many years road maps were free in the United States, and over 150 million were produced annually. They reached an advanced state

of quality, and their design was perhaps more closely tied to their function than any other class of map.

There is no limit to the number of classes of maps that can be created by grouping them according to their dominant subject matter. Thus, there are soil maps, geological maps, climate maps, population maps, transportation maps, economic maps, statistical maps, and so on without end. Such categories are useful only in that any one such class will have numerous similarities in the cartographic treatment of the substantive material and the associated problems. It is a mistake, however, to think that all such maps are alike. There is likely to be more difference in the cartography between a large-scale map of surface bedrock in an area and a small-scale map of plate tectonics in the Atlantic basin—both geological maps—than there would be between a soil map and a vegetation map of a given area. Cartography is independent of subject matter.

SELECTED REFERENCES

Balchin, W. G. V., "Graphicacy," *The American Cartographer, 3,* 33–38 (1976).

Board, C., "Maps as Models," in R. J. Chorley and P. Haggett (eds.), *Models in Geography,* Methuen and Co. Ltd., London, 1967, pp. 671–725.

Board, C., "Cartographic Communication and Standardization," *International Yearbook of Cartography, 13,* 229–236 (1973).

Greenhood, D., *Mapping,* University of Chicago Press, Chicago, 1971.

Keates, J. S., "Symbols and Meaning in Topographic Maps," *International Yearbook of Cartography, 12,* 168–181 (1972).

Kishimoto, H., *Cartometric Measurements,* published by the author, Zürich, 1968.

Koláčný, A., "Cartographic Information—A Fundamental Term in Modern Cartography," *The Cartographic Journal, 6,* 47–49 (1969).

Monkhouse, F. J., and H. R. Wilkinson, *Maps and Diagrams: Their Compilation and Construction,* 3rd ed., Methuen and Co. Ltd., London, 1971 (University Paperbacks).

Morrison, J. L., "Changing Philosophical-Technical Aspects of Cartography," *The American Cartographer, 1,* 5–14 (1974).

Morrison, J. L., "The Science of Cartography and Its Essential Processes," *International Yearbook of Cartography, 16,* 84–97 (1976).

Muehrcke, P., "Trends in Cartography," in P. Bacon, *Focus on Geography, Key Concepts and Teaching Strategies,* National Council for the Social Studies, Washington, D.C., 1970, pp. 197–225.

Robinson, A. H., and B. B. Petchenik, *The Nature of Maps, Essays toward Understanding Maps and Mapping,* University of Chicago Press, Chicago, 1976.

Thompson, M. M., and G. H. Rosenfield, "On Map Accuracy Specifications," *Surveying and Mapping, 31,* 57–64 (1971).

Thrower, N. J. W., and J. R. Jensen, "The Orthophoto and Orthophotomap: Characteristics, Development and Application," *The American Cartographer, 3,* 39–56 (1976).

Bibliographies of Cartography

U.S. Library of Congress, Geography and Map Division, *The Bibliography of Cartography,* G. K. Hall and Company, Boston, 1973 (5 vols.).

Institut für Landeskunde and Deutsche Gesellschaft für Kartographie, *Bibliotecha Cartographica,* Bad Godesberg, Federal Republic of Germany.

THE HISTORY OF MAPMAKING

The art and science of cartography has no doubt been practiced for at least as long as there has been communication by written language. The history of mapmaking is a record not only of the human struggle to understand the environment, from the immediate surroundings to the whole earth, but it also reflects people's attitudes, beliefs, and priorities at various times. A general awareness of the changes and developments during the long history of cartography provides the cartographer with an understanding of the roots of the métier, but perhaps even more important, it provides a background that serves properly to emphasize the profound developments of the present day.

The history of cartography is a field of study to which a great many scholars have devoted their lives. It has a very large literature, ranging from the very scholarly to the popular. In our brief review we can only touch the high points, and the interested student can refer to the bibliography for additional reading.

No one knows when the first cartographer prepared the first map, which was no doubt a crude representation of locations drawn in soft earth or sand or scratched on a rock. Certainly the spatial concern was relatively narrow. Perhaps the oldest authentic map that survives is a clay tablet nearly 5000 years old showing mountains, water bodies, and other geographic features in Mesopotamia. It is thought that portions of the valley of the Nile were carefully mapped in ancient times in order to recover property lines after the annual floods. Figure 2.1 is an early, large-scale Mesopotamian city plan.

The maps of primitive peoples of recent times have been studied, and the evidence suggests that mapmaking is a common and well-developed skill, albeit with major differences from the maps made by more advanced groups. All peoples seem to develop an appreciation of the spatial distribution of topographic phenomena and employ whatever materials are at hand to represent vital matters such as routes and hunting grounds. Eskimos use driftwood, pebbles, and bone; South Sea islanders use reeds, shells, and leaves; and

CHAPTER 2
CARTOGRAPHY PAST AND PRESENT

FIGURE 2.1 *An ancient plan (map) of a city in Mesopotamia on a clay tablet. (Courtesy of Norman J. W. Thrower.)*

Indians use birch bark and sand (Fig. 2.2). Although it is difficult to generalize from the meager data, it appears that primitive peoples are especially conscious of topological relationships (relative positions, etc.) and employ a different system of distance scaling than that normally used in most modern mapping. Distances are often thought of in terms of relative travel time, regardless of the actual earth space involved. Both this method of scaling and their topological concerns are eminently practical.

When people began to observe the earth from a scientific-philosophical point of view, they began to view things differently, and their maps reflect the change. This occurred both in the Western and Eastern worlds.

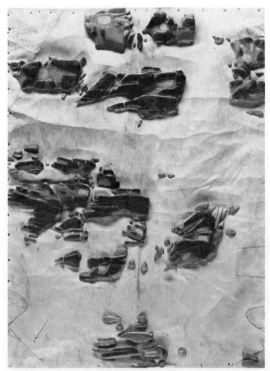

FIGURE 2.2 *A primitive map, carved from drift-wood, colored and tied with gut to sealskin, showing islands, reefs, lakes, swampy ground, tidal areas, and so on, in an area of 70 mi. ² Made by an Eskimo who had never seen any other map. (From the Library of Congress collection.)*

The Beginnings of Western Cartography

The jump from a concern with the local area to the contemplation of the character of other lands or even the nature of the whole earth is very large. No one knows when this took place in the Western world; but we do know that by the time of Aristotle (384–322 B.C.), the earth was recognized as being spherical from evidence such as differences in the altitudes of stars at different places, from the fact that shorelines and ships seemed to "come over" the horizon as one moved across the sea, and even from the assumption that the sphere was the most perfect form.

By the second century B.C. the system of describing positions on the earth by latitude and longitude and the division of the circle into 360° had become well established. The development

of civilization allowed more frequent travel, and a greater interest in faraway places was accompanied by increased thought about ways of presenting the relationships of areas on maps. Estimates of the size of the earth were made by the ancient scholars Eratosthenes (ca. 276–195 B.C.) and Posidonius (ca. 130–50 B.C.) from angular observations on the sun and stars in the eastern Mediterranean area (see Fig. 3.1). The methods used were entirely correct, but the required assumptions and the precision of their observations were not. Nevertheless, although we cannot be absolutely sure of their results in terms of modern units of measure, the estimates seem not to have been very far off. Both apparently overestimated the size by only 12 to 15%.

From the occasional references to maps in the classical Greek literature we may infer that mapping was not considered to be anything unusual. Consequently, although we may assume that many maps were made during the time of the classical Greeks, *none* of the actual maps appears to have survived. Fortunately for our understanding of the cartography of the ancient period, there is, however, a record that apparently clearly reflects the advanced stage to which cartography had developed by the end of the Greek period; this record is the writings of Claudius Ptolemy (ca. 90–160 A.D.).

In the ancient Carthaginian trading city of Tyre on the southern coast of present Lebanon, a cartographer-geographer, Marinus, a contemporary of Ptolemy, wrote and made maps. Unfortunately, Marinus' materials are lost, but Ptolemy set out to correct Marinus and, by good fortune, what is thought to be Ptolemy's writings have survived. Ptolemy lived and worked in Alexandria, Egypt, which had become the intellectual center of the Western world. There was a great library at Alexandria as well as a community of scholars. Ptolemy brought together into a book, called simply *Geography,* all that was apparently then known concerning the earth. The *Geography* included, among other things, a treatise on cartography.

He described how maps should be made; he gave directions for dealing with the problem of presenting the spherical surface of the earth on a flat sheet; he recognized the inevitability of deformation in the process; and, in many other ways, he described what was known about cartography at that time. Ptolemy also made a list of numerous places in the world he had learned about from the writings of others and from travelers, and he gave his best estimate of their positions in latitude and longitude terms. To illustrate this he is thought to have provided a series of maps of local and larger areas (Fig. 2.3); this is not certain, however, because if he did, they have since been lost. He did give a detailed account of how they were made. Ptolemy's writings were lost to the Western world for more than 1000 years, but fortunately they were preserved in the Arab world and later came to light again in Europe. From his descriptions the Ptolemaic maps were reconstructed, and they had a profound influence on European geographical and cartographical thinking during the Renaissance.

It is worth observing here that Ptolemy was a compiler of maps, he was not a surveyor. Although there is no clear distinction even today, a compiler obtains data to be mapped from a variety of sources, ranging from survey maps and official records to the resources of libraries; the topographic surveyor obtains data from field measurements and air photographs. The compiler, like Ptolemy, usually is concerned with preparing small-scale maps of large areas, while the surveyor works toward producing large-scale maps of small areas. The skills, methods, and problems involved are quite different.

The Medieval Period
The Greeks and Romans apparently were technologically primitive and the classical period is "classical" only in philosophical and intellectual

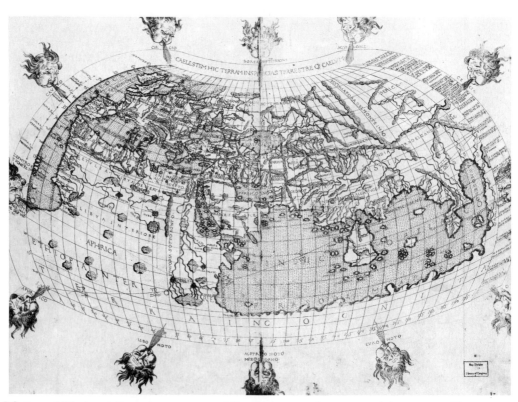

FIGURE 2.3 One of "Ptolemy's" maps as constructed from his written directions and descriptions. The area shown extends from the Atlantic on the west to beyond the closed Indian Ocean. This world map reflecting the knowledge of the second century A.D. was better than any other when Ptolemy's works were reintroduced to Europe in the fifteenth century. (From the Library of Congress collection.)

matters. So far as we can tell, the writings of Marinus and Ptolemy's *Geography* were unique, and there is no reference to any other such activity for many centuries. The Romans did survey, and there are references in Roman literature to administrative and engineering types of maps but, as the Roman Empire declined, concern with the general theory and practice of cartography all but died out.

Ptolemy's concern in the second century A.D. with projections and systematic ways of making maps to show the actual geography was followed in succeeding centuries by a quite different attitude. The map became a didactic device to illustrate scriptural theory about the nature of the earth, and objective geographical thinking about faraway places was replaced by fancy and whimsy. Maps of the "known world" (called *mappa mundae* in Latin) were produced but, whereas during the Greek period these had been based on observation and reason, they now came to be but media for preserving the results of fanciful speculation and literal interpretations of biblical passages. Quite a variety were prepared containing various symbolic representations of the earth, such as rectangular maps probably based on the scriptural reference to the "four corners of the earth." But even more common were circular maps (also probably because of a biblical reference), with the holy city of Jerusalem at the center. These maps reflected the threefold division of the earth among Shem, Ham, and Japheth, the sons of Noah. They are called T in O maps, because they were designed with the Mediterranean as the upright part of the T, the Don and Nile rivers as the crosspiece, and the whole inside a circular ocean (Fig. 2.4).

The farthest area (known at all) was, of course, the Orient. It became traditional to locate Paradise in the difficult to reach, far eastern area and to put it at the top of the map. From this practice we have derived the term *to orient* a map—that is, to turn it so that the directions indicated are understood by the reader—but today to orient a map means to arrange it so that map and earth directions correspond or, alternatively, so that north is at the top.

On these maps often were placed mythical places, beasts, and dangers such as the kingdom of the legendary Gog and Magog, who were nonbelieving menaces to the Christian world. This

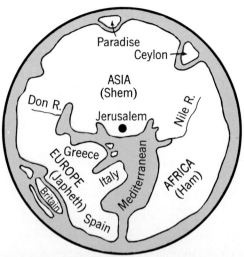

FIGURE 2.4 *The layout of the T in O maps. Jerusalem was usually located at the center and Paradise in the top central section; the term "to orient" a map comes from this. The known world was surrounded by the river Oceanus. Compare with Fig. 2.5.*

kind of cartography hung on for a long time, and some of the later maps that follow this general pattern are very detailed and ornate. One of them is the map from the Hereford Cathedral in England (Fig. 2.5).

The earlier part of the medieval period in the Western world is called the Dark Ages, and it was not until after the first millenium of the Christian era that enlightment began to show in the field of cartography and geography. Three advances can be noted: (1) the contributions of the Arabs, (2) the growth of interest in faraway places, and (3) the remarkable development of portolan charts.

The Arabs had preserved the writings of Ptolemy, had translated them, and had made several attempts to measure the size of the earth in the ninth century A.D. The culmination of Arab cartography was reached in the work of Idrisi (1099–1164 A.D.), an Arab scholar at the court of King Roger of Sicily. His world maps, made in 1154 and 1161 A.D., were a great advance over contemporary non-Arab works (Fig. 2.6).

The first few centuries of the second millenium A.D. saw a change taking place in intellectual standards that was reflected in the cartog-

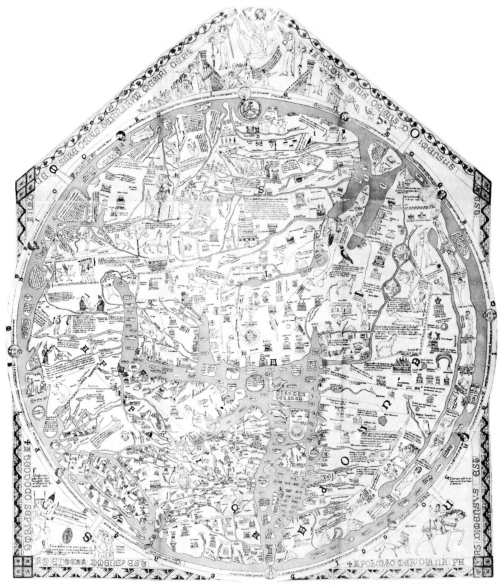

FIGURE 2.5 *The Hereford world map. This map, more than 3 m wide, made in the thirteenth century, illustrates the degree to which cartography had degenerated from the time of Ptolemy, 1000 years earlier. The map is oriented with east at the top and Jerusalem at the center. (From the Library of Congress collection.)*

raphy. The transformation began slowly; but the several Crusades, the travels of missionaries and merchants, such as the Polos, and the generally increased movement of peoples and goods began to awaken an interest in the outside world; the fancy and superstition characteristic of the Dark Ages began to recede. With the translation into Latin, and then the printing of Ptolemy's writings and maps (items that had lain dormant for 1000 years) in the fifteenth century, a new interest in geography and cartography developed in Europe.

One of the more remarkable occurrences in

the history of mapmaking is the sudden appearance in the thirteenth century of quite accurate sailing charts called portolan charts. These charts, prototypes of our modern nautical charts, may have come into use as a consequence of the employment of the magnetic compass around 1200 A.D. They were the products of the experience of a large amount of open-water and coastwise sailing in the Mediterranean and adjacent areas (Fig. 2.7). Characteristically, they were covered by a systematic series of unlabeled, intersecting rhumb lines radiating from compass roses. The lines and shapes of the bounding coasts of the seas were surprisingly accurate but, unfortunately, this accuracy did not seem to extend to maps of the land behind the coasts until much later. Although many portolan charts must have been made, none earlier than the late thirteenth century has survived.

The medieval period in cartography may be said to have come to an end during the fifteenth century. During that notable century, there were many important developments but, before turning to them, it is appropriate to compare the course of cartography in the Eastern world.

Cartography in the Eastern World

The oldest civilization in Asia is that which developed in China, from where it spread to other parts of the continent. There is evidence of astronomical knowledge as early as the first historic dynasty, the Shang Dynasty, which ended in the eleventh century B.C. There are literary references to geography and maps at a very early date (B.C.) but, as in Europe, the basic principles of cartography were not enunciated until much later.

Two names in the early history of Chinese cartography are equal in importance to that of Ptolemy and Marinus in the Western world. Chang Heng was a contemporary of Ptolemy and, like Ptolemy, he was an astronomer, but not apparently a geographer. He is said to have originated one of the distinctive characteristics of Chinese cartography, the use of a plane square grid on which the map is drawn. Phei Hsiu (224–271 A.D.), Minister of Works during the Chin Dynasty, prepared a manual outlining the principles of geographical description and mapmaking, including elements such as scale, the square grid, distance, direction, and elevation.

The manual is lost, but the principles were kept alive.*

Technologically China was more advanced than the Western world. The compass was used much earlier there, paper was invented in China in the second century A.D. (not really used in the Western world until about 1000 years later), and the first printing of a map in China occurred about 1155 A.D., 300 years earlier than in Europe.

Probably the most distinctive feature of Chinese cartography up to relatively recent times is the square grid that appears on their maps. The lines are not latitude and longitude lines, and there is no real evidence of the use of projections by the early Chinese mapmakers. The square grid is just that, a rectangular grid with a specified earth distance for the interval. It was a useful tool for working out distances and, in a sense, was a sort of scale.

The relative accuracy of Chinese cartography is noteworthy. One of the most remarkable instances is a map entitled Map of the Footsteps of Yü the Great (Yü Chi Thu) that was carved on stone in about 1137 (Fig. 2.8). It has a square grid with a scale of 100 li (58 km or 36 mi) to each square. Another notable name in Chinese cartography in Chu Ssu Pen (1273–1335), who established a scientific mapping tradition and prepared a relatively accurate detailed map of China that was ultimately made into an atlas.

By the seventeenth century, European influences began to make themselves felt in Chinese cartography, particularly through the work of Jesuit missionaries. Thereafter the mapmaking of the Eastern and Western worlds ceased to be very distinctive.

The Renaissance in Western Cartography

The Age of Discovery, centering on the monumental achievements of Columbus, da Gama, Cabot, Magellan, Elcano, and others at the end of the fifteenth and the beginning of the sixteenth centuries, kindled such an interest in the rapidly

---

* Recent discovery of three 2d. century B.C. Chinese maps of extraordinary quality add further evidence of early development in that region. See Mei-Ling Hsu, "The Han Maps and Early Chinese Cartography," Annals, Association of American Geographers, forthcoming.

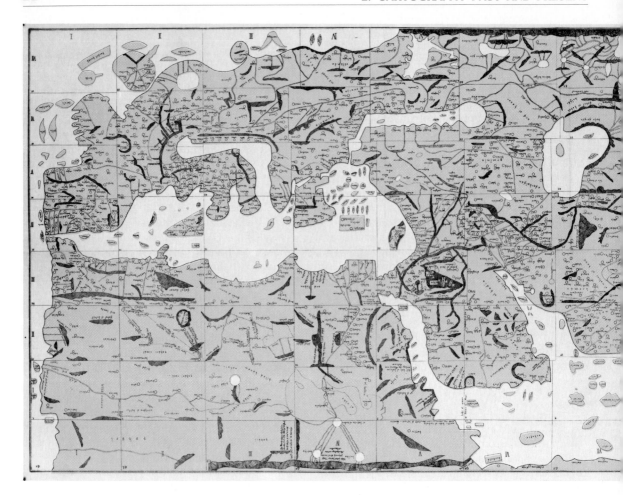

expanding world that map publishing soon became a lucrative calling. By the latter half of the sixteenth century, the profession was generally in good standing and was well supported, even though its products were still far from being first-class examples of objective scientific thought.

One of the circumstances that contributed greatly to the rapid advance of cartography was the invention in Europe shortly after 1450 of printing and engraving, which soon made possible the reproduction of maps in numerous copies. Previously, each map had to be laboriously hand drawn. Great map-publishing houses such as Mercator, Blaeu, Hondius, and others in Holland and France rose and flourished. Their maps were primarily general maps containing not much more than coastlines, rivers, cities, and occasional crude indications of mountains. Fancy and intricate craftsmanship was popular, and the maps were richly embellished with ornate scrolls, compass roses, and drawings of people, animals, and ships. Except for religious and some navigational data, the mapping of any geographic information beyond what we today call base data was unknown. The mapping of most other kinds of geographical distributions had to wait for the inquiring minds who would want such maps.

The Early Period of Modern Cartography

Between 1600 and 1650 a new and fresh attitude toward cartography arose. For the first time since

FIGURE 2.6 *The smaller of two world maps by Idrisi made in 1161* A.D. *as derived in reduced form from the original 73 sheets. The original is oriented with south at the top. Here it is turned upside down so that the geographical relationships can be more easily recognized, since we are now used to seeing maps with north at the top. (From the Library of Congress collection.)*

the ending of the classical era 1500 years earlier, accuracy and the scientific method became fashionable again. This attitude, replacing the dogmatic and unscientific outlook that was more or less dominant during the long Dark Ages, made itself evident in a number of ways in cartography.

In the second half of the seventeenth century the French Academy was founded with the avowed purpose of improving charts and navigation. This, of course, included cartography among its important concerns. Precise navigation in the open ocean was becoming a serious problem, and its solution depended on an accurate determination of the size and shape of the earth and on the development of a practical method for determining the longitude. The necessity for increased

mobility in military actions also made desirable the development of land survey methods.

The French Academy began by measuring accurately the arc along a meridian and, with the new technique of triangulation, it began to position accurately the outlines of France (Fig. 2.9). Because of differences noted in the lengths of degrees along the meridian and the behaviour of the pendulum at various latitudes, the question arose as to the precise shape of the earth. During the first half of the eighteenth century, French expeditions were sent to Peru and Lapland to measure other arcs along meridians. Their determinations settled once and for all that the polar radius was shorter than an equatorial radius.

Of particular interest in the history of cartog-

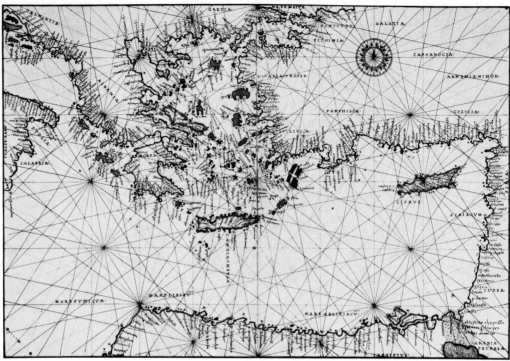

FIGURE 2.7 *One of the later portolan charts. Note the detail and relative accuracy of the delineation of the coasts. From Plate 12 of the manuscript atlas prepared by Battista Agnese of Venice about 1543, now in the Library of Congress. The maps are drawn on vellum and decorated in colors and gold. (From the Library of Congress collection.)*

raphy was Edmund Halley's publication in 1701 of what we would now call a thematic map (Fig. 2.10). On it he showed the distribution, as known at that time, of the declination of the compass by means of lines passing through points of equal declination. Now known as isogonic lines, these Halleyan lines or curves led to the widespread use of this kind of line symbol in the following century.

In the middle of the eighteenth century the French initiated a detailed topographic survey of their country at a scale approximately 1¼ mi to the inch and almost completed it prior to the end of the century. Harrison's chronometer for longitude determination was perfected in England in 1765; and there were many other evidences of curiosity about the earth. But perhaps the most notable and significant cartographic trend was the realization by many that their fund of knowledge about the land behind the coastlines was quite

erroneous. Even the administrators and rulers of countries, particularly in Europe, became aware that it was impossible to govern (or fight wars) without adequate maps of the land. Soon other great national topographic surveys of Europe were established, such as that of England in 1791, and the relatively rapid production of topographic maps followed.

The problem of representing the landsurface form arose and, almost as quickly, devices such as the hachure and contour were developed. By the last half of the nineteenth century, a large portion of Europe had been covered by topographic maps. These maps were expensive to make, however, and did not have a wide distribution. But they were the foundation on which all future cartography of the land was to be based.

As briefly described above, as early as the thirteenth century nautical charts were in use to

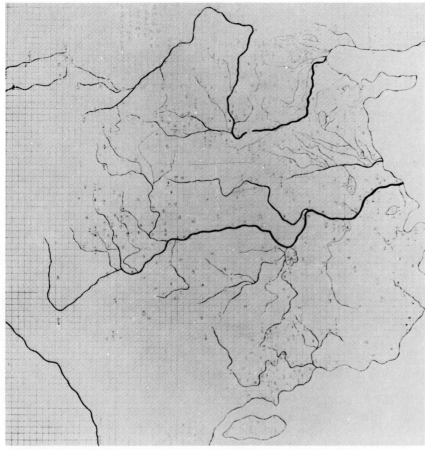

FIGURE 2.8 *A map of China carved in stone about 1137 by an unknown cartographer of the Sung Dynasty showing the characteristic square grid. The map is remarkably accurate, as comparison with a modern map will show. (Courtesy of Norman J. W. Thrower.)*

assist the mariner. Their production was largely in private hands but, by the early sixteenth century in both Spain and Portugal, official organizations were established to supervise chart making. It was not until the nineteenth century that the great modern governmental hydrographic offices were established to produce the detailed charts needed for safe use of the oceans. These counterparts to the great land surveys produced navigational charts, bathymetric maps ("topographic" maps of the sea bottom), and a variety of special purpose oceanic maps.

The Introduction of the Metric System

Before the beginning of the nineteenth century, every country had its own unique system of weights and measures. This caused much difficulty, because no one knew precisely the relationship of one national unit to another (e.g., the French toise to the English yard). Consequently, it was difficult to compile maps and to convert the scales of maps.

Earlier attempts by scholars to institute a new unit of length were given considerable impetus by the proposal in 1791 of the French Academy and

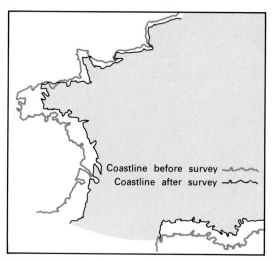

FIGURE 2.9 *The outlines of France before and after the first accurate triangulation (essentially completed by 1740). Similar corrections of other areas could be made after they were accurately triangulated.*

by the National Assembly's approval of its plan to employ the length of a meridian quadrant in fixing a universal unit.* The Revolution in France first delayed and then promoted what was to become the metric system, in which the basic unit of distance was the meter, defined as 1/10,000,000 part of the arc distance from the equator to the pole. This provided a kind of natural universal unit to which every other unit could be compared. Soon thereafter, in the early nineteenth century, map scales began to be stated as fractions or proportions in which one unit on the map represented so many of the same units on the earth, for example, 1:100,000. When map scales are stated in this way, conversions are easy, because such a proportion is independent of any one kind of unit of measurement. The use of the metric system led to "round number" scales such as 1:25,000 instead of odd scales such as 1:63,360 (1 in. on the map represents 1 mi on the earth).

All the world, except for a few areas including

* Georg Strasser, "The Toise, the Yard and the Metre-The Struggle for a Universal Unit of Length," Surveying and Mapping, 35, 25–46 (1975).

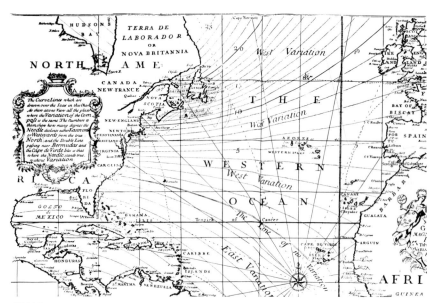

FIGURE 2.10 *A portion of Halley's isogonic chart of 1701 showing "curve lines" that supposedly connect points with the same declination of the compass. Note the compass rose in the lower right with its radiating rhumb lines. (From the Library of Congress collection.)*

the United States, have officially adopted the metric system with its easy decimal relationships. It is used in all scientific inquiry and, even though not officially in civilian use it has been legal in the United States for many years. The military agencies adopted it years ago, and most maps made in the United States now include graphic scales that show metric units. It is to be expected that the United States will officially "go metric" within a very few years.

The Rise of Thematic Cartography

Until the nineteenth century most maps simply showed where places, rivers, coasts, and boundaries were located. Beginning slowly in the eighteenth century, in addition to the general small-scale and large-scale topographic maps and navigational charts, a second great class, the thematic map, was added to the repertoire of cartography by the early nineteenth century. Thematic cartography had sketchy beginnings and had to wait for the broadening of scientific inquiry and the study of people and their institutions, which grew rapidly in the nineteenth century. This required much mapping. As we saw in Chapter 1, the thematic map is quite different from the general atlas map, the navigational chart, and the topographic map. It is a kind of graphic "geographical essay," since its main objective is to communicate specific geographical concepts regarding the structural character of particular distributions; it is concerned with densities, relative magnitudes, gradients, spatial relationships, movements, and the many interrelationships, and aspects among the distributional characteristics of the earth's phenomena.

Most important to the development of thematic cartography was the branching out of science into a number of separate fields; this was in contrast to its previous state, which was a kind of all-inclusive complex of physical science, philosophy, and general geography. The more exact studies such as physics, chemistry, mathematics, and astronomy had progressed far; but earth and life scientists who were concerned with certain classes of earth distributions, such as the physical geographer, geologist, meteorologist, and biologist, and the investigators whom we now call "social scientists," were just beginning their studies. The taking of censuses and the recording of all sorts of geographical data from weather observations to crime rates was generally initiated during the early part of the nineteenth century, and this also had a significant effect on the development of thematic cartography. Most of the rapidly growing earth, life, and behavioral sciences needed maps and, in general, they required the smaller-scale, compiled, thematic maps.

The Growth of Modern Cartography

Many factors helped to promote an acceleration in cartography during the nineteenth century. They include the development of lithography, which made possible the easy and inexpensive duplication of drawings; the invention of photography; the development of color printing; the rise of the technique of statistics; the growth of mass transport, such as the railroad; and the rise of professional scholarly societies.

By the beginning of the twentieth century, the thirst for knowledge about the earth had led to remarkable strides in all aspects of mapmaking. The colored lithographic map was fairly common; a serious proposal to map the earth at the comparatively large scale of 1:1,000,000 had been made; and some great map-publishing houses such as Bartholomew in Great Britain, Justus Perthes in Germany, and Rand McNally in the United States had come into being.

Cartography has probably advanced more, technically, in the past three quarters of a century than during any other period of comparable length. It is probably correct to say that the number of maps made since 1900 is greater than the production during all previous time, even if we do not count the many millions made for military purposes. Many factors have combined to promote this phenomenal growth; we will mention only a few.

The development of the airplane was very significant to modern cartography. It operated as a catalyst in bringing about the demand for more mapping; at the same time, the airplane made possible photogrammetry (mapping from photographs). The need for smaller-scale coverage of larger areas, such as the aeronautical chart, promoted larger-scale mapping of the unknown areas. Furthermore, the earth seen from the vantage point of an airplane in flight appears somewhat like a map, and the air traveler frequently develops in interest in maps. The rapid developments connected with space exploration and satellite technology is having a similar effect, but at a different level. Remote sensing, mapping of the moon and planets, and precise measurements of the earth and of relative positions on it from orbital observation are just a few of the developments.

Photography, in which rapid advancements were made during the latter half of the nineteenth century, has also had enormous significance for mapping. The potential of photographs for photogrammetric uses was evident almost immediately after the invention of the camera, and the advantages of aerial over terrestial photographs were obvious. Today, photogrammetric methods are extensively used in the preparation of nearly all published large-scale topographic maps. Furthermore, it is now possible to remove the perspective displacement in an air photograph, and a new kind of map with a photographic base, called an orthophotomap, is becoming increasingly available.

The union of photography with the printing processes in the latter part of the nineteenth century provided a relatively inexpensive means of reproducing an image on almost any material. From then on technical developments in the photolithographic and photoengraving fields have been rapid and continuous, and they have become inextricably woven into the methods of cartography. Today, although costs are considerable, high-speed, multicolor presses are capable of handling any kind of cartographic problem. Modern map construction with its scribing, open window negatives, photo lettering, and so on, is as different from the mapmaking of 50 years ago as is the automobile from the horse and buggy.

The rapidly increasing population of the earth, the phenomenal growth of urban centers, and the associated technological and industrial development have also had a powerful impact on cartography. The need to protect the environment and the demand for environmental information by planners and administrators has greatly promoted all classes of cartography. The urban, rural, resource, industrial, or other kinds of planner-engineers are usually concerned with relations in the spatial dimension; for this they must have maps.

Probably the most profound changes in the broad field of cartography occurring at present are a result of continuing development of techniques associated with automation and computer applications. The hardware, consisting of computers, digitizers, plotters, printers, and the cathode ray tube together with the associated software, the programs for accomplishing particular tasks, are literally revolutionizing cartography. However, the remarkable mechanical and technical advances simply make the solutions to problems easier. Cartography is a body of theory and method for dealing with problems of recording, analyzing, and communicating geographical information; the computer, air photography, the scribe coat, and the other innovations are only aids to better, more economical, faster mapmaking.

THE PRESENT FIELD OF CARTOGRAPHY

Mapmaking is a very old activity but, in the Western world, it did not attain the status of a profession until the fifteenth century. At that time, with the burgeoning of exploration, the translation of Ptolemy's *Geography* into Latin, and the application to maps of the newly developed printing processes, a class of scholar-technician mapmakers came into being. As we noted earlier, in Spain and Portugal, countries deeply committed to overseas colonial development and trade, official bureaus were established to promote chart making and navigation methods. Later, large commercial trading institutions in several countries, notably Holland and England, maintained similar, but private, mapping offices. After the introduction of topographic mapping in the eighteenth century, government agencies concerned with official large-scale mapping of the land were established. Today, most countries support large government agencies with primary responsibilities for official mapping and charting. Many other agencies, organizations, and individuals are active in cartography. The great technical advances in the processes of mapmaking, the growing need for education and research in cartography, and the importance of maps as documents, analytical tools, and communicative devices has caused cartography to become a broad, diverse field.

Major Divisions of Cartography

To separate the field of cartography into divisions is as difficult as trying to classify the kinds of maps.

Nevertheless, as we saw when discussing the various classes of maps, there are meaningful distinctions to be made.

Cartography is usually thought to consist of two classes of operations. One is concerned with the preparation of a variety of general maps used for basic reference and operational purposes. This category includes, for example, large-scale topographic maps of the land, hydrographic charts, and aeronautical charts. The other division has to do with the preparation of an even larger variety of maps used for general reference and educational (in the broad sense) purposes. This category includes the usually small-scale thematic maps of all kinds, as well as atlas maps, road maps, maps to accompany the written text in books, and planning maps. Within each category there is also considerable specialization such as may occur among the survey, design, drafting, and reproduction phases of a topographic map. All such divisions and activities blend one into another, however, and sharp compartmentalization rarely occurs.

The first-mentioned category of cartography works primarily from data obtained by field or hydrographic survey or by satellite and photogrammetric methods. Things such as the shape of the earth, height of sea level, land elevations, precise distances, and detailed locational information are of fundamental concern. Complex electronic and photogrammetric instruments and remote sensing are an integral part of this sort of cartography. Generally, this group includes the great national survey organizations, the oceanographic and aeronautical charting agencies, and most military mapping organizations, which altogether employ thousands of cartographers. They produce an enormous number of maps. For example, the National Ocean Survey in the United States, only one of the several federal mapmaking agencies, issues more than 40 million copies of charts each year. The fundamental bases for all mapping come from the work of these governmental departments.

The other category, which includes thematic cartography, draws on the basic work of the first group but is mostly concerned with the communication of general information and with the effective graphic delineation of relationships, generalizations, and geographical concepts. The specific subject matter may be drawn from history, economics, urban planning, rural sociology, engineering, and many other areas of the physical and social sciences—there is no limit. Uppermost, however, is the concern for a properly designed, clear, legible, graphic communication to assist in understanding and interpreting the social and physical complex on the earth's surface. The data for such maps, ranging from the thematic to the small-scale reference maps in an atlas, are usually compiled.

It should not be inferred that the practical considerations that tend to separate the cartographers basically concerned with large-scale maps, survey, and photogrammetric methods from those concerned with small-scale maps and compilation necessarily create a well-defined void between the two groups. The contrary is true. The fundamental problems of each group are conceptually similar. For example, the cartographic concepts on which the delineation of landforms by contours depends are the same as those on which the delineation of an abstract statistical surface by isopleths depends. Most of the principles on which cartographic techniques are based are equally applicable in either division of the field. Instead, the major distinction is in the methods of acquiring the data to be mapped; these are commonly different. On the other hand, the cartography, that is, the conception, the designing, and the execution of the map, as distinct from the gathering of the data, is fundamentally the same in both divisions.

Other Cartographic Activities

The major conceptual-operational divisions of the field outlined above do not take into account some other fundamental areas of cartographic activity that, in the United States, tend to cut across the large-scale, survey-oriented and the small-scale, compilation-oriented groups. There are several such areas: private mapmaking companies, commercial survey and mapping companies, planning agencies, and cartographic information departments.

Commercial cartography is made up of a large number of companies, from small, local organizations to large, complex organizations with national and international markets. In the United States there are more than 100 such firms, and they employ over 1000 cartographers; the majority of them are technicians, and the rest of them are compilers, editors, and researchers. The products of such companies are very diverse, ranging from street and plat maps to specialized atlases and road maps. The maps are either prepared from material supplied by the consumer, such as a street plan for a town, or they are compiled by research staffs in major mapmaking companies or in the cartographic departments of publishing companies or encyclopedia publishers.

Many commercial survey and mapmaking companies do contract mapping for industries, planning departments, and state and municipal agencies. This often involves air survey. The work ranges from cadastral-type mapping for private and public appraisal or tax assessment needs to site analyses for urban and industrial development. Some of these companies are quite sophisticated in their application of advanced electronic and computer aids to surveying and cartography.

Most cities, many states, and even some regional groupings of areas have planning departments that make quantities of large-scale and small-scale maps. An example of a large-scale map would be the detailed mapping of an area to be redeveloped; population and land use maps for zoning and other aspects of general urban planning are representative of smaller-scale maps. Cartographers employed by planning agencies often work with air photography.

The enormous number and diversity of maps that have been made in recent years, the great variety of activities that require maps, the need to provide reference assistance to those who need to know what is available, and the necessity for maintaining collections of maps has led to the rapid growth of cartographic information services. Many cartographers are involved in this facet of cartography, which ranges from the map divisions of large libraries and the numerous smaller ones in federal mapping agencies, universities, and colleges to the map information bureaus maintained by the federal and state governments. The professional activities in this area of cartography are less concerned with mapmaking than they are the communication of information about maps, their availability, reliability, cost, and so on, although such organizations often publish index maps showing the variety of coverages, scales, and types of maps. A great deal of cataloging and bibliographic work is involved. The two main federal

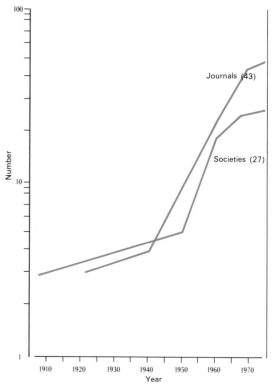

FIGURE 2.11 Cumulative numbers of cartographic societies and journals. Note that the period of fastest growth has been since 1940. The y-axis is a logarithmic scale so that the slope of the curve shows the relative rate of growth. (From data compiled by John A. Wolter.)

units in this area are the Geography and Map Division of the Library of Congress* and the National Cartographic Information Center (NCIC), a cooperative activity of federal mapping agencies maintained by the U.S. Geological Survey, which grew out of the former Map Information Service.†

Societies and Journals. Since the middle of the 1940s, the field of cartography has grown rapidly in all aspects. The establishment of profes-

---

* Walter W. Ristow, "Cartographic Information Services of the Library of Congress," The American Cartographer, 1 (2), 125–130 (1974).

† The postal address: National Cartographic Information Center, U.S. Geological Survey, 507 National Center, Reston, VA 22092.

sional societies has been particularly notable. A main function of professional societies is to disseminate information; one way to accomplish this is by the publication of a journal. Consequently, the number of journals has grown along with the societies. There are nearly 30 societies around the world and more than 40 journals are being published (Fig. 2.11). The major journals are listed in the bibliography.

Because cartographers have much in common with geographers, surveyors, and photogrammetrists, some societies include these interests. For example, in the United States the American Congress on Surveying and Mapping (ACSM) encompasses surveyors as well as cartographers. The Cartography Division of that organization is one of three "separate" divisions and has been in existence since soon after the founding of the ACSM in 1941. There are also cartographic associations in English-speaking Austrailia, Britain, Canada, and New Zealand and in most of the other leading countries of the world.

Cartography is represented at the international level by the International Cartographic Association (ICA), an organization in which countries, not individuals, hold membership. More than 50 countries now are members. The ICA is affiliated with the International Geographical Union, and so far the two organizations have scheduled their general assemblies together every 4 years.

SELECTED REFERENCES

Bagrow, L., *History of Cartography,* revised and enlarged by R. A. Skelton, C. A. Watts and Co. Ltd., London, 1964.

Brown, L. A., *The Story of Maps,* Little, Brown and Company, Boston, 1949.

Kember, I. E., "Some Distinctive Features of Marine Cartography," *The Cartographic Journal, 8,* 13–20 (1971).

Leverenz, J. M., "The Private Cartographic Industry in the United States, Its Staff and Educational Requirements," *The American Cartographer, 1,* 117–123 (1974).

Ristow, W., *Guide to the History of Cartography,* Library of Congress, Washington, D.C., 1973.

Robinson, A. H., and H. Wallis, "Humboldt's Map of Isothermal Lines," *The Cartographic Journal, 4,* 119–123 (1967).

Robinson, A. H., "The Genealogy of the Isopleth," *The Cartographic Journal, 8,* 49–53 (1971).

Robinson, A. H., and B. B. Petchenik, *The Nature of Maps, Essays toward Understanding Maps and Mapping,* University of Chicago Press, Chicago, 1976, p. 108–123.

Strasser, G., "The Toise, the Yard and the Metre—The Struggle for a Universal Unit of Length," *Surveying and Mapping, 35,* 25–46 (1975).

Thrower, N. J. W., *Maps and Man,* Prentice-Hall, Inc., Englewood Cliffs, NJ, 1972.

Tooley, R. V., *Maps and Map Makers,* 2nd ed., Bonanza Books, New York, 1952.

A select list of cartographic serials that publish articles in English and the country of their present publication

The American Cartographer (USA)

The Canadian Cartographer (Canada)

The Cartographic Journal (United Kingdom)

Cartography (Australia)

International Yearbook of Cartography (Federal Republic of Germany)

New Zealand Cartographic Journal (New Zealand)

Bulletin, Society of University Cartographers (United Kingdom)

Surveying and Mapping (USA)

Imago Mundi: A Review of Early Cartography (United Kingdom)

Bulletin, Special Libraries Association, Geography and Map Division (USA)

Cartographica (Canada)

World Cartography (United Nations, New York)

In recent times careful mapping has been extended to the surface of the moon and to some of the planets. Mapping other celestial bodies is made more difficult by their distance and consequent difficulty of their observation, but the basic methods involved in locating and representing on a plane the relative planimetric (horizontal) and hypsometric (vertical) positions of places at a reduced scale are essentially the same, regardless of the object being mapped, because celestial bodies are spherical, large, and require a coordinate system. Although the following discussion refers to the earth, the principles apply to the mapping of any similar body.

THE EARTH

The Shape of the Earth

If we disregard the relatively minor irregularities of the earth's outer shell, such as its continents with their irregular land surface form, the shape of the earth may be defined roughly as spherical. In reality, the earth is a complex geometric figure. The shape assumed by our plastic planet, spinning on its axis through space, is the result of the interaction of several internal and external forces such as gravity, the centrifugal force of rotation, and variations in the density of its rock constituents.

The interaction of tectonic and gradational

CHAPTER 3
THE SPHEROID, COORDINATE SYSTEMS, AND MAP SCALE

forces has produced other irregularities such as mountains, plains, and ocean basins. This class of irregularities, so noticeable to the human eye, is relatively so small, however, that they are significant only to the cartographic problem of the delineation of the land form. For example, on a globe with a diameter of 30 cm (about 1 ft), mountains and ocean basins would scarcely be noticeable; the maximum deviations from average sea level would be less than 0.25 mm (0.01 in).

Maps are representations of the complex shape of the earth on a plane; accordingly, it is necessary to transfer systematically the geometric relations from the one shape to the other.* If this is to be done accurately it is apparent that the characteristics of both shapes must be known. Furthermore, in order that the transformation may be done systematically, we must assume that the earth shape is a simple, solid form. It is important to understand that this part of the mapping process involves three steps: (1) the determination of the regular geometric figure that most closely approximates the actual size and shape of the irregular earth; (2) the systematic transfer of positions on the irregular earth to the surface of the "approximate" earth; and (3) the transformation of that three-dimensional surface to a plane surface. There are many complications involved in making observations of the surface of the earth, but the transformations to the approximate earth and to the plane are accomplished mathematically.

Human being's first ideas of the earth around them probably included little beyond that which they could see, since the view was limited by the horizon; consequently, the surface appeared flat. The idea of sphericity was not generated until the philosophers of the pre-Christian era applied reason to the problem, and 500 years or more before the birth of Christ the earth was theorized to be spherical. By the time of Claudius Ptolemy (second century A.D.), the spherical earth was generally recognized.

Since then the concept of sphericity has dominated, although some people from time to time,

thought the earth to be flat. There are still some, but views from space are making it difficult to believe! Supposed fears about falling off the edge of the earth to the contrary, it is likely that since the beginning of the Christian era, most people have known that the earth is spherical.

The Geoid

The figure of the earth is unique and can only be described as being a geoid, meaning earthlike. It is the shape that would be approximated by an undisturbed mean sea level in the oceans and the level of the water in a series of sea level canals crisscrossing the land. It is properly defined as an equipotential (i.e., gravity potential everywhere equal) surface to which the direction of gravity is everywhere perpendicular. Because of variations in the density of the earth's constituents and because the constituents are irregularly distributed, the geoid generally rises over the continents and is depressed in oceanic areas. It also shows various other bumps and hollows that depart from an "average smoothness" by as much as 60 m. The geoid is a very significant factor in mapping because all observations on the earth are, of course, made on the geoid. Because the geoid is irregular and therefore the direction of gravity is not everywhere toward the center of the earth, great efforts must be expended so that determinations of position will be consistent. The determination of the precise figure of the earth is part of the responsibility of the science of geodesy.

The Spheroid

In addition to the irregularities caused by the variations in density and distribution of the earth materials, the geoid is deformed from approximating a sphere by the rotation of the earth. Because it spins on an axis, the earth is bulged somewhat in the equatorial area and flattened a bit in the polar regions. The actual amount of the flattening is about 21.5 km (13.5 mi) difference between the polar and equatorial radii; the equatorial radius is, of course, the larger.

For precise mapping of large areas, such as topographic mapping, a regular geometric figure must be used. As explained above, observations on the geoid are likely to be inconsistent with one another because of differences in the direction of gravity; these can be reconciled by transferring the

---

*Strictly speaking, we may map on surfaces other than a plane. The representation on a reduced globe and that of a three-dimensional terrain model are both maps. Ninety-nine percent of all maps are on a plane.

data to the regular form that most closely approximates the geoid. This is an ellipsoid of revolution, that is, a figure produced by an ellipse rotated around its minor axis.

The amount of the polar flattening (ellipticity or oblateness) is given by the ratio $f = (a - b)/a$ where a is the equatorial semiaxis and b is the polar semiaxis. It is usually expressed as $1/f$ and, for most spheroids, the value of $1/f$ is close to 297.

Because of the earth's oblateness, a line extending around the earth that passes through the poles will not be a circle but will be slightly oval in shape, with the flatter portion in the polar sector and the more rapid curvature in the equatorial. Since a considerable amount of navigation is based on observations aimed at finding the angle between some celestial body and the horizon plane (or a perpendicular to it—the vertical) at a point, it is apparent that complications result from this departure from a true sphere. Consequently, whenever maps are being prepared for navigation or for determining or plotting exact courses and distances from one place to another, it is necessary to take the oblateness into account. In most cases of very small-scale mapping it may safely be ignored.

The Size of the Earth

Before the beginning of the Christian era both Eratosthenes (about 250 B.C.) and Posidonius (about 100 B.C.) calculated the size of the earth and apparently came close to the figures we now accept. Both proceeded by finding the number of degrees in an arc of assumed length on the surface. Eratosthenes estimated that the distance between Alexandria and Syene (modern Aswan) in Egypt was 5000 stadia. He determined the arc distance to be 7°12' by observing the angle above the southern horizon of the noon sun at Alexandria to be 82° 48' when the noon sun was supposed to be vertical at Syene (Fig. 3.1). Arc distance 7° 12' is $^{1}/_{50}$ of a full circle, so 50 times the assumed distance between Alexandria and Syene, 5000 stadia, equaled 250,000 stadia. Assuming that the stadium is about 185 m, this means that Eratosthenes found the earth's circumference to be 46,250 km (approximately 28,740 mi), which is only about 15% too large. Each of Eratosthenes' measurements and assumptions was somewhat

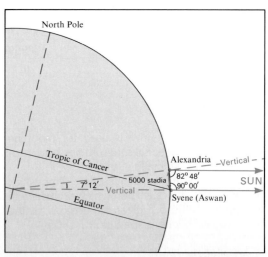

FIGURE 3.1 *Eratosthenes' method of calculating the circumference of the earth. See text for explanation.*

erroneous but, fortunately, the errors tended to compensate one another.

The early estimates by Eratosthenes and Posidonius were reported by others and recorded by Ptolemy who, recognizing the errors, but not their compensation, accepted "corrected" values that reduced the earth's circumference by nearly a fourth. Unfortunately, or fortunately, depending on one's viewpoint, Ptolemy's convictions were generally accepted. If Columbus had known the true length of a degree of arc on the earth, he probably would not have dared set sail to find the Indies by going west; in that case, perhaps the Americas might have been discovered instead by explorers from China.*

In recent times the dimensions of the spheroid have been calculated with great care from various astro-geodetic arcs measured with precision (Table 3.1). In the United States

---

*When Columbus added the (overestimated) "known" distance from western Europe to eastern Asia and the distances westward to the Azores and eastward from Asia to Japan and then divided the sum by his too-small degree, the result when subtracted from 360° meant that the remaining distance westward to the Orient from the Azores was hardly more than the length of the Mediterranean Sea. To the end of his days, Columbus believed that he had reached the Indies and the Orient—hence the West Indies and the term Indians applied to the natives!

TABLE 3.1

Name	Equatorial semiaxis—*a*		Polar semiaxis—*b*		1/*f*
	m	ft	m	ft	
Clarke 1866	6,378,206	20,926,062	6,356,584	20,855,121	294.98
International	6,378,388	20,926,488	6,356,912	20,856,029	297

Clarke's 1866 and the International Spheroid

Clarke's spheroid of 1866 is used for mapping purposes. An international spheroid has been proposed but is not generally in use.

Given a spheroid, various other dimensions of the earth may be calculated which are useful in cartography. Table 3.2 gives dimensions of the earth based on Clarke's spheroid of 1866.

Areas on the Earth

As the spherical earth complicates the cartographic problem of representing on a plane such concepts as distance and direction, the allside curved surface likewise makes difficult the reckoning and representing of areas. To arrive at the area of a polygonal segment of the surface of a sphere is relatively easy, but the earth is not a sphere and, for various other reasons, the establishment of exact position is difficult. If positions are doubtful, then the shape of the spherical segment is in doubt and thus the area of it is open to question. Besides, most of the areas in which we would be interested, aside from small land holdings, are extremely irregular (e.g., continents or countries with complex boundaries or coastlines). Consequently, one easy way to determine the actual size of such areas is to map them first and then compute or measure on a map in some manner the area enclosed. Of course, such a map must be one in which the transformation from the spherical surface to the plane surface has been made so that areas are uniformly represented as to size anywhere within the map.

The Great Circle

The shortest distance between two points is a straight line; however, on the earth it is obviously impractical to follow this straight line through the solid portion of the planet. The shortest course over the surface between two points on a sphere is the arc on the surface directly above the straight line. This arc is formed by the intersection of the spherical surface with the plane passing through the two points and the center of the earth (Fig. 3.2). The circle established by the intersection of such an extended plane with the surface divides the earth equally into hemispheres and is termed a *great circle*.

Great circles bear a number of geometrical relationships with the spherical earth that are of considerable significance in cartography and map use.

1. A great circle always bisects another great circle.
2. An arc of a great circle is the shortest course between two points on the spherical earth.
3. The plane in which any great circle lies always bisects the earth and hence always includes the center of the earth.

TABLE 3.2

	Kilometers	Statute miles (U.S.)
Equatorial diameter	12,756.4	7,926.5
Polar diameter	12,713.2	7,899.6
Equatorial circumference	40,075.4	24,901.7
Radius of a sphere of		
equal area*	6,370.9	3,958.7
Area of the earth (approximate)	510,900,000 km^2	197,260,000 mi^2

Dimensions of the Earth

* The sphere of equal area is the true sphere that has the same surface area as the earth spheroid. All radii of a sphere are identical.

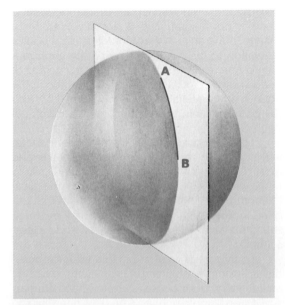

FIGURE 3.2 *The trace of the intersection of a plane and a sphere is a great circle whenever the plane includes the center of the sphere. Any two points on a sphere can be connected by a great circle, such as A and B, and this is the shortest route over the surface between the two points.*

Because a great circle is the shortest distance between two points on a spherical surface, air and sea travel, insofar as is possible or desirable, move along such routes. Radio signals and certain other electronic impulses tend to travel along great circles; therefore many maps must be made on which great circles are shown by straight lines or simple curves.

✳ COORDINATE SYSTEMS

To locate points relative to one another, it is necessary to employ the concepts of direction and distance. These can only be specified in terms of some system; primitive peoples probably did so in relative terms, using aids such as the directions of the rising and setting sun, forward and backward, left and right, and so on, and they probably expressed distance in terms of travel time—all of these being reckoned with respect to one's location. Any universal or general system must, however, be established in relation to some unique

reference or starting point. If such a point is designated, then the location of every other point can be stated in terms of a defined direction and distance with it.

Plane Coordinates

On a limitless plane surface there is no natural reference point; that is, every point is like every other point. An arbitrary system of location on a plane surface has long been used by establishing a "point of origin" at the intersection of two conveniently located, perpendicular "axes." The plane is then divided into a grid by an infinite number of equally spaced lines parallel to each axis. The position of any point on the plane with reference to the point of origin may then be stated by indicating the distance from each axis to the point, measured in each case parallel to the other axis, and expressed to any desired precision. In the familiar rectangular coordinate system (e.g., cross-section paper) the "horizontal" distance is

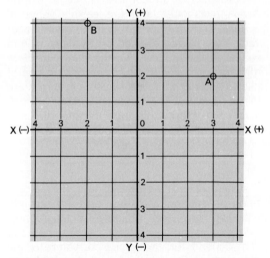

FIGURE 3.3 *A plane coordinate system. The origin if 0. The abscissa values are −X 0+X and the ordinate values are −Y 0+Y. The position of point A is 3, 2; the position of point B is −2, 4. In the geographical coordinate system, Y values correspond to latitude north and south and X values to longitude east and west of the origin (0° latitude and 0° longitude). In designating position in a plane coordinate system, the X value is always given first; on the earth's coordinate system, latitude is usually given first.*

called the X value or *abscissa,* and the distance perpendicular to it is called the Y value of the *ordinate* (Fig. 3.3).

Spherical Coordinates

On a sphere a similar (but much older) coordinate system is used. A spherical surface is an allside curved one (i.e., it slopes away equally in every direction from every point), and the use of parallel straight lines is impossible. On a motionless spherical surface there is also no natural reference point, but celestial bodies are not motionless with respect to other bodies. On the earth, for example, two convenient reference points are established by the poles; these are the two points where the axis of rotation intersects the spherical surface. If we imagine the earth before us with the axis vertical, the ordinate values, corresponding to the Y values in the rectangular coordinate system, are called latitude and the abscissa values are called longitude. The arrangement anywhere away from the poles of these two sets of perpendicular coordinates establishes the cardinal directions on the earth. A rectangular and the earth's spherical coordinate system have much in common; both systems may be represented by two sets of grid lines (called the graticule on the earth) that are perpendicular to one another, but only in one of the sets (latitude) that make up the graticule are the lines parallel. On a spherical surface we may conveniently specify distance in any direction (along a great circle arc, of course) by degrees, minutes, and seconds of arc.

Latitude

The system of locating oneself in a north-south position on the spherical earth was first put into practice by the Greek philosopher-geographers before the beginning of the Christian era. The system depends on the regular curvature of the earth's surface. One's latitude may be defined as the angle between a normal (perpendicular) to the surface and the plane of the equator at that place. This is accomplished by observing the altitude (angle above the horizon) of some celestial body and then, with the aid of an ephemeris, calculating the desired angle in the north-south direction.*

* *An ephemeris is an astronomical almanac that contains tables that show the daily positions of celestial bodies.*

Because the earth is spherical, change of position along a north-south line is accompanied by a change in the angular elevation of celestial bodies in relation to the horizon plane on the earth in a one-to-one relationship; that is, for each degree of north-south arc distance, the elevation above the horizon of a celestial body will change by 1°. Any star or the sun can be observed, and the result will be the same.

The foregoing simplifies the problem somewhat, because the earth rotates on its axis and most of the celestial bodies therefore seem also to move while the observer is moving from one place to another. The fundamental fact remains that position north-south can be determined by measuring the angle between the horizon and a celestial body.

To utilize this relationship in a spherical coordinate system was natural. The ancients imagined an infinite number of circles around the earth parallel to one another (Fig. 3.4). The one dividing the earth in half, equidistant between the poles, was named, as might be expected, the equator. The series north of the equator was called north latitude, and the series south of the equator was

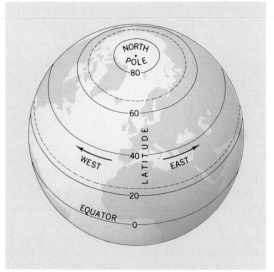

FIGURE 3.4 *The parallels of latitude (distance north-south) coincide with the directions east-west. (From Trewartha, Robinson, Hammond, and Horn,* Fundamentals of Physical Geography, *3d ed., McGraw-Hill Book Company, 1977.)*

called south latitude. To determine which circle one was on and, hence, the arc distance north or south of the equator required only the observation of the angle between the horizon and some known celestial body such as the sun, Polaris, or some other star.

No change has been made in the geographical coordinate system since it was first devised nearly 2200 years ago.

The Length of a Degree of Latitude. In the usual system of angular measurement, a circle contains 360°; and a half-circle contains 180°. Consequently, there are 180° of latitude from pole to pole. The quadrant of the circle from the equator to each pole is divided into 90°, and the numbering starts from 0° at the equator and goes by degrees, minutes, and seconds to 90° at each pole. Latitude is always designated as north or south.

Degrees of latitude are very nearly the same length, but not quite. Because the earth approximates an oblate spheroid, a north-south line (a meridian) has more curvature near the equator and less near the poles. Therefore, to observe a 1° difference in the altitude of a celestial body requires a lesser traverse along the meridian in the equatorial regions and a greater traverse in the polar regions. Consequently, degrees of north-south arc on the earth are not quite the same lengths in units of uniform surface distance, but vary from about 110.6 km (68.7 mi) near the equator to about 111.7 km (69.4 mi) near the poles. This difference of about 1 km in 110 is of little significance in small-scale maps, but is important on large maps of small areas. A complete table of lengths is included in Appendix C.

Longitude

Only position north-south on the earth is established by the latitude component of the geographical coordinate system. The transverse component, longitude, or distance east-west is provided by an infinite set of meridians, arranged perpendicular to the parallels. Unlike the equator in the latitude system, there is no meridian with a natural basis for being the starting line from which to reckon distance east-west in degrees, minutes, and seconds of longitude. From a given meridian, selected as a starting line, east-west position is designated

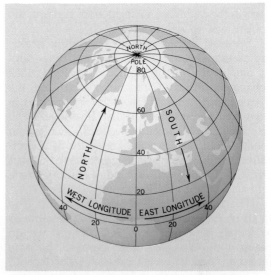

FIGURE 3.5 *The meridians of longitude (distance east-west) coincide with the directions north-south. (From Trewartha, Robinson, Hammond and Horn, Fundamentals of Physical Geography, 3d ed., McGraw-Hill Book Company, 1977).*

by the angular distance along the parallel circle in the latitude system (Fig. 3.5).

Prior to the middle of the eighteenth century, only latitude could be easily reckoned with any precision. Distance east-west depends on time differences and, for easy reckoning, requires that one know the times of day at the two places at the same instant. Without accurate timepieces that can be carried or instantaneous communication, this can only be done by very elaborate astronomical observation and calculation. Through the years this led to considerable error in east-west location, and it was one of the contributing factors to the glorious error of the fifteenth century, the idea that the distance separating Europe from Asia westward was less than half its actual value.

When the determination of longitude became critical for navigation, generous prizes for its solution were offered, and a variety of suggestions were put forward; the suggestions ranged from the observations of a celestial timepiece, such as the behavior of the satellites of Jupiter, to the employment of the variations (declination) of the magnetic compass. When the chronometer (a very accurate clock) was developed by Harrison and others in the middle of the eighteenth century, the problem was solved.

Because all parallels are concentric circles, they all rotate at the same angular speed—360° per day or 15° per hour. By carrying a clock showing the accurate time somewhere else, the difference between that time and local suntime in hours, minutes, and seconds can be converted to the longitude difference between the two places merely by using arithmetic. Today this is accomplished by radio time signals broadcast at regular intervals, as well as with the aid of chronometers.

The Length of a Degree of Longitude. The length of the equator is very nearly the same as the length of a meridian circle but, as we go toward the poles, all other parallels become smaller and smaller circles; yet each is divided into 360°. Therefore, each east-west degree of longitude becomes shorter with increasing latitude and is finally reduced to nil at the poles. The relationship between the length of a parallel (the circumference of a small circle) and the circumference of a great circle (e.g., the equator or a meridian circle) is the circumference of the great circle multiplied by the cosine of the latitude of the parallel; stated another way, the length of a degree of longitude = cosine of the latitude × length of a degree of latitude.

A table of cosines (see Appendix D) will show that

$$\cos\ \ 0° = 1.00$$
$$\cos 30° = 0.87$$
$$\cos 60° = 0.50$$
$$\cos 90° = 0.00$$

Thus, at 60° north and south latitude, the length of a degree of longitude is half that of a degree of latitude. Table 3.3 illustrates the decreasing lengths of the degrees of longitude from the equator toward the pole. A more complete table is included in Appendix C.

The Prime Meridian. The meridians are all alike, and any one can be chosen as the meridian of origin from which to start the numbering for longitude. The choice became, as might be expected, a problem of international consequence. Numerous countries, each with characteristic national ambition, wished to have 0° longitude

TABLE 3.3

Lengths of Degrees of Parallels		
Latitude	Kilometers	Statute Miles
0°	111.321	69.172
10°	109.641	68.129
20°	104.649	65.026
30°	96.448	59.956
40°	85.396	53.063
50°	71.698	44.552
60°	55.802	34.674
70°	38.188	23.729
80°	19.394	12.051
90°	0	0

within its borders or as the meridian of its capital. For many years each nation published its own maps and charts with longitude reckoned from its own meridian of origin. This, of course, caused much confusion.

During the last century, many nations began to accept the meridian of the observatory at Greenwich near London, England, as 0° and, in 1884, it was agreed on at an international conference. Today this is almost universally accepted as the prime meridian, but some maps still show two sets of meridians, one based on a local prime meridian and the other on the Greenwich system. Since longitude is reckoned as either east or west from Greenwich (to 180°), the prime meridian is somewhat troublesome because it divides both Europe and Africa into east and west longitude. The choice of the meridian of Greenwich as the prime meridian establishes the "point of origin" of the geographical coordinate system in the Gulf of Guinea. The opposite of the prime meridian, the 180° meridian, is more fortunately located; its position in the Pacific provides a convenient place for the international date line. Days on earth must begin and end somewhere, and only a few deviations from 180° are needed in that sparsely populated region to keep from separating inhabited areas into different-day time zones.

Rectangular Coordinates

The geographical coordinate system is useful for large areas, and the measurement of distances and directions in angular measure in degrees, minutes, and seconds can hardly be improved on. But for small areas it is cumbersome. With the increasing range of artillery in World War I it became more and more difficult to arrive at accurate azimuth (bearing or direction) and range (distance). To simplify the problem, the French constructed a series of local plane, rectangular coordinate grids on maps. Since the formulas of plane geometry are far simpler than those of spherical geometry, other nations quickly followed suit and, between World Wars I and II, a great many systems of plane rectangular coordinates were devised and put into use. By now the use of rectangular grid systems is almost universal.

The procedure is as follows: first, a map is made by transforming the spherical surface to a plane (by a system of map projection), preparing the map on the plane, and placing a rectangular plane coordinate grid over the map. The two sets of straight, parallel grid lines are equally spaced and perpendicular to one another. To locate a position we need only specify the X and Y coordinates to whatever degree of precision we desire in whatever earth distance units are used, usually in a decimal system. This is much simpler than degrees, minutes, and seconds of latitude and longitude.

To simplify the reckoning of position, only the upper right-hand part of a plane coordinate system is used (Fig. 3.3) so that both sets of coordinates are positive, and therefore there will be no repetition of numbers east and west or north and south of the axes. Normally, the origin of the numbering is assumed to be outside the map area to the lower left.

The reading of a grid reference proceeds in the normal way a point is located on cross-section paper. In rectangular map coordinates the X value is always given first and is called an *easting;* the Y value is called a *northing.* A rule of thumb is that when using grid references we must always "read right–up." Reference to Fig. 3.6 will show that point P can be given an easting value of 145 and a northing value of 201 by decimal subdivision of the squares. The grid reference would simply be 145201. With lesser precision it would be 1420, and with greater precision it would be 14562011. Grid references always contain an even number of digits; the first half refers to the easting and the second to the northing. If each square in Fig. 3.6 represented 1 km² (1000 m on a side), the reference 145201 would be a statement of location to within a 100 m square, that being the location of the southwest corner of the square. Adding an ad-

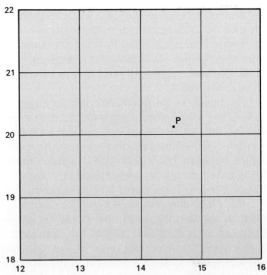

FIGURE 3.6 *A portion of a plane rectangular grid. If the squares are 1 km on a side, then point P may be located to within a 10 m² with the grid reference 14562011.*

ditional digit would narrow the position to a 10 m square.

Normally, rectangular coordinates are only used on large-scale maps, since the distortions that result from the transformation of the spherical surface to the plane make small-scale maps undesirable for detailed reference and calculation. Small-scale reference maps sometimes employ a letter-number indexing system to help locate map data, but that is not a rectangular coordinate system.

Direction

Directions on the earth are entirely arbitrary, since a spherical surface has no edges, beginning, or end. By definition, then, north-south is along any meridian and east-west is along any parallel; because of the arrangement of the graticule, these two directions are everywhere perpendicular except, of course, at the poles. The directions determined by the orientation of the graticule are called geographic or *true* directions as distinguished from two other kinds of direction, *grid* direction and *magnetic* direction.

It is apparent that when a rectangular grid is placed over the graticule of a map, in most places the "north" direction of the grid will not coincide with true north as specified by the graticule (see Fig. 4.25). Therefore, most detailed topographic maps specify the discrepancy in degrees and minutes between grid north and true north at the center of the sheet.

The needle of the magnetic compass aligns itself with the total field of magnetic force; in most parts of the earth this is not parallel with the meridian.* Consequently, there is usually a difference between true north and magnetic north that is called magnetic declination. Like the difference between true north and grid north, the amount of this difference with true north is usually given on detailed maps. Furthermore, the magnetic field changes slowly so that the declination value is likely to be correct only for the date the map was issued. Often a statement of the amount of annual change in declination is included.

Prior to the regular use of the magnetic compass, mariners in the Mediterranean area often specified directions by typical winds; these were shown diagramatically on a wind rose, a star-shaped device with 8 or 16 labeled points. From this developed a compass card in which the 360° circle was divided into 32 points (1 point = 11¼°). It became common practice on charts to include a similar representation, called a compass rose, to help the navigator reckon position and plot the course. Divisions finer than a point later became necessary, and today the compass rose on charts usually consists of a circle divided in degrees (Fig. 3.7).

The direction of a line on the earth is called many things: bearing, course, heading, or azimuth. Their meanings are essentially the same, differing largely in the context in which they are used. The two of importance in cartography are azimuth and bearing.

The Azimuth. As is apparent from studying a globe, the directions on the earth, established by the graticule, are likely to be constantly different if we move along the arc of a great circle. Only on a

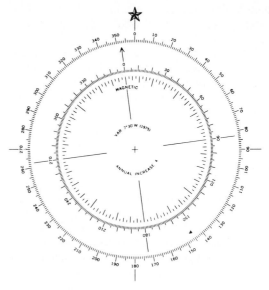

FIGURE 3.7 A typical "compass rose" from a modern nautical chart published by the U.S. National Ocean Survey. The rose is positioned on the chart so that 0°/360° shows true north. Note how magnetic north and annual variation are indicated.

meridian or on the equator does direction remain constant along a great circle. It is convenient to be able to designate the "direction" the great circle has at any starting point toward a destination. This direction is reckoned by observing the angle the arc of the great circle makes with the meridian of the starting point. The angle is described by the number of degrees (0 to 360°), reading clockwise usually from north.

In these days of radio waves and air transport directions, routes of movement along great circles are of major importance. Hence many maps are constructed so that directional relations are maintained as far as possible.

The computation of azimuths in the geographical coordinate system is quite involved and is ordinarily not needed except in geodetic work. In plane rectangular coordinates the grid azimuth (Az_G) from A to B is

$$\tan Az_G = \frac{E_2 - E_1}{N_2 - N_1}$$

in which E_1 is the easting value for A and E_2 is the easting value for B; N_1 is the northing value for A

---

* Contrary to the belief of the uninformed, the compass needle does not point directly toward the magnetic pole(s), except in the sense that if one were to "follow the compass" he would ultimately arrive at the magnetic pole, but it would be by a devious route.

and N_2 the northing value for B. Az_G will be found in a table of tangents such as Appendix D, and it must then be employed as an angular measure clockwise from grid north. If B lies in the northeast quadrant *from A*, then the tangent value $= Az_G =$ the grid azimuth; if B is in the southeast quadrant, then $Az_G = 180° -$ tan; if B is in the southwest quadrant, then $Az_G = 180° +$ tan; and, if B lies in the northwest quadrant, then $Az_G = 360° -$ tan.

The Rhumb Line or Loxodrome. A bearing is the direction from one point to another, usually expressed in relation to the compass rose either in a fashion such as northeast or as north 45° east. A great circle is the most economical route to follow when traveling on the earth. A pilot may do this by following a radio beam, but it is practically impossible otherwise, except when travel is along a meridian or the equator, because directional relations constantly change along all other great-circle routes. This is illustrated in Fig. 3.8. Because a course of travel must be directed in some manner, such as by the compass, it is not only inconvenient but impracticable to try to change course at, so to speak, each step.

A line of constant bearing is called a rhumb line or loxodrome. Meridians and the equator are rhumbs as well as great circles, but all other lines of constant bearing are not great circles. As a matter of fact, such rhumb lines are complicated curves, making equal oblique angles with all meridians; if one were to continue along an oblique rhumb line one would spiral toward the pole but in theory never reach it.

In order for ships and aircraft to approximate as closely as possible the great circle route between two points, movement is directed along lines of constant bearing that approximate it. Such a course is planned to begin on the great circle and shortly return to it, then depart and return again, as shown in Fig. 3.9. This procedure is similar to following the inside of the circumference of a circle by a series of short straight-line chords. It is not the same, of course, since the great circle is the "straight line" and the rhumb lines of constant bearing are curved lines and are actually longer routes.

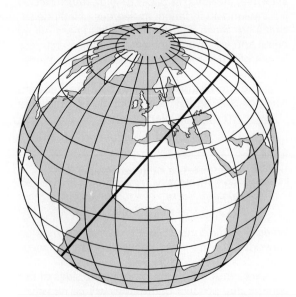

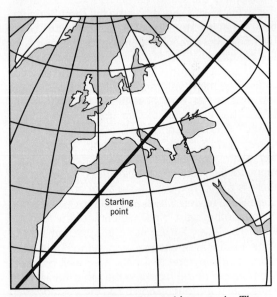

FIGURE 3.8 *How azimuth (direction) is given. The drawings show a great circle on the earth's graticule. The drawing on the right is an enlarged view of the center section of the drawing on the left. The azimuth, from the starting point, of any place along the great circle to the northeast is stated as the angle between the meridian and the great circle, reckoned clockwise from the meridian. Notice that the great circle intersects each meridian at a different angle.*

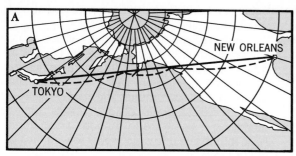
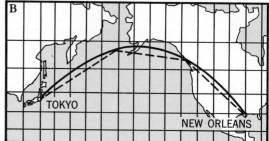

FIGURE 3.9 Two maps showing the same great circle arcs, and rhumb lines. (A) This map is constructed so that the great circle arc appears as the shortest distance between Tokyo and New Orleans, that is, a straight line, while the rhumbs appear as somewhat longer "loops." This is their correct relationship. (B) In this map the representation has been reversed by constructing the map in such a way as to make the rhumbs appear as straight lines, which "deforms" the great circle arc into a curve.

Orientation. A map sheet when looked at or held in the hand has a top and a bottom; if the sheet represents part of the earth in some way (picture or map) it is natural to think of what is shown in the upper part of the representation as being farther away. In the allegorical medieval *mappae mundi* ("world maps") Paradise was logically located at the top and placed in the most remote part of the known earth. This was the Orient, or the Far East. From this came the term "orientation," having to do with determining one's bearings and arranging things with the points of the compass.

Several centuries ago it became the practice to arrange maps with north instead of east at the top, and it has now become so strongly established that we think of "up north" and "down south." Australia and New Zealand being "down under," and "upper" Michigan and "lower" California are examples of the unconscious adjustment to this convention. Needless to say, since there is no up or down along a spherical surface, there is no reason why a map cannot be oriented in some other way. Since we think of the top of a sheet as "away" from us, it is apparent that in some instances orienting a map in the direction of interest or movement, if any, may well promote the purpose of the map. On the other hand, conventional north orientation helps the average map user because of familiarity with the convention.

Distance

Distances on the earth's surface are always reckoned along arcs of great circles unless otherwise

qualified. Because no map, except one on a globe, can represent the distances between all points correctly, it is frequently necessary to refer to a globe, to a table of distances, or to calculate the length of the great-circle arc between two places. A piece of string or the edge of a piece of paper can be employed to establish the great circle on a globe. If the scale of the globe is not readily available, the string or paper may be transferred to a meridian and its length in degrees of latitude ascertained. Since all degrees of latitude are nearly equal and are approximately 111 km (69 mi), the length of the arc in conventional units can be determined.

The arc distance D on the sphere between two points A and B, the positions of which are known, can be calculated by the formula

$$\cos D = (\sin a \sin b) + (\cos a \cos b \cos P)$$

in which

D = arc distance between A and B
a = latitude of A
b = latitude of B
P = degrees of longitude between A and B

When D is determined in arc distance, it may be converted to any other convenient unit of measure.

Note. If A and B are on opposite sides of the equator, the product of the sines will be negative. If P is greater than 90°, the product of the cosines will be negative. Solve algebraically.

The plane grid distance D_G, between two points A and B on a rectangular coordinate system can be calculated by the formula

$$D_G = \sqrt{(E_A - E_B)^2 + (N_A - N_B)^2}$$

in which E_A and E_B are the eastings of points A and B, respectively, and N_A and N_B are the respective northings.

There are, of course, many different units of distance measurement that have been used on maps in the past, and it is sometimes necessary to refer to encyclopedias or other sources to obtain data for conversion. Most maps today use metric units, and even nautical charts are converting depth measures to meters from the traditional 6-ft fathom. Civilian mapping agencies in the United States are now beginning to change to the metric system, but the conversion is costly and will take time. The relations between measures of the International (metric) System and the U.S. Customary System are given in the conversion table in Appendix A.

SCALE

Since maps must necessarily be smaller than the areas mapped, their use requires that the ratio or proportion between comparable measurements be expressed on the map. This is called the map scale and should be the first thing of which the map user becomes aware. The scale of a map may be shown in many ways: it can be specifically indicated by some statement or graphic device, and it may be shown indirectly by the spacing of the graticule and even subtly by the size and character of the marks on the map.

Scale is an elusive thing in maps because, by the very nature of the necessary transformation from the sphere to the plane, the scale on a map must vary from place to place and will commonly also vary even in different directions at a point.

Statements of Scale

The scale is commonly thought of as being an expression of a distance on the map to distance on the earth ratio with the distance on the map always expressed as unity. The map scale may be expressed in the following ways.

Representative Fraction (RF). This is a simple fraction or ratio. It may be shown either as 1:1,000,000 or 1/1,000,000. The former is preferred. This means that (along particular lines) 1

mm or 1 cm or 1 in. on the map represents 1,000,000 mm, cm, or in. on the earth's surface. It is usually referred to as the "RF" for short. The unit of distance on both sides of the ratio must be the same.

Verbal Statement. This is a statement of map distance in relation to earth distance. For example, the RF 1:1,000,000 denotes a map on which 1 mm represents 1 km or about 1 in. to 16 mi. Many older map series were commonly referred to by this type of scale, for example, the 1-in. or 6-in. maps of the British Ordnance Survey (1 in. to 1 mi, 6 in. to 1 mi).

Graphic or Bar Scale. This is a line placed on the map, often in the legend box or margin of the sheet, that has been subdivided to show the map lengths of units of earth distance. One end of the bar scale is usually subdivided further so that the user may measure distances more precisely (Fig. 3.10).

Area Scale. This refers to the ratio of areas on the map to those on the earth. When the transformation from the sphere to the plane has been made so that all area proportions on the earth are correctly represented, the stated scale is one in which 1 unit of area (square centimeters, square inches) is proportional to a particular number of the same square units on the earth. This may be expressed, for example, either as $1:1,000,000^2$ or as 1 to the square of 1,000,000. Usually, however, the fact that the number is squared is assumed and not shown.

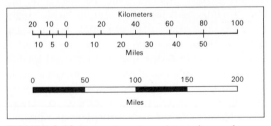

FIGURE 3.10 Examples of graphic or bar scales. Often, the left end of the bar is subdivided in smaller units in order to provide easier estimation of precise distances.

Scale Factor

It is not possible to transform the spherical surface to a plane without "stretching" or "shrinking" differentially the spherical surface in the process. This means that the stated scale, the RF, will be a correct statement of the scale only at selected points or along particular lines; elsewhere the actual map scale will be either larger or smaller than the given RF. This is true to some degree in *all* flat maps. The statement of the relation between the given RF and the actual scale value is called the *Scale Factor* (SF).

Perhaps the simplest way to appreciate the concept of the SF is to imagine the necessary reduction and transformation of the spherical surface as being accomplished in two stages: (1) the reduction of the earth to a globe of a selected scale, and (2) the transformation of that spherical globe to the plane of the map. The stated RF of the map will then be the RF of the globe and is called the *principal* (or nominal) RF. The real RF will be the *actual* scale on the map and will, of course, vary from place to place.

The SF may be computed by the following formula.

$$SF = \frac{\text{actual scale}}{\text{principal scale}}$$

This expresses the SF as a ratio related to the principal scale as unity. A SF of 2.000 would mean that the actual scale was twice the principal scale, which would be the case if, for example, the actual scale were 1:15,000,000 and the principal scale were 1:30,000,000. (The student must remember that the larger the denominator of the RF, the smaller the scale.) Similarly, a SF of 0.5000 would show that the actual scale was half that of the principal scale, as would be the case if, for example, the actual scale were 1:60,000,000 and the principal scale were 1:30,000,000.

Scale factors of the magnitudes used for the foregoing illustrations occur only on small-scale world maps. On large-scale maps they will vary only slightly from unity. For example, on large-scale maps employing the transverse Mercator projection, the SF magnitudes within a 6° longitude zone may vary only from 0.99960 to 1.00158.

Determining the Scale of a Map

Sometimes maps are made that do not include a statement of scale. This is ordinarily poor practice but, nevertheless, it occurs. More often it is necessary to determine the scale for a particular part of the map since, as observed previously, the scale can never be the same all over a flat map. The map scale along a particular line may be approximated by measuring the map distance between two points that are a known earth distance apart and then computing the scale. Certain known distances of the graticule are easy to use, such as the distance between parallels (average of 111 km, 69 mi) or the distance between meridians (see Appendix C). Care should be exercised that the measurement is taken in the direction the scale is to be used; frequently the distance scale of the map will not be the same in all directions from a point.

If the area scale is desired, a known area on the earth (see Appendix C) may be measured on the map with a planimeter and the proportion thus determined. Remember that area scales are conventionally expressed as the square root of the number of units on the right of the ratio. Thus, if the measurement shows that 1 square unit on the map represents 25,000,000,000,000 of the same units on the earth, it would not be recorded that way but as $1:5,000,000^2$ or merely by the square root, 1:5,000,000, which approximates the linear scale.

Transforming the Map Scale

Frequently the cartographer is called on to change the size of a map, that is, to reduce or enlarge it. The mechanical means of accomplishing this are dealt with in a later chapter, but the problem of determining how to change it in terms of scale is similar to the problem of transforming one type of scale to another. If the cartographer can develop a facility with this sort of scale transformation, no difficulty will be experienced in enlarging or reducing maps.

There is, of course, no difficulty in transforming decimal metric scales. U.S. Customary units are more bothersome. The essential information is that 1 mi (statute) = 63,360 in. With this information we can change each of the linear scales (RF, graphic, stated) previously described to the others. Examples follow.

If the RF of the map is shown as 1:75,000:

EXAMPLE 1.

To determine the stated scale:

Metric.

The cm/km scale will be:

1. 1 cm (map) represents 75,000 cm (earth), and
2. 75,000 cm = 0.75 km
3. 1 km/0.75 km = 1.333
4. Therefore 1.333 cm represents 1.0 km

U.S. Customary.

The in./mi scale will be:

1. 1 in. (map) represents 75,000 in. (earth), and
2. 1/75,000 = x/63,360, and
3. x = 0.845
4. Therefore, 0.845 in. represents 1.0 mi

EXAMPLE 2.

To construct a graphic scale, a proportion is established as:

Metric.

1. 1.333 cm:1.0 km::x cm:10 km
2. x = 13.33
3. 13.33 cm represents 10 km, which may be easily plotted and subdivided

U.S. Customary.

1. 0.845 in.:1.0 mi::x in.:10 mi
2. x = 8.45 in.

3. 8.45 in. represents 10 mi, which may be easily plotted and subdivided

EXAMPLE 3.

To determine the RF:

Metric.

If the graphic scale shows by measurement that 1 cm represents 50 km, then:

1. 1 cm represents 50 × 100,000 cm, or
2. 1 cm to 5,000,000 cm and, therefore,
3. The RF is 1:5,000,000

U.S. Customary.

If the graphic scale shows by measurement that 1 in. represents 75 mi:

1. 1 in. represents 75 × 63,360, or
2. 1 in. to 4,752,000 in. and, therefore,
3. The RF is 1:4,752,000

Changing the scale of a map that has an area scale is accomplished by converting the known area scale and the desired area scale to a linear proportion.

Common Map Scales

Maps are made with an infinite variety of scales. An experienced map reader learns to associate an approximate level of generalization and accuracy with particular scales, and the RF then becomes a kind of index of precision and content. It is helpful to translate the RF mentally into common units of

TABLE 3.4

Map scale	One centimeter represents		One kilometer is represented by		One inch represents	One mile is represented by
1:2,000	20 m		50	cm	56 yd	31.68 in.
1:5,000	50 m		20	cm	139 yd	12.67 in.
1:10,000	0.1	km	10	cm	0.158 mi	6.34 in.
1:20,000	0.2	km	5	cm	0.316 mi	3.17 in.
1:24,000	0.24	km	4.17 cm		0.379 mi	2.64 in.
1:25,000	0.25	km	4.0	cm	0.395 mi	2.53 in.
1:31,680	0.317	km	3.16 cm		0.500 mi	2.00 in.
1:50,000	0.5	km	2.0	cm	0.789 mi	1.27 in.
1:62,500	0.625	km	1.6	cm	0.986 mi	1.014 in.
1:63,360	0.634	km	1.58 cm		1.00 mi	1.00 in.
1:75,000	0.75	km	1.33 cm		1.18 mi	0.845 in.
1:80,000	0.80	km	1.25 cm		1.26 mi	0.792 in.
1:100,000	1.0	km	1.0	cm	1.58 mi	0.634 in.
1:125,000	1.25	km	8.0	mm	1.97 mi	0.507 in.
1:250,000	2.5	km	4.0	mm	3.95 mi	0.253 in.
1:500,000	5.0	km	2.0	mm	7.89 mi	0.127 in.
1:1,000,000	10.0	km	1.0	mm	15.78 mi	0.063 in.

Common Map Scales and Their Equivalents

measure. For example, on a map at a scale of 1:1,000,000, 1 mm represents 1 km and $\frac{1}{16}$ in. represents about 1 mi. Table 3.4 contains a listing of some of the more common map scales.

SELECTED REFERENCES

Aeronautical Chart and Information Center, *Geodesy for the Layman,* St. Louis, 1959.

Bowditch, N., *American Practical Navigator,* H.O. Pub. No. 9, 1966 Corrected Print, U.S. Navy Hydrographic Office (now Naval Oceanographic Office), U.S. Government Printing Office, Washington, D.C., 1966.

King-Hele, D., "The Shape of the Earth," *Scientific American, 217,* 67–76 (1967).

Maling, D. H., "The Figure of the Earth and the Reference Surfaces Used in Surveying and Mapping," in *Coordinate Systems and Map Projections,* George Philip and Son Ltd., London, 1973, pp. 1–17.

Proudfoot, M., *The Measurement of Geographic Area,* U.S. Bureau of the Census, Washington, D.C., 1946.

Robinson, A. H., "The Elusive Longitude," *Surveying and Mapping, 33,* 447–454 (1973).

U.S. Naval Observatory, *The American Ephemeris and Nautical Almanac,* U.S. Government Printing Office, Washington, D.C., issued annually.

Because of natural limits on human observation, people need help in seeing both very small and very large objects of interest. With the aid of enlarging optical and electronic devices we can observe things that are extremely small or very far away. Cartography is a similar system for bringing sections of, or all, the earth and other celestial bodies into view so that we may study their character. Instead of enlargement, reduction is the operation that makes observation possible.

Since most celestial objects are apparently essentially spherical, a simple way of attacking the problem is to prepare a globe map. Except for the graphic symbolism on a globe map, all that has been changed is the size; all other geometric relationships (relative distances, angles, relative areas, azimuths, rhumb lines, great circles, etc.) are retained without change. A globe is, therefore, a "naturally accurate" map.

A globe map, being on a spherical surface, has, on the other hand, a considerable number of practical disadvantages, one of them being the very fact that it is a three-dimensional round body only less than half of which can be observed at any time. In addition, it is cumbersome to handle, difficult to store, expensive to make and reproduce, and it is not easy to measure on its three-dimensional surface. Consequently, for most of the purposes for which we use a map, the globe map is less desirable than one that has included in its preparation a transformation of the spherical surface. Most of the disadvantages inherent in using the spherical form for the map are eliminated by transforming the surface to a plane; a substance in the form of a plane is easy to handle, all of it can be observed at once, it is relatively cheap to prepare and reproduce, and it is easy to measure and draw on it with ordinary instruments.

The actual process of transformation is called projection, and the term "projection" stems from

CHAPTER 4
MAP PROJECTIONS AND PLANE COORDINATES

the fact that some means of transformation can be accomplished by geometrically "projecting," with lines or shadows, the homologous points from the sphere to a plane surface. Actual geometric projection from the sphere to the plane includes only a few of the possibilities, however, and there are a larger number of possibilities for the retention of significant earth relationships that can be worked out mathematically. These are also called map projections, and no useful purpose is served by attempting to distinguish between geometric and mathematical map projections.

Map Projections and Cartography

Every flat map must employ a system of projection. Consequently, the methods of transformation from the sphere to the plane are of central concern, not only to the mapmaker, but to all who use maps, because it is an indisputable geometric fact that it is impossible to make the transformation without in some manner modifying the natural surface geometric relationships. But it is also a fact that there are innumerable transformations that can retain, on the plane, one or several of the spherical relationships. In addition, and of great significance, is the fact that a given system of transformation can produce both some desirable and some undesirable geometric characteristics. A good example of the last point is provided by Mercator's projection.

The sixteenth century was an exciting time of exploration and greatly expanded sea travel. Two new continents had been found; one of Magellan's ships had succeeded in circling the globe; the earth's land areas were beginning to take shape on the world map; and ships were setting out "to all points of the compass." One of the major trials of early navigators was that, although they had a rough idea of where lands were and had the compass to help them, they had no way of determining the bearing of courses that, with any degree of certainty, would take them to their destinations.

The Flemish mathematician and cartographer, Gerardus Mercator, provided a solution in 1569. It was pointed out earlier that a mariner must sail along a rhumb line, a line of constant bearing, because there is no other course that can be readily traveled. The solution was to transform (mathematically) the spherical earth's surface to a plane in such a way that a straight line on the resulting chart, *anywhere in any direction,* was a rhumb line. Thus the mariner need only draw a straight line (or a series of straight lines) from the starting point to the destination and, if appropriate allowances were made for drift, winds, and compass declinations, the mariner had a reasonably good chance of arriving somewhere near the destination. Parallels and meridians are rhumb lines, but all other rhumbs are parts of complex curves on the globe; by making them all appear as straight lines, for purposes of *navigation,* Mercator's projection produced a desirable geometric condition.

The navigator is not much concerned with the larger shapes and relative sizes of land areas, but teachers and students of the earth are. In Mercator's projection, in order to straighten out the rhumbs, the spherical surface is severely strained, so to speak, so that on the flat map larger areas in middle and higher latitudes appear misshapen and grossly enlarged (Fig. 4.5). On a Mercator chart showing both Greenland and Arabia, Greenland appears enormously larger whereas, in fact, on a globe they appear about equal (Fig. 4.4). For *general* purposes, then, Mercator's projection produced quite undesirable geometric conditions.

The fact that a system of projection can be suitable for one purpose and quite unsuitable for another suggests that the essential problem of the mapmaker is the analysis of the geometrical requirements of the proposed map and the *selection* of the system of transformation that will best or most nearly meet these needs. But that is not all. A cartographer "understands" maps, and either for personal work or for other people one must often measure things such as distances, angles, areas, or directions, and perhaps simply interpret mapped geographical phenomena. An understanding of projections and their attributes is necessary for accurate interpretation and use of maps.

In the four centuries since Mercator introduced his projection, the world has changed tremendously. In terms of human travel distances have been reduced a thousandfold; people have investigated and mapped an untold number of subjects; and all branches of learning, including mathematical cartography, have progressed immensely. The development of map projections has kept pace with the developments in other fields and, as the needs arose for ways of presenting particular geographic relations, a means of transforming the spherical surface to the plane to accomplish the purpose usually became available.

Obviously, a textbook of this sort cannot go deeply into a subject as mathematically complex as map projections. Nevertheless, in order that the student of cartography, and the user of maps generally, may have an understanding of the fundamentals involved in the subject of transformation from the sphere to the plane, a relatively non-mathematical description of the more basic elements follows.

INTRODUCTION TO MAP PROJECTION

Scale and Surface Transformation

When a spherical surface is mapped on a globe of any size, only reduction takes place. As pointed out in Chapter 3, the scale factor (SF) remains 1.0 everywhere on the globe surface, and relative distances, angles, or areas are preserved. When the sphere (globe) surface is transformed to a plane surface, there will be a systematic variation from place to place in the value of the SF.

To visualize what happens, imagine a pattern of equally distant points on the globe surface, and the establishment of corresponding or homologous points on the plane. The system employed to specify the positions of the points on the plane constitutes the method of projection. Since the two surfaces are what is called not *applicable*—cannot be transformed one to the other without stretching or tearing—distance relationships among the points on the plane must be modified. Since angular and areal relationships are functions of relative distances, alterations of these relationships are bound to occur. Consequently, it is impossible to devise a system of projection for two nonapplicable surfaces such that any figure drawn on the one will appear exactly the same on the other. Nevertheless, by suitable arrangement of the SF variations, it is possible (1) to retain some angular relationships, *or* (2) to retain relative sizes of like figures. If these particular qualities are not wanted in a projection, but some other geometric attribute of the spherical surface is desired, such as straight line azimuths from one point to all others,

then most angular relationships will usually be changed, and areas on the two surfaces will not have a constant ratio to each other.

In order to understand these stated facts as they apply to the transformation of a spherical surface to the plane surface of a map, it is necessary to realize, from the calculus of finite differences, that SF values may occur at a point and may be different in different directions at a point.

To demonstrate the first of these propositions, imagine an arc of 90°, as in Fig. 4.1, projected orthographically to a straight line tangent at a. If $a, b, c, \ldots j$ are the positions of 10° divisions of the arc, their respective positions after projection to the line tangent at a are indicated by $a, b', c', \ldots j'$. Line aj' therefore represents line aj. It may be seen from the drawing that the intervals on the straight line, starting at the point of tangency (a), become progressively smaller as j' is approached. If SF = 1.0 along the arc then, on the projection of the arc, the SF is gradually reduced from 1.0 at a to 0.0 at j'. The rate of change is graphically indicated by the diminution of the spaces between the points. Since it is a continuous change, it is evident that every point on aj' must have a different SF.

In order to visualize that scale at a point may be different in different directions, imagine a rectangle a, b, c, d, as in Fig. 4.2A, and its orthogonal projection resulting from rotating it around

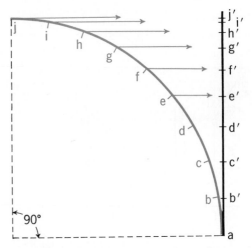

FIGURE 4.1 Orthographic projection of an arc to a tangent straight line.

the axis *ad* so that side *ad* coincides in *abcd* and its projection *ab'c'd*.

Figure 4.2B shows rectangle *abcd* with its projection superimposed with *ad* of each coincident. If the SF of *abcd* is assumed to be 1.0 and the length *ad* is the same in each rectangle, there has been no change in scale in that direction. However, since the length *ab'* is half the length *ab* and since it is evident from the method of projection that the change has been made in a uniform fashion, then the SF along *ab'* must be 0.5 or half the scale that obtains along *ad*. By projection line *ac* has become line *ac'*. The ratio of lengths *ac'* to *ac* constitutes the SF along *ac'*, and it is evident that it is neither the 1.0 ratio along *ad* or the 0.5 ratio along *ab'*; it is somewhere in between. Any other diagonal from *a* to a position on the side *bc* would have its corresponding place of intersection on *b'c'*. The ratio of lengths of similar diagonals on the two rectangles would be different for each such line. Hence, the scale at point *a* in rectangle *ab'c'd* is different in *every* direction.

The values of the SF in different directions at each point on a map projection and the changes in SF from point to point provide the bases for analyzing, to a considerable degree, what the system of map projection has accomplished for, or done to, the geometric realities of distances, directions, angles, areas, and so forth, on the sphere.

The Law of Deformation

At any point on a spherical surface there are, of course, an infinite number of directions and therefore an infinite number of paired perpendicular directions, such as N-S with E-W, NE-SW with NW-SE, and so on. When the spherical surface is transformed to a plane, generally the angular relation among the directions at each point will have been changed by the system of transformation; that is, a pair of directions that is perpendicular on the globe will not necessarily be perpendicular on the projection.

The law of deformation, formulated by M. A. Tissot, states that whatever the system of transformation, *there is at each point of the spherical surface at least one pair of perpendiculars which will be retained as perpendicular on the projection,* although all the other angles at that point may be altered from their original position.* If angles have been preserved by the system of projection (called *conformality*), there will be an infinite number of pairs of perpendicular directions retained as perpendiculars at each point, and at each point the SF will be the same in all directions (although usually different from point to point). This is a special case. On the other hand, when conformality has not been preserved in the transformation, then only one pair of perpendiculars at each point will be retained as perpendicular on the projection and the SF will normally be different in the two directions. By analysis of the SF differences in the directions of these perpendiculars, we may assess some of the kinds of deformation that occur on any map projection.

Tissot's Indicatrix. To calculate the amounts of angular and area deformation that occur at points on a map projection, Tissot employed a device that he called the indicatrix. For this he proposed that each point on the sphere be represented by a circle, the radius of which is unity (SF = 1.0). In a system of transformation the SF in the two directions of the perpendiculars retained as perpendicular will ordinarily not equal 1.0 and, except in conformal map projections, the SF will be different in the two directions. When the latter is the case, the indicatrix circle becomes an

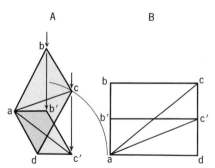

FIGURE 4.2 Projection of a rectangle to another rectangle with one side held constant. (A) The reduced, perspective drawing shows the geometric relation of the two rectangles. (B) Shows the relation of the two rectangles when they are each viewed orthogonally (i.e., perpendicularly to their surface).

---

* A proof of the law of deformation is given in Appendix I.

ellipse, the major and minor axes of which are given by the values of the SF.* By an analysis of the geometric changes that result from the transformation of the original circle on the sphere to an ellipse on the plane, we can determine the amount of angular and areal deformation that has occurred at that point. To demonstrate how this may be done, it is convenient to use finite lengths to represent the SF magnitudes, even though these are values that occur at infinitely small points and thus have no actual dimension.

Figure 4.3 represents a *point* on the sphere at O. The SF in every direction at O is considered unity, so that the point O can be represented by a circle with $OM = 1 =$ the radius r. In the system of projection the directions, OB and OA are the directions of the perpendiculars on the sphere that

---

* *In conformal projection the SF is the same in all directions at a point; consequently, the indicatrix remains a circle on the plane. If we consider a circle as a special form of the ellipse in which major and minor axes are the same, we may say that the indicatrix is always an ellipse. This will simplify wording.*

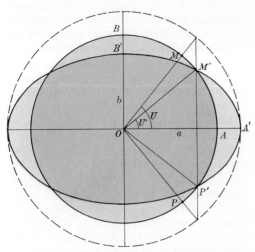

FIGURE 4.3 *The indicatrix (the ellipse) has been constructed as an equal-area representation of the circle to which the following indices apply: OM = OB = OA = r = 1; Scale factors in the directions OA' = 1.25, and OB' = 0.80; MOA = U = 51°21'40"; M'OA' = U' = 38°39'35"; U − U' = 12°42'05"; 2ω = 25°24'10". (Modified from Marschner.)*

are retained as perpendiculars in the projection. The SF in the direction OA is designated as a and that in OB as b. Since $OB = OA = OM$, a and b in the original circle $= 1.0$. In the ellipse resulting from projection, $a = 1.25$ and $b = 0.80$. These are also the directions of maximum changes in scale in the indicatrix, and the values for a and b are all that are needed to determine for any point on any projection the amount of either the angular or areal alteration that has occurred at that point.

ANGULAR CHANGE. To analyze first the angular deformation that has taken place at O, it is necessary to understand that all points on the circumference of the circle will have their counterparts on the periphery of the ellipse. Point B has been shifted to point B', and A to A' on the ellipse. It is evident that no angular change in these directions has taken place, since angle BOA = angle B'OA'. All other points on the arc between B and A will, when projected to the ellipse, be shifted a greater or lesser amount in their direction from O. Each such point on the circle will have a corresponding position on the ellipse. The point subjected to the *greatest* deflection is identified in the circumference of the circle with M, and it has its counterpart in the periphery of the ellipse in point M'. The angle MOA on the sphere thus becomes M'OA' in the projection. If angle $MOA = U$ and angle $M'OA' = U'$, then $U - U'$ denotes the *maximum* angular deformation within one quadrant. The value of $U - U'$ is designated as *omega* (ω). If an angle such as MOP were to have its sides located in two quadrants and if they were to occupy the position of maximum change in both directions, then the angle in question would be changed to M'OP' and would thus incur the maximum deflection for one quadrant on both sides. Consequently, the value of 2ω denotes the possible *maximum angular* change that may occur at a point. All other angular deformations at O would be less than 2ω. Since the values of ω will range from 0° in the directions of the axes to a maximum somewhere between the axes, it is not possible to state an average angular deformation.

AREA CHANGE. Changes in the representation of areas may or may not be a corollary to the transformation of the circle into an ellipse by the projection system. If there has been a change in the surface area, its magnitude can be readily established by comparing the areal "contents" of the original circle with that of the ellipse. The area of a circle is $r^2\pi$, whereas the area of an ellipse is $ab\pi$, where a and b represent the semimajor and semiminor axes, respectively. Therefore, since the axes of the ellipse are based on the original circle whose radius was unity, and since π is constant, the product of ab compared to unity expresses how much the areas have been changed. The product of ab is designated as S.

Deformation in Map Projections

Regardless of the system employed to transform the spherical surface to a plane, the geometrical relationships on the sphere cannot be entirely duplicated. Angles, areas, distances, and directions are subject to a variety of changes, and there are many other specific spatial conditions that may or may not be duplicated in map projections, such as parallel parallels, converging meridians, perpendicular intersection of parallels and meridians, poles being represented as points, and so on. The major alterations, however, are those having to do with angles, areas, distances, and directions. In order to provide the student with an understanding of the basic consequences of these alterations, a brief resumé of their characteristics follows.

Angular Representation.

The compass rose appears the same everywhere on the globe surface (except at the poles); that is, at each point, the cardinal directions are always 90° apart, and each of the intervening directions is everywhere at the same angle with the cardinal directions.

It is possible to retain this property of angular relations to some extent in a map projection. When it is retained, the projection is termed *conformal* or *orthomorphic,* and both words imply "correct form or shape." It is important to understand that these terms apply to the directions or angles that obtain *at points*. The attribute of correct shape is not meant to apply to areas of any significant dimension; no projection can provide correct shape to areas of any great extent.

On a sphere or a globe map, by definition the SF is 1.0 in every direction at every point. In the transformation process deformation of some kind must occur, and the SF must vary from point to point. It is possible, however, to arrange the stretching and compression so that at each point on a conformal projection $a = b$ in all directions. When this condition obtains, all directions around each point will be represented correctly and the parallels and meridians will intersect at 90°. It must be emphasized that this desirable quality is limited to directions at points and does not necessarily apply to directions from one place to another. It is also important to realize that just because a projection shows parallels and meridians as crossing one another with right angles, it does not necessarily mean it has the property of conformality.

Area Representation.

It is possible to retain in a map projection the representation of relative sizes of surface areas. When this characteristic of the surface of the sphere is retained, the projection is said to be *equivalent* or *equal-area.* The property of equivalence is obtained by arranging the SF in the directions of the perpendiculars retained on the projection so that the product of $ab = S = 1.0$ everywhere. If this is so, all areas of figures on the projection will be represented in correct relative size.

On such a projection the SF can be the same in *all* directions at only one or (at the most) two points or along one or two lines. At all other places it will be different in different directions from each point. Hence, angles around all such points will be deformed. Since the scale requirements for conformality and for equivalence in a map projection are contradictory, it is apparent that no projection can be both conformal and equivalent. Thus all conformal projections will present similar earth areas with unequal sizes and all equivalent projections will deform most earth angles.

The representation of areas and angles are the most important for the majority of cartographic representations. Two other aspects of representation need be considered, however, in order that the student may have a sound understanding of what must happen when the one surface is transformed to another. One of these concerns distances.

Distance Representation.

Needless to say, any map projection represents all distances "correctly," provided we know the scale variations involved. As generally understood, however, distance representation is a matter of maintaining consistency of scale; that is, for finite distances to be represented "correctly," the scale must be *uniform* along the extent of the appropriate line joining the points being scaled and must be the same as the principal scale on the nominal globe from which the projection was made, that is, have an SF of 1.0. The following are possible.

1. Scale may be maintained in one direction, for example, north-south or east-west, but only in one direction. When this is done, the lines (whether or not they are parallels or meridians) along which the SF remains 1.0 are called *standard*.

2. Scale may be maintained in all directions from *one* or *two* points, but only from those points. Such projections are called *equidistant*.

Direction Representation. Just as it is impossible to represent all earth distances with a consistent scale on a projection, so also is it impossible to represent all earth directions correctly with straight lines on a map. It is true that we can arrange the SF distribution so as to obtain straight rhumb lines or great circles. But no projection can show *true direction* in the proper sense that all great circles will be shown as straight lines that will have the same angular relations to the graticule of the map that they have with the sphere. For example, the oft-stated assertion that Mercator's projection "shows true direction" applies only to the fact that constant bearings are shown as straight lines. Such a statement is erroneous in that true direction on a sphere is along a great circle, not along a rhumb except, of course, when the two coincide (Fig. 4.4).

When directions are defined properly as great-circle bearings and, if we think of a correct direction as being that shown by a great circle as a straight line on the map having the proper azimuth reading with the local meridian, certain representations are possible.

1. The course of great circle arcs between all points may be shown as straight lines for a limited area, although the angular intersections of all the great circles with the meridians (azimuths) will not be shown correctly. To do this causes such a strain, so to speak, on the transformation process that it is not possible to extend it to even an entire hemisphere.

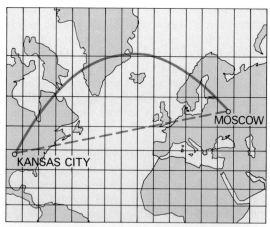

FIGURE 4.4 *The great circle and the rhumb from a point in the United States to a point in Russia as they appear on a Mercator projection. The great circle shown as a solid line is the "true direction" (i.e., most direct from the one point to the other, not the rhumb).*

2. Straight great circles with correct azimuths may be shown for all directions from *one* or, at the most, *two* points. Such projections are called *azimuthal*.

The Analysis and Depiction of Deformation

In order to compare one projection with another in terms of the amount and distribution of the deformation, we may employ various approaches. Some are entirely graphic and simply show a sort of graphic index of the amount and location of the deformation. Of generally more utility are the quantitative measures 2ω and S.

On all conformal projections, $a = b$ everywhere on the projection and, when $a = b$, the value of 2ω is $0°$. Hence there is no angular deformation at points on a conformal projection but, because the values of a and b vary from place to place, the product of ab, that is, S, will also vary from place to place. Consequently, all conformal projections exaggerate or reduce relative areas, and S at various points provides an index of the degree of areal change.

On equivalent projections the scale relationships at each point are such that the product of ab always equals 1.0. Any difference between the values of *a* and *b* will produce a value of 2ω greater than 0°. Consequently, all equivalent projections deform angles, and the value of 2ω at various points provides an index of the degree of angular deformation.

On all projections that are neither conformal or equivalent, *a* will neither equal *b* nor will the product of $ab = 1.0$. Therefore, on such projections, both the values of S and 2ω will vary from place to place.

The distributional pattern of the values of S and 2ω that occur at various points on a projection may be shown by isarithms of equal value. Many projections may differ in detail but be derived from the same basic geometric model, such as a cone or a cylinder. Commonly, all the projections in such a class will have similar patterns of deformation.

It is also possible to derive the mean value of the deformation for either the entire projection or for only a portion of the projection, such as the land area. A comparison of mean values for several similar projections is helpful in evaluating their relative merit.

Tissot's indicatrix is limited, in its analytical function, to values at a point. It does not provide much help in depicting another kind of deformation, which exists in all projections, that of changes in the distance and angular relation among widely spaced points or areas such as continents. To date, this kind of deformation has not been found to be commensurable; consequently, graphic means have been employed to help show the alteration that takes place with respect to the larger spatial relations.

Various devices have been employed to this end, such as a head plotted on different projections to illustrate elongation, compression, and shearing of larger areas (Fig. 4.5). The deformation of directions is effectively demonstrated by plotting various great circles in different parts of a projection (Fig. 4.6). Another device has been the covering of the globe with equilateral triangles and then reproducing the same triangles on the different projections.

The Arrangement of the Graticule

The surface of a sphere is the same everywhere insofar as its general geometry is concerned. At any point, the surface curves away in all directions

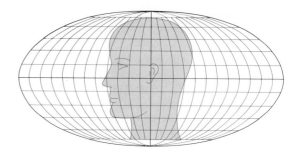

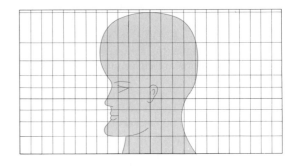

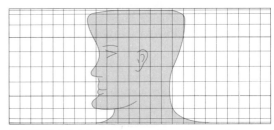

FIGURE 4.5 *A head portrayed on Mollweide's projection and the outline then plotted on Mercator's projection (center), and (bottom) the cylindrical equal-area projection with standard parallels at 30 degrees, to illustrate deformations. Because the profile looks most natural on Mollweide's projection does not mean that projection is "better." The natural profile could have been drawn on any one and then plotted on the others.*

at the same rate. In order to have something to provide positive identification of location on such a limitless, uniform space, we use the earth's coordinate system. It is so useful that we find it difficult to think of the earth's surface without automatically including the graticule. Consequently, we

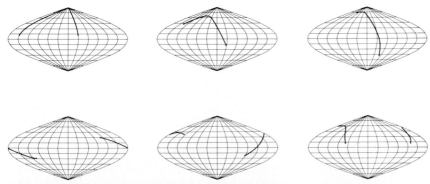

FIGURE 4.6 *Selected great circles on the sinusoidal projection showing the departure from a straight line. Each uninterrupted and interrupted arc is 150 degrees long. (From a report produced for the Office of Naval Research, Geography Branch, by W. Tobler.)*

tend to conceive of a map projection as a representation of the graticule. Furthermore, because the concept of cardinal directions is so important to people in their thinking about the earth, it has become conventional to represent the surface in such a way as to present these directions to good advantage. This is commonly done by making significant directions appear as straight lines or arcs, such as east-west parallels or north-south meridians. This necessitates organizing the chosen system of projection in a particular or conventional way.

It is important for the student to realize that however the system of projection may be arranged or oriented with respect to the graticule, any system of projection is merely one of transforming a spherical *surface* to a plane *surface*. The graticule has nothing to do with it except to provide a handy series of reference points. In whatever way the system of projection may be positioned with respect to the sphere, the characteristics of the projection would still be the same. The pattern of deformation and the amounts of deformation would not change. Figure 4.7 shows how the graticule appears when the sinusoidal projection is centered at various places. They are all the same system of projection.

On the other hand, in working with a spherical surface, we must have reference points. The coordinate system provides these. Therefore, since the graticule is useful to establish convenient

reference points (because it is well known and because the cardinal directions are important), it is common for a transformation system to be applied to the sphere in such a way that the graticule is displayed in a simple fashion. This may be accomplished by arranging the system so that deformation is symmetrical around well-known lines, such as a meridian or one or two parallels. Such lines, when a SF of 1.0 is held constant along them, are standard lines. They constitute reference lines that define the principal scale and from which the scale departs in other parts of the map.

The concept of standard parallels and standard meridians will be used in the next section to help describe the conventional form of projections. Remember, however, that the arrangement of the transformation system so that standard great circles coincide with the equator or meridians, or so that the standard small circles coincide with parallels, is merely a great convenience; it is not a necessity.

MAP PROJECTIONS

There are an infinite number of ways to transform a spherical surface to a plane and, of course, the variety of cartographic objectives is unlimited. Fortunately, a number of systems of map projection combine several useful characteristics, with the result that relatively few map projections are in common use. The versatility of the computer and

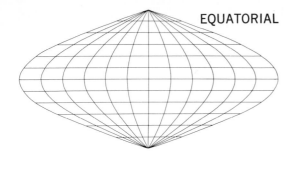

EQUATORIAL

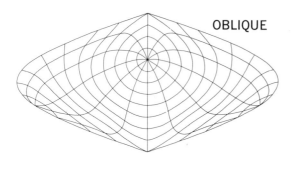

OBLIQUE

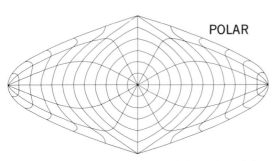

POLAR

FIGURE 4.7 *Different centerings of the sinusoidal system of projection produce different appearing graticules. Nevertheless, the pattern of the deformation is the same on all, since the same system of transformation is employed; therefore, they are all the same projection. (From a report produced for the Office of Naval Research, Geography Branch, by W. Tobler.)*

plotter have made it much easier to obtain a good match between map objective and projection, but the advantage of familiarity has tended to restrict the choice of transformation systems to those that are better known.

The number of diverse factors that may influence the choice of a map projection is surprising. The geographer, historian, and ecologist are likely to be concerned with sizes of areas. The navigator, meteorologist, astronaut, and engineer are generally concerned with angles and distances. The atlas mapmaker often wants a compromise, the illustrator is usually limited by a prescribed format, and the maker of a series of maps is interested in how sheets may be made to fit together well.

Because the selection from among map projections necessarily involves evaluation of deformation, there is a tendency to think of map projections as necessary but poor substitutes of the globe surface. On the contrary, in most instances projections are advantageous for reasons in addition to the fact that it is cheaper to make a flat map than a globe map. Map projections enable us to map distributions and derive and convey concepts that would be either impossible or at least undesirable on a globe. Furthermore, the notion that one projection is by nature better than another is unwarranted. There are some projections for which no useful purpose is known, but there is no such thing as a bad projection—there are only poor choices.

The Choice of Map Projections

There are many factors to be kept in mind when choosing a system of projection. No specific formulas can be given that will lead to the right selection because each map is a complex compound of objectives and constraints. A few generalizations can be drawn that will exemplify the complex nature of this fundamental choice that the cartographer must make.

One of the more important factors involved in selecting a map projection is the manner in which the inherent deformation is arranged with respect to the area covered by the graticule. Certain general classes of projections have specific arrangements of the deformation values, and the knowledge of these patterns helps considerably in choosing and using a particular system.

Maps to be made in series, such as sets for atlases or even topographic series, have different requirements from those made as individual maps. For example, a most useful attribute of some aspects of some projections is the fact that any portion can be cut from the whole and provide a segment that is in itself a relatively good selection for a smaller area, with good symmetry and deformation characteristics. Most projections in which the meridians are straight lines that meet the parallels at right angles satisfy this requirement.

A great many maps demand more from the map projection than merely one or a combination of the special properties of projections (equivalence, conformality, and azimuthality). Projection attributes such as parallel parallels, localized area deformation, and rectangular coordinates frequently are significant to the success of a map. For example, a map of a distribution that does not require equivalence may have a concentration of the information in the middle latitudes. In that case, a projection that to some extent expanded the areas of the middle latitudes would be a great help by allowing relative detail in the significant areas. Any small-scale map of temperature distri-

butions over large areas is made more expressive if the parallels are parallel; it is even more expressive if they are straight lines that allow for easy north-south comparisons. A map for which indexing of places is contemplated is more easily done with rectangular coordinates than with any other kind.

The overall shape of an area on a projection is likewise of great importance. Many times the shape and size (format) of the page or sheet on which the map is to be made is prescribed. By using a projection that fits a format most efficiently, a considerable increase in scale can often be effected, which may be a real asset to a crowded map.

The Classification of Projections. The usual categorization of projections is based on general geometric characteristics. Conceptually, the spherical surface is transformed to a "developable

CYLINDRICAL   AZIMUTHAL   CONIC

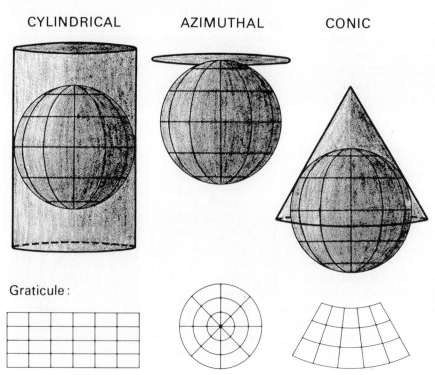

Graticule:

FIGURE 4.8 *Some of the developable surfaces to which the earth's surface may be "projected" and the appearances of the graticule when conventionally arranged.*

surface," which is a geometric form capable of being flattened, such as a cone or a cylinder (both of which may be cut and laid out flat) or a plane (which is already flat). Conventionally, the axis of the earth is aligned with the axes of the cylinder and cone (see Fig. 4.8) so that the graticule lines will be simplified. In a projection based on a cone, meridians converge in one direction and diverge in the other; on the opened-up cylinder, meridians are straight parallel lines. Projections on a plane are not so conventionally aligned, and no generalizations can be made about their appearance. Such a constructional grouping of projections results in categories called cylindrical, conic, azimuthal (plane), pseudocylindrical, and miscellaneous (those based on no geometric form). Whatever the terminology employed, the grouping is not strictly a classification but a listing. Other classifications or listings of projections include the parametric, the appearance of the graticule, and the relation of the spherical surface to the plane (secant, tangent, transverse, or oblique, for example).

The subject of classification of projections has been much investigated and general proposals for its accomplishment have been made, none of which seem to have gained general acceptance. The topic is of considerable interest from a theoretical point of view, but it is too complex to be dealt with here.*

The Basic Attributes of Projections. The primary elements in the choice of projections are generally their major property or properties: azimuthality, conformality, and equivalence. Each property is generally of sufficient significance so that it is usually the first distinguishing characteristic with which the cartographer begins to make a choice. The major properties have been defined earlier.

Secondary elements of considerable significance are the amounts and patterns of distribution of the deformation. *Mean deformation*, either maximum angular or area (2ω or S), is the weighted arithmetic mean of the values that occur over the projection. When derived for similar areas on different projections, a comparison of the mean deformation values provides one index of the relative efficiency of the forms of projection.

The pattern of deformation is the arrangement of either S or 2ω values on a projection. It is most easily symbolized and visualized by thinking of the quantities as representing the Z values of an S or 2ω third dimension above the projection. Then isarithms (contours) of this surface will show the arrangement of the relative values and, by their closeness, the gradients or rates of change. Certain classes of projections have similar patterns of deformation. These are diagrammed in Figs. 4.9 to 4.11. Within the shaded areas the heavy lines are the standard lines; the darker the shading, the greater the deformation.

An azimuthal pattern occurs if the transformation takes place from the sphere to a tangent or an intersecting (secant) plane. The trace of the intersecting plane and sphere will, of course, be circular. Lines of equal deformation are concentric arcs around the point of tangency or the center of the circle of intersection (see Fig. 4.9).

A conic pattern results if the initial transfor-

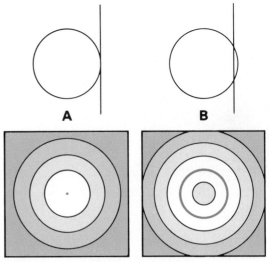

FIGURE 4.9 *Azimuthal patterns of deformation. (A) The pattern when the plane is tangent to the sphere at a point. (B) The pattern when the plane intersects the sphere. The trace of the intersection will, of course, be a small circle.*

* Refer to D. H. Maling, Coordinate Systems and Map Projections, *George Philip and Son Ltd., London, 1973, pp. 82–108.*

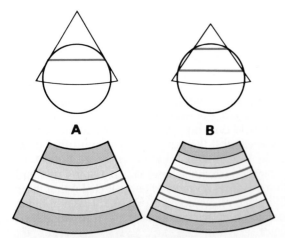

FIGURE 4.10 *Conic patterns of deformation. (A) The pattern when the cone is tangent to one small circle. (B) The pattern when the cone intersects the sphere along two small circles.*

mation is made to the surface of a true cone tangent at a small circle or intersecting at two small circles on the sphere. Lines of equal deformation parallel the standard small circles (Fig. 4.10).

A cylindrical pattern occurs on all map pro-

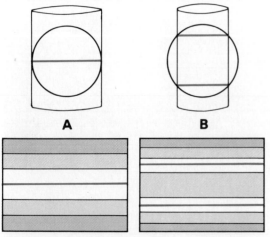

FIGURE 4.11 *Cyclindrical patterns of deformation. (A) The pattern when the cylinder is tangent to a great circle. (B) The pattern when the cylinder is secant and the standard lines are parallel small circles (they need not be parallels of the graticule).*

jections that, in principle, are developed by first transforming the spherical surface to a tangent or secant cylinder. In all cases the lines of equal deformation are straight lines parallel to the standard lines, the least deformation being along the line of tangency or intersection (see Fig. 4.11). Deformation will increase in all these instances away from the standard lines (or point). The greatest gradient will usually be in the direction normal (perpendicular) to the standard line.

Conformal Projections

Maps that are to be used for analyzing, guiding, or recording motion and angular relationships require the employment of conformal projections. In these categories fall the navigational charts of the mariner or aviator, the plotting and analysis charts of the meteorologist, and the general class of topographic maps. Many of the uses to which topographic maps are put do not require conformality but, since topographic maps serve a wide variety of engineering purposes, they have been made more and more often on conformal projections.

Because there is no angular deformation at any point on a conformal projection, the notion is widespread that shapes of countries and continents are well presented. Although it is correct to state that small areas on conformal projections retain good shapes, it is also true that in order to retain angular representation it is necessary to alter the area relationships. Thus, on small-scale conformal projections the area scale varies greatly from point to point; consequently, large areas may be imperfectly represented with respect to shape.

It is difficult to deal with the concept of deformation on a conformal projection because, in a sense, there is nothing deformed, since all angular relationships at each point are retained (i.e., $a = b$ everywhere). All that changes is the SF, and one point is as "accurate" as another; only the scales are different. Thus we can only refer to the principal scale as "correct," and the other parts of the projection will be merely relatively enlarged or reduced.

There are four conformal projections in common use: Mercator's, the Transverse Mercator, Lambert's conic with two standard parallels, and the stereographic. Much the best known is Mercator's projection because of its long utility to the mariner. The stereographic projection, familiar to the classical Greeks, is also widely used. In recent times a conic conformal, Lambert's conic with two standard parallels, and a transverse form of Mercator's have become more popular among cartographers.

Mercator's projection (Fig. 4.12) is probably the most famous map projection ever devised. It was introduced in 1569 by the famous Flemish cartographer specifically as a device for navigation, and it has served this purpose well. Conceptually, it is a cylindrical projection, but it must be mathematically derived. The theoretical cylinder is made tangent to the equator. It has the property that all rhumb lines (lines of constant bearing) appear as straight lines, an obvious advantage to someone trying to proceed along a compass course. Except for the meridians and the equator, great circles do not appear as straight lines, so Mercator's projection does not show "true direction"; but such courses can be easily transferred from a gnomonic projection that does so. The great-circle course can then be approximated by a series of straight rhumb lines. It is apparent that Mercator's projection enlarges (not distorts) areas at an increasing rate toward the higher latitudes,

so it is of little use for purposes other than navigation.

In the normal form of Mercator's projection the standard line is the equator along which the SF is 1.0. The rate of change of the SF values is relatively small for the lower latitudes, meaning that a zone along the equator is exceedingly well represented. For example, even at 10° latitude the SF would only be about 1.016. Since the earth is spherical and the equator is like any other great circle, it is apparent that we can twist Mercator's projection (or the sphere) 90° so that the standard line becomes a meridian (a great circle) that takes the place of the equator (Fig. 4.13). When this is done, the projection is called the Transverse Mercator.

The Transverse Mercator projection is conformal but, because most of the parallels and meridians are curves, it does not have the attribute that all rhumbs are straight lines. Consequently, since scale exaggeration increases away from the standard meridian, it is useful for only a small zone along that central line (Fig. 4.13). Lines of equal-scale exaggeration (parallels of the graticule in the normal aspect of Mercator's projection) are

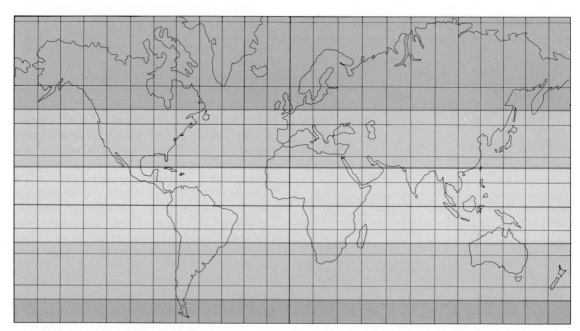

FIGURE 4.12 *Mercator's projection. Values of lines of equal area exaggeration over the principal scale at the equator are 25 and 250 percent.*

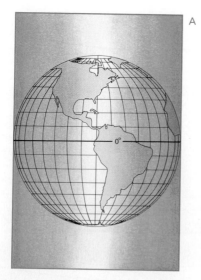

A

B

FIGURE 4.13 (A) *The conceptual cylinder for the normal form of Mercator's projection is arranged parallel to the axis of the sphere, resulting in the equator (0°) being tangent and thus the standard line. (B) To develop the Transverse Mercator projection, the cylinder has been turned 90 degrees, resulting in a meridional great circle being tangent and a meridian, in this case 90°W, becoming the standard line.*

less than 1.0 between the standard parallels and more than 1.0 outside them. Area deformation between and near the standard parallels is relatively small; thus the projection provides exceptionally good directional and shape relationships for an east-west latitudinal zone. Consequently, the projection is much used for air navigation in intermediate latitudes and for meteorological charts.

The stereographic projection (Fig. 4.21) also belongs in the azimuthal group. The deformation (in this case, area exaggeration) increases outward from the central point symmetrically. This is an advantage when the area to be represented is more or less square or of continental proportions.

Equivalent Projections

For the majority of maps used for instruction and for small-scale general maps, the property of equivalence commands high priority. The relative extent of geographical areas, such as nations, water bodies, and regions of geographical similarity (vegetation, population, climate, etc.) are matters of obvious significance. Such relationships can appear properly only on equivalent or near-equivalent projections.

Many of our general impressions of the sizes of regions are acquired subconsciously. Because nonequivalent projections have been so frequently used for instructional and small-scale general maps in the past, most people think Greenland is considerably larger than Mexico (nearly the same size) and that Africa is smaller than North America (Africa is more than 2 million mi^2 larger). For some maps the property of equivalence is more than a passive factor. Some kinds of symbolization require equivalence in order to portray relative densities. An example is illustrated in Fig. 4.15.

The choice among equivalent map projections depends on two important considerations.
1. The size of the area involved.
2. The distribution of the angular deformation.
There are a great many possibilities from which to choose; if the cartographer will but keep these two elements in mind, he will rarely make a bad choice.

As a general rule, the smaller the section to be represented, the less significant is the choice of an equivalent projection. Because a great many

small circles parallel to the standard central meridian (Fig. 4.23). In recent years it has become popular as a projection for topographic maps and as a base for plane coordinate systems.

Lambert's conic projection with two standard parallels in its normal form has concentric parallels and equally spaced, straight meridians that meet the parallels at right angles (Fig. 4.14). The SF is

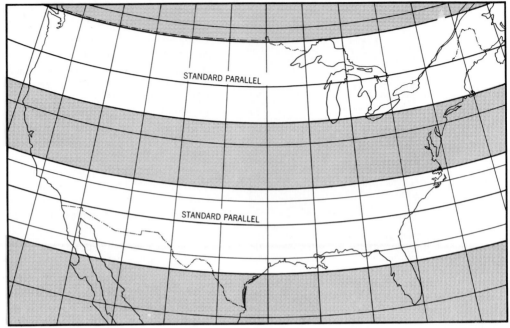

FIGURE 4.14 *Lambert's conformal conic projection. Values of lines of equal area exaggeration are 2 percent.*

ways have been devised to transform the surface of a sphere to a plane while maintaining the requirement for equivalence ($ab = 1$ everywhere), there are a relatively larger number of well-known

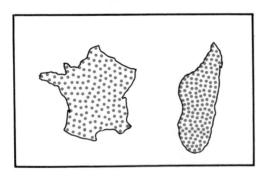

FIGURE 4.15 *These two areas France, left, and Madagascar (Malagasy Republic), right, are nearly the same size on earth. On the nonequal-area projection from which these two outlines were traced, France appears larger than it should be in comparison to Madagascar. The same number of dots has been placed within each outline, but the apparent densities are not the same.*

equivalent projections than azimuthal and conformal projections. Because the entire surface area of the earth can be plotted within the bounding lines of a plane figure of almost any shape, there is also a large number of such "world projections," many of which are equivalent. A representative selection is shown in Fig. 4.16.

It is apparent that for a world map the pattern of deformation is a matter of paramount concern. The "better" portions of such projections can also be used for maps of continents or even for areas of smaller extent.

A few representative types of common equivalent projections are shown, along with brief notes on their employment. Most of the illustrations include isarithms of 2ω to show the overall pattern of deformation.

Albers' conic projection (Fig. 4.17) has two standard parallel small circles (conventionally these are made parallels of the graticule) along which there is no angular deformation. Because it is conically derived, deformation zones are arranged parallel to the standard lines. Any two small circles in one hemisphere may be chosen as

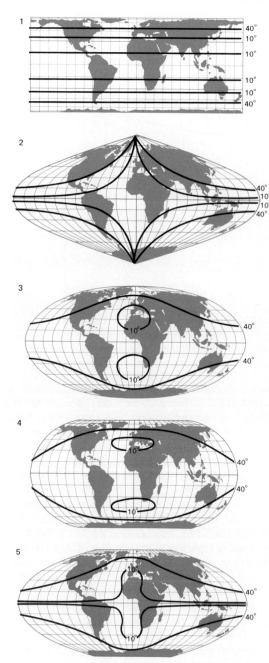

FIGURE 4.16 *A few of the many equivalent world map projections. Identified by number they are: 1-Cylindrical equal-area with standard parallels at 30°N and S latitude; 2-Sinusoidal projection; 3-Mollweide's projection; 4-Eckert's No. IV projection; and 5-Flat polar quartic authalic projection.*

standard, but the closer together they are, of course, the better will be the representation in their immediate vicinity. Because of the low deformation value and its neat appearance with straight meridians and concentric arc parallels that meet the meridians at right angles in its conventional form, this is a good choice for a middle-latitude area of greater east-west extent and a lesser north-south extent. Outside the standard parallels, the SF along the meridians is progressively reduced. Parallel curvature ordinarily becomes undesirable if the projection is extended for much over 100° longitude. In recent decades Albers' projection has replaced other projections (notably the polyconic) as the common choice for many maps of the United States. Its obvious superiority for maps on which to plot and study geographical distributions has led to its selection as the standard base map by many government agencies such as the Bureau of the Census.

Lambert's equal-area projection (Fig. 4.21) is azimuthal as well as equivalent. Since deformation is symmetrical around the central point, which can be located anywhere, the projection is useful for areas that have nearly equal east-west and north-south dimensions. Consequently, areas of continental proportions are well represented on this projection. It is limited to hemispheres.

The cylindrical equal-area projection (Fig. 4.16) is capable of variation like Albers' conic, that is, the projection has two parallel standard small circles, usually parallels of latitude. The two small circles (parallels) may "coincide," so to speak, and be a great circle (the equator), or they may be any two others so long as they are homolatitudes (the same parallels in opposite hemispheres). Deformation is arranged, of course, parallel to the standard small circles. Although for a variety of reasons this projection "looks peculiar" to many people, it does in fact provide, when standard parallels just under 30° are chosen, the least *overall* mean deformation of any equal-area world projection.

The sinusoidal projection (Fig. 4.16), when conventionally oriented, has a straight central meridian and equator, along both of which there is no angular deformation. All parallels are standard and a merit of this projection is that they are equally spaced, giving the illusion of proper spacing so that it is useful for representations

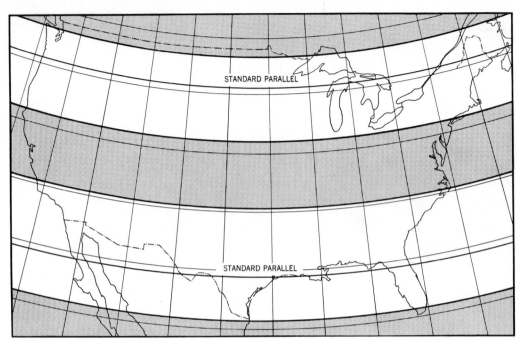

FIGURE 4.17 *Albers' conic projection. Values of lines of equal maximum angular deformation are 1 degree.*

where latitudinal relations are significant.* The sinusoidal is particularly suitable when properly centered for maps of less-than-world areas, such as South America, for which area the deformation distribution is especially fortuitous.

Mollweide's projection (Fig. 4.16) does not have the excessively pointed polar areas of the sinusoidal, and thus it appears a bit more realistic. In order to attain equivalence within its oval shape, it is necessary to decrease the north-south scale in the high latitudes and increase it in the low latitudes. The opposite is true in the east-west direction. Shapes are modified accordingly. The two areas of least deformation in the middle latitudes make the projection useful for world distributions when interest is concentrated in those areas.

Eckert's IV (Fig. 4.16) projection, when conventionally arranged, has a pole represented by a line half the length of the equator instead of by a

point. Accordingly, the polar areas are not so compressed in the east-west direction as on the preceding two projections. This takes place, however, at the expense of their north-south representation. As in Mollweide's, the equatorial areas are stretched in the north-south direction. Deformation distribution is similar to that of Mollweide's.

The flat polar quartic projection (Fig. 4.16) has a line one-third the length of the equator for the pole (when conventionally arranged). This provides a better appearance for the polar areas than either the sinusoidal or the Mollweide. It is becoming more and more popular for world maps.

Several projections for world maps have been prepared by combining the better parts of two. The best known of these is Goode's Homolosine (Fig. 4.18), which is a combination of the equatorial section of the sinusoidal and the poleward sections of Mollweide's; thus, it is equal-area.* The two projections, when constructed to

---

* It is an illusion because the distance between parallels is properly measured along a meridian, not simply perpendicular to the lines representing the parallels on the map.

* Mollweide's projection is sometimes called the homolographic, hence the combined form homolo + sine.

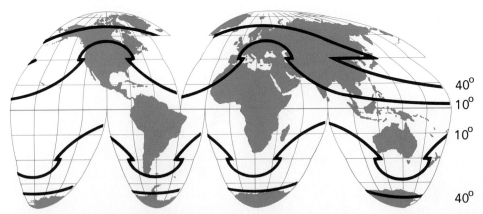

40°
10°
10°
40°

FIGURE 4.18 *Goode's Homolosine projection is an interrupted union of the sinusoidal projection equatorward of approximately 40 degrees and the poleward zone of Mollweide's projection. Values of black lines of equal maximum angular deformation are 10 and 40 degrees.*

the same area scale, have one parallel of identical length (approximately 40°) along which they may be joined. It is usually used in interrupted form (see below), and has been widely employed in the United States. Its overall quality, as shown by a comparison of mean values of deformation, is not appreciably better than Mollweide's alone.

Interruption and Condensing. When either the continents or the seas are the areas of primary interest, it is possible to display one or the other (lands or seas) to better advantage on an equivalent world projection by (1) interrupting the projection, and (2) condensing the map. Interruption allows one to repeat the better parts of the projection, as in Goode's Homolosine projection (Fig. 4.18). Condensing is merely "cutting out" the unwanted sections of the projection in order to attain greater scale for the areas of interest (Fig. 4.19).

Azimuthal Projections

As a group, azimuthal projections (sometimes called zenithal) have increased in prominence in recent years. The advent of common air travel, the development of radio electronics, the mapping of other celestial bodies, and the general increase in scientific activity all have contributed to this development. It has occurred because azimuthal projections have a number of useful qualities not shared by other classes of projections.

All azimuthal projections are "projected" on a plane that may be centered (i.e., made tangent or secant) to the sphere anywhere. A line perpendicular to the plane at the center point of the projection will necessarily pass through the center of the sphere. Consequently, an azimuthal projection is symmetrical around the chosen center. The variation of the SF in all cases changes from the center at the same rate in every direction. If the plane is made tangent to the sphere, there is no deformation of any kind at the center; if it is made secant the deformation will be least along a small circle. Furthermore, all great circles passing through the center will be straight lines and will show the correct azimuths from and to the center in relation to any point. It should be emphasized that only azimuths (directions) from and to the *center* are correct on an azimuthal projection.

At the center point all azimuthal projections with the same principal scale are identical, and the variation among them is merely a matter of the scale differences along the straight great circles that radiate from the center. Figure 4.20 illustrates this relationship. The fact that the deformation is arranged symmetrically around a center point makes this class of projections useful for areas having more or less equal dimensions in each direction, or for maps in which interest is not localized in one dimension. Because any azimuthal projection can be centered anywhere and still present a reasonable-appearing graticule, the class is more versatile than others.

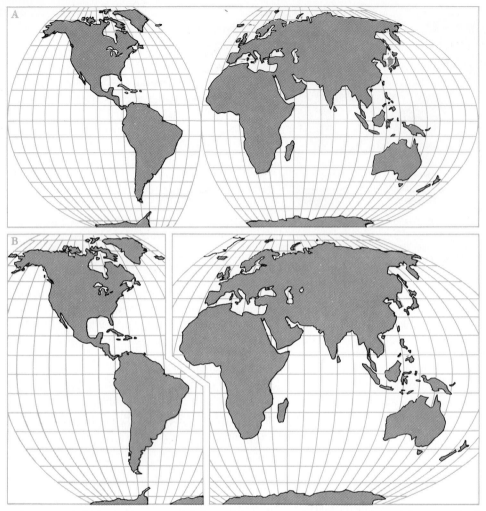

FIGURE 4.19 *A shows an interrupted flat polar quartic equivalent projection of the entire earth. By cutting out unwanted areas, as in B, additional scale is obtained within a limiting width. The areas of concern could just as well be the ocean areas.*

There are, of course, an infinite number of azimuthal projections possible,* but only five are well known: the stereographic, Lambert's equal area, the azimuthal equidistant, the orthographic, and the gnomonic (Fig. 4.21).

The conformal stereographic and Lambert's equal-area projections (Fig. 4.21) were considered in the sections that treated those two major

properties. The stereographic, however, is unique among projections; it has the attribute that any circle or circular arc on the sphere will plot as a circle or circular arc on the projection.* It is pos-

---

* *Any vertical air photograph is a perspective azimuthal projection.*

* *Since all great circles through the center of the projection are, of course, straight lines, we must define these as circles with a radius of infinity to make this statement not open to argument! In any case, any small circle on the earth within the limits of the projection can truly be drawn with a compass on the projection. It may be done by locating the ends of a diameter along a straight line radial*

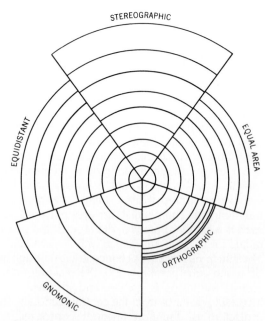

FIGURE 4.20 Comparison of azimuthal projections with a common center, in this case a pole on the earth. In all cases meridians (great circles) are simply straight lines radiating from the center. Note that the only variation is in the spacing of the parallels; in other words, the only difference among the projections is the scale away from the center.

sible, therefore, to plot the ranges of radiating objects, such as radio waves or aircraft, merely with a compass, as directed in the footnote.

The azimuthal equidistant projection has the unique quality that the linear scale is uniform along the radiating straight lines through the center. Therefore, the position of every place is shown in simple relative position and distance in relation to the *center*. Directions and distances between points whose great circle connection does not pass through the center are not shown correctly.

Any kind of movement that is directed toward or away from a center, such as radio impulses and seismic waves, is well shown on this projection. The projection has an advantage over many of the other azimuthal projections in that it is possible to show the entire sphere (Fig. 4.22). Most azimuthals are limited to presenting a hemisphere or less. Since the bounding circle on this world projection is the point on the sphere opposite to the center point (antipode), shape and area deformation in the periphery are excessive.

The orthographic projection looks like a perspective view of a globe from a considerable distance, although it is not quite the same. For this reason it might almost be called a visual projection in that the deformation of areas and angles, although great around the edges, is not particularly apparent to the viewer. On this account it is useful for preparing illustrative maps wherein the sphericity of the globe is of major significance.

The gnomonic projection has the unique property that all great-circle arcs are represented as straight lines anywhere on the projection. The projection is useful in marine and air navigation, since the navigator need only join the points of departure and destination with a straight line on a gnomonic chart and the location of the great-circle course is displayed. Because compass directions constantly change along most great circles, the navigator transfers the course from the gnomonic graticule to one on a conformed projection, such as Mercator's projection, and then approximates it with a series of rhumbs, which are straight lines on the Mercator chart.

Other Systems of Projection

At least 250 different systems of projection have been devised and described in the literature of mathematical cartography. New ones are proposed with regularity either to fit a particular need or merely because the transformation of a spherical surface to a plane with given constraints is an intriguing mathematical problem. It is impossible in this textbook to do more than describe briefly the widely used projections. To range further, refer to technical treatises, some of which are listed in the bibliography.*

(great circle) through the center, and then finding the construction *center* of the circle by halving the diameter. The actual center of the circle on the earth and the construction center for that circle on the projection do not coincide except at the center of the projection.

* A most useful reference and introduction to the complexity of the great variety of projections is D. H. Maling, "The Terminology of Map Projections," International Yearbook of Cartography, 8, 11–64 (1968).

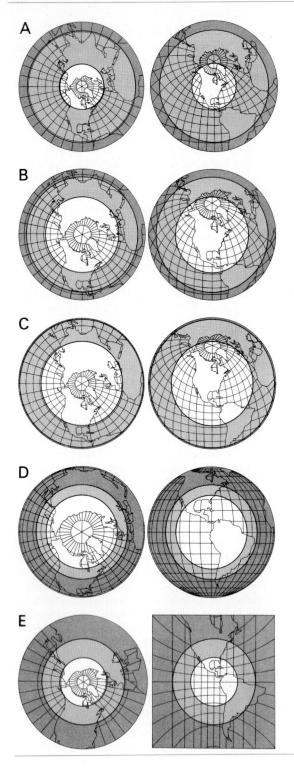

Most of the well-known and widely used projections are those described above, which have specific properties, but there are a few that do not have specific properties and yet have particular attributes that make them useful for particular purposes. Those include the polyconic, the plane chart, and the simple conic with two standard parallels.

The polyconic projection was first widely used in the United States by the Coast and Geodetic Survey (now the National Ocean Survey), and it was subsequently adopted as the projection for the Standard topographic series of the U.S. Geological Survey. It has a straight, standard central meridian. The parallels are arcs of circles, and each is a standard parallel, drawn with the proper radius for its cone and hence with its own center. Thus the parallels are not concentric. Although the SF along each parallel is 1.0, the scale along the curved meridians is greater and increases with increasing distance from the central meridian. The projection is, therefore, neither conformal nor equivalent. On the other hand, when it is used for a small area, bisected by a central meridian, both these qualities are closely approached.

For the mapping of a large area on a large scale, the development of each map sheet on its own polyconic projection is therefore an acceptable solution to the projection problem. Each such small section will fit perfectly with the adjacent ones to the north or south but, because of the curved meridians, they will not fit together east-west. The variations of the SF within the 7½′ and 15′ quadrangles of the standard topographic maps of the United States is insignificant and usually less than that which results from paper shrinkage or expansion. Although it is admirably suited for individual maps covering small areas, it is clearly not suited for maps of larger areas.

A modified polyconic projection is used for the so-called Millionth Map or International Map

FIGURE 4.21 *The five well-known azimuthal projections. (A) Stereographic. (B) Lambert's equal-area. (C) Azimuthal equidistant. (D) Orthographic. (E) Gnomonic. In A the zones of areal exaggeration are less than 30%, 30 to 200%, over 200%, In B, C, D, and E, zones of maximum angular deformation are 0 to 10 degrees, 10 to 25 degrees, over 25 degrees.*

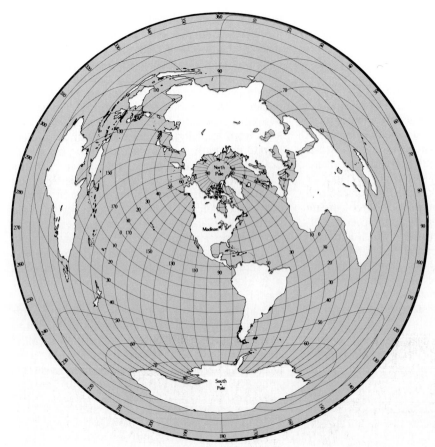

FIGURE 4.22 *An azimuthal equidistant projection of the world centered on Madison, Wisconsin, prepared for geophysical uses. Note the deformation near the outer edge. (University of Wisconsin Cartographic Laboratory.)*

of the World. It was changed by making the meridians straight instead of curved and by making two of them, instead of one, standard on each sheet. Whereas in the ordinary polyconic all parallels are standard, in the modified polyconic only the bottom and top one in each sheet are standard. This makes it possible to fit sheets together east-west as well as north-south at the slight expense of some linear and area scale variation. The diagonal sheets cannot fit.

The plane chart, sometimes called the equirectangular projection, is one of the oldest and simplest of map projections. It is useful for city plans or base maps of small areas. It is easily constructed and, for a limited area, has small defor-

mation. All meridians and any chosen central parallel are standard. The projection may be centered anywhere.

There are several possible arrangements of scale that may be used to produce a conic projection with two standard parallels. These projections are similar in appearance to Albers' and Lambert's conic projections, but they do not have the special properties of those two. Such a projection does not distort either areas or angles to a very great degree if the standard parallels are placed close together and provided the projection is not extended far north and south of the standard parallels. This kind of projection is frequently chosen for areas in middle latitudes of too large an extent for an equirectangular projection and for maps not requiring the precise properties of equivalence or conformality.

PLANE COORDINATE SYSTEMS

As described in Chapter 3, the use of rectangular coordinate systems has become widespread during this century. Most large-scale topographic maps show one or more systems of rectangular coordinates, and some areas (e.g., Great Britain) have adopted a national system of rectangular coordinates. Although it is simple enough to draw a rectangular grid on any map and thereby create a rectangular coordinate system, for precise referencing and directional relationships it is necessary to develop a plane coordinate system on a map projection, preferably one that is conformal. Three such projections are used: the Transverse Mercator projection, Lambert's conic with two standard parallels, and the stereographic.

The UTM System

Although individual countries may develop particular systems suitable to their needs, one system that has become widely used is the Universal Transverse Mercator (UTM) grid system. The grid system and the projection on which it is based have been widely adopted for topographic maps, referencing of satellite imagery, and other applications that require precise positioning. As described earlier in this chapter, the conformal Transverse Mercator projection (also called Gauss-Krüger) results from applying Mercator's system of projection to the earth, with a meridian being the standard great circle in place of the equator. In the simplest case this would mean that the SF along the meridian would be 1.0 and that the SF magnitudes would increase at right angles away from the central meridian. To balance the scale variation in the UTM system, the projection is made secant so that the standard lines are small circles paralleling the central meridian spaced 180 km east and west of it. Figure 4.23 shows this relationship and how the SF range is thereby reduced.

In the UTM grid system the area of the earth between 80°N and 80°S latitude is divided into north-south columns 6° of longitude wide called zones. These are numbered from 1 to 60 eastward, beginning at the 180th meridian. Each col-

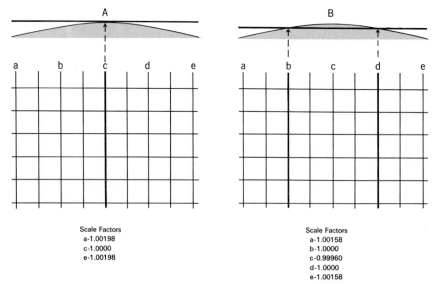

Scale Factors
a-1.00198
c-1.0000
e-1.00198

Scale Factors
a-1.00158
b-1.0000
c-0.99960
d-1.0000
e-1.00158

FIGURE 4.23 *Instead of a single line being standard, as in A, two are made standard, as in B, on the transverse Mercator projection system used for UTM rectangular grid reference purposes. The standard lines (small circles) b and d in B are 180 km east and west of the central meridian. The east-west zone shown here would be approximately 725 km wide or about 6.5° longitude near the equator.*

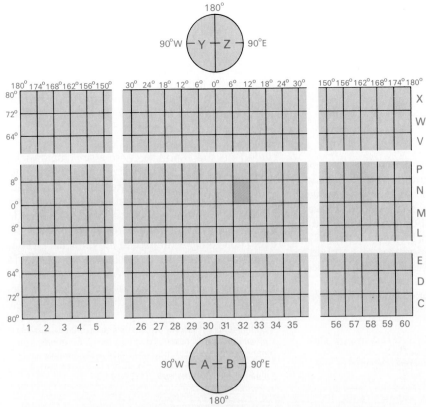

FIGURE 4.24 *The system of UTM grid zone designations. Each 6° latitude by 8° longitude quadrilateral is identified by its column number and row letter (I and O are omitted). The tinted zone is 32N.*

umn is divided into quadrilaterals 8° of latitude high. The rows of quadrilaterals are assigned letters C to X consecutively (with I and O omitted) beginning at 80°S latitude (Fig. 4.24). Each 6° east-west by 8° north-south quadrilateral is designated by the appropriate number-letter combination. As always, in giving a grid reference, one reads right, up. Each quadrilateral is divided into 100,000-m zones designated by a system of letter combinations.

Within each zone the meridian in the center of the zone is given an easting value of 500,000 m. The equator is designated as having a northing value of 0 for the northern hemisphere coordinates and an arbitrary northing value of 10 million m for the southern hemisphere.

The UPS System

The Universal Polar Stereographic (UPS) grid system is based on the conformal stereographic projection centered at the pole. Like the UTM, in order to decrease the SF variation, it is made secant with the standard small circle being about 81°. The SF at the center (pole) is 0.994, at 81° latitude (approximately) it is 1.000, and it increases to 1.0016 in the vicinity of 80° latitude.

In the UPS grid system each circular polar zone is divided in half by the 0°–180° meridian. In the north polar zone the west half (west longitude) is designated grid zone Y, the east half, Z. In the south polar zone the west half is designated A, the east half B (Fig. 4.24).

In the polar areas the northings and eastings

of a grid system must be arbitrarily assigned. In both zones the 2 million m easting coincides with the 0°–180° meridian line. The 2 million m northing coincides with the 90°E–90°W meridian line. Grid north is parallel to true north along the 0° meridian and, therefore, also to true south along the 180° meridian. The UPS zones are divided into 100,000-m squares like the UTM.*

State Plane Coordinates

In order to provide the convenience of plane coordinates and a way to ensure the permanent recording of the location of original land survey monuments, the U.S. Coast and Geodetic Survey (now the National Ocean Survey) worked out a system of plane coordinates for each of the states on either the Transverse Mercator or Lambert's conic projection that are specifically tied to locations in the national geodetic survey system. To keep the inevitable scale variation to a reasonable

minimum, a state may have two or more overlapping zones, each of which has its own projection system and grid. The units used are feet. The large-scale topographic maps published by the U.S. Geological Survey now carry tick marks showing the locations of the 10,000-ft grid.*

Figure 4.25 shows both the 1000-m UTM grid lines of Zone 16 and the 10,000-ft grid lines of the Wisconsin Coordinate System South Zone, extended over the southern third of the area included on the Madison, Wisconsin Quadrangle, 1:62,500, U.S. Geological Survey topographic map. The three different orientations (the graticule, that is, the net of parallels and meridians, the UTM grid, and the grid of the Wisconsin Coordinate System South Zone) result from each set being differently arranged. In the graticule all lines are true north-south or east-west. In the UTM Zone 16 the meridian of 87°W longitude is central and is the north-south axis about which the rec-

---

*A complete description of the UTM and UPS grid systems is given in Department of the Army, Grids and Grid References, TM 5-241-1, Headquarters, Department of the Army, Washington, D.C., 1967.

* A complete description of the state plane coordinate systems is given in Hugh C. Mitchell and Lansing G. Simmons, The State Coordinate Systems, U.S. Coast and Geodetic Survey, Special Publication No. 235, U.S. Government Printing Office, Washington, D.C., 1945.

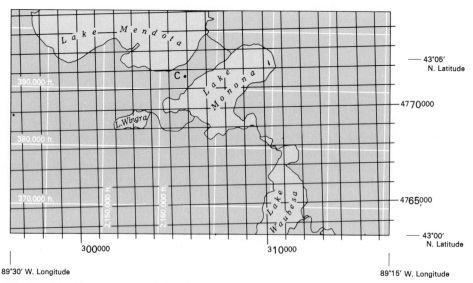

FIGURE 4.25 *Three systems of coordinates that appear on the Madison Quadrangle, USGS, 1:62,500. The bordering ticks show the graticule; the white lines show the 10,000-ft grid lines of the Wisconsin Coordinate System, South Zone; and the interior black lines show the 1000-m UTM grid lines.*

tangular grid is arranged.* In the Wisconsin Coordinate System the north-south axis of the rectangular grid is 90°W longitude. Accordingly, in the area shown on Fig. 4.25 it will be seen that the UTM grid deviates slightly from a north-south, east-west orientation in a fashion opposite to the arrangement of the Wisconsin Coordinate System.

On published topographic maps the positions of the grid lines on the two sets are shown only by hard-to-distinguish ticks around the margins of the map. Point *C* in Fig. 4.25 is the State Capitol. Grid references of *C* would be read as follows.

1. UTM SYSTEM. Normally the small initial digits in the UTM grid shown on a map (3 in the eastings and 47 in the northings on Fig. 4.25) are not used. In the UTM Grid System the reference 0571 would locate Point *C* within a 1000-m square and 058716 would locate it within a 100-m square, these designations being the locations of the south-west corners of the respective squares. Adding yet another digit to each set would locate it within a 10-m square, and adding a fifth digit to the easting and the northing would locate it within a 1-m square. Naturally, extremely accurate, larger-scale maps would be needed to obtain that much precision.

2. WISCONSIN COORDINATE SYSTEM. On the published topographic map 1/100 in. represents 52 ft so that the reading of a coordinate position to that precise value would be quite useless, since the paper the map is printed on would be subject to greater distortion with

changes in humidity. State plane coordinates are surveyed to the fraction of a foot in the field, and then are fully given. Point *C* in Fig. 4.25 is located at approximately 2,164,600 ft east and 392,300 ft north in the Wisconsin Coordinate System, South Zone.

SELECTED REFERENCES

Balchin, W. G. V., "The Choice of Map Projections," *Empire Survey Review, 12,* 263–276 (1954).

Dahlberg, R. E., "Evolution of Interrupted Map Projections," *International Yearbook of Cartography, 2,* 36–53 (1962).

Department of the Army, *Grids and Grid References,* TM 5-241-1, Headquarters, Department of the Army, Washington, D.C., 1967.

Keuning, J., "The History of Geographical Map Projections Until 1600," *Imago Mundi, 12,* 1–24 (1955).

Maling, D. H., *Coordinate Systems and Map Projections,* George Philip and Son Ltd., London, 1973.

Marschner, F. J., "Structural Properties of Medium and Small Scale Maps," *Annals,* Association of American Geographers, *34,* 1–46 (1944).

Mitchell, H. C., and L. G. Simmons, *The State Coordinate Systems,* U.S. Coast and Geodetic Survey, Special Publication No. 235, U.S. Government Printing Office, Washington, D.C., 1945.

Steers, J. A., *An Introduction to the Study of Map Projections,* 14th ed., University of London Press, London, 1965.

Stewart, J. Q., "The Use and Abuse of Map Projections," *The Geographical Review, 33,* 589–604 (1943).

Tobler, W. R., "Geographic Area and Map Projections," *The Geographical Review, 53,* 59–78 (1963).

Tobler, W. R., "A Classification of Map Projections," *Annals,* Association of American Geographers, *52,* 167–175 (1962).

---

* The 100,000-m squares of the military UTM grid system are not used in the application of the UTM system on U.S. Geological Survey topographic maps. The numbering is simply based on the equator and the central meridian of each zone, as earlier described.

Inherent in the cartographic process are several basic operations, all of which are interdependent to some degree. Some are universal aspects of knowing such as the recognition of classes of phenomena and their measurement, but some are unique to cartography. One of these is the reduction of reality (map scale) in order to bring more of an area into view. Another is the transformation of the spherical surface to a plane (map projection) in order that the map can be geometrically faithful to the area it represents. A third is the denotation of the components of reality by various marks (map representation) in order to characterize and distinguish among them. A fourth is the acquisition of data and its transfer to the map in proper positional relationship (map compilation) in order to display their spatial relationships; this, of course, is the fundamental objective of cartographic communication.

As we have seen in the preceding chapters, even conceptually straightforward procedures such as reduction and spherical transformation involve a considerable variety of possibilities and complications. The same is true of representation, compilation, and symbolization; the following chapters will examine several aspects of these operations in detail, as they are related to particular kinds of data and communication objectives. In this chapter we will focus on the essential nature of geographical data, their measurement, their representation, and their compilation as an introduction to the more specialized treatments that follow in later chapters. We must begin with a consideration of how the data we map exist and how we conceive of them for mapping purposes.

CHAPTER 5
DATA ORDERING, REPRESENTATION, AND COMPILATION

ORDERING SPATIAL DATA

Anything that is anywhere, either tangible, such as a road, or intangible, such as religious adherence, is a spatial phenomenon having location and is capable of being mapped. Because the components of reality are infinitely varied and complex, in order to prepare effective maps it is necessary first to analyze the essential nature of the spatial data. This involves several approaches. Foremost, of course, is the locational relationship of the data, or their geographical ordering; this display is what cartography is all about. Geographical ordering is an inherent quality of all spatial data, and we can take it for granted, even though establishing position is not always easy. Equally important is a systematic ordering of the conceptual and measurable characteristics of spatial data. Without such an organized approach to their qualitative and quantitative characteristics, there would be nothing but confusion.

The recent extension of the study of spatial phenomena and their mapping to other celestial bodies, such as the moon or Mars, has produced some minor problems in terminology. Ever since people began to study their surroundings, near and far, that meant the earth, of course. As time passed, the systematic study of spatial relationships in general became known as *geography* (Gr. *geo,* earth + *graphein,* to write). Now that we are studying (and mapping) other celestial bodies, it is a bit awkward to use earth-oriented terms such as geographical or geological. Nevertheless, we will on occasion continue to use the terms geographical and geography to refer simply to spatial phenomena and their systematic study in the general sense.*

Classes of Phenomena

We can recognize four basic categories of spatial phenomena: *positional, linear, areal,* and *volumetric.*

Positional Data. A point is simply a nondimensional location, a place. Conceptually, posi-

tional or point data are things that exist at individual places. There is a great variety, ranging from the location of a depth sounding to a road intersection. At a higher level of abstraction, we may conceive of a city at a location. Even though the city may cover a considerable area, it can be endowed with the attribute of being at a different place than another city. Even some summary characteristic of a considerable region, such as the average annual production of a state or country, can be thought of as being represented by some central location within the boundaries. The distinguishing character of positional data is the conception of its existence at a single location, however abstract that conception may be.

Linear Data. Some kinds of geographical phenomena are linear, the distinguishing characteristic being the quality of one-dimensionality. Even though a phenomenon may have significant width, such as a road or a river, its course and relative length are its dominant attributes that allow us to think of it simply as a line. There are many kinds of linear data, ranging from the invisible boundary between two areas of different administration or the coastline separating land from water, to the route followed by spreading ideas.

Areal Data. Conceptually areal data are two-dimensional and the focus of interest is the areal extent of the phenomenon. Even though a region may be long and narrow, as an area, its linearity is not its prime attribute. Like the other classes of geographical data, the conceptual range of areal data is wide. Attributes such as national sovereignty, dominant language or religion, climatic type, and soil character all fall in the class of areal data.

Volumetric Data. Geographical volumetric data are three-dimensional in concept. Such data may range from a mental construct (e.g., the population of a city as a "quantity" of people), or it can be tangible (e.g., the volume of precipitation that falls on an area or the tonnage of freight moved by rail). We conceive of geographical volumes in different ways. The population of a city or the gross national product of a country are simply sums, whereas many geographical volumes are

---

* When more specific reference is necessary, one should use terms such as selenography (moon) and areography (Mars). See, for example, R. M. Batson, "Cartography of Mars: 1975," The American Cartographer, 3 (1), 57–63 (1976).

thought of as quantities spread out over some datum, and extending above or below it, such as land or water in relation to sea level. Conceptually, a volume can be quite abstract, such as geographical density, that is, so many units of some phenomenon per unit area.

We tend not to be very systematic in the way we deal with spatial data because we often put the same item in different categories, depending on how we may be thinking about it. For example, we may conceive of New York City as positional (a place, in contrast to Philadelphia), or as an area (a particular administrative region, as contrasted with an adjacent region), or as a "volume" of humanity. Nevertheless, any geographical phenomenon may be placed in one of these four categories, and some phenomena may be transformed conceptually from one to another.

Continuity and Smoothness

Some geographical arrays are discrete or discontinuous in that a distribution may be composed of individual items at particular locations, and the intervening areas are empty of the item. Such would be the case, for example, of individual houses, industrial plants, cities, or routes of movement. In contrast, other distributions are continuous in that no area is empty. It is inconceivable, for example, that temperature or the surface of the solid earth (above and below sea level) cannot exist anywhere. Spatial data that intrinsically are discrete can be transformed conceptually into continuous data. For example, people are discrete, so that population would have to be classed as discontinuous; but if numbers of people are related to the areas that they occupy by applying the density concept (number of persons per square kilometer,) the ratio becomes continuous, since all areas then must have a density value, even those where the value is zero.

Geographical distributions are also either smooth or nonsmooth. Smooth phenomena are those in which the differences from place to place are transitional, not abrupt. For example, atmospheric pressure is found to change gradually from place to place and the pressure between two places close together will be intermediate. On the other hand, some things, such as land-use categories and gross national products, change abruptly at the boundaries between classes.

Commonly, areal data tend to be nonsmooth and volumetric data tend to be smooth; however, just as in the case with continuity, some kinds of distributions can conceptually be either. For example, density can be thought of as volume data in which the value changes are abrupt at the boundaries of the areal units used in its calculation. But, more abstractly, it can also be conceived as forming a smooth, undulating, statistical surface with sloping ups and downs from place to place.

Measurement of Phenomena

When dealing cartographically with positional, linear, areal, and volumetric data, it is necessary, of course, to determine the locations of the data. This provides the spatial measurement or geographical ordering that is the fundamental function of the map; but that is not enough. It is also necessary to differentiate within the classes of data; otherwise a map that only distinguished among the four classes of data and displayed their locations would not be very useful. The best differentiation method is to apply a system of measurement that classifies and distinguishes within and among the data according to four levels of precision. Such a method is called a scaling system, and the four levels, in increasing order of descriptive efficiency, are named *nominal, ordinal, interval,* and *ratio.*

Nominal Measurement.
Nominal measurement is employed when we distinguish among things only on the basis of their intrinsic character. The classes that may be involved are unlimited but, in nominal scaling, the distinctions are conceived as being based only on qualitative considerations without any implication of a quantitative relationship.

Examples of nominal differentiation of positional data are Chicago, a gravel pit, and the north magnetic pole; an example of areal data is a land-use class; an example of linear data is a river or road; and an example of volumetric data is a maritime air mass or foreign-born population.

Although we can conceive of particular geographical volumes on a nominal scale, they cannot be mapped as volumes without employing a higher order of measurement (ordinal, interval, ratio). Because the ordinary map is only two-dimensional, a volume without any quantitative attribute can only be treated as positional, linear, or areal data. For example, a volume of foreign-born population might be mapped as existing in a city (positional data) or in an area (areal data); a lake or an air mass would have to be mapped simply as occurring over an area (areal data); and a volume of freight moved over a given railroad line could only be mapped as a line.

Ordinal Measurement. Ordinal scaling involves nominal classification; it also differentiates within a class of data on the basis of rank according to some quantitative measure. Rank only is involved; that is, the order of the data categories is given, but not any definition of the numerical values. For example, we can differentiate major ports from minor ports, intensive from extensive agriculture, among small, medium, and large cities, hot and cold temperatures, and so on. It enables the map reader to tell that point, line, area, or volume data are larger or smaller, more or less important, younger or older, and so on, but it does not provide any specific magnitude assignment.

Interval Measurement. Interval scaling adds the information of distance between ranks to the description of class and rank. To employ interval scaling, we must use some kind of standard unit (which may be quite arbitrary) and then express the amount of difference in terms of that unit. For example, we differentiate among temperatures by using a standard unit, the degree (degrees Celsius or Fahrenheit), among city sizes by using a person (as a unit of population) as a standard unit, and among differences in elevation by employing standard units of linear measure such as the meter or foot.

Although interval measurement of positional, linear, areal, and volumetric data provides more information than nominal and ordinal scaling, one must be careful not to infer more than is warranted by the nature of the unit and the scale to which it applies. For example, one could not say that 40°F is twice as much temperature as 20°F.

Ratio Measurement. To provide magnitudes that are intrinsically meaningful requires the employment of an interval scale in which the intervals begin at a zero point that is *not arbitrary*, as are the zeros of the Fahrenheit and Celsius temperature scales. Elevation above a datum, barometric pressure, the Kelvin thermometric scale, depth of snow or precipitation, populations of cities, and tons of freight are all ratio measurements. Most measures pertaining to length, area, and volume are ratio measures.

Ratio measures relating to the concept of geographical density (number of items per unit of area), although they are absolute magnitudes, do not have complete geographical validity, since the quantities refer to areas instead of to points.*

From the point of view of mapmaking there is no difference between the treatment of geographical data measured on interval and ratio scales. In both instances a range is being displayed and, from the point of view of representation, it is immaterial whether or not the scale begins at an arbitrary zero.

REPRESENTATION

In the usual sense a map is a visual representation of selected spatial phenomena; in order to display the data we employ an almost unlimited variety of marks. By relating identifiable graphic characteristics of the marks to chosen attributes of the data, we assign nominal and quantitative meanings to the marks, and they become symbols. By arranging the marks planimetrically in the map plane, we endow them with geographical meaning. Throughout the long history of cartography many systems of symbolization have been devised; a good part of this textbook describes and illustrates cartographic symbolism.

A graphic system of communication demands that the various marks be distinguishable, just as the letters of the alphabet used in a language must appear different so that we do not get mixed up as to the sounds they represent. Furthermore, by the systematic use of graphic similarities and contrasts among the marks, we are able to express likenesses and differences in the data they represent. Therefore, in preparation for

---

* See Phillip C. Muehrcke, "Concepts of Scaling from the Map Reader's Point of View," The American Cartographer, 3 (2) 123–141 (1976), for a fuller discussion of the relation between systems of measurement and the interpretation of mapped data.

the study of compiling and symbolizing geographical data, it is necessary first to survey the basic categories of marks and how they may be graphically modulated.

The Visual Variables

Although positional distinction among data is the primary purpose of a map, if that were all a map could reveal, there would be little communication. In order to represent the chosen characteristics of spatial data in meaningful fashion, we must make use of variations in the graphic qualities of the marks. These graphic qualities, analogous as communicative media to the sounds of speech, have been called the visual variables.* These are the perceptual dimensions of graphic character that can be systematically modulated to convey meaning. There are seven. One is *position* in the visual field, which in cartography is largely prescribed by the geographic ordering of the data and the frame of the map. The other six are *shape, size, color, value, pattern,* and *direction.*

Just as there is an almost unlimited number of ways of combining the sounds of speech into the words of language, graphic communication has an enormous capability for combining the visual variables.

1. **Shape.** Shape is the graphic characteristic provided by the distinctive appearance of (a) a regular form, such as a circle or triangle; (b) the outline of an irregular area, such as a state or island; or (c) the contour of a linear feature, such as a river or coast (Fig. 5.1).
2. **Size.** Marks vary in size when they have dif-

J. Bertin, Semiologic graphique, Gauthier-Villars, Paris and Mouton, Paris-The Hague, 2d edition, 1973, pp. 42–97.

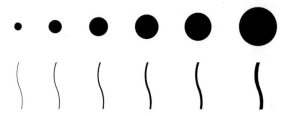

FIGURE 5.2 *Contrast of the quality of size.*

ferent apparent dimensions—diameter, area width, height (Fig. 5.2).

3. **Color.** Color is a complex visual perception and will be discussed at greater length in a later chapter. Suffice it here to point out that the common use of the term "color" refers to hue. When we say that things are different colors, we are usually describing characteristics such as blue, green, red, and so on.
4. **Value.** As a graphic quality, value refers to the relative lightness or darkness of a mark, whether or not hue is involved. Surfaces that reflect physically measurable amounts of light are said to have a tone (gray). Perceptually, a given tonal surface may appear different under different circumstances; when referring to the sensation of tone, it is better to use the term value, which refers to the perceptual scale of light (high value) to dark (low value) (Fig. 5.3).
5. **Pattern.** Any graphically distinctive arrangement of marks will herein be called a pattern.* As used in connection with systematic graphic representation, of necessity we restrict the meaning of the term "pattern" to a regular arrangement of component marks (Fig. 5.4). A fuller discussion of patterns will be found in Chapter 13. This use of "pattern" as a kind of mark should not be confused with geographical arrangements such as patterns of climate or occupance.
6. **Direction.** Direction refers to the orientation of an elongated individual mark or the parallel arrangements of marks (pattern) that are positioned distinctively with respect to some frame of reference. Such a frame may be the map

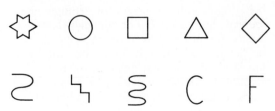

FIGURE 5.1 *Contrast of the quality of shape may be accomplished in many ways, only a few of which are suggested here.*

The meaningful organization of things has acquired a variety of terms, such as pattern, texture, grain, and structure, none of which has a distinct definition in the abstract.

FIGURE 5.3 *Contrast of the quality of value. The complete range is from white (high value) to black.*

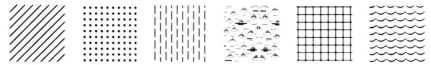

FIGURE 5.4 *Contrast of the quality of pattern.*

frame, graticule, or circular subdivision (Fig. 5.5).

Although this list includes the major and common ways that map marks may be varied, there are some less common ways that, nevertheless, may be of great significance in certain instances. Examples are focal quality, the attribute of being sharply distinct or, conversely, fuzzy; color saturation, which is more or less of a hue in a unit area; and position in a series, which may give distinctiveness to a mark relative to other marks.

The elemental visual variables are not easy to define with words, since words are a foreign language as far as graphic communication is concerned. Furthermore, the words that we must use do not have specific meanings out of context. It is worth reiterating that here we are using the visual variable terms in the restricted sense of how graphic marks may be made to appear different from one another.

Classes of Representation

It is by now abundantly apparent that there is an unlimited variety of spatial data that can be mapped and a considerable array of marks with which to represent them. Furthermore, the marks can be used singly or in combination and one kind of mark may be used to represent quite different phenomena. In order to consider the ways in which representation can be employed in symbolizing data, it is helpful to classify the systems of graphic representation on a functional basis. We can recognize three classes of symbols: *point, line,* and *area*.

1. **Point Symbols.** Sometimes called spot symbols, point symbols are individual marks used to represent positional data, such as a city, a spot height, the centroid of some distribution, or a conceptual volume at a location, such as the population of a city. Even though the mark may cover some map space, if conceptually it refers simply to a location, it is a point symbol. A few representative examples are shown in Fig. 5.6.

2. **Line Symbols.** Sometimes called band symbols, line symbols are individual linear marks used to represent a variety of geographical data. Just because a linear mark may be used does not mean that the class of data being represented is linear; for example, contours are lines used to represent elevations and depths (point data) from which volumes may be determined. Also, it should be noted that the lines of a graphic pattern are not linear symbols in the sense used here. Examples of some linear symbols are shown in Fig. 5.6.

3. **Area Symbols.** Sometimes called field symbols, area symbols are marks—color, tone, pattern—within a map area used to indicate

FIGURE 5.5 *Contrast of the quality of direction.*

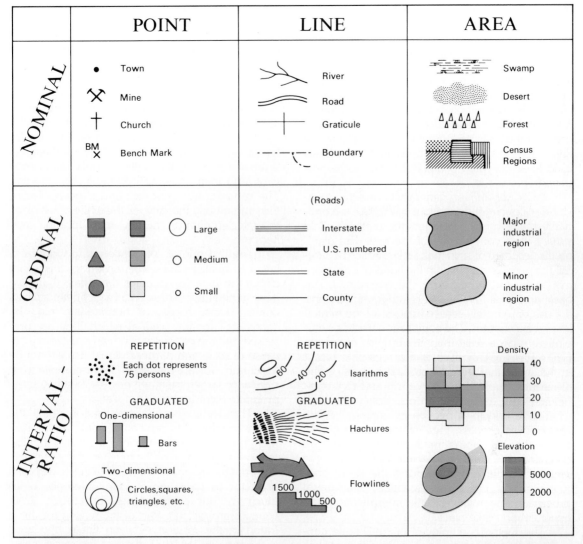

FIGURE 5.6 *Some examples of the three classes of representation (point, line, area) and how they might be used to portray nominal, ordinal and interval-ratio data.*

that that region has some common attribute, such as an administrative jurisdiction or some measurable characteristic. When used this way, an area symbol is graphically uniform, that is the pattern, color, or tone is the same over the entire area being represented by it.* Some examples are shown in Fig. 5.6.

* *Strictly speaking, a second, nonuniform kind of area symbol is the application of smooth tonal variations to*

COMPILATION

In cartography the term "compilation" refers to the assembling and fitting together of the diversity of geographical data that is to be included in a

portray three-dimensional form (e.g., of the land surface) in the manner of a photograph. Less tangible variations from place to place, such as ratios, can also be represented this way. Ordinarily, in cartography this technique of representation is called shading.

map. The "fitting together" means locating the various data in their proper relative horizontal positions (planimetry) according to the map projection system and the map scale being employed.

As suggested in Chapter 2, one of the more important distinctions in mapmaking is the one between the processes employed to produce large-scale topographic and special reference maps and the methods used for small-scale general and thematic cartography. One of the major differences lies in the methods and techniques used to acquire and compile the data to be portrayed.

Large-scale maps, generally considered to include scales larger than about 1:75,000, are ordinarily made by photogrammetric methods, field survey, or a combination of the two. The planimetric accuracy of such maps is controlled as carefully as possible and, within the limits of definition, map scale, and human error, they are correct. The basic principles involved in photogrammetry and the use of remotely sensed images will be treated in the next chapter. The compilation methods employed in large-scale mapping by way of precise field survey are essentially noncartographic except in the sense that the end product is a map. The disciplines involved are surveying and civil engineering and, except for an occasional reference where the principles overlap or impinge on the cartographic process, they will not be treated in this text-book. Some medium-scale maps are compiled by tracing selected data from the larger-scale maps and reducing the result photographically. These are essentially mechanical processes, and relatively little interpretation and generalization take place.

The methods employed in small-scale general and thematic mapping are quite different from those used in large-scale mapping. Often, small-scale mapping involves maps of relatively large areas; this means that data may need to be acquired from diverse sources. Additionally, thematic mapping very often requires that the primary subject matter of the map be presented against a background of locational information, which is called the *base data*. This base material is ordinarily compiled first; the accuracy with which it is done largely determines the accuracy of the final map. This is because of the practical requirement that the thematic cartographer must compile

much of the subject-matter data by using the base material as a skeleton on which to hang it. These data, usually consisting of coasts, rivers, lakes, and political boundaries, are generally available from the large-scale general reference maps.

The Compilation Process

The compiling of data requires using many maps (as well as other sources) from which to gain the desired information. The maps may be on different projections; they may differ markedly in level of accuracy; the dates of publication may vary; and their scales will probably be different. The cartographer must pick and choose, discard this, and modify that and, all the while, must place the selected data on the new map, locating each item precisely.

The first rule of compilation is to work from larger to smaller scales. The reason for this is that all but the largest scale maps show data that have been generalized. These data (e.g., linear, such as coasts, rivers, roads, or boundaries) may be "accurate" for the scale at which they are presented, but the generalization made necessary because of scale and purpose would usually not be "accurate" for a larger scale. If we compile from smaller to larger scale, we may be building inappropriate error into the compilation.

The techniques of compilation (i.e., the means employed to position the data on the new map) range from those that are largely mechanical to those that consist of the transfer of data by eye. It is possible by photography or with the help of a projector to transform a scale difference of as much as four or five times, but much larger changes are difficult. Part of the reason for this is found in the relation between linear and areal scale: the size (area) of maps of the same region will vary as the square of the ratio of their linear scales. For example, a region shown in 1 cm^2 at a scale of 1:50,000 will occupy only 1 mm^2 (1/100 cm^2) at a scale of 1:500,000. Therefore, although positions of individual points or simple lines could be easily determined, anything that involves complications in areal spread would be reduced to an almost indecipherable complexity.

The process of compilation often requires that much of the data be transferred by eye. The graticule shown on the projection in each case constitutes the guide lines and all positions must be estimated. The eye is remarkably discriminating and, with practice, can position data with all the precision that is ordinarily necessary. Very often the use of mechanical means for reduction is prevented by a change in projection. Do not be concerned about the accuracy of such a process; 90% of all small-scale maps have been compiled in this manner.

When the projections differ between those of the sources being used and those of the map being compiled, it is necessary for the cartographer to become adept at imagining the shearing and twisting of the graticule from one projection to another and to modify the positioning accordingly. The difficulties occasioned by map projection differences between the sources and the compilation can be largely eliminated by making the graticules comparable. That is to say, the same interval on each will greatly facilitate the work (Fig. 5.7).

Compilation is most easily undertaken by first outlining on the new projection the areas covered by the source maps. This outlining is similar to the index map of a map series. The sheet outlines may be drawn, and the special spacing of the graticule on each source (5°, 2°, etc.) may be lightly indicated.

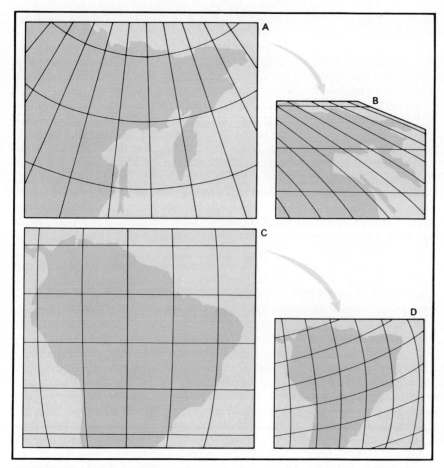

FIGURE 5.7 Shearing and shape changes in the compilation procedure. Maps B and D are derived from A and C.

The Significance of Base Data to Thematic Maps

The importance of including on the finished map an adequate amount of base data cannot be over-emphasized. Nothing is so disconcerting to a map reader as to see a large amount of detail presented on a map and then be confronted with the realization that there is no "frame" of basic geographic information to which the distributions can be related. The most important objective of thematic mapping is to communicate geographical relationships. Since few people are able to conjure up an acceptable "mental map" and project it on the thematic map, it is incumbent on the cartographer to provide it.

The amount and detail with which the base data are shown will, of course, vary from map to map. The usual thematic map must have on it the coastlines, the major rivers and lakes, and at least the basic civil divisions. The graticule, in most cases, should also be indicated in some fashion. The objectives of the map will dictate the degree of detail required, but it is a rare map that can be made without these kinds of information to aid the reader in obtaining the relationships presented.

Compiling Base Data. It is important to appreciate that there are different kinds of surveys according to the definitions and assumptions made by the survey organization. Cadastral survey is commonly done for a limited area; because the curvature of the earth's surface is relatively insignificant over a small area, it may not be taken into account. The lines of such a plane survey are determined from ground observations and are normally mapped as observed instead of first being referred to a spheroid. Topographic maps, on the other hand, are normally made by referring the ground observations to a spheroid; consequently, the two kinds of surveys usually do not match.

Most large-scale plane survey and cadastral maps do not show much physical environmental data. If compilation requires the union of physical and cultural data, the cartographer may be hard put to resolve the differences. For example, if we wish to make a map showing up-to-date information concerning (1) the streams, lakes, and swamps, and (2) the roads of a region, we may find the first category on topographic maps (of different dates!), but not the second; the roads will probably be available from county or state road maps, but these may not show the drainage details. The two sources will be essentially "accurate" according to the definitions used for their mapping, but they will not match one another. In general, the practical significance of these kinds of problems varies according to the scale of the map being compiled; the smaller the scale, the less the difficulties, since positional discrepancies diminish and the desirability of generalization increases with reduction in scale.

Determining the Map Scale

All maps are, of course, constructed to a scale. In actual practice many thematic maps are prepared in a size to fit a prescribed format, the format being the size and shape of the sheet on which the map will appear. The format may be a whole page in a book or atlas, a part of a page, a separate map requiring a fold, a wall map, or a map of almost any conceivable shape and size. Whatever the format may be, the map must fit within it.

The shapes of earth areas may vary considerably when mapped, depending on the projections on which they are plotted. Hence, an important concern of the cartographer regarding map scale may be the projection on which the map will be made. When the projection choice has been narrowed to those that are suitable, the variations in the shapes of the mapped area on the different projections can be matched against the format to see which will provide the best fit and maximum scale.

The easiest way to do this is to establish the vertical and horizontal relationship of the format shape on a proportion basis and then compare the proportion against representations on the various projections. We can usually find examples of projections or can easily calculate critical dimensions from the tables or directions for their construction. In this instance the actual dimensions as to length and breadth are not the critical matter; the proportion between them in relation to the format is important. When the projection has been selected that best fits the purpose and format of the proposed map, the scale of the finished map may be determined. Normally it is good practice to use either a round number RF or a simple stated scale (map distance to earth distance).

The cartographer must also decide at this stage whether to compile on an already drawn projection or base map, or to construct a projection and compile the entire map. Most projections may be employed for alternative east-west areas by simply tracing the graticule and renumbering the longitude. We must be careful, however, not to use a projection that has been copyrighted or patented unless permission is obtained. As a general rule, it is far better to construct the projection anew to fit precisely the objective of the map. Even without computer assistance, most projections are not difficult to construct except in special phases, and it is poor practice to produce an inferior map "projection-wise" in order to save a few hours' time. Frequently mapped areas such as continents or countries have appeared, however, on most of the appropriate standard projections; if they are available there is no reason why a good projection should not be used. We should always, however, carefully check such a projection to be certain that it has been properly constructed.

Compilation Scale in Relation to Finished Scale. If the map is to be scribed it is normal practice to compile at the finished scale. If carefully done the compilation itself will be the basis for the image that can be made to appear in the scribe coat.* Similarly, if the map is to be reproduced from an original drawing on plastic or paper by a process that does not allow changing scale, compilation will have to be done at the finished scale. On the other hand, if reproduction is to be by a process that allows reduction, it is desirable to compile and draft at a larger scale and reduce the artwork to the finished scale.

Sometimes the complexity of the map or other characteristics makes it desirable to compile at a larger scale even for a map to be scribed. In this case reduction would be used to place the image on the scribe coat.

In general an ink drawing can be reduced with advantage: at the larger scale drafting is easier, reduction "sharpens up" the lines, and so on.

The amount of reduction will depend on the process of reproduction and on the complexity of the map. It will also depend on whether definite specifications have been determined for the reproduction, as may be the case with maps of a series. In general, ink-drafted maps are made for from one-quarter to one-half reduction in the linear dimensions. It is unwise to make a greater reduction because the design problem then becomes difficult.

Changing Map Scale
The scale of a base map or a source map may be changed by optical projection, photography, or similar squares.*

A variety of projection devices are available, and most cartographic establishments have one. They may operate by projection from overhead onto an opaque drawing surface or from underneath onto a translucent tracing surface. A source map is placed in the projector and, by adjusting the enlargement or reduction, the projection of the image to the drawing surface may be changed in scale to that of the map being compiled. When properly aligned, the desired information may then be traced. Needless to say, if the projections of the source map and the map being compiled are very different, such a projector cannot be used.

Changing scale by photography is also a very easy process. It may be accomplished either by photographic enlargement or reduction through the use of the conventional film process or by several kinds of photocopy processes. It is necessary to specify the reduction or enlargement for some types of photocopy work by a percentage ratio, such as a 50 or 75% reduction. Care should be exercised in specifying the percentage change, since 50% reduction or enlargement means one-half in the linear dimension and, in some processes, is the most that can occur in one "shot." In such instances a change of more than 50% would require repeating the process; to reduce something by 75% in the linear dimension would then require two exposures of 50% each, not one at 50% and another at 25%. Available photocopy

---

* Scribing is a construction process by which an image may be fixed in the coating (scribe coat) on a sheet of plastic. The cartographer, being guided by the image, then removes the coating with special tools where lines, and the like, are to appear (see Chapter 17).

* The pantograph, a mechanical device for changing scale, is rarely used in cartography today, because it is relatively cumbersome and slow.

paper is often limited to 18×24 in., and anything larger may need to be done in sections. It is difficult to match the sections, since the paper ordinarily changes shape unevenly in the developing and drying.

Photographic enlargement or reduction using film negatives may be made somewhat more precisely, since the image may be either projected or viewed in the camera and, therefore, dimensions can be scaled more exactly. All the cartographer need do is to specify a line on the piece to be photographed and then request that it be reduced to a specific length. The ratio can be worked out exactly, and the photographer needs only a scale to check the setting. Any clearly defined line or border will serve as the guide. If none is available, the cartographer may place one (with light blue pencil) on the drawing.

In some instances we are forced to change scale by a method called "similar squares." This involves drawing a grid of squares on the original and drawing the "same" squares, only larger or smaller, on the compilation. The lines and positions may then be transferred by eye from one grid to the other (see Fig. 5.8). With care it is quite an accurate process; it is the same as compilation.

Occasionally, a cartographer is called on to produce an oversized chart or map that involves great enlargement. In most cases, extreme accuracy is not required. If the outlines cannot be sketched satisfactorily because of their intricacies,

it is possible to accomplish an adequate solution by projection onto a wall. A slide or film transparency may be projected to paper affixed to the wall and the image traced thereon. If an opaque classroom projector with a large projecting surface (that is sufficiently cool in operation) is available, it may be used directly so that the necessity of making a transparency or slide is eliminated.

Compilation Procedure

The compilation process has as its objective the preparation of a composite that contains all the base data, the lettering, the geographical distribution(s) being mapped, and so on. This becomes the guide for the construction of the map, either by scribing or by ink drawing. The mechanical processing of the compilation is not of concern in this chapter (see Chapter 17), but it is important that the cartographer proceed at this stage in a manner that will make the ultimate map construction as easy as possible.

Perhaps nothing helps the compiling procedure so much as a translucent material with which to work. A tracing medium of some sort (e.g., plastics of various kinds and tracing papers) enables the compiler to accomplish a number of things in addition to the convenience of being able to trace some data. The compiler may lay out lettering for titles, and the like, and move the layout around under the compilation worksheet. If the compiler wishes to draw a series of parallel lines,

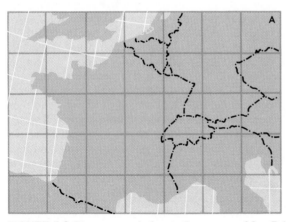
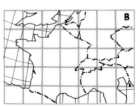

FIGURE 5.8 *Changing scale by similar squares. Map B has been compiled from Map A.*

letter at an angle, or place dots regularly, some cross-section paper need only be placed under the tracing paper. The use of tracing paper, however, occasions the problem of maintaining size, since paper contracts and expands with changes in humidity and temperature. Plastic materials are more stable. If registry, (i.e., the matching of several compilation or final drawings) is a problem, a dimensionally stable material should be used.

The Worksheet. The composite that results from the compilation process is called the worksheet. For a simple map the worksheet contains everything, but for complex maps there may be a need for more than one worksheet, each carefully registered (accurately fitted and positioned) with its companions. When completed, a worksheet is comparable to a corrected, ''rough draft'' manuscript: all that is then necessary is for the final flaps (the artwork) to be prepared by scribing or tracing.

The various marks on the worksheet may be done by pencil, ball-point pen, or any satisfactory medium. It will be of great help in compilation if each kind of line that is to be shown differently on the final map is put in a different color. Lettering, if its positioning is no problem, may be roughly done; if the positioning is important, the lettering should be laid out with approximate size and spacing. If the first try does not work, it may be erased and done over. Borders and obvious line work need only be suggested by ticks.

When the worksheet and the compilation have been completed, the image may be transferred to a scribe sheet for processing or ink drafting may be done on translucent material directly over the worksheet. If the map is simple and the artwork is to be prepared by the cartographer, things such as the character of the lines will be planned; but if it is to be drafted by someone else, the cartographer must prepare a sample sheet of specifications to guide the draftsman. Even for the cartographer-produced map, this is wise. This is simple to do if each category is in a different color or otherwise clearly distinguished. Separation drawings for small maps may be made easily from a single worksheet, and they will register.

It is a common experience for a cartographer to find that many of the elements included in one map might easily be used for another. Base materials such as boundaries, hydrography, and even lettering may not vary much, if at all, from one to another in a series of maps. Consequently, the cartographer can save considerable future effort by anticipating possible subsequent use of the compilation efforts; the cartographer should prepare the worksheet and plans with these possibilities in mind. Modern reproduction methods make it relatively easy to combine different separation drawings, even when printing in one color (see Chapter 16).

Coastlines

The compiling of coasts for very small-scale maps is not much of a problem, since they usually require so much simplification that detail is of little consequence. This is not the case when compiling at medium scales where considerable accuracy of detail is necessary.

Perhaps the major problem facing the cartographer is the matter of source material. It is well to bear in mind that some coasts will be shown quite differently on different maps, yet both may

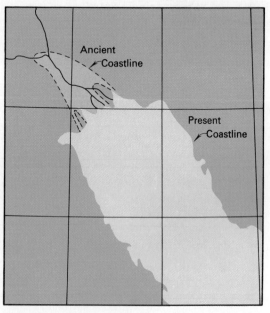

FIGURE 5.9 *Major changes in coastlines occur over long periods of time, and these changes may be significant even on small-scale maps. A portion of the Persian Gulf.*

be correct. Hydrographic charts are made with a datum, or plane of reference of mean low water, whereas topographic maps are usually made with a datum of mean sea level. The two are not the same elevation, and it is to be expected that there will therefore be a difference in the resulting outline of the land. In parts of the world that experience high tidal ranges or where special planes of reference are used, the differences will be greater. Another difficult aspect of dealing with such source materials is that the coloring of the various charts and maps of the same area may be inconsistent. Marsh land, definitely not navigable, is likely to be colored as land on a chart, and a compiler would assume it to be land by its appearance. On the other hand, low-lying swamp on a topographic map is likely to be colored blue, as water, and only a small area may be shown as land. On

many low-lying coasts the cartographer may be faced with a decision as to what is land; the charts and maps do not tell him.

Through the years some coasts change outline sufficiently so that it makes a difference even on medium-scale maps. Figure 5.9 shows the north coast of the Persian Gulf in the past and at present; Fig. 5.10 shows a portion of the Atlantic coast. If we were making maps of an historical period we would endeavor to recreate the conditions at the period of the map. This problem is particularly evident on coastal areas of rapid silting, which in many parts of the world seem to be important areas of occupancy.

In a number of areas of the world, notably in the polar regions, the coastlines, like many other elements of the map base, are not well known, and they vary surprisingly from one source to an-

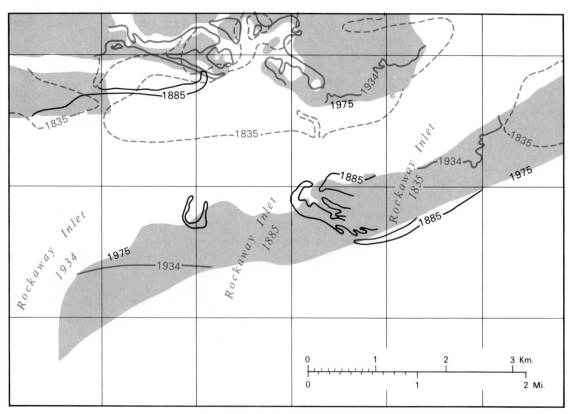

FIGURE 5.10 *Frequent changes have occurred in some areas. The various lines show the positions of the shoreline of Rockaway Inlet, Long Island, at several periods. (Modified and updated from Deetz, Cartography, Special Publication 205, U.S. Coast and Geodetic Survey.)*

other. On some maps a particular region may appear as an island; on others, as a series of islands; and on still others, as a peninsula. On simple line maps a broken or dashed line suffices for unknown positions of coastlines, but it becomes a larger problem when the water is to be shaded or colored; no matter what type of line is used to delineate the coast, the tonal or color change outlines it clearly.

Administrative Boundaries

Compiling political boundaries is sometimes a remarkably complex problem, since the boundaries must be chosen for the purpose and date of the map. The problem becomes more difficult as the area covered by the map increases. Almost all man-made boundaries change from time to time, and it is surprising how difficult it is to search out the minor changes. The problem is compounded in some kinds of thematic maps. For example, a population map of the distribution of people in central Europe prior to World War II, but also showing present boundaries, raises these problems: (1) international boundaries today, and (2) census division boundaries as of the dates of the enumerations in the various countries. The major difficulties are twofold. The first is that of finding maps that show enumeration districts and that also show latitude, longitude, and other base data so that the boundaries may be transferred to the worksheet in proper planimetric position. The second is that of placing later international boundaries in correct relation to the earlier enumeration boundaries.

It is not uncommon for official civil division boundary maps to be without any other base data, even projection lines. Such a condition should impress on the cartographer who uses such deficient maps the need to provide base data for the users of the maps.

Hydrography

The compiling of rivers, lakes, and other hydrographic features as part of base data is very important. These elements of the physical landscape are in some instances the only relatively permanent interior geographical features on many maps; they provide helpful ''anchor points'' both for the compilation of other data and for the map reader's understanding of the ''place correlations'' being communicated.

The selection of the rivers and lakes depends, of course, on their significance to the objective of the map. On some maps the inclusion of well-known state or lower order administrative boundaries makes it unnecessary to include any but the larger rivers. Maps of less well-known areas require more hydrography, since the drainage lines, which indicate the major landform structure of an area, are sometimes a better-known phenomenon than the internal boundaries. Care must be exercised to choose the ''main stream'' of rivers and major tributaries. Often this depends not on the width, depth, or volume of the stream, but on some economic, historical, or other element of significance.

Just as coastlines have characteristic shapes so do rivers, and these shapes help considerably on the larger-scale maps to identify the feature. The braided streams of dry lands, intermittent streams, or meandering streams on flood plains are examples. On small-scale thematic maps it frequently is not possible to include enough detail to differentiate among stream types, but the larger sweeps, angles, and curves of the stream's course should be faithfully delineated. Likewise, the manner in which a stream enters the sea is important. Some enter at a particular angle, some enter into bays, and some break into a characteristic set of distributaries. Swamps, marshes, and mud flats are also commonly important locational elements on the map base.

Sources and Credits

The sources to which a cartographer may turn to obtain geographical information are legion; and it is not our purpose to suggest and evaluate them. It is useful, however, to divide the great variety into two classes with respect to the necessity for citation.

A considerable share of the geographical information compiled as base data—things such as coastlines, rivers, and boundaries—may be classed as general knowledge in the sense that they are well-known facts. They have been mapped many times, and it is quite unnecessary to cite the source of such information. On the other hand, if the delineation differs significantly from the usual way the phenomenon has been mapped because some very new source has been used, such as a new survey, remotely sensed imagery (such as LANDSAT, see Chapter 6), or as a result of recent boundary changes, it is helpful to the map user for that to be indicated.

In all instances of thematic maps and large-scale specialized reference maps, it can be assumed that the data being mapped will not be common knowledge. Consequently, the map user should be given the source and, if it is important, the date of the special information. Indications such as, "U.S. Bureau of the Census, 1970," "Land-use data from provisional 1972 map by Special Project Committee . . . , etc.," or "Zoning boundaries as approved by the County Board in 1976," will provide the user with necessary information.

Many relatively large-scale compiled maps have been put together from a variety of sources in a kind of patchwork fashion. Instead of individual phenomena being compiled from several sources that cover the whole map area, one part of the map may be derived from modern topographic maps, another part from older compiled maps, and so on. If there is any significant difference among the quality of the sources, then a coverage diagram should be prepared. An example of such a diagram is shown in Fig. 5.11. Many kinds of information can be shown on coverage diagrams, such as the quality of source maps, dates of censuses, or names of compilers or investigators.

Whenever specialized information has been obtained from a nonpublic source, such as a research foundation paper, a scholarly publication, or an industrial report, it is common courtesy to acknowledge the source. Judgment is always involved. The information may be widely known and may appear in substantially the same form in several sources. In that case citation may be unnecessary but, whenever there is any doubt, the source should be acknowledged.

Copyright

A large proportion of published material is protected by copyright, the name given to the legal right of a creator to prevent others copying his or her work. The application of statutory copyright to maps is not very well understood because the law is primarily aimed at the protection of literary

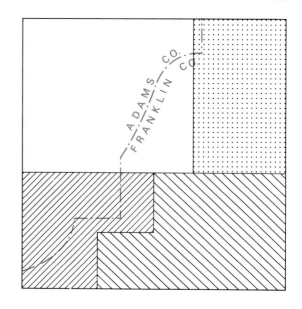

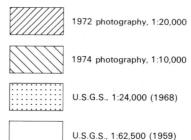

1972 photography, 1:20,000

1974 photography, 1:10,000

U.S.G.S., 1:24,000 (1968)

U.S.G.S., 1:62,500 (1959)

FIGURE 5.11 Coverage diagram showing the kind and date of sources used in compilation.

works, but it encompasses a large number of classes of materials ranging from musical compositions to photographs, and including maps. Furthermore, the copyright law of the United States has just been revised. We cannot, of course, offer legal advice, but a few general observations are provided so that the student will appreciate both the complexity of the situation and the necessity for complying with the provisions of the law.

As a rule, no map copyrighted in the United States can be *copied* without obtaining permission from the copyright holder. The term "copied" is used here in its literal sense, meaning "reproduced exactly." One must not, therefore, either through photography or tracing, copy exactly the way some other mapmaker has represented geographical data. Although the exact position of the coastline of the United States, for example, cannot be copyrighted, the way someone has generalized it can. Furthermore, the exact manner in which some cartographer has processed and symbolized statistical data for an area is protected.

Most publications, including maps, produced by agencies of the U.S. government are in the public domain; that is, they are not copyrighted and can be used freely. One ought always to give credit where it is due but, in the case of specialized publications containing judgments and opinions of named authors, it is not only courteous but wise to request permission from the issuing office, since the material may have been copyrighted by the authors separately or some of it may have come from copyrighted sources.

Although governmentally produced survey maps in the United States of the topographic variety, census materials, and the like may be used freely as sources of data, this is most definitely *not* the case with such materials produced by other governments. In most instances reproduction of such materials without permission is strictly prohibited. Since the United States is a party to the international Universal Copyright Convention, and since this and other agreements make it obligatory for citizens to comply with the copyright laws of the agreeing nations, one must be very careful.

Generally, whenever one wants to copy or reproduce material copyrighted by others for any purpose other than private study, written permission to do so must be obtained from the holder or holders of the copyright.*

SELECTED REFERENCES

Bond, B. A., "Cartographic Source Material and its Evaluation," *The Cartographic Journal, 10,* 54–58 (1973).

Chang, K., "Data Differentiation and Cartographic Symbolization," *The Canadian Cartographer, 13,* 60–68 (1976).

Dobson, M. W., "Symbol-Subject Matter Relationships in Thematic Cartography," *The Canadian Cartographer, 12,* 52–67 (1975).

Jenks, G. F., "Contemporary Statistical Maps— Evidence of Spatial and Graphic Ignorance," *The American Cartographer, 3,* 11–19 (1976).

Kaminstein, A. L., "Copyright and Registration of Maps," *Surveying and Mapping, 13,* 182–184 (1953).

Kaminstein, A. L., "Maps, Charts and Copyright," *Special Libraries, 51,* 241–243 (1960).

Muehrcke, P. C., "Concepts of Scaling from the Map Reader's Point of View," *The American Cartographer, 3,* 123–141 (1976).

Robinson, A. H., "The Cartographic Representation of the Statistical Surface," *International Yearbook of Cartography, 1,* 53–61 (1961).

Robinson, A. H., "Scaling Non-numerical Map Symbols," *Technical Papers,* 30th Annual Meeting, American Congress on Surveying and Mapping, pp. 210–216 (1970).

Wright, J. K., "'Crossbreeding' Geographical Quantities," *Geographical Review, 45,* 52–65 (1955).

Wright, J. K., "Map Makers are Human: Comments on the Subjective in Maps," *Geographical Review, 32,* 527–544 (1943).

---

* There are many complications involved, such as the so-called fair use doctrine, which applies to certain scholarly uses of copyrighted materials, but it is difficult for individuals to obtain clear legal opinions on copyright matters. The best advice is always to request written permission; it is usually given.

Humans have built-in senses by which we can experience conditions of our environment. We "remotely sense" some of these conditions with hearing and sight. For example the ear reacts to sound waves travelling through the atmosphere, enabling us to hear thunder, wind and other sounds, and the normal eye is sensitive to some of the electromagnetic energy emitted and reflected by objects. In order to increase our range of perception, we have learned to fabricate and utilize other kinds of sensors which are sensitive to electromagnetic energy. Although remote sensing has long been an important part of mapping, espe-cially in large-scale mapping, it is becoming increasingly so with the modern developments of electronic processing and satellites.

To understand modern remote sensing, we must first look at the characteristics of electromagnetic energy and the various ways it can be sensed. Later we will consider its cartographic applications.

Electromagnetic Energy
Energy is emitted from any object that has a temperature above 0°K, with the amount of energy emitted increasing with a rise in temperature. The

CHAPTER 6
REMOTE SENSING AND THE USE OF AIR PHOTOGRAPHS

theoretical black body is used as a standard to compare emittance of radiation.* There is a specific spectral distribution of emitted energy; however, the amount of energy is not the same at all wavelengths. For very short and very long wavelengths the emittance is low; for some wavelengths in between, it reaches a maximum value, depending on the temperature of the black body (Fig. 6.1). This peak in the energy distribution curve is shifted to shorter and shorter wavelengths as the temperature of the black body increases. If, for example, a piece of iron were heated enough, its radiation curve would first intercept the long waves of the visible spectrum and appear to be a dull red. Upon further heating the peak of the curve would shift to shorter wavelengths and the

color would change to orange, then yellow, and finally white.

Three components of the waves of radiant energy are the wavelength, wave frequency, and wave velocity (Fig. 6.2). Since the velocity is constant (speed of light), the length of the wave and the frequency have a reciprocal relationship. In radio communication a particular band of radiation is identified by its wave frequency. However, in remote sensing, it is customary to refer to the wavelength to indicate a particular part of the spectrum. Because of the variety of substances on the earth and the great range of conditions, the waves of reflected or emitted electromagnetic energy vary in length from the very short gamma rays and X rays* to radio waves, which can be many kilometers in length.

*The black body is the theoretical material that transforms heat energy into radiant energy with the maximum rate permitted by thermodynamic laws.

*Lengths of X rays are expressed in angstroms. One angstrom = one hundred-millionth of a centimeter.

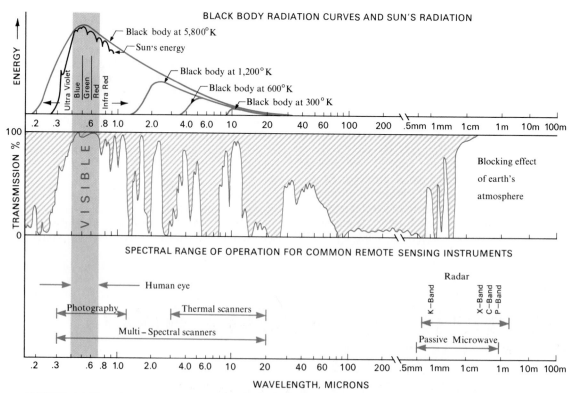

FIGURE 6.1 The electromagnetic spectrum showing the common bands of energy, energy curves for various black body temperatures, and windows where remote sensing can be accomplished. (After Scherz)

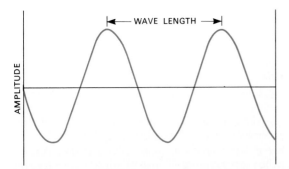

FIGURE 6.2 In remote sensing, wavelength is used to indicate a particular band of energy.

For convenience, the continuum of electromagnetic energy is often divided into bands or regions such as ultraviolet, visible, infrared, and microwave. Boundaries of these bands are not precise and should be thought of only as zones of transition. In Fig. 6.1 one can see that the visible band extends from about 0.4 to 0.7 micrometers (μm),* which means that our eyes can detect only a tiny portion of the entire spectrum. Photography can record wavelengths from about 0.3 to 1.2 μm or about three times the range the human eye can see. To sense wavelengths longer than 1.2 μm, instruments other than photographic cameras are used.

A photograph is the result of a chemical emulsion being acted on directly by reflected energy, whereas the imagery produced by other remote sensing is the result of the detection of emitted or reflected energy and the conversion of the detected signals into a picturelike format. Remote sensors can detect and record energy in many parts of the spectrum that are invisible to the human eye. These new unfamiliar views of our landscape can give analysts helpful new insights.

The numerous remote sensing devices that detect and record energy may be divided into active and passive systems. Passive sensors detect natural energy, either reflected or emitted; active systems generate the energy that is directed at and subsequently received from a target. Probably the most popular of the active systems is radar, which operates in the microwave portion of the spec-

trum. An example of passive sensing is photography, which uses reflected solar energy, usually in the visible portion of the spectrum. When artificial sources of illumination such as flares are used, photography becomes an active system.

Effects of the Atmosphere. Materials in the atmosphere such as carbon dioxide, dust, pollen, pollutants, moisture and ozone block, absorb, and scatter incoming solar radiation. These changes affect the levels of electromagnetic energy that can be recorded at the surface of the earth. Further changes occur when the energy is reflected or emitted from the earth back through the atmosphere to a remote sensing device.

The blocking effect of the earth's atmosphere is not of the same magnitude for all wavelengths. The bands of the spectrum where atmospheric attenuation is slight are called *windows* and are the regions utilized for remote sensing (see Fig. 6.1).

Absorption is the major factor in altering reflected infrared energy. Although the principal window is a band from 8 to 14 μm, the middle infrared, from 3.5 to 5.5 μm, is commonly used because the instruments designed for these shorter wavelengths are less costly to construct. For passive microwave and radar sensing there are windows between 1 mm and 1 cm. Beyond the wavelength of 1 cm, the atmosphere is relatively "transparent" and accommodates various types of radio equipment.

Photography

For more than 100 years remote sensing has been used to obtain geographical information. The aerial photograph was the first remote sensing device utilized to inventory and map characteristics of the earth. Although some highly sophisticated systems have recently been developed to record earth data, the photograph, with its great resolving power, will undoubtedly continue to be widely used as a means of remote sensing.* Even though most of these newer sensors detect energy outside the visible portion of the spectrum and record that

Micrometer (μm) = one millionth of a meter.

Resolving power is the degree to which a photographic emulsion or a detecting device can record and display small objects or small distances between objects.

information in some form of imagery, many conventional photo-interpretation techniques and procedures are still applicable in its analysis.

In the past a major problem was how to obtain the needed number of photographs, but today the problem is how best to handle the tremendous volume of photographs and imagery that is available. Although statistical and computer techniques are used to convert some of the remotely sensed data to usable form, it appears that for the foreseeable future people will still play an important role in interpreting both photographs and other sorts of imagery.

As early as 1840 it was suggested that photographs be used for the purpose of mapping; during the following decades, a variety of schemes were devised to procure photography. Balloons, kites, rockets, and even carrier pigeons were used to send cameras aloft.

Pigeons carried breast-mounted cameras with automatic timing devices. When released several kilometers from their loft, they returned at a relatively constant speed and on a straight course, thus providing the means for obtaining a series of photographs along the entire route. With balloons and kites, cameras of considerable size could be sent aloft to provide photography for civilian and military uses.

The first photographs from an airplane were made by Wilbur Wright over Italy in 1909. During the early days of World War I, airplanes were in use; however, they were not used to procure photography until authorities were convinced of the intelligence-gathering value. At first, hand-held ground cameras were used but, by 1915, cameras specially designed for aerial photography were in use. Although facilities were rather primitive by today's standards, the military forces were able to produce thousands of prints daily. The usefulness of aerial photography was well established by the end of World War I, but little had been done to establish training procedures; after the armistice, military photo interpretation virtually came to a standstill.

In the civilian community advances continued to be made in the commercial and scientific uses of aerial photography and, during the 1930s, government agencies made extensive use of aerial survey. The Agriculture Adjustment Administration systematically produced photography of agri-

cultural activities over most of the United States, and photography was used by the U.S. Geological Survey for topographic mapping and by the Forest Service for timber inventories. State and local agencies recognized the value of aerial photography and began using it for planning purposes.

The greatest stimulus for photo interpretation probably came with World War II. Prior to the war it was recognized by many military leaders that intelligence gathering by use of photo interpretation might well have great influence on the outcome of a future conflict. Germany made good use of this source of intelligence during its early offensive by photographing most military and transportation facilities across all of western Europe.

The English were forced to rely on photo interpretation for intelligence gathering after the retreat from Dunkirk in 1940. Procedures and methods they developed were invaluable to the United States, since it had little or no photo interpretation capability upon entering the war.

The mass of photography that had to be handled required that a great number of interpreters be trained; many of these people retained an interest in photo interpretation and contributed research to the field. Likewise, the urgent need for maps during the war resulted in greater use of photogrammetric* methods, both because of the speed with which maps could be made and because the photographs provided a good source of information for inaccessible areas.

Improvements in aircraft, cameras, filters, and film emulsions have enabled aerial photography to maintain its role as an important source of information about the earth. Multilens cameras are able to record several simultaneous views of the same scene using different film-filter combinations, thereby obtaining views in various parts of the visible and near-visible spectrum. Panoramic cameras, by using a narrow slit that pans from side to side and exposes film held in the form of an arc, can produce high-resolution photography of a large area in a single exposure. Strip cameras, by having rolled film move past a slit at a rate coordinated with the speed of the aircraft, can

---

Photogrammetry is the art or science of obtaining reliable measurements by means of photography.

record a continuous image of the landscape along the flight line. Because of the great resolving power of black and white films and recent improvements in the fidelity of color films, one can usually expect more landscape detail for mapping from them than from other kinds of sensors.

Black and White Photography. Photography, which records energy between 0.3 and 1.2 μm, includes all of the visible spectrum and extends into the near infrared band. Panchromatic emulsions, which are used for most black and white photography, record wavelengths from about 0.3 to 0.7 μm, essentially the visible spectrum. Energy originating in the sun and modified by the earth's atmosphere is reflected from objects on the landscape and passes through a lens onto the light-sensitive emulsion of the film. The amount of light reaching a particular area of the film determines the density (opaqueness) of that area after the film is developed. An object that reflects a great deal of light causes its area of the film to be dense, and an object that reflects less light produces an area that is less dense. When the processed film, or negative, is used with sensitized photographic paper to make a contact print (or positive), the dense areas hold back light and produce light tones on the print, whereas the less dense areas permit more light to be transmitted to produce darker tones.

An object on the landscape is visible when its tone is different from that of its background. If its tone is the same as the tone of the background, the object will not be visible and might be detected only if it casts a shadow.

In the visible and near-visible portion of the spectrum haze results from the scattering of the short wavelengths of blue light. To penetrate the haze, a filter that blocks the shorter wavelengths is used. If one wants to photograph only certain selected wavelengths, filters can be used that block all of the visible portion of the spectrum except the wanted wavelengths, which are allowed to pass through.

During manufacture, the emulsion of a film can be formulated so that it reacts only to particular wavelengths, and its sensitivity can be extended into the near infrared, energy that is not visible to the eye, but that can be recorded as an image on the film. The tone of an object resulting from infrared exposure may be quite different from that produced by reflected light. On panchromatic film an object may not be visible because its tone is the same as its background, whereas that same object and background might have contrasting tones on infrared film and be quite discernible. Since most of the shorter wavelengths are eliminated, black and white infrared photography is very effective in penetrating haze and can be used successfully on days when ordinary film would be unsatisfactory.

The greater contrast in images on infrared film makes it especially useful for some kinds of interpretation. For example, it is a favorite of foresters, since different types of forest cover can be more easily distinguished by variations in tone. Because water absorbs infrared radiation more effectively than it does visible light, it registers as a dark gray (or black) on infrared film, and so infrared photography is a convenient means of delineating hydrographic features.

The making of infrared images is considered photography but, in a strict sense, it is not. "Photo" means "of or produced by light," and infrared radiation is not in the visible or "light" part of the spectrum. On the other hand, infrared images (unlike those produced by other sensors) are made as photographs are, by energy passing through a lens directly onto sensitized film. Placing infrared sensing in the category of photography allows for a neat distinction between it and thermal infrared sensing, which uses the longer wavelengths.

Color Photography. Although black and white photography continues to be the workhorse for landscape interpretation and mapping, color photography is becoming more and more important. It is much more expensive and, until recently, the resolution of its pictures was comparatively poor. But now color prints and transparencies are challenging black and white prints in quality; they show landscape detail with ever improving clarity.

Early experiments in color photography were conducted during the latter part of the nineteenth century, but it was not until 1935 that Kodachrome film was placed on the market. In 1942 Kodacolor Aero Reversal Film, a negative positive reversal film was introduced; however, slow film speeds and problems in processing discouraged extensive use of color film. Color infrared or "false color" film, which was developed for detection of artificial camouflage, has since been found invaluable in the plant sciences. In the past 30 years vast improvements have been made in film speed and resolution.

The human eye can distinguish only a limited number of gray tones but a very large number of colors (hues). Since many elements of the landscape have special or unique colors associated with them, it follows that in some cases interpretation of a scene is made easier and more reliable when one is able to view it in color. Identification of objects involves consideration of the size, shape, tone, texture, and location of features. However, to determine the condition of objects, such as diseased or distressed vegetation, one may find color the only clue.

Color film has three separate layers of continuous-tone (panchromatic) emulsion, each formulated during manufacture to be sensitive to a particular portion of the visible spectrum (Fig. 6.3). The two kinds of color film, color reversal and color negative, differ in that the former produces a color transparency that has the same tones and values as the original scene, while the latter yields a negative in which the tones and values are reversed. The color negative can be used to produce either a positive transparency or a color print on paper.

Reversal (color) processing makes use of the residual-silver phenomenon* to form a positive image directly from the original film exposed in the camera. In this process the film is developed in black and white developer, producing a negative

*Development of a negative leaves residual silver halide that was not used to form the negative image and that has the gradation of a positive.

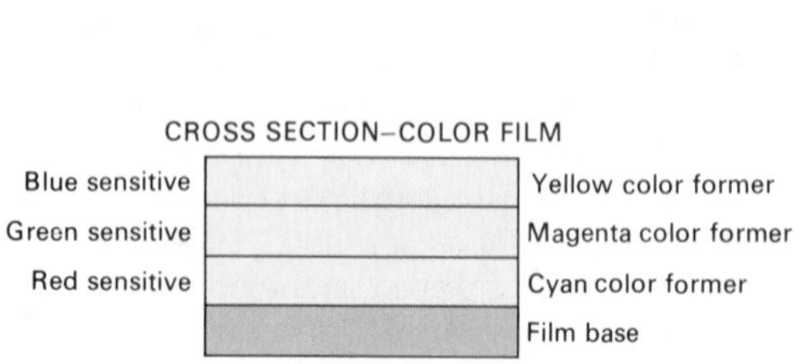

CROSS SECTION—COLOR FILM

Blue sensitive		Yellow color former
Green sensitive		Magenta color former
Red sensitive		Cyan color former
		Film base

FIGURE 6.3 Color film consists of three emulsion layers, each sensitive to different wavelengths of light.

silver image in each of the three emulsion layers. Next the film is exposed to white light or is chemically treated to fog the silver halide that was not utilized during the first development. The film is then placed in the color developer, where the most recently exposed silver halide oxidizes the developer to form positive images of yellow, magenta, and cyan. Combinations of varying amounts of these dyes produce the colors of the original scene. The final step is a bleaching process in which the silver is converted to salts that are soluble in hypo. Since the dye is insoluble, it remains to record the colored image.

Negative color films incorporate the dyes in the emulsion at the time of manufacture. After development, the silver image, which is formed along with the dyed image, is removed by bleaching. The remaining dyes are complimentary to the colors and negative in relation to the tonal gradations of the original scene. To produce a color print, white light is directed through the negative onto a three-layer emulsion on paper. The dyes permit selected wavelengths to pass and expose the appropriate layers on the paper, producing an image in the proper colors.

False Color Photography. During World War II, false color or camouflage detection film was developed; this allowed interpreters to distinguish easily dead vegetation or artificial camouflage materials from live vegetation. It is a reversal film that differs from conventional color film in that the three layers of the emulsion are sensitive to infrared, green, and red instead of to blue, green, and red. The combinations of sensitivities and dyes produce photographs in which the colors are false for most objects (Fig. 6.4). The green-sensitive layer is dyed to produce yellow, the red-sensitive layer produces magenta, and the infrared-sensitive layer produces cyan. An object that reflects infrared or longwave red appears red on the final print, those items that reflect visible red light appear green, and those reflecting green light are blue.

With conventional color photography the color of an object appears approximately the same as it does to the human eye; therefore if an item is painted the same color as its background, it will not be visible. On false color film objects painted with infrared-absorbing paints pho-

CROSS SECTION—FALSE COLOR FILM

Green sensitive	Yellow Color former
Red sensitive	Magenta color former
Infrared sensitive	Cyan color former
	Film base

FIGURE 6.4 *The special combination of sensitivity and dyes in the three layers produces a print on which the colors are not natural.*

tograph blue or purple, whereas live, healthy deciduous vegetation with its high reflection in the infrared appears red (magenta). Military equipment such as guns and vehicles that have been painted the color of vegetation or sites that have been covered with dead vegetation can easily be distinguished from live, healthy vegetation on the basis of their color on the photography.

False color film is being used in a number of investigations including identification of land use, crops, geologic formations, forest species, vegetation disease, and water pollution. In urban land use mapping, for example, the film is useful because the utilization of longer wavelengths permits better penetration of haze. Also, there is sharp contrast between vegetation, which appears reddish, and cultural features, which appear blue.

Nonphotographic Sensors

Shortly after World War I experiments were conducted in the use of nonphotographic sensors. The early work was to a great extent concerned with the detection of military targets by utilization of the infrared portion of the electromagnetic spectrum. Improvement in sensors has depended to a great extent on development of more efficient detectors. Nonphotographic sensors are now utilized in the ultraviolet, visible, infrared, and microwave portion of the spectrum.

Along with the improvement in sensing devices during the past 100 years, we see a parallel development of new vehicles to carry the equipment aloft. The aircraft used for photographic missions during World War II were far superior to those used in World War I. Since the 1940s, airplanes have been produced that can carry an arsenal of sensing equipment to extremely high altitudes. The 1957 launch of Sputnik I paved the way for the use of spacecraft to carry detecting de-

vices far beyond the earth's atmosphere. Systematic use of orbital observations began in 1960 with the launch of Tiros I, and photography produced by automatic cameras in an orbiting spacecraft became available in 1961. Although the pictures were made to monitor the attitude of the spacecraft, they served as a stimulus for future planned photographic missions. Manned space flights permitted selection of targets while in orbit, producing unique and valuable pictures.

Scanning Devices. Instead of recording in a wide-angle field the way a camera does, scanning devices detect the energy from one small element of the landscape at a time. As the vehicle (aircraft or satellite) moves, the scanner collects the energy from the landscape in a series of scanlines that are perpendicular to the line of flight. Rotating mirrors reflect the radiation onto detectors. These spinning mirrors are synchronized with the speed of the sensor platform so that as it moves along the flight line, a new swath is scanned that is adjacent to the previous one (Fig. 6.5). The data that is col-

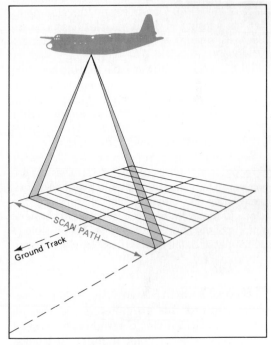

FIGURE 6.5 *Scanners detect energy along a scan line that is perpendicular to the line of flight.*

lected in the form of electronic signals from narrow strips of the landscape is stored on magnetic tape and can later be processed to produce a picturelike format.

Because of unwanted motions in aircraft or space vehicles, a long series of perfectly parallel scan lines is practically impossible. Constant scale along a flight line then is unlikely and, because of the nature of scanning, scale along the scan line is also not accurate. The mirror, rotating at a constant speed, senses radiation from equal angular intervals. Larger ground areas are sensed toward the horizon than are sensed directly under the vehicle. As a consequence of these variations along both the flight and scan lines, known ground locations are used to calculate an approximate scale.

Ultraviolet Sensing. Unless an external source of energy is used, sensing in the ultraviolet is limited to daylight hours, since it depends on reflected energy from the sun. Because of the blocking effect of the earth's atmosphere (mainly by the ozone layer) most of the energy in wavelengths shorter than 0.28 μm do not reach the surface of the earth. Consequently, developments in remote sensing in the ultraviolet has not progressed as rapidly as they have in the longer wavelengths.

Film emulsions are available that are sensitive to wavelengths down to 0.29 μm and that provide a high degree of resolution. Photography in these short wavelengths requires special lenses with a high quartz content, since most ordinary lenses are opaque to wavelengths shorter than 0.36 μm.

Scanning devices can also be used to detect and record in the ultraviolet. The impulses can be recorded on tape and used later to produce a photolike image. By electronically filtering out the "noise" in the data that is caused by atmospheric scattering of the short wavelengths, it is possible to produce better detail than is possible with ultraviolet photography.

Thermal Infrared Sensing. The radiation distribution curve for the sun shows an energy peak in approximately the visible portion with most of the remaining radiation in the shorter wavelengths of the infrared. The earth, because of much lower temperatures, radiates energy at longer wavelengths. Curves for most objects on the earth peak near 10 μm, although the curves vary with the temperature and the nature of the material of the object. Devices used to sense thermal radiation operate in wavelengths from about 3 to 14 μm and therefore are not dependent on light. Sensors operating in the range of 3 to 5 μm, however, receive both thermal and reflected radiation and, during daylight hours, will receive roughly equal amounts of each.

The thermal sensor is a scanning device that uses a detector to receive the energy from rotating mirrors. This energy, which is received from along a scan line, is converted to electrical impulses proportional to the intensity of the energy received. The signal can be stored on tape or used to produce a display on a television screen. Ordinary film can be exposed in synchronization with the display on the tube to produce imagery of the thermal radiation.

Since the temperature of the scanning device itself causes it to emit infrared radiation in the same band in which it is operating, the scanner will affect the data. To overcome this problem, scanners are cooled to extremely low temperatures and enclosed in a heat-proof box. Scanners that operate in the band from 3 to 5.5 μm are less expensive, since they need to be cooled to only about $-200°C$, whereas those operating in the longer wavelengths must be cooled nearer to absolute zero.

Tonal differences on thermal imagery are a result of temperature differences created by variations in the nature of the material or the form of the surface (see Fig. 6.6). Tones can vary relative to each other, depending on diurnal temperature changes. For example, two similar rocks may appear quite unlike if sensed at different times (e.g., day and night).

The day and night capability of thermal sensing is an advantage over photography but, on the other hand, the resolution of landscape detail is much inferior. Thermal sensing is used successfully to detect and delineate the edges of forest fires; water pollution can easily be detected by locating warm water discharges; and it can also be used for inventorying livestock and wild animals because of its day and night capability and its ability to discriminate between an animal and its background on the basis of temperature differences.

FIGURE 6.6 Infrared imagery (left) obtained at night, and a conventional aerial photograph (right) display an almost complete reversal in tone. The arrows point to outcrops of a Pliocene shale formation. (Images courtesy of U.S. Geological Survey and National Aeronautics and Space Administration.)

Microwave Sensing. Microwave sensing is concerned with wavelengths of the electromagnetic spectrum from about 0.1 cm to 100 cm where atmospheric attenuation is negligible. These electromagnetic waves are, of course, invisible to the eye and are detected by an antenna. At these longer wavelengths (compared to infrared and thermal infrared), there is very little energy available, and very sensitive equipment is necessary to detect the naturally radiated energy in this part of the spectrum.

PASSIVE MICROWAVE SENSING. The microwave radiometer, a passive sensor, depends on energy that is emitted, transmitted, or reflected naturally from a surface. The emitted energy is related to temperature, and the transmitted energy has its origin in the subsurface of an object. During daylight hours, the reflected component is present. The passive microwave sensors are comparable to the thermal sensors, except they operate in longer wavelengths and sense with an antenna instead of with a heat-sensitive detecting device. Since somewhat better spatial resolution is possible in the shorter wavelengths (with a given antenna size), the passive sensors operate in that portion of the microwave spectrum from approximately 20 to 100 cm.

A microwave radiometer with a narrow beam antenna can be attached to a scanning device that moves along a path transverse to the flight line. Signals received are amplified and stored on magnetic tape. Computers can assign colors to different levels of recorded energy to produce a false color image or they can be displayed as numerical readouts. The weak signals would be no particular problem if the antenna were able to gather energy from a target for a long period of time. Because of the speed with which the scanning radiometer must operate, large areas must be sensed to gather enough energy for a satisfactory reading. Having to sense a relatively large area results in poor spatial resolution, which places some limitations on the uses of microwave imagery. Improvements in these sensors, however, promise much better imagery in the future.

Presently, data obtained by passive microwave can be used to determine soil moisture conditions, since one component of the energy originates in the subsurface. Other possible cartographic applications include the mapping of ocean surface conditions, inventory of the water content of snowfields, location of ice water boundaries, and various kinds of geologic exploration.

RADAR. Radar operates in the microwave portion of the spectrum with wavelengths from approximately 1 cm to 1 m and is an active system, furnishing its own source of energy. These wavelengths give radar the capability of penetrating clouds and haze and, since the system depends on no outside source of energy, it can operate during daylight or darkness. This day or night, all-weather capability can be successfully used for mapping in parts of the world where weather conditions have prevented the use of photography and some other types of sensing devices. The photolike format of the imagery permits the use of many of the photo-interpretation techniques for analysis, however many unknown aspects of the character of the images are currently under investigation (Fig. 6.7).

Radar transmits a particular band of electromagnetic energy toward an object and detects a portion of the energy reflected from the object (backscattered). As well as establishing the direction to the target, the system also determines the

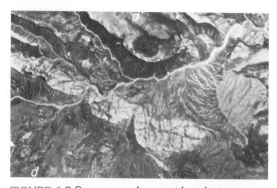

FIGURE 6.7 *Same area shown with radar image* (*above*) *and photograph*(*below*). (*Courtesy of the Westinghouse Electric Corporation*)

distance by measuring the elapsed time between transmitted pulse and the return of the reflected energy. Since each pulse yields such a small amount of information, thousands of transmissions per second are necessary to produce a sample that can be converted to the photolike image.

Radar developed during the early 1940s was of the PPI (Plan Position Indicator) type, which has a rotating antenna that provides a scan of a full 360°. The echo received is played back instantly in the form of a spot of light on a cathode ray tube. Reflecting objects such as storms, aircraft, or ships can be located in terms of direction and distance. The system became sophisticated enough during the war to supply information for aiming anticraft guns at moving targets.

To produce photolike imagery, a system called side-looking airborne radar (SLAR) has been developed. Instead of an antenna sweeping through a full circle, an antenna, which transmits and receives a narrow beam, scans a swath of the terrain perpendicular to and to the side of the flightline (see Fig. 6.8). The relative intensity of the signal reflected from an object is converted to an image, and the location on the landscape is determined by the time elapsed between transmitting and receiving a given pulse. As the aircraft moves forward, a new strip of land is imaged. By recording the strengths of the signals from these narrow strips on a roll of film, an image of the terrain is produced.

Resolution of radar imagery is best at close range and it can be improved by the use of longer antennae or by operating with shorter wavelengths. However, clouds and the atmosphere tend to absorb the shorter wavelengths. Also, there is a practical limit to the length of antennae that can be carried by aircraft. For good resolution conventional SLAR must operate in low-level, short-range situations.

To improve resolution at greater distances, *synthetic-aperture* radar has been developed. A relatively short antenna transmits and receives signals from a given target at regular intervals as the aircraft moves along the flightline. The backscattered signals are stored on film and later combined in a special way to produce the radar image with which we are familiar. The effective length of the antenna can be the distance the aircraft travels

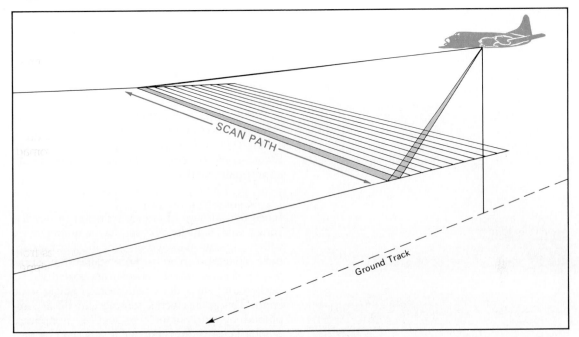

FIGURE 6.8 Radar scans a swath of land only on one side of the aircraft.

along the flight line while a particular target is within view. An antenna only 1 m long can have an effective length of hundreds of meters.

Visual characteristics of size, shape, texture, tone, and shadows used in the interpretation of aerial photographs are also used in the analysis of radar imagery. Tonal differences are related to the variations in the signals returned to the antenna. The nature of this reflected energy is dependent on the properties of the transmitted energy and the properties of the surfaces being sensed. The electromagnetic energy can vary in wavelengths, polarization, and direction. Properties of the surfaces that affect the reflected energy include roughness, slope, water content, dielectric and conducting properties, and attitude relative to the transmitter. Smooth surfaces and water, for example, appear dark; sand and rocks tend to be light tone.

Lack of a return signal causes a dark tone and, since the electromagnetic energy has little ability to penetrate solid objects, buildings, landforms, and vegetation can block the signal and cause shadows. Because of the angle at which the

ground surface is being "illuminated," the shadow effects are similar to low-sun angle photography. This enhances the impression of relief and is useful in the analysis of the form of the land. Shadows should be oriented so they do not fall away from the observer to prevent the illusion of inverted relief. Usually surveys are made in the same "look" direction throughout the flight to produce consistency in shadow direction. To operate while flying in both directions of the traverse, the systems are arranged so they can view from either side of the aircraft as required.

Electromagnetic energy is composed of waves and vibrations that can be made to oscillate in a horizontal, vertical, or some other plane. When energy vibrates in a particular plane, it is said to be polarized. Radar can send signals horizontally or vertically, and the returning signal exhibits polarization. Images can be produced with the same polarization (HH) as was sent or with the opposite polarization (HV). Using different polarization modes produces images that can be quite different (Fig. 6.9). The amount of backscattering varies with the wavelength. Shifts

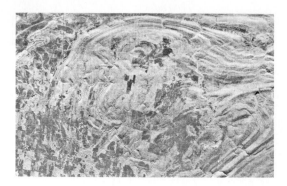

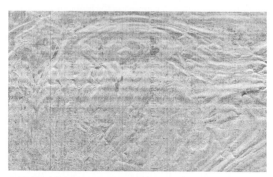

FIGURE 6.9 *Top image produced with same polarization. Bottom image produced with opposite polarization. (Courtesy of the Westinghouse Electric Corporation)*

from band to band, then, can also produce images with different characteristics. Radar that operates in two bands and at two polarizations produces four different images simultaneously. Like the multispectral scanner or multilens camera, this provides a considerable amount of additional data for the analyst.

The geometry of the radar image is quite different from that of a vertical or oblique aerial photograph. As the image is constructed, scale adjustments are made that produce consistent scale if the landscape is flat. In mountainous terrain, however, complications occur. Since radar is a distance-ranging device, the face of a mountain may be perpendicular to the pulse front of the beam, and little or no difference in time interval occurs from the return from the foot and the summit; facing slopes would be shortened and

backslopes elongated. The top of a tall tower could be imaged before the base and it would appear to lean toward the observer. It would appear to overlie the foreground instead of the background, as is the case on oblique photography.

With the aircraft operating at 6,000 m (20,000 ft) or more, scales of the imagery range in the area of 1:200,000 to 1:400,000. At an elevation of 6,000 m (20,000 ft), a strip of terrain 16 to 20 km (10 to 12 mi) wide can be covered. Stereoscopic coverage is possible by flying with appropriate overlap.

Multispectral Sensing

Since the amount of energy reflected or emitted varies with wavelength, sensing simultaneously within separate narrow bands of the electromagnetic spectrum may provide data that show greater differentiation between an object and its background than does data from a single wide band. This multispectral sensing can be accomplished by photographic as well as nonphotographic sensors, and separate images can be produced in black and white or they can be combined to produce color renditions. Transparencies from various wavelengths used in projectors equipped with color filters can be reconstituted into images in which the strengths of the colors can be varied. Often combinations can be found that make the desired information more evident.

Tonal variations on imagery indicate that changes are present; however, the eye is often unable to discern boundaries between tonal classes because of the subtle gradations. Instruments are available that can measure the optical densities of the film emulsion to establish these boundaries, or they can be established by electronic means from taped data. Bright colors can be assigned to the image classes, thereby enhancing the densities to make them strikingly visible. See Figs. CI.2 and CI.3.*

Multilens Photography. Although experiments in multispectral black and white photography were conducted as early as 1861, not until a 100 years later was the technique used for

---

Figures labeled "CI" refers to the color insert.

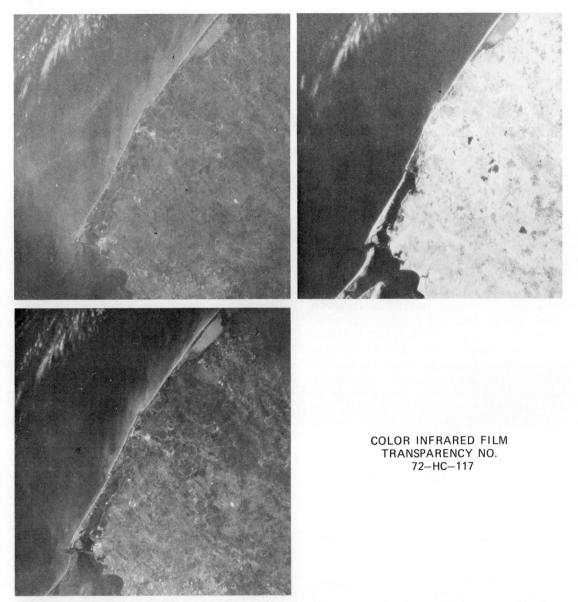

COLOR INFRARED FILM
TRANSPARENCY NO.
72—HC—117

FIGURE 6.10 Multilens photography from space. Differences in tones on the three images are caused by the particular wavelengths used. (Images courtesy National Aeronautics and Space Administration)

aerial survey of terrain features. Since the late 1960s, missions of both space vehicles and aircraft have made extensive use of photography made simultaneously in more than one wave band.

Cameras have been developed that have two or more lenses, each of which can be used with a different filter to expose a different kind of film. The result can be a series of photographs of the same landscape exposed at the same time, each recording reflected energy in a different portion of a visible or near-visible part of the spectrum (Fig 6.10). Since the exposures are made simultaneously, variations in illumination are nonexistent and the interpreter has pictures that differ only because of the particular wavelength used. Disadvantages of multilens photography are that the image format is smaller than with ordinary photography and there are many more images for an interpreter to attempt to comprehend.

Multispectral Scanner. Optical multispectral scanners can divide the ultraviolet, visible, and infrared portion of the spectrum into as many as 18 separate channels in order to study reflectance and emittance phenomena.* Particular wavelengths can be selected that are more important for a specific study, and various wavelengths can be used for comparison (Fig. CI.1). Since the images in the various bands are procured simultaneously, as they are in the multilens camera, the effect of changing illumination is not a problem.

Except for an additional separation assembly, the multispectral scanner operates much as the thermal scanner. Spinning mirrors again collect the data and pass it on to the detectors. However, between the mirror assembly and the detectors there is a prism or diffraction grating (separation assembly) that splits the incoming energy into narrow bands. By strategically placing special detectors at specific geometric positions, each small wavelength band is directed at the proper detector.

Characteristics of Images

Unlike a drawn map in which the selected features of the landscape are represented by devised symbols, remotely sensed images are records in various tones of gray or colors, of all those objects the particular device is able to resolve. Identifying objects on imagery is usually more difficult than recognizing symbols; there are, however, several characteristics of images that can be used as aids in the process of identification.

Tone and *color* variations result from differences in the reflective and emissive capabilities of objects, which are recorded on imagery in a variety of gray values on a black and white rendition or the combinations of hue, value, and chroma on a color image. Grays are described in terms such as light, medium, and dark and,

in interpretation, prepared gray and colored scales can be used for comparison with images. Because of differences in the amount of electromagnetic energy available during an image acquisition mission or because of variations in the processes used to form the image, a given object may have variations in tone or color on several images on one set of coverage.

With conventional photography, tone differs even more, depending on the kind of film used. On panchromatic film, for instance, water is usually a medium to medium dark tone; however, if the sun's rays are reflected directly into the camera, an extremely light or almost white tone is the result (Fig 6.11). In infrared photography, on the other hand, water is recorded in a very dark tone with the sun's reflection not affecting it.

Moisture, especially in soils, causes tone to be darker, and marshland tends to be darker than the surrounding areas. Human activity in an area, on the other hand, tends to cause lighter tones. Rural schools are often distinguishable from rural churches because of the lighter tones around a school that are caused by childrens' play.

On positive thermal imagery higher temperatures are generally recorded as lighter tones. Nu-

FIGURE 6.11 *When the sun's rays are reflected from water directly into the camera (lower left), the tone becomes very light.*

Although it is possible to sense 18 separate bands, it has been found that much of the information is redundant, consequently, only 6 or 9 channels are usually used.

merous natural conditions can, however, have a pronounced effect on the appearance. Clouds produce a patchy warm (light) and cool (dark) pattern, and rain can produce streaks parallel with the scan lines. Downwind from obstructions, wind streaks in the form of light-tone plumes can obscure landscape detail. Diurnal variations in the tones of features can be expected, since objects cool and heat at different rates.

The tone of an individual feature can vary with the band used to sense the data, since the amount of energy reflected or transmitted by a feature varies with the wavelength. By producing images in carefully selected narrow bands, more pronounced tonal contrasts in object and background can be produced. *Multispectral image analysis* denotes the use of images from more than one spectral region.

Density slicing, where a slice or segment of a gray scale is selected to emphasize a feature, can be accomplished both by photographic and electronic methods. For example, on near infrared black and white pictures, water usually has the darkest tone, since it absorbs infrared radiation. By controlling the exposure time one can photographically remove all gray tones that are lighter than water; the result is that water is black and everything else is white. This rendition could save a cartographer a great deal of time in delineating, hydrographic features.

Colors may be substituted for gray scales by both optical and electronic means (see Fig. CI.2.) Because this system can provide a great number of colors, there is a marked improvement in discrimination over that which can be realized with tones of gray.

Tones of objects on radar imagery depend to a great extent on the polarization and the strength of the signal returned to the receiving antenna. Shadows are very dark, since they result from a lack of signal return.

Texture is related to the arrangement of tonal repetitions in objects and may be described by terms such as stippled, granular, or mottled. Texture might also be related to the roughness or smoothness of a surface, and it is especially valuable in the analysis of radar and imaging microwave radiometry.

The size of objects necessary to produce a texture depends on the scale of the imagery. On large-scale aerial photography individual trees of a forest might be discernible, whereas at a smaller scale the crowns may be so small that only a texture is produced. Information concerning tree size can often be determined, since large crowns produce a coarse, cobbled texture, and smaller saplings appear smoother. Grass, because of its tiny blades, usually produces a smooth texture on a photograph, and brush is somewhat coarser.

Shape or the general form (which can include the three-dimensional stereoscopic view) may be the single most reliable evidence for identification. Buildings often have characteristic shapes because of their use or because of tradition of design. Churches are often in the shape of a cross and usually have a spire. Thermal electric plants are usually housed in structures that have three distinct levels: the lowest for administration, the second for generating equipment, and the highest for coal-burning equipment; in addition, a large smokestack will usually be visible. Shape is also useful for evaluating or determining the significance of features. Distinctive shapes of landforms, for example, often provide clues to the underlying structure or erosional processes.

Shadows can aid an interpreter because they provide a view more like the ground view of an object; they can be a hinderance because they tend to hide detail. Although the dark shadows on radar imagery enhance the three dimensional effect of relief, they mask ground features completely.

On aerial photography the lengths of shadows can be used to determine the heights of objects. The heights of some sample objects are first measured on the ground, and their shadows are then measured on the photograph. A shadow factor, which is the height in meters per unit of measurement of the shadow, is then computed and applied to the measured lengths of shadows of objects for which heights are desired.

Aerial photography is usually flown within 2 hr of noon to avoid long shadows and to reveal ground features more clearly. Sometimes, however, long shadows can be an aid to interpretation, since they can enhance shapes. Photography to be used in geologic investigations is sometimes procured at low sun angles to accentuate small surface irregularities.

Occasionally, when viewing the imagery with the naked eye, shadows falling away from the viewer cause an illusion in which the relief on the photograph appears reversed. Streams appear to be on ridges and ridges appear to be valleys. This probably happens because we are accustomed to frontal overhead sources of illumination which cast shadows below and toward us. In any event, the reversed effect should not be evident if a picture is oriented so that the shadows fall toward the viewer.

Size of images is extremely important because it can be a deciding clue when distinguishing between objects alike in shape. Measuring the object may be necessary before the interpreter can make an accurate identification. There are a variety of scales available for measuring, and the interpreter should select one that best meets present needs. If we wish to convert to ground distances in the metric system we should, of course, use a scale graduated in metric units. A scale graduated in thousandths of a foot is useful in that a measurement need only be multiplied by the denominator of the representative fraction to convert the photo distance to ground distance in feet.

The representative fraction assigned to photography can be calculated by using the formula

Representative Fraction (RF)
$$= \frac{\text{Camera Focal Length (Cf)}}{\text{Elevation of Camera (H)}}$$

Example: $RF = \dfrac{.5}{10,000}$ or $\dfrac{1}{20,000}$

or 1 : 20,000.

Since the elevation of the camera is related to only one datum on the photo, the RF represents only a general scale.

Pattern, the arrangement on the landscape of both physical and cultural features, is often distinctive and may be useful for recognition and evaluation. Drainage patterns, for example, can be evidence for geological interpretation. Field and road patterns can be useful for work in agriculture and economic studies. Pattern, like texture, has a relationship to scale in that some features may form discernible patterns only at certain scales. For example, at large and medium scales, trees in an orchard may form a pattern; at much smaller scales the reduction in size of images could change the pattern to a texture. Then, at an even smaller

scale, it is possible that the distribution of many orchards would form a pattern.

Site, the location on the landscape, can contribute to identification, since many features are found in characteristic places. A particular type of vegetation may, for example, appear only along streams, only on ridges, only on north slopes, or only on south slopes.

Associated features are commonly found adjacent to or in conjunction with the object under investigation. We would expect to find heavy industry adjacent to some form of heavy transportation such as rail or water. Thermal electric plants and paper mills need large supplies of water and are usually located on or near lakes or rivers.

Identification should not be made on the basis of just one of an object's characteristics; all the evidence available should be used. One characteristic may be more obvious or dominant, but consideration of all the characteristics will produce more reliable results.

Stereoscopy. Stereoscopic study, although it limits the field of vision, permits the interpreter to view features in three dimensions and usually also supplies magnification. Since the vertical view of the landscape is itself somewhat foreign to most people, restricting that view to two dimensions adds a serious limitation. In the study of things such as slope or drainage, it is almost imperative that stereoscopic methods be used.

Several kinds of stereoscopes are available, but the two most commonly used are the lens type (Fig. 6.12) and the mirror type (Fig. 6.13). Without the binoculars attached, the mirror type permits the larger field of vision. The folding lens type is more portable, of course and, for an experienced interpreter, is probably more useful. Proper use of a stereoscope will result in a better three-dimensional view and will avoid undue eye strain.

When first attempting to view photographs with a stereoscope, we should:

1. Locate and mark the principal point of each photograph. This point is located where lines connecting opposite fiducial marks intersect (Fig. 6.17).
2. Locate and mark on each photograph the image of the overlapping photograph's principal point (conjugate principal point). A line

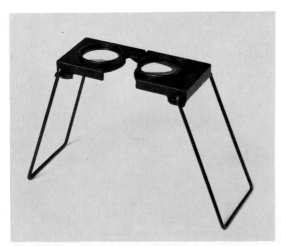

FIGURE 6.12 *Lens type stereoscope. (Courtesy of Gordon Enterprises.)*

connecting the principal point and the conjugate principal point on each photograph is the flight line.

3. On a flat surface overlap the photographs so images on the two photographs match and the flight lines are superimposed. The photographs must be oriented so the flight line extends from left to right relative to the observer. While keeping the flight lines superimposed, separate the photographs approximately the distance between the viewer's eyes (roughly 5 cm).

4. Place the stereoscope on the photographs so each lens is directly above the object to be viewed.

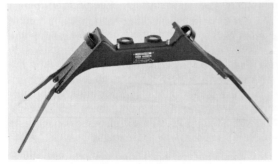

FIGURE 6.13 *Mirror-type stereoscope. (Courtesy of Gordon Enterprises.)*

5. While viewing through the stereoscope, each eye must see only that image directly beneath the lens. It might be helpful to imagine that the image is some distance away to assist in keeping each eye looking at its respective image. If the images do not merge and form a good three-dimensional model, we may need to make minor adjustments in the amount of separation of the photographs, the separation of the lenses of the stereoscope, or the alignment of the pictures.

After a short period of practice we need no longer follow these steps, since we will be able to judge the alignment and spacing and will be able to achieve the stereoscopic view in a few seconds.

The ability to use a stereoscope requires binocular vision. Some students with a weak eye are unable to experience the three-dimensional view, and others may assume they see it because of apparent relief caused by shadows. The stereogram (Fig. 6.14) can be used to determine whether or not the student has the three-dimensional view. The two blocks of letters are arranged so that most of the letters will appear on the plane of the paper when viewed through the stereoscope. Some scattered letters, which form three words, appear to float some distance above the plane of the page. The ability to read these three words indicates that the viewer has achieved a stereoscopic view.

Keys as an Aid in Interpretation. An interpretation key is an assembly of reference materials prepared to assist the interpreter in identifying and, in some cases, determining the significance of objects on images. A key contains information, much of it usually in graphic form such as stereograms or pictures of ground views, that illustrates the recognition characteristics of particular objects or groups of objects in a prescribed area. Since distinguishing characteristics of objects are not always the same in every part of the world, it usually is not possible to prepare a key that has worldwide application.

There has been disagreement among photo interpreters concerning the value of keys. Keys, of course, cannot replace the interpreter's use of reasoning, but they do have some important uses. An interpreter often finds it necessary to compile information in some unfamiliar field; keys can be

```
E S P B O E Y V E S O D        E S P B O E Y V E S O D
P S A T B L P M G M I D        P S A T B L P M G M I D
F  EM L EN MS ENTS             F E ML E NM S ENTS
M T Q H O G I P M M T F        M T Q H O G I P M M T F
Y U E S G J G A Z G J P        Y U E S G J G A Z G J P
B G S O P F S E F T F M        B G S O P F S E F  T F M
O T H P Z P R P O F P V        O T H P Z P R P O F P V
J T C P F A P R G H G T        J T C  P F A  P R  G H G T
F L T S F L P T A N T O        F L T S F L P T A N T O
O G M R D A P P H R Y M        O G  M R  D A P  P H  R Y  M
F B E R P F L A E Y J L        F B E R P F L A E Y J L
G O E C P H E A Z G A Z        G O E C P H E A Z G A Z
```

FIGURE 6.14 *When viewed stereoscopically, three words should appear to float some distance above the plane of the page.*

used to assist in identification. They are also useful for training interpreters or for would-be interpreters to train themselves. An experienced interpreter may find keys helpful to refresh his or her memory or to introduce image characteristics in a new geographical area.

Cartographic Presentation

For precise mapping, image positions must be related to some kind of reference system that specifies location on the earth's surface. The reference system might be a transformation of the geodetic coordinates (latitude and longitude) or it may be a mathematically related rectangular system such as the *State Plane Coordinate System* or the *Universal Transverse Mercator* (UTM). Because the latter is based on the metric system, has a relatively small scale factor, and covers the whole earth from 80°S latitude to 84°N latitude, it has become preferred for mapping and geographical data files. When only small areas of the earth are involved, a local *XY* coordinate system is sometimes applied. Sensor records can be analyzed in this local arrangement and later transformed to a standard system.

The positions of points on imagery depend on the relative location and attitude of the sensor (exterior orientation), the interior geometry of the sensor system (interior orientation), and errors caused by processing. In the case of conventional aerial photography, exterior orientation is constant for all points on a particular photograph, since all points were imaged at the same instant with the camera at a particular location and orientation. With scanning devices, however, the exterior orientation can vary on each scan line or at each point along the line, so a continuous record of the location and attitude of the device is neces-

sary. Sensor platforms orbiting in space have an advantage over aircraft in that they are not subjected to atmospheric conditions, and corrections in attitude can be made very slowly and can be more accurately measured.

There are two approaches to presenting remotely sensed data in cartographic form. A coordinate system can be distorted to fit the image, or the sensed data can be warped to fit a reference system. The latter method is considered more useful; however, it is also more difficult to perform.

Distortion of the coordinate system simply involves determining the reference system coordinates of a few points and computing the location of reference grid lines, which can then be plotted on the presentation. On weather satellite pictures a perspective graticule (latitude and longitude) can be produced from information about the spacecraft position and the camera pointing angle. Political boundaries can then be plotted as base data to assist in relating the picture to the earth. The main disadvantage of fitting a reference system to an image is its incompatibility with other image and graphic presentations of the same area.

The simplest method of fitting sensor data to a reference system is by visually comparing the sensor image to a reference image that already has the desired geometry. By comparing the sensor image with a topographic map, for instance, the cartographer is able to transfer, by eye, the data from the image to the map. More sophisticated systems include optical devices such as a reflecting projector or a camera-lucida.*

Methods using optical, electronic, and digital

A camera-lucida is an instrument that permits the eye to receive two superimposed images.

rectification have also been developed; these methods employ both the image geometry and the exterior orientation data to fit the image to a reference system. The systems are extremely complex.

Although numerous methods have been suggested and devised to relate image points to earth reference systems, operational procedures for accurate planimetric mapping from remotely sensed data are available only for conventional photography. Probably most work has been in the area of producing topographic maps from side-looking imaging radar because of the all-weather capability of securing the data. Although it has been demonstrated that topographic mapping at a scale of 1:250,000 is feasible, the systems are not yet operational. Radar presentations are usually semi-controlled mosaics in which different image scales have been corrected, but displacements caused by topographic displacement remain.

Satellite imagery can be useful in thematic mapping. Because of the relatively small scales used, data can readily be transferred to a map base without great concern for precise planimetric position. Categories of information such as open water or certain types of vegetation can be isolated by either photographic or digital processing.* Imagery is also useful for updating or making corrections on medium-scale charts as well as on thematic maps.

Sources for Imagery

Although much of the information from sensors in the early satellites was gathered for the purpose of investigating weather conditions, equipment used in later missions was designed to record other phenomena.

The most ambitious civilian undertaking in terms of an earth inventory is the mission of LANDSAT.† The vehicle is equipped with two imaging systems—a four-band Multispectral Scanner (MSS) and three Return Beam Vidicon‡

(RBV) cameras operating in different spectral bands. Because of a switch problem in the control circuit, the RBV was closed down after about 2 weeks of operation. The MSS, however, has been supplying complete coverage of the United States and Canada every 18 days, and nearly 80% of the earth's land areas have been imaged at least once.

LANDSAT-2, which carries the same equipment as LANDSAT-1, was placed in orbit in 1975. The MSS equipment on the two satellites is able to provide repetitive coverage of North America every 9 days. At this writing, the RBV on LANDSAT-2 has not been turned on.

LANDSAT-C, which is scheduled to be launched by 1978, is to be equipped with two RBV systems that will operate in the same band. The alignment of the cameras will result in two images that will cover a ground scene of nearly 100 km². The MSS will differ from those on LANDSAT-1 and LANDSAT-2 in that a fifth band, which will operate in the thermal (emissive) spectral region (from 10.4 to 12.6 μm), will be added.

The National Cartographic Information Center* has been established to collect and disseminate information relative to the cartographic holdings (including imagery) of the many federal agencies as well as state, county, and private organizations. Sources for particular imagery can be learned by proper inquiry to this office; however, time may be saved by dealing directly with a holding agency.

The Earth Resources Observation System (EROS) Program, administered by the Geological Survey, operates a Data Center to provide access to aerial photography acquired by the U.S. Department of Interior, NASA's† LANDSAT imagery, and photography and imagery from Skylab, Apollo, and Gemini spacecraft and from research aircraft.‡ A central computer complex, which controls a data base of more than 5 million images, performs searches of specific geographical areas

---

*Alden P. Colvocoresses, "Evaluation of the Cartographic Application of ERTS-1 Imagery," The American Cartographer, 2(1), 5—18 (1975).

†Launched in 1972, the satellite was originally called ERTS-1, for Earth Resources Technology Satellite. In 1975 the name was changed to LANDSAT-1.

‡The RBV cameras used on LANDSAT store the images on photosensitive surfaces that are then scanned to produce video outputs.

*National Cartographic Information Center, U.S. Geological Survey, 507 National Center, Reston, VA 22092.

†NASA—National Aeronautic and Space Administration.

‡To place an order or secure information contact:
User Services Unit
EROS Data Center
Sioux Falls, SD 57198

to retrieve coverage information. To request a search, which is performed free of charge, one is advised to use the inquiry form available from the center. The search will provide a computer listing and detailed description of the characteristics of all images and photographs available over or close to the area of interest.

Mosaics of LANDSAT imagery of the United States have been prepared by the Soil Conservation Service. They are not available from the Data Center, but should be ordered from:

> Cartographic Division
> Soil Conservation Service
> Federal Center, Building No. 1
> Hyattsville, MD 20782

The Agricultural Stabilization and Conservation Service* (ASCS) of the Department of Agriculture has probably the largest collection of aerial photography. Both contact prints and a variety of enlargements are available on either paper or on a polyester base. Since local ASCS offices (state and county) hold the most recent coverage for their areas, a personal visit is recommended to view the photography before placing an order.

Defense Meteorological Satellites are equipped with sensors that provide data in the visible/near infrared (0.4 to 1.1 μm) and the infrared (8.0 to 13.0 μm). Positive transparencies of both high and low resolution are available for both the visible and infrared and can be secured on loan from:

> DMSP Satellite Data Library
> Space Science and Engineering Center
> 1225 W. Dayton Street
> Madison, WI 53706

Photocopies of portions of the transparencies in an 8×10-in. format can be purchased at the above address.

NASA also supports the Technical Application Center,† which disseminates Skylab, Gemini, and Apollo earth-oriented photography and also produces educational slide sets from satellite photography.

Aerial Photography Field Office
Administrative Service Division
ASCS-USDA
2505 Parley's Way
Salt Lake City, UT 84109
†Technology Application Center, University of New Mexico,
Albuquerque, NM 87131

Handling of Imagery. Most interpretation is still done with paper prints of photographs or other images, although the use of films (transparencies) is increasing. Whether the film is used in rolls or as single frames, care should be exercised so they are not scratched or otherwise damaged. A clear material such as acetate can be placed over films to protect them while in use.

Whether we are working with film or paper prints, adequate storage such as a file cabinet is required to keep the materials in good condition and readily accessible. Each picture should be identified by an overall numbering system or by the systematic use of a number that may be a part of the marginal information. To be able to select individual images, some sort of index showing location is necessary. Plotting outlines or even just the identification numbers of individual prints in their correct locations on topographic sheets results in a useful index (Fig. 6.15). However, even a chart showing flight lines and relative locations of prints could be adequate.

The marginal information on aerial photography usually includes the date when the photograph was produced and an identifying number for each print. In some instances, information such as the altitude of the aircraft, time of day, and camera specifications are also included. The date of the photography is important to the interpreter, because the images of many features of the landscape reflect changes with changing climatic conditions. This is especially true of cultivated crops, but it is evident also in natural vegetation. Abrupt changes in images in several prints of a flight line may be accounted for by reference to the date. If, for example, a number of photographs do not conform to the specifications, those portions of a flight line may have been rephotographed at a later time. Consequently, on the later photos, trees may reflect a different amount of light because of changes in leaves or because the trees could even have dropped their foliage.

THE USE OF AIR PHOTOGRAPHS

The cartographer relies more and more on photogrammetric methods to furnish planimetric and hypsometric positions for map preparation. Compared with field methods, photographs can pro-

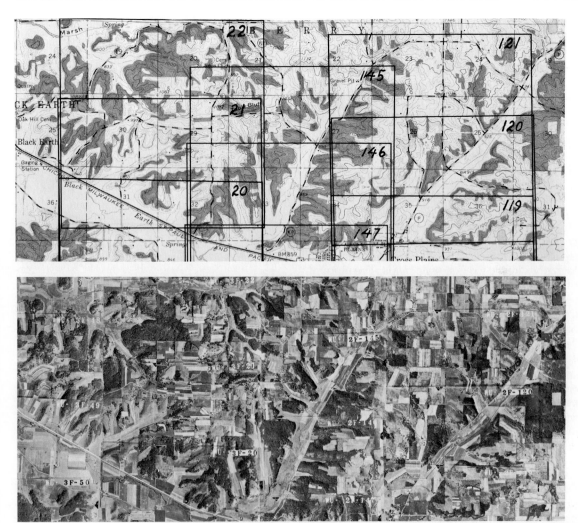

FIGURE 6.15 Topographic map used as an index for photography compared to a photo mosaic index.

vide such information much more rapidly and at a reduced cost. With limited ground control, photogrammetric methods can provide accurate large-scale maps for areas where it is difficult to conduct extensive field surveys because of terrain or climatic conditions. This advantage, of course, is especially important for military operations.

Most detailed large-scale maps made from aerial photographs are produced by government agencies or by commercial organizations working under contract. The high cost of precise stereoplotting instruments prohibits their use by individ-

uals or small commercial firms. Some, however, can be found at educational institutions where they are used mainly for instruction and research.

Even without expensive equipment, cartographers can make use of photographs for extending control. Equipment that is not particularly costly can furnish control that is precise enough for the preparation of smaller-scale maps or for map revision. Space will limit discussion here to a description of a system of extending control using radial line plotting and to a discussion of some plotting instruments.

Geometry of Aerial Photographs

The perspective projection of the image of the land surface on an aerial photograph causes the scale relationships to differ from those that would occur on an orthogonal projection of the land to a map. Disregarding things such as the distortion produced by the lens, paper, and film, the scale of a truly vertical photograph of perfectly flat terrain would be nearly the same as that of an accurate map. The occurrence of relief, however, causes variations in scale to appear because of the perspective view of the camera. Relative to one particular level of the terrain, higher points will be displaced away from the center of the photograph and lower points will be displaced toward the center. These differential variations in scale preclude our merely tracing information from photographs directly to large-scale maps. The amount of displacement can be measured, and the disadvantage of not being able to trace information directly for large scale map production is far outweighted by our ability to use displacement to determine distances above or below a chosen datum.

Consideration of the complex geometry of an aerial photograph will be limited here to the aspects of vertical photographs with no tilt. The optical or lens distortion which, on a typical 9 × 9-in. photograph, might be in the magnitude of a displacement of a fraction of a millimeter, can usually be ignored except for determination of elevations with stereoscopic plotting instruments. Distortion resulting from changes in the dimensions of the film base are usually very slight, but paper distortion of prints can be considerably more bothersome.

Scale. The precise scale ratio between two points on a vertical aerial photograph usually differs from that of the general or average scale. As observed earlier, the general scale is the ratio of the focal length of the camera to the elevation of the camera with respect to some specific elevation on the landscape; it follows that this ratio will not be correct for any other elevation or datum. Each of the infinite number of horizontal planes has its own specific ratio or scale. Figure 6.16 illustrates how scale varies with differences in the vertical positions of points on the landscape. The true map locations of two towers on the land

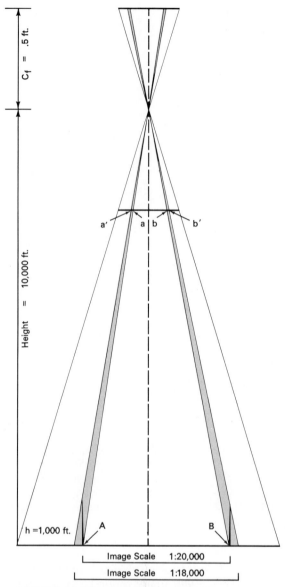

FIGURE 6.16 *Because of the perspective projection of an aerial photograph, the locations of images on the film (and print) are determined by their vertical positions and their distances from the point directly beneath the camera.*

surface at the same elevation is at points A and B. On the photograph, at a scale of $1:20,000$ these locations will be at a and b. Because of the perspective in the photograph, the tops of the towers will appear in the photograph at *a'* and *b'*, which are clearly farther apart and therefore at a larger scale than *a* and *b*, for example, a scale of 1:18,000. In this particular case, we could plot the positions of the towers at a scale of 1:20,000 because the bases as well as the tops of the towers would be visible on the photograph. On the other hand, if the problem involved a hill, the base would not be visible and, therefore, we could not plot the position of a hill directly from the photograph at a scale of 1:20,000.

Displacement. Displacement because of relief occurs at a radial direction from the nadir, the point on the plane of the photograph located by the extension of a vertical line through the center of the camera lens which, on a truly vertical photograph, coincides with the principal point or geometric center. Straight lines across a photograph connecting opposite fiducial marks intersect at this principal point (Fig. 6.17). No displacement occurs at the principal point, then, when it is aligned with the perspective center of the camera

lens, but image distance from the center of the photograph to any other point will depend on (1) the relative vertical location of that point, and (2) the distance of it from the principal point.

The amount of displacement changes directly with the vertical departure from a chosen datum and the distance from the principal point, and inversely with the height of the camera. A comparison of similar triangles in Fig. 6.18 indicates this relationship; as either *h* or *d* increases, *rd* also increases; and, as *H* increases, *rd* decreases. Photographs made at low altitudes, then, show more displacement than those made at high altitudes.

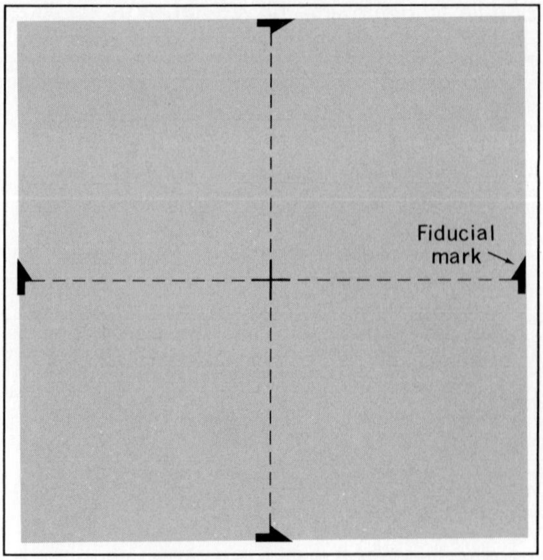

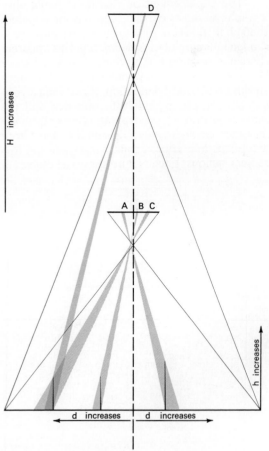

FIGURE 6.17 *Fiducial marks are located in the focal plane of the camera. They are etched on the surface of the glass, which is in contact with the film or, in the case of the open-type focal plane, they are projections of metal that extend into the negative area.*

FIGURE 6.18 *As the distance from the center increases and as the height of the object increases, the amount of displacement (rd) is greater. As the altitude of the camera increases, displacement is less.*

Although the latter have scale relationships that more closely approximate a map, the low-altitude photographs with the greater displacements turn out to be more useful for the determination of elevations.

We perceive depth in our vision in a variety of ways. With one eye we must rely on sizes of objects, clarity of detail, or whether one object appears in front of another. When using both eyes, each eye sends its own signal to the brain. Because our eyes are separated, each sees an object (if it is within approximately 600 m) from a different angle, and the subsequent signals to the brain cause the sensation of depth.

The displacement of objects on aerial photographs produces parallax, which is the apparent change in position of an object because of a change in the point of observation. This apparent change in position is the principal reason for our being able to view two photographs and produce an illusion of a third dimension. By viewing an object on one photograph with one eye and the same object on an overlapping photograph with the other, we are in effect viewing the object from two points that approximately represent the two camera stations. Each picture shows all objects in

perspective from each camera station, and our eyes each send a signal to our brain causing the image to take on an apparent third dimension.

On a photograph with no tilt, the parallax is a linear element used for determination of elevation. It is parallel with the axis of the camera stations and is associated with the height of the object, the focal length of the camera, the distance between the camera stations, and the distances from the camera stations to the object. The algebraic difference of the parallax on two overlapping photographs is used to determine elevations using stereoscopic plotting instruments.

The parallax difference can be measured graphically; however, the error is likely to be somewhat larger than when using plotting instruments. Two overlapping photographs should be carefully aligned, as shown in Fig. 6.19. The principal points and conjugate principal points (image of the principal point of the overlapping photograph) of both photographs must fall on a straight line. The average distance b between principal points and conjugate principal points is then computed. The parallax difference is the difference of the distance xx' and yy' and can be used in the following formula to determine the difference in elevation

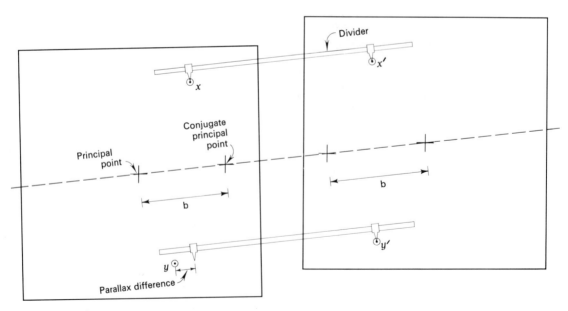

FIGURE 6.19 *Graphically measuring parallax difference. The locations of one of the points on the two photographs is indicated by x and x', and the second point indicated by y and y'.*

between the two points.

$$\Delta e = \frac{H}{b} \times \text{parallax difference}$$

where H is the average height of the camera above the terrain.

Radial Line Plotting

On a truly vertical photograph, azimuths from the principal point are correct to any point on the photograph; this condition permits us to perform triangulation directly from the photographs. Since the direction from the principal point to its conjugate principal point is correct on each of two overlapping photographs, it follows that by superimposing these two lines we produce a baseline with ends from which all angles are correct. On pieces of overlay material we can mark the principal points of each photograph, and draw lines through the conjugate principal points and from the principal points through a third point c (Fig. 6.20). When the overlays are placed so the baselines are superimposed, we have produced a triangle in which the angles at a and a' are correct. If these two angles are correct, the angle at c must, of course, also be correct (Fig. 6.21), the elevation of c having no effect on the horizontal angles. The three points have been located in a true planimetric relationship, and the scale of the three sides of the triangle is the same.

In radial line plotting each point is relocated by the amount of its displacement and all points are then located at a common scale. The actual scale of the plot bears no special relationship to the general scale of the photography and can be either larger or smaller, depending on the distance between the principal points of the two photographs (Fig. 6.21). In this situation scale can be determined if we know the ground (or map) distance between two points in the plot.

Usually a radial line plot is prepared at a predetermined scale by using control points established in the field or from an accurate map. Ideally, we should plot three control points at the desired scale in the area of overlap of the first two photographs. The principal points are then either separated or moved closer together to cause radial lines to intersect at the plotted control points.

Hand Template Method. By using additional photographs, we are able to extend control over larger areas by using the following procedure (Fig. 6.22).

1. Locate the principal points on all photographs and mark them by pricking a small hole, point A (Fig. 6.22).
2. Locate and mark all conjugate principal points while viewing stereoscopically, point b (Fig. 6.22).

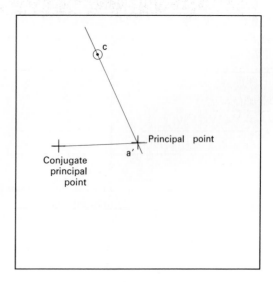

FIGURE 6.20 *The angles at points a and a' are correct.*

3. Select at least three points (called pass points) on each photograph in the area of side lap of the adjacent flight line, points *c* Fig. 6.22). Although it is desirable, pass points need not be features that ordinarily would be symbolized on a map but can be any images that are easily recognized on adjacent photographs.

4. Mark all control points to be used in the plot. [Three control points in the area of overlap of the first two photographs establish the scale, points *d* (Fig. 6.22); however, more control points are recommended for other photographs to help hold the entire plot to the predetermined scale.]

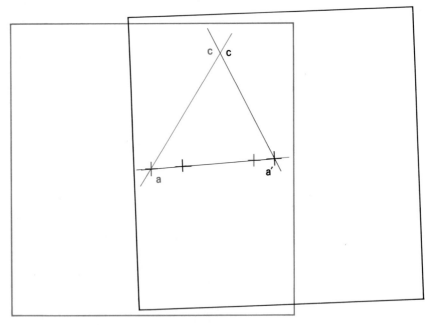

FIGURE 6.21 *The angles at points a and a' are correct; therefore, the angle at c must be correct.*

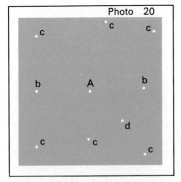 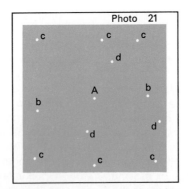 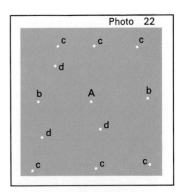

FIGURE 6.22 *The principal point, conjugate principal points, and pass points have been marked and can subsequently be transferred to an overlay.*

5. Templates are prepared on a clear or translucent material that has been cut into pieces slightly larger than the photographs. (Photographs can then be taped face down to the overlay material, thereby preventing the tape from touching the emulsion.)

6. To make templates of the overlay material (Fig. 6.23):
   a. Mark the principal points.
   b. Draw lines from the principal points through the conjugate principal points (baselines).
   c. Draw short lines or rays from each principal point through each pass point on that photograph extending ½ in. or so on either side. If the intended scale of the plot is to depart considerably from the general scale of the photographs, the lines through the pass points must be longer. Extend them away from the center for larger scale and toward the center for smaller scales.
   d. Draw short segments of radials from the principal point of each photograph through features that have been selected for establishing new control points.

7. Identify each template by adding the picture number in any convenient location on the photograph.

8. Separate templates from the photographs

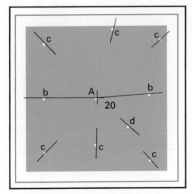 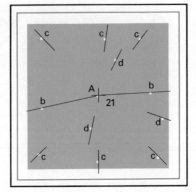 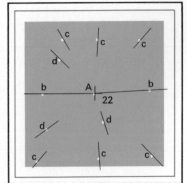

FIGURE 6.23 *Prepared template with principal points marked, base lines plotted, and rays to pass points drawn.*

and arrange in groups according to flight lines.

9. Prepare a base sheet for template assembly by plotting existing control on a translucent material (Fig. 6.24).

10. Begin laying the templates by selecting two photographs (preferably near the center of the plot) that have the greatest number of control points and placing them on the base sheet with the baselines of the two photographs coincident. Adjust the photographs by separating the principal points or moving them closer together until the radials of the control points intersect at the plotted positions on the base. Tape the template to the bases. The principal points of the first two photographs are now in their correct position relative to all the control plotted on the base. The pass points, likewise, are in their correct positions (Fig. 6.24)

11. Working along the flight line, lay each consecutive template on the plot, keeping the baselines superimposed and moving the template until its radials through the pass points intersect at those points established by the prior templates. As we proceed, the additional plotted control should be used to hold the scale and adjust for accumulated errors.

12. The adjacent flight line is laid in the same manner, using established pass points and control points from the previous flight line.

13. When all the templates are in place, the assembly may be turned over and, since the base sheet is translucent, any points where radials intersect can be selected and marked on the reverse side of the base sheet.

The base sheet with the original control points and those new ones established by the radial line plot can now be used to compile the map features. Photographic images should be adjusted to fit the control by the use of an instrument that will change the scale of the photograph and, if possible, correct for small amounts of tilt.

The accuracy of the radial line plot depends on several factors. If any of the photographs have considerable tilt, the angles at the principal point

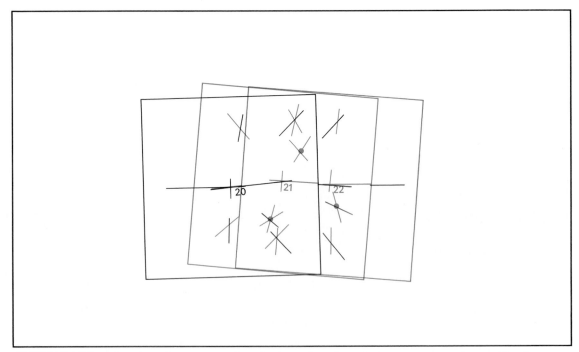

FIGURE 6.24 Templates have been arranged on the base sheet (on which map or ground control has been plotted) so that appropriate rays intersect at the control points and establish the scale for the plot.

will not be correct and rays will not intersect at the correct positions. Sometimes the paper on which the prints are made expands or shrinks in one direction as compared to another causing differences in relative positions. The care with which the plot is prepared is also reflected in the accuracy of positions. Errors in the plot usually appear in the form of triangles where rays should, of course, intersect at a point. Sometimes it is necessary to pick up several templates and adjust them slightly for discrepancies that appear.

Slotted Template Method. Slotted templates, mechanical substitutes for the hand template, are pieces of material (which need not be translucent) in which slots substitute for the lines drawn on the hand templates. A machined round stud, which fits the slot precisely, can be slid within the slot (Fig. 6.25). Since the stud can slide in all the templates having slots for a common point, the templates tend to settle at the correct position. Discrepancies then tend to "shake out" as the plot proceeds.

FIGURE 6.26 *Radial arm template set. (Courtesy of Gordon Enterprises.)*

Since the templates need not be translucent, rigid and stable material can be used. Selected points are recorded on the base map by inserting a sharp pin into the hole in the center of the stud and piercing the base material. Special punches are needed to cut the slots and, since the slots are cut for a specific photograph, the templates cannot be used for other plots.

Radial Arm Template Method. Another mechanical solution, using adjustable radial arms, permits reuse of the equipment. Flat metal arms of various lengths with holes at one end and a slot at the other again replace the lines in the hand template. The holes in the arms are placed over a bolt secured at the principal point with the middle of the slots falling on the conjugate principal point and pass point studs. When all radial arms are in place, a nut is tightened to hold the template rigid and the radial arms are used in the same manner as slotted template (Fig. 6.26). When the plot has been completed, the template can be disassembled and the parts used again.

Plotting Instruments

Plotting equipment for map compilation from aerial photographs falls in two general groups, monocular and stereoplotting instruments. Monocular instruments are used chiefly for small-scale mapping or for revising existing mapping. With simple stereoplotting instruments parallax can be measured and at least generalized contour

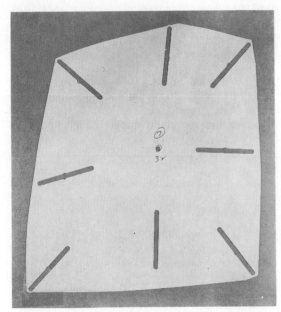

FIGURE 6.25 *In the slotted template method, the centers of the slots are the radials from the principal point to the pass points, and to the conjugate principal points.*

lines or form lines can be drawn. Photogrammetric plotting instruments used in the preparation of large-scale accurate topographic maps are sophisticated machines that provide very precise measurements and plot the map data in orthogonal projection.

Monocular Instruments

Two general types of monocular instruments are those that use the camera-lucida principle and those using optical projection. Both types are essentially devices by which photographic detail can be transferred to a base map on which control has been plotted. Changes in scale can be made to facilitate fitting the detail to the control. Some instruments also have the capability of making adjustments for tilt.

The sketchmaster uses the camera-lucida principle, and the eye receives two superimposed images, one from the base map and the other from the photograph. The image from the photograph is reflected from a semitransparent mirror, and the base map is visible through the mirror. The instrument is adjustable so that we can change the scale of the photograph to fit the base. Foot screws on each of its three legs can also be adjusted to remove small amounts of tilt from the photograph (Fig. 6.27).

Another machine, useful for medium- and small-scale compilation from photographs, is the vertical reflecting projector. The image from opaque copy is reflected from mirrors through a lens to a table on which the map is placed. Usually the projectors are designed so that we are able to change the size of the image up to three or four times, permitting tracing at the desired scale. On some projectors the table on which the base map is placed can also be tilted slightly to allow for tilt in the photograph. Neither an instrument that employs the camera-lucida principle or one that uses optical projection has the capability of removing the effect of displacement used by relief.

Stereoscopic Plotting Instruments

Stereoscopic plotting devices differ from monocular instruments in that they employ two overlapping photographs enabling us to make parallax measurements in order to determine elevations.

One of the simpler instruments is the stereocomparagraph, which consists of three major parts: (1) a stereoscope, (2) a device for measuring parallax, and (3) a drawing attachment (Fig. 6.28). In the stereocomparagraph a pair of photographs is viewed stereoscopically so as to produce a three-dimensional image or "model" of the terrain. The measurement of parallax is accomplished by adjustment of a "floating dot." This dot is actually made up of the images of two dots each seen with only one eye. The single dot may be adjusted "vertically" in space by changing the horizontal spacing of the two dots that form its image. The tracing arm is linked to the dot. Changing the apparent vertical position of the dot does not affect the position of the tracing arm, but horizontal movement does.

FIGURE 6.27 *The sketchmaster can be adjusted for different scales and to compensate for tilt. (Courtesy of Gordon Enterprises)*

FIGURE 6.28 *With the stereocomparagraph, parallax can be measured for calculating differences in elevation. (Courtesy of Gordon Enterprises.)*

By viewing the photographs stereoscopically and adjusting the two dots, the floating dot can be brought into apparent contact with the surface of the earth in the three-dimensional stereo model. Then, by keeping the elevation constant and moving the tracing arm so that the dot remains in contact with the surface of the model, a contour line may be traced. We may move the dot to points of different elevations and, by adjustment, the floating dot can be brought into apparent contact with the surface of the model. Readings from the micrometer showing the position of the dot provide parallax measurements from which differences in elevation can be calculated.

The stereocomparagraph makes no corrections for tilt. Furthermore, the plot produced is still in the perspective projection of the photograph; therefore, each contour line is projected at a different scale. For precise maps we must compensate for tilt and provide means for compiling the information on the map sheet in an orthogonal projection at a consistent scale.

Some photogrammetric plotters or plotting assemblies operate on the basis of systems that, in a sense, reverse the rays that formed the camera picture and, by projection, produce the image of a three-dimensional model of the terrain on a mapping table. When the projectors are properly oriented so as to adjust the projected images to fit already plotted horizontal control, the apparent model may be used to delineate planimetric detail and plot contours.

In such an assembly, two or more projectors are mounted so that they can be moved in any direction and thereby duplicate the relative position and orientation of the camera at the instants the pictures were taken (Fig. 6.29). Overlapping photographs in the form of glass positives are placed in alternate projectors, and one photograph is projected through a red filter and the other through a cyan (greenish blue) filter. A cartographer, wearing spectacles with one cyan and one red lens, sees the image of one photograph with one eye and the other photograph with the other eye, resulting in a three-dimensional view. A stereo pair superimposed in red and cyan is called an anaglyph.

A small tracing stand with a white disc or platen has an illuminated aperture in its center (Fig. 6.30). This platen "disappears" in the projected image of the model because it becomes part of it, so to speak, but the tiny illuminated aperture appears to float in space like the dot in the stereocomparagraph. The platen, and thus the dot, can be moved vertically, and its height within the model is indicated on a micrometer scale.

When the floating dot is placed in contact with the surface at a given elevation, the tracing

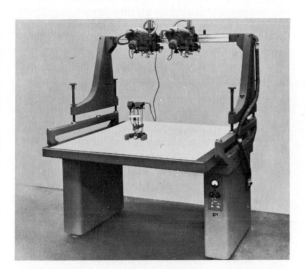
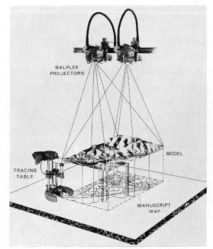

FIGURE 6.29 (*Left*) *A Balplex Plotter used for very precise maps.* (*Right*) *Illustration of the basic operation of the Balplex Plotter.* (*Courtesy of Bausch and Lomb, Incorporated, Special Products Division.*)

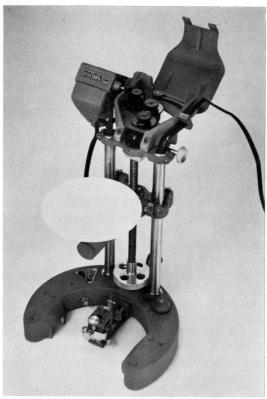

FIGURE 6.30 *Balplex Universal Foot Reading Tracing Table. (Courtesy of Bausch and Lomb, Incorporated, Special Products Division.)*

nating images of the left and right photographs. The operator, viewing through shutters synchronized with those of the projectors, sees the right-hand photograph only with the right eye, and the left with the left eye, thus producing the stereoscopic three-dimensional model.

Advantages of this system over anaglyphic systems are as follows.

1. Much more light reaches the platen.
2. Because the two images may be matched in sharpness, resolution is greatly improved.
3. The operator sees no subimage (intended for the other eye), since the images are completely separated.
4. Color photographs can be used with the system. This capability, along with recent improvements in the resolution qualities of color film, suggest that color photography may be used much more for map production in the future.

Orthophotoscope

The perspective image of a photograph can be changed to an orthogonal* projection by use of a device called an orthophotoscope. The new picture may be made free of tilt, and thus all points may be placed in their correct locations relative to one another. Scale, then, is constant and all angles are true. The usefulness of such a photograph is quite obvious. Areas measured on it will be correct, and angles on the photograph will match angles in the field. Because of the great amount of detail on photographs, in some instances the orthophotograph can be more useful than a topographic map for work in the field. For example, an image on a photograph can often be correlated with the corresponding ground feature much more easily than can a map symbol. Furthermore, the correct planimetry of an orthophotograph permits precise location of features that are not and often cannot be symbolized on a map.

Although there are now a variety of instruments that can produce orthophotographs, most operate on the basic principle of the first one developed in 1953. The projection of a three-dimensional stereoscopic image produced by a

table may be moved, keeping the dot in contact with land. A pencil or scribing instrument built into the tracing stand directly beneath the dot traces the contour in its correct planimetric position. Planimetry, such as the location of a road, may be plotted from the model by tracing the feature and at the same time continually adjusting the floating dot so that it is always in contact with the surface.

A development by the U.S. Geological Survey alleviates some of the problems inherent in systems that require colored filters to produce the three-dimensional model. A system known as the Stereo Image Alternator permits an operator to view the two photographs without the need for colored spectacles.

Rotating shutters mounted in front of each projector interrupt the light at a rate of sixty times per second. They are synchronized to flash alter-

---

The new image is orthogonal in theory; in practice it can only closely approach a truly orthogonal projection.

stereoscopic plotter is exposed to a film through a very small opening moved across the model. As the tiny aperture moves along a narrow strip, the film being exposed remains stationary in its horizontal position but is moved in the vertical dimension to keep the aperture "in contact with the surface" of the three-dimensional image. The operation is much the same as tracing correct planimetry from a model by continually adjusting the floating dot to keep in contact with the surface of the stereo model. After the aperture has moved across the model once, it is moved sideways a distance equal to the width of the opening.

Orthophotographs can, of course, be assembled to form a mosaic which, in turn, can be overprinted with map symbols to produce an orthophotomap. Photomaps, prepared from mosaics of conventional photographs, have long been used as map substitutes, but the displacement inherent in the photos caused scale discrepancies that limited their usefulness. The orthophotomap has the advantage of the photographic image and provides accurate scale as well.

SELECTED REFERENCES

Avery, Thomas Eugene, *Interpretation of Aerial Photographs,* Burgess Publishing Company, Minneapolis, 1977.

Barrett, Eric C., and Leonard F. Curtis, *Environmental Remote Sensing,* Crane, Russak and Company, Inc., New York, 1974.

Barrett, E. C., and L. F. Curtis, *Introduction to Environmental Remote Sensing,* John Wiley & Sons, New York, 1976.

Colwell, Robert N., "Some Significant Elements in the New Remote Sensing Panorama," *Surveying and Mapping, 34,* 133–142 (1974).

Estes, John E., and Leslie W. Senser, *Remote Sensing,* Hamilton Publishing Company, Santa Barbara, CA, 1974.

Holz, Robert K., *The Surveillant Science: Remote Sensing of the Environment,* Houghton Mifflin Company, Boston, 1973.

Manual of Photogrammetry, American Society of Photogrammetry, Falls Church, VA, 1966.

Manual of Color Aerial Photography, American Society of Photogrammetry, Falls Church, VA, 1968.

Manual of Remote Sensing, American Society of Photogrammetry, Falls Church, VA, 1975.

Rudd, Robert D., *Remote Sensing,* Duxbury Press, North Scituate, MA, 1974.

Thrower, Norman J. W., and John R. Jensen, "The Orthophoto and Orthophotomap: Characteristics, Development and Applications," *The American Cartographer, 3,* (1) 39–52 (1976).

Veziroglu, T. Nejat, *Remote Sensing,* Hemisphere Publishing Corporation, Washington, D.C., 1975.

Williams, Richard S., and William D. Carter, *ERTS 1: A New Window on Our Planet,* Professional Paper 929, U.S. Geological Survey, Washington, D.C., 1976.

Data handling in cartography can be conveniently broken down into three stages: (1) data capture, (2) data manipulation, and (3) data symbolization and plotting. The first stage, data capture or the obtaining of the raw data for the map, is accomplished in diverse manners. Some data are captured by questionnaires and others are obtained by photographic means as discussed in the previous chapter; coordinate digitizers are now being used to "capture" data from existing maps and charts. The sources of data for cartographers are also varied; some result from primary data where the cartographer personally collects the data, but most are secondary data where the cartographer utilizes data existing in some already-collected form. These secondary data, when initially collected by a census taker, surveyor, or satellite, present the cartographer with two evaluation problems: (1) the appropriateness of the data for a given mapping purpose, and (2) the resolution of conflicting data acquired from two or more sources. Moreover, when primary data are being acquired, it must be done with full realization of the required level of reliability.

Obtaining appropriate data for the map the cartographer wishes to make is an important first step toward the success of the communication, but perhaps more important is the second stage, the manipulations the cartographer performs on the data. Improper manipulations can destroy the excellence of good data and mask the shortcomings of poor data.

More and more of a cartographer's effort is devoted to mapping data that require statistical manipulation in order to obtain a variety of descriptive measures. Much of the data with which a cartographer works consist of samples obtained in various ways, and the realistic portrayal of these data cannot be accomplished unless the cartographer has a clear understanding of the reliability of the data and the relative appropriateness of the various statistical measures as descriptive devices.

CHAPTER 7
PROCESSING CARTOGRAPHIC DATA

Furthermore, in the manipulation of the data, in the selection of categories, in the use of nominal, ordinal, interval, and ratio scales, and in the planning of the symbolization, the cartographer must use the techniques of the statistician. The accomplished cartographer, therefore, finds it necessary to be familiar with some of the basic concepts of statistical method.

It is clearly not possible in an introductory textbook, where the main theme is cartography, not statistics, to go into much detail concerning the variety of statistical techniques that the cartographer might find useful. Nevertheless, the preparation of distribution maps usually requires familiarity with the general concepts involved. Acquire a good book on statistical method to use as a reference. A number of titles are included in the bibliography.

The third stage in data handling, its symbolization and plotting, requires a clear understanding of the various scales used for measuring. The various scales of measurement provide criteria that enable the cartographer to select symbolization in harmony with the established hierarchy of importance in the map communication. It is therefore vitally necessary that the cartographer be familiar with the appropriate descriptive statistical measures for each scale of measurement.

This chapter introduces the reader to the use of various simple statistical techniques that the cartographer commonly must employ in manipulating the data that are used in mapping. Of foremost importance are the types of quantities the cartographer encounters, their scales of measurement along with the associated appropriate descriptive statistical measures, and other common statistical manipulations.

BASIC STATISTICAL CONCEPTS AND MANIPULATIONS

Prerequisite to refining and processing data is the decision regarding the type of presentation to be employed for the cartographic communication. This does not mean that the cartographer must first select a specific form of symbolization; instead, it means that one must settle on a hierarchy of importance for the different classes of data and select the basic conceptual form in which it is to be displayed, for example, whether each data set is to

be conceived as a discrete or a smooth distribution. After these basic decisions are made, the cartographer may then proceed to solve a variety of problems concerning the data.

When data are obtained from a variety of sources, it usually is necessary to equate them so that they provide comparable values. For example, different countries use different units of measure such as long tons or short tons, U.S. gallons or Imperial gallons, hectares or acres, and so on. Frequently the units must be further equated to bring them into strict conformity. If, for example, we were preparing a map of energy reserves, it would not be sufficient to change only the volume or weight units to comparable values, but it would also be necessary to bring the figures into conformity on the basis of their BTU ratings. It is also frequently necessary to process the statistics so that unwanted aspects are removed. A simple example is provided by the mechanics of preparing a density map of rural population based on county data. The total populations and areas of incorporated divisions as well as the totals for counties may be provided in census tables; if so, the areas and populations of the incorporated divisions must be subtracted from the county totals.

Another illustration is provided by the well-known regional isothermal map. If the objective of the map is to reveal relationships among temperatures, latitudes, and air masses, for example, the local effects of elevation must be removed from the reported figures. This involves ascertaining the altitude of each station and the conversion of each temperature value to its sea-level equivalent.

After the statistical data have been made comparable, the next step is to convert them to mappable data. This may, of course, not be necessary for many maps, such as the isothermal map, because the data need only to be plotted and isarithms drawn. On the other hand, ratios, per hectare yields, densities, percentages, and other indices must be calculated before plotting.

A calculator that can add, subtract, multiply, and divide is of constant use to a cartographer working with distribution and statistical maps. If a great number or if more difficult calculations are required, it is well worth the initial tabulating time to prepare the data for subsequent machine processing. An increasing amount of data of all kinds is becoming available in machine usable form, and the cartographer need only obtain copies of the necessary information. The entire processing can be done by programming, and the result can be printed tabulations or machine-produced plots. Effort spent in initial search for data in the form easiest to process will often save considerable total expenditure of time.

Absolute and Derived Quantities

One of the most important changes in cartography during the past 200 years has been the introduction of thematic cartography. People became conscious of the variety of phenomena on the earth even before the flowering of the Classical Era, and it was only natural that they should also have been interested in variations in amount from place to place. Their observations were generally so incomplete and unreliable, however, that it was not possible to map the results in any consistent fashion until comparatively recently. With the development of accurate enumeration and observation and the rise of statistical method, ways of mapping quantitative distributions quickly developed. Early thematic cartography during the first part of the nineteenth century was largely concerned with these kinds of maps. With very few exceptions (one is the isarithm), most of the methods of portrayal by point, line, and area symbols were devised and employed since 1800 to portray ordinal and interval-ratio categories of geographical distributions. Each year there is a greater production of these kinds of thematic maps, and it is to be expected that the rate of increase will continue to rise. Consequently, it is imperative that the cartographer become adept at selecting appropriate methods of communicating quantitative data.

All maps fall in one of two classes: they portray either (1) absolute, observable qualities or quantities, or (2) calculated (i.e. derived) qualities or quantities. Examples of the first class are maps showing categories of land use, the production or consumption of goods, or the elevations of the land surface above sea level. The qualities or quantities are simply those observed concerning a single class of data, and they are expressed on the map in absolute terms according to some measurement scale, for example, output of hydroelectric power per state or numbers of beef cattle per county. In this group many combinations are possible, and several kinds of values may be presented at once. In no case are the data expressed as a relationship, such as per capita consumption of services or percentage change in population density.

In the second group of maps are those that portray derived values (e.g., averages, percentages, and/or densities); on these, the mapped values express either some kind of summarization or some sort of relationship between two or more sets of data. Examples of this second group include: areas of suitable soil where percolation tests have indicated acceptability for septic tank use, per capita income, or categories of similar ground slope. This group, showing derived rather than absolute values, includes four general classes of relationships.

Averages. The first, and probably the most common, are the maps of averages, such as are obtained by the reduction of large amounts of statistical data, for example, weather observations, well records, or numbers and sizes of farms. In this group there are several common kinds of averages called measures of central tendency such as the mode, the median, and the mean. These will be considered in more detail later in this chapter, but it is worth pointing out here that generally the most important in cartography is the mean. It is usually symbolized by $\overline{X}$ in the following equation

$$\overline{X} = \frac{\Sigma X}{N}$$

where ΣX is the summation of all the X values, and N is the number of occurrences of X.

Ratios. The second class includes all those maps of *ratios,* such as proportions, percentages, or rates, in which some element of the data is singled out and compared to the whole. These are illustrated by maps such as those showing the percentage of rainy days, the proportion of all cattle that are beef cattle, mortality per 1000 persons, or the rate of growth or decline of some phenomenon. In this group the numerical value mapped will ordinarily be the result of one of the following basic kinds of operations:

$$\text{a ratio} = \frac{n_a}{n_b}$$

$$\text{a proportion} = \frac{n_a}{N}$$

$$\text{a percentage} = \frac{n_a}{N} \times 100$$

where n_a is the number in one category, n_b the number in another category, and N the total of all categories.

As used by the cartographer, such ratios, proportions, and percentages commonly have the characteristics of a spatial average, since 12 persons per square kilometer is a ratio obtained by dividing the total number of people by the total number of square kilometers. (This kind of ratio is the basis of the familiar density concept treated below.) Rates and ratios, of course, need not be quantities related to the land. For example, assume the numerator was the total number of dairy cows in a county: we could use as the denominator quantities such as the number of farms, the number of farm operators, or the total number of cattle. These would result in the ratios of the number of dairy cattle per farm or per farm operator, or the proportion of dairy cattle to total cattle.

Maps of these kinds of relative quantities are made to show variations from place to place in the relationship mapped, and they are usually prepared from summations of statistical data either over area or through time. The appropriateness of the quantity obviously depends on the specific use to which it is being put, but a few words of caution are in order. Percentages, rates, and ratios, when mapped on the basis of enumeration units, are usually assumed by the map reader to extend more or less uniformly throughout the enumeration unit. If the phenomenon does not in fact so occur, then the ratio mapped may be quite misleading, just as it may if too few of the items occur. Thus, a value of 100% of farms with tractors could be the result of only one farm with one tractor in a large area.

Quantities that are not comparable should never be made the basis for a ratio. For example, the number of tractors per county is next to meaningless, and we ought not even calculate let alone map the number of tractors per farm by dividing the total number of tractors by the number of farms in a county unless the farm sizes (or some such significant element) are relatively comparable. Common sense will usually dictate ways to insure comparability.

Densities. The third class of maps of related quantities consists of those commonly called *density* maps. These maps are employed when the major concern is focused on the relative geographical spacing of discrete phenomena. Examples are maps of the number of persons (or trees, cows, etc.) per square kilometer, or the average spacing between phenomena such as service stations or mail collection points. The density (D) is computed by

$$D = \frac{N}{A}$$

where N is the total number of phenomena occurring in an enumeration unit, for example, a county, and A is the area of the unit. The average spacing between phenomena, another way of looking at density, is computed by

$$\overline{S} = 1.0746 \sqrt{\frac{A}{N}}$$

where $\overline{S}$ is the average spacing of the items or the mean distance between them in linear values of the same units used for A when an hexagonal spacing is assumed, this being the most economical of area. If we assume simply a square arrangement, the interneighbor interval, as it has been called, is simply the square root of the reciprocal of the population density expressed in suitable linear units.

The density class is more closely related to the land than the first two, and the significant element in the relationship is area. Thus, for example, 5000 persons in an area of 100 km² is a density of 50 persons per square kilometer; if arranged hexagonally within the 100 km², each individual would be about 150 m from neighbors and, if arranged rectangularly, paired persons would be slightly closer. In many instances, a density value derived from the (1) total number within, and (2) the total area of a statistical unit is not so significant as one that expresses the ratio between more closely related factors. For example, the relation of the number of people to productive area in predominately agricultural societies is frequently found to be more useful than is a simple population to total area ratio. If the data are available we can easily relate population to cultivated land, to productive area defined in some other way, or to that spatial segment that is important to the objective of the analysis.

When working with densities and average spacings, the cartographer is limited in the detail he or she can present by the sizes of the statistical units for which the enumeration of the numbers of items has been made. Generally, the larger the units (such as townships, counties, or states), the less will be the differences among the values. In many cases the initial data must be supplemented by other sources in order to present a distribution as close to reality as possible.

ESTIMATING DENSITIES OF PARTS. When preparing density maps, it is not uncommon for the raw data to have been gathered for enumeration districts within which there is considerable unevenness of distribution. This is not apparent from the raw data, of course. If supplementary data exist that allow us to estimate with reasonable accuracy the density in one part, the value to be assigned to the other can easily be calculated.

Assume, for example, an enumeration unit with a known average density of 100 per square kilometer. Assume, further, that examination of other evidence has shown that this unit may be divided into two parts, m, comprising 0.8 of the entire area having a relatively low density, and n, comprising the remaining 0.2 and having a relatively high density. If, then, we estimate that the density in m is 10 per square kilometer, a density of 460 to the square kilometer must be assigned to n in order that the estimated densities m and n may be consistent with 100, the average density for the enumeration unit as a whole.

The figure 460 for the density in n is obtained by solving the following equation:

$$\frac{D - (D_m a_m)}{1 - a_m} = D_n$$

or

$$\frac{100 - (10 \times 0.8)}{0.2} = 460$$

where D is the average density of the unit as a whole, D_m the estimated density in m, a_m is the fraction of the total area of the unit comprised in m, $1 - a_m$ the fraction comprised in n, and D_n is

the density that must accordingly be assigned to n.

D_m and a_m are estimated approximately. It is not necessary to measure a_m accurately, since the amount of error in a rough estimate is likely to be less than the amount of error in the best possible estimate of D_m.

Study of neighboring areas sometimes gives a clue to a value that may reasonably be assigned to D_m. For example, other maps may show what would appear to be similar types of distribution prevailing over D_m, a part of the unit, and over the whole of E, an adjacent unit. It would be reasonable, therefore, to assign to D_m a density comparable with the average density in E.

Having assigned estimated but consistent densities to two parts of an area, we may then divide each or one of these parts into two subdivisions and work out densities for these two in the same manner; the process may be repeated within each subdivision.

The method is merely an aid to consistency in apportioning established values within the limits of territorial units for whose subdivisions no statistical data are available. The method can easily be applied in the mapping of any phenomena for which statistics are available only by larger units. The table in Appendix F enables us to solve the fundamental equation without either multiplication or division. Figure 7.1 illustrates the refinement that can be made.

Potentials. The fourth class of distribution maps comprises those called *potential* maps. These maps assume that the individuals comprising a distribution, such as people or prices, interact or influence one another, directly with the numbers or magnitudes involved and inversely with the distance between them. Because this assumption is derived from the physical laws governing the gravitational attraction of inanimate masses, it is called a gravity concept. It has been applied to a variety of economic and cultural elements.

The value of the potential at any point is the sum at that point of the influence of all other points on it plus its influence on itself. The potential P of place i of phenomenon X will be

$$P_i = X_i + \sum \frac{X_j}{D_{ij}}$$

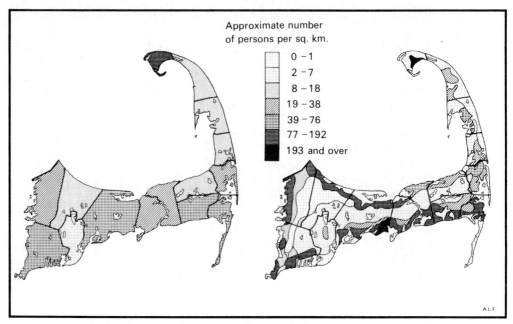

Approximate number
of persons per sq. km.

	0 – 1
	2 – 7
	8 – 18
	19 – 38
	39 – 76
	77 – 192
	193 and over

FIGURE 7.1 The map on the left shows the density according to whole census enumeration units, while the map on the right shows the refinement that can be developed by the system described in the text for estimating densities of parts. (Redrawn from The Geographical Review, published by the American Geographical Society of New York.)

where X_j is the value of X at each place involved and D_{ij} is the distance between place i and j. The preparation of a potential map requires that this summation be repeated for each place. The resulting map values will be in numbers of units of X at particular points. It is apparent that the relationship can easily be amended by inserting a constant in the equation or by defining in special ways the distance between the places.

Scales of Measurement and Their Appropriate Descriptive Statistics

Often in cartography we are called on to map summaries of distributions. There are different ways to summarize data, and these differences must be kept clearly in mind. First, we may summarize data through time. For example, a series of climatic observations at one station may result in an average January temperature, or the decennial percentage change in population for an enumeration area can be calculated. We may also summarize data over area. An example would be the ratio of beef cattle to total farm animals in 1974 by

county for a state. Furthermore, in summarizing we may combine time and space by determining the percentage change in the ratio of beef cattle to total farm animals by county from 1964 to 1974. One set of statistical manipulations is appropriate for data summarized over time and another for data summarized over area. The confusion is slightly compounded by the fact that each scale of measurement (nominal, ordinal, interval, and ratio) calls for particular statistical techniques. In each case, the cartographer may be interested in either a measure of central tendency or a measure of variation; in some instances, both measures may be of concern.

The cartographer's concern with the nature of averages and their quality, as shown by their indexes of variation, extends beyond the simple facts. In communicating geographical characteristics by way of a map, the cartographer has numerous opportunities to vary the manner of presentation so as to portray the character of the distribution properly. For example, distribution maps may "look" very precise, or they may be made to look general, or titles and legends may be designed to draw the map reader's attention to the critical characteristics of the data so that it will not mislead. In order properly to communicate geographical facts, we must understand the nature of the distributions with which we work and summarize them accordingly.

There are various measures of central tendency and a larger number of indexes of variation. One is more appropriate than another, depending primarily on the variable being mapped. The study of these measures and their calculation is more appropriate in a course in statistical method or from a book that has that as its primary aim. Only the more important measures will be mentioned here, and any extended discussion will be limited to the mean and the standard deviation, generally the most used and significant in quantitative distribution mapping.

Table 7.1 shows the most appropriate measure of central tendency with which to summarize a distribution scaled in a certain way, together with a commonly used associated index of variation.

Nominal Scaling. In a nominal distribution the mode is the class that occurs most frequently. For example, if we were mapping the distribution of forests, grasslands, or shrublands, on the basis of data acquired by small unit areas, we would expect that each small mapping unit (area) would contain some of each class, such as grassland on the uplands, forest in the river valleys, and so on. We would assign to each such small unit area the quality of the modal class, that is, the quality that occurred with the greatest areal frequency (covers the greatest amount of area in the small unit area). For example, note the category grassland in Fig. 7.2. Often this decision is not easy for the cartographer. The determination of the mode of an area can be very elusive unless the distribution being

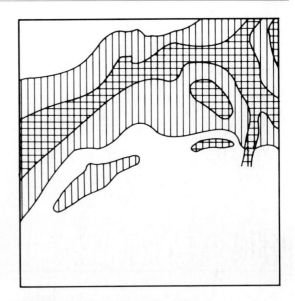

Forest

Shrub

Grassland

FIGURE 7.2 A vegetation-cover map illustrating the areal mode for nominally scaled data. The category grassland covers more of the area shown than either of the other two categories. If the scale of this map were greatly reduced, the entire area could be summarized by its areal mode, and mapped as grassland.

mapped is visible. Even then it is possible that two or more categories equally comprise an areal unit or that the variation within a unit area is so great that less than 25% occurs in any single class. The smaller the unit areas, the less likely this is to be a problem.

Similarly, we might wish to map the predominant occurrence of some phenomenon that varied through time at a series of places, such as the predominant fuel used to generate electricity. The modal class would be the class that occurred most often (e.g., coal in Fig. 7.3). In this case the event itself has no areal significance and, the mode is simply the most frequently occurring value. The data are summarized through time for locations, and these locations are then mapped.

The variation ratio (v) is a statistic that indi-

TABLE 7.1

	Measurement Scales and Appropriate Descriptive Statistics	
Scaling Method	Measure of Central Tendency	Index of Variation
Nominal	Mode	Variation ratio
Ordinal	Median	Decile range
Interval	Mean	Standard deviation
Ratio	Mean	Standard deviation

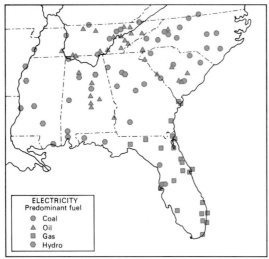

FIGURE 7.3 *Predominant fuel used to generate electricity in the southeastern United States. The modal class (i.e., the one occurring with greatest frequency) is coal.*

cates how representative the mode is of the distribution. This index of variation gives the proportion of nonmodal cases; v varies from a value of near 1, which indicates that nearly all units are nonmodal, to 0, which indicates only the occurrence of the modal class. Therefore, the nearer v is to zero, the better the quality of the mode as a summarizing statement. The variation ratio is calculated by

$$v = 1 - \frac{f_{modal}}{N}$$

where f_{modal} is the frequency or number of occurrences in the modal class and N is the total of all occurrences. For example, the data in Fig. 7.3 have a variation ratio of 0.61. This indicates that the mode in this example is not a good representation of the distribution.

If we are concerned with areal frequency, the calculation of the variation ratio must be weighted by area and becomes

$$v = 1 - \frac{a_{modal}}{A}$$

where a_{modal} is the area occupied by the modal class (the class occurring in the largest area) and A

is the total area. From Fig. 7.2 the areally weighted variation ratio is 0.43.

The areally weighted variation ratio resembles the variation ratio, since it indicates the proportion of area occupied by the nonmodal classes. This ratio varies from nearly 1 to 0 and, as it approaches zero, it indicates that the areal mode itself is a good descriptor of the distribution. Accordingly, the areal mode of Fig. 7.2 is a better descriptor than is the mode of Fig. 7.3.

Ordinal Scaling. The median is the point on an ordinal scale that is in the middle, that is, that neither exceeds nor is exceeded in rank by more than half the total observations. Suppose we were to rank in each county the quality of the farms on a four-category scale from top (A) to bottom (D). The median quality would be the rank in which half the farms were above it and the other half were below it in rank. For example, if the data for 46 counties showed that there were 42 of A class, 1 of B class, 2 of C class, and 1 of D class, the median class would be A; if the data showed 10 of A, 12 of B, 3 of C, and 21 of D, the median class would be C. Obviously, the median itself may not be characteristic of the distribution, as the latter example demonstrates.

The median may also be areally weighted. When this is done, the median signifies that rank below which and above which half the area under consideration lies. The method of calculation requires that the observations be ranked along with their respective areas. Then, beginning at either the lowest or highest rank, the areas associated with each observation are progressively summed. The areally weighted median rank is the rank of that observation whose associated area, when it is added to the accumulating sum, makes the sum equal to half of the total area. An example is given by the data in Table 7.2 and the map in Fig. 7.4. The areal median divides the counties between number 38, Grant, and number 39, Sauk. As a result, the areal median is the rank category "rural." If a map were made of the data in Table 7.2 and the areal median were used as the class limit, approximately half of the area of the state would be above the median (all counties in Table 7.2 from Milwaukee to Grant), and half of the area of the state would be below the median (all counties in Table 7.2 from Sauk to Washburn). The areally

TABLE 7.2

				Area	Area
Rank	**Percent Urban**	**Observation (County)**	**Category**	**Area (km²)**	**Area Decile**
1	100.0	Milwaukee	Highly urban	619	1
2	81.6	Brown	Highly urban	1355	1
3	80.2	Waukesha	Highly urban	1440	1
4	78.2	Winnebago	Highly urban	1176	1
5	77.2	Dane	Highly urban	3100	1
6	76.1	Racine	Highly urban	873	1
7	74.9	La Crosse	Urban	1215	1
8	74.9	Rock	Urban	1867	1
9	73.3	Douglas	Urban	3393	1
10	71.5	Kenosha	Urban	707	2
11	69.2	Eau Claire	Urban	1681	2
12	68.6	Outagamie	Urban	1642	2
13	67.5	Ozaukee	Urban	609	2
14	61.1	Sheboygan	Urban	1311	2
15	60.2	Manitowoc	Urban	1526	2
16	57.4	Ashland	Urban	2686	2
17	57.1	Fond du Lac	Urban	1875	2
18	55.0	Lincoln	Urban	2331	2
19	52.2	Jefferson	Urban	1461	3
20	52.2	Wood	Urban	2103	3
21	49.6	Marathon	Rural	4103	3
22	49.4	Portage	Rural	2098	3
23	47.0	Washington	Rural	1109	3
24	46.9	Langlade	Rural	2222	3
25	45.8	Dodge	Rural	2310	4
26	44.7	Calumet	Rural	816	4
27	43.4	Marinette	Rural	3595	4
28	41.8	Green	Rural	1518	4
29	38.7	Dunn	Rural	2222	4
30	38.7	Walworth	Rural	1450	4
31	37.7	Monroe	Rural	2370	4
32	36.5	Kewaunee	Rural	857	5
33	36.3	Crawford	Rural	1518	5
34	35.4	Waupaca	Rural	1945	5
35	34.5	Chippewa	Rural	2655	5
36	33.7	Door	Rural	1272	5
37	33.6	Oneida	Rural	2885	5
38	32.8	Grant	Rural	3025	5
39	32.0	Sauk	Rural	2176	6
40	31.4	Green Lake	Rural	919	6
41	29.8	Richland	Rural	1513	6
42	28.9	Columbia	Rural	2015	6
43	28.4	St. Croix	Rural	1906	6
44	28.1	Oconto	Rural	2606	6
45	25.8	Rusk	Rural	2357	6
46	23.4	Pierce	Highly rural	1531	6
47	21.4	Barron	Highly rural	2243	7
48	21.4	Jackson	Highly rural	2590	7
49	20.4	Taylor	Highly rural	2536	7
50	20.3	Price	Highly rural	3284	7
51	19.9	Shawano	Highly rural	2328	7
52	18.8	Juneau	Highly rural	2059	8
53	16.9	Iowa	Highly rural	1971	8
54	15.2	Vernon	Highly rural	2085	8
55	9.1	Clark	Highly rural	3165	8
56	0.3	Waushara	Highly rural	1627	8
57	0.0	Adams	Highly rural	1753	8
58	0.0	Bayfield	Highly rural	3818	8
59	0.0	Buffalo	Highly rural	1844	9
60	0.0	Burnett	Highly rural	2176	9

TABLE 7.2 (Continued)

Rank	Percent Urban	Observation (County)	Category	Area (km²)	Area Decile
61	0.0	Florence	Highly rural	1267	9
62	0.0	Forest	Highly rural	2616	9
63	0.0	Iron	Highly rural	1932	9
64	0.0	LaFayette	Highly rural	1665	9
65	0.0	Marquette	Highly rural	1184	9
66	0.0	Menominee	Highly rural	938	10
67	0.0	Pepin	Highly rural	614	10
68	0.0	Polk	Highly rural	2419	10
69	0.0	Sawyer	Highly rural	3297	10
70	0.0	Trempealeau	Highly rural	1914	10
71	0.0	Vilas	Highly rural	2246	10
72	0.0	Washburn	Highly rural	2113	10

weighted median can also be obtained from a cumulative frequency graph mentioned later in this discussion.

An appropriate descriptive statistic to measure variation on an ordinal scale is the quantile range, such as the decile range.* The decile range

*A quantile is obtained by dividing a distribution into segments; quartiles divide it into 4 categories, deciles into 10 categories, and so on, to centiles (or percentiles) that divide it into 100 categories.

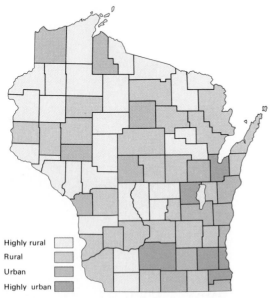

Highly rural

Rural

Urban

Highly urban

FIGURE 7.4 Map showing an ordinal ranking of the counties of Wisconsin, depending on degree of urbanization. Data from Table 7.2, column labeled "category."

(d) is simply a statement of the number of ranks included between the first and the ninth deciles. It is obtained by $d = d_9 - d_1$, in which d_9 is the rank below which 90% of the cases occur, and d_1 is the rank below which 10% of the cases occur. In the illustration of farm quality above, in the first ranking the decile range would be 0, indicating that the median (A) is an excellent statement of the average; in the second illustration, the decile range would be 3, showing that the median C is not a good statement of the average.

This measure can also be areally weighted; in such instances the weighted decile range indicates the number of ranks included between the first and the ninth deciles (where each decile is equal to 10% of the total area). Table 7.2 is divided by deciles such that each decile contains 10% of the total area of Wisconsin. In this example the first areal decile, below which 90% of the area of Wisconsin falls, separates Douglas and Kenosha counties, both of which rank in class 2—urban. The ninth areal decile, below which 10% of the area is located, falls between Marquette and Menominee counties, both of which rank in class 4 —highly rural. Thus, class 4 minus class 2 gives a decile range of 2, indicating that the areally weighted median is not an especially good summary statistic for the data in Table 7.2.

Interval and Ratio Scaling. The arithmetic mean is perhaps the most frequently used average in cartography. Most of the maps of temperature, pressure, precipitation, income, yield, and the other elements common in physical and human geography are based on means derived in one way or another. The mean is obtained by summing the values of the items and dividing by the number of items. The formula was presented earlier in this chapter.

The mean can also be areally weighted. In distributions, whenever the values (X) are in any way related to areal extent, they must be weighted for their frequency: this is most easily done by multiplying each X value by the area of its extent, summing these products, and dividing the sum by the total area. The general expression for any areally weighted mean is, therefore,

$$\overline{X} = \frac{\Sigma aX}{A}$$

in which ΣaX represents the sum of the products of each X value multiplied by its area, and A is the total area, that is, Σa.

In the geographic analysis of problems employing maps, we are often required to obtain means from mapped data in which totals are difficult if not impossible to obtain. For example, suppose we wished to obtain the average elevation above sea level from an ordinary contour map of elevation. Figure 7.5 may be used as an illustration, and it is assumed to be a perfectly conically shaped island, whose form and elevations (Z) are shown by evenly spaced contours at a 30-m interval. It is clear from inspection that different elevations occur more or less frequently. In this in-

stance, the geographic frequencies are the relative areas of the occurrences; consequently, it is necessary to obtain the areas of the spaces between the contours. This can be done most accurately by measuring them with a planimeter or, if the contours have been digitized, by computer computation (see Chapter 12). The results obtained are expressed in any convenient square units, such as square centimeters. The measurements obtained from the original drawing of Fig. 7.5 (before reduction for printing) are shown in the first two columns of Table 7.3. The mean elevation may be determined in two ways: by calculation or by graphic analysis. Using the areally weighted mean formula given above, the data in Table 7.3 yield

$$\overline{Z} = \frac{9329.25}{102.65} = 90.9 \text{ m}$$

The same result can be obtained graphically by constructing a cumulative frequency graph. Additionally, a cumulative frequency graph can be used to obtain a value for an areally weighted median using interval or ratio data instead of ordinal data. A cumulative frequency graph is constructed by plotting the values of the distribution on the Y axis of an arithmetic graph in the order of their values against the extent of their progressively cumulated areas on the X axis. The paired data from the first and fifth columns of Table 7.3 may be graphed, as shown in Fig. 7.6, by plotting

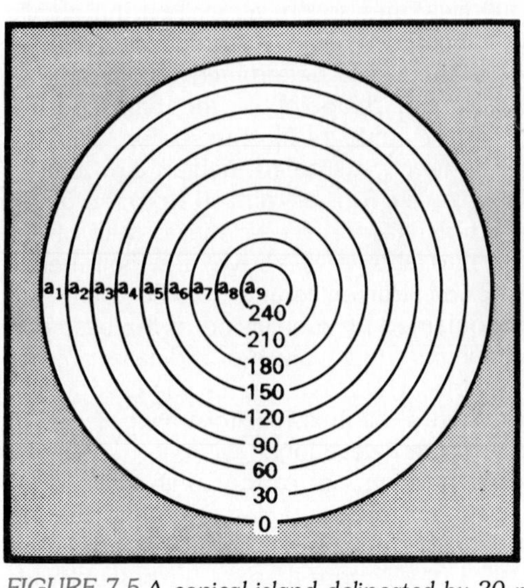

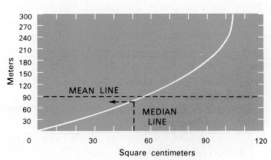

FIGURE 7.5 A conical island delineated by 30-m contours.

FIGURE 7.6 A cumulative frequency graph. The horizontal dashed line is the top of a rectangle of the same area as that included between the curve and the baseline. The height of the rectangle on the Y-axis is the geographical mean. The vertical dashed line is erected at the midpoint of the cumulated areas which are plotted along the base. The value of the curve at the point of intersection with the vertical dashed line is the geographical median.

TABLE 7.3

Elevation Classes (m) (Z)	Midvalue of Elevation Classes (m) (Z_m)	Map Areas of Elevation Classes (cm²) (a)	(aZ_m)	Cumulated Areas (cm²) ($a_1 + a_2 + a_3$ . . .)
Calculation of the Mean Elevation of the Island Shown in Figure 7.5				
0–30	15	21.42	321.30	21.42
30–60	45	18.84	847.80	40.26
60–90	75	16.45	1233.75	56.71
90–120	105	14.06	1476.30	70.77
120–150	135	11.55	1559.25	82.32
150–180	165	8.84	1458.60	91.16
180–210	195	6.39	1246.05	97.55
210–240	225	3.81	857.25	101.36
240–270	255	1.29	328.95	102.65
Total		102.65	9329.25	

the value of the upper-class limit of the first elevation category, Z_1 (30), on the Y axis and the area (21.42) on the X axis. The next pair, Z_2 (60) and $a_1 + a_2$ (40.26), is similarly plotted. The addition of each area value to the sum of the preceding areas (after arranging them in the ascending order of Z values) assures that the curve, which results from joining the plotted points with a smooth line, will rise to the right. The area under the curve (bounded by the curve, the baseline, and an ordinate erected at the value of the total area) may then be measured and expressed in square units. When this total is divided by the length of the baseline from the origin to the total area (measured on the graph in linear terms of the same units used to express the area under the curve), the quotient is the height of a rectangle with the same area as that under the curve, one side of which is the length of the baseline. The height of the rectangle when plotted on the Y axis of the graph is the mean height of the curve and is the areally weighted mean.

In Fig. 7.6 the area under the curve (on the original drawing, before reduction) was 13.42 cm². The length of the base as plotted was 8.74 cm. Division of the area by the base (13.42/8.74) resulted in a quotient of 1.535 cm. This value, when measured on the Y axis from the base, gives a rectangle with a height having an elevational value of 90.65 m, which is the areally weighted or geographic mean. In Fig. 7.6 this height is shown by the dashed horizontal line. The difference of approximately 0.25 m between the mean determined arithmetically (90.9 m) and the mean determined graphically (90.65 m) results from the graphic process and the consequent rounding of numbers, as well as the fact that midvalues of classes were used in one computation, whereas upper-class limits were used in the other. This kind of graph is also occasionally called a hypsographic curve because it has been used to determine mean land heights (mean hypsometric values) exactly as illustrated here. It may, of course, be applied to any other kinds of data that are expressed on an interval scale and that have actual or assumed continuous areal extent.

The cumulative frequency graph may also be employed to determine areally weighted percentiles, quartiles, and so on, as well as the areally weighted median, of a series scaled on an interval basis. The median is determined by erecting an ordinate at the midpoint of the baseline and reading the Y value of its intersection with the curve. The areally weighted median elevation of the island on Fig. 7.5, as derived from the graph in Fig. 7.6, is close to 81 m; that is, half of the map area as shown is above and half below that elevation. Percentile and quartile values are similarly obtained by subdivision of the base and by determining the Y values of ordinates erected at such points. The curve can also be used to find the total area of all those sections lying above or below any given elevational value.

Many problems of cartographic analysis include data with open-end classes, that is, the top or bottom class may be only available on a "less than . . ." or a "more than . . ." basis. This is the case with the series of elevations shown by the contours on the island in Fig. 7.5, wherein the contour of highest value is 240 m. By definition, the area inside that contour is above (i.e., more than) 240 m, but by exactly how much is not shown. In other cases, the neat line or map edge may cut through a statistical unit area and produce a similar result. When the limits of all classes are not given, the mean value cannot be accurately ascertained, but judicious estimation and interpolation will usually make significant errors unlikely.

THE STANDARD DEVIATION. The measure of variation appropriate for interval or ratio scaled data is called the standard deviation. The standard deviation is perhaps the most important index of concern to cartographers. It is especially important because of (1) its descriptive utility, (2) its use in the evaluation of the reliability of measures computed from samples, and (3) its wide employment.

Many kinds of phenomena, when scaled on an interval or ratio basis, show a similarity in the frequency with which particular values occur in the series. The values that occur with the greatest frequency are usually those near the mean of the series. The greater the difference (or deviation) of a value from the mean of its series, the less frequently that value is likely to occur.* When an

---

Although a great many phenomena show a "normal" distribution of values, some do not. A description of other kinds of distributions is more appropriately a topic for study in statistics.

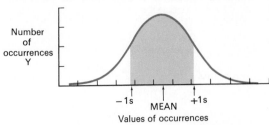

FIGURE 7.7 *A frequency curve of a normal distribution. The number of occurrences are plotted on the Y-axis and the values on the X-axis. The shaded area shows the proportion (68.27%) of the occurrences within one standard deviation on either side (plus or minus) of the mean.*

"ideal" series is graphed by plotting the frequencies against the values, the result is a *normal curve* of a *normal distribution*. It is illustrated in Fig. 7.7 by an ordinary (not cumulative) frequency graph on which the frequency of occurrence of the values is plotted on the Y axis against the values on the X axis. Of additional significance in mapping data that have been obtained by taking samples is the fact that if a series of random samples is taken from a normal statistical "population," the means of the individual samples also tend to have a normal distribution.

The standard deviation is a way of describing the dispersion of the values in a normal distribution. When data approximate a normal distribution, the cartographer, by using the standard deviation, has the advantage of being able to divide the distribution easily into classes that should contain any selected percentage of the observations. The standard deviation specifies the range on either side of the mean value of a normal distribution, which includes slightly more than two-thirds (68.27%) of all the values in the series; no matter how the original observations are distributed, the range of two standard deviations on either side of the mean will include at least three-quarters (75%) of the observations. For example, the areally weighted standard deviation of the elevations on the island in Fig. 7.5 is 63.47 m. This means that one could expect that about two-thirds of the island has an elevation within 63.47 m of the mean; or, about two-thirds of the island lies above approximately 27.43 m (90.9 − 63.47 m) and below approximately 154.37 m (90.9 + 63.47 m); that is, about two-thirds lies between those two elevations. The use of fractional standard deviations (1) to set class limits so that approximately equal numbers of observations fall in each class, or (2) to obtain other criteria is becoming more common in cartography.

The standard deviation is the square root of the mean of the squared deviations from the mean. The difference or deviation of each item in the distribution is squared, and the sum of these squares is known as the *variance*. The *standard deviation* is the square root of the variance. The standard deviation may be easily computed by the following formula:

$$s = \sqrt{\frac{\Sigma(X - \overline{X})^2}{N}}$$

in which s is the standard deviation, and $(X - \bar{X})$ is the difference between each value and the mean. When working with data, if we do not wish to determine the differences between each value and the mean, we may use the following formula:

$$s = \sqrt{\frac{\Sigma X^2}{N} - (\bar{X})^2}$$

As pointed out in connection with the determination of the mean, when data are in any way related to areal extent, it is necessary to take this into account. The easiest way to do this is to use the areal extent as the frequency. The expression for an areally weighted standard deviation then becomes

$$s = \sqrt{\frac{\Sigma a Z^2}{A} - \left(\frac{\Sigma a Z}{A}\right)^2}$$

in which $\Sigma a Z^2$ is obtained by first squaring each Z value, multiplying it by the area it represents, and summing the products. The term $(\Sigma a Z/A)^2$ will be recognized as the square of the areally weighted mean. As an illustration, the areally weighted standard deviation of the elevations of the island shown in Fig. 7.5 is calculated below. The necessary data are shown in Table 7.4, and it will be seen that midvalues are employed.

$$s = \sqrt{\frac{1,261,424.25}{102.65} - \left(\frac{9329.25}{102.65}\right)^2}$$

$$= \sqrt{12,288.59 - 8259.91}$$
$$= \sqrt{4028.68}$$
$$= 63.47 \text{ m}$$

The standard deviation is widely used in both descriptive and inferential statistics. From the preceding discussion, the descriptive utility of the standard deviation should be obvious. Its importance is demonstrated in practice by its widespread use. Cartographers often map data that are dependent on standard deviations when they map distributions such as rainfall variability, probability of temperature extremes, or potential corn yields. In addition to this descriptive utility, the utility of the standard deviation for setting class limits for distributions cannot be understated. It has been suggested by some that only by the use of the mean and standard deviation can class limits be set that will allow comparability of distributions through time for statistics about distributions such as birth rates and death rates, diseases, or income. Today's cartographer must be familiar with the concept of standard deviation, its calculation, and its correct uses.

THE STANDARD ERROR OF THE MEAN. Many maps are made of distributions wherein the data are obtained by sampling. For example, a map of average temperatures is based on temperature records for a particular period and, there-

TABLE 7.4

Calculation of the Standard Deviation of Elevations on the Island Shown in Fig. 7.5

Midvalue of Elevation Class (m) (Z_m)	Map (or actual) Area of Class (cm²) (a)	(aZ_m)	(Z_m^2)	(aZ_m^2)
15	21.42	321.30	255	4819.50
45	18.84	847.80	2025	38151.00
75	16.45	1233.75	5625	92531.25
105	14.06	1476.30	11025	155011.50
135	11.55	1559.25	18225	210498.75
165	8.84	1458.60	27225	240669.00
195	6.39	1246.05	38025	242979.75
225	3.81	857.25	50625	192881.25
255	1.29	328.95	65025	83882.25
Total	102.65	9329.25		1,261,424.25

fore, is only one sample of the time through which such temperatures have occurred. The particular sample available will ordinarily not provide the true mean because it contains only a portion of the total of all possible values. The likelihood of the mean being a different value can be inferred from the standard deviation of the values in the sample. This inference is made by calculating the standard deviation of the mean. It is usually called the standard error of the mean and is symbolized by $s_{\bar{x}}$. It is obtained by dividing the standard deviation of the sample values by the square root of the number of the sample values that were used in calculating the mean. Its expression is

$$s_{\bar{x}} = \frac{s}{\sqrt{N}}$$

where s is the standard deviation of the values of X, and N is the number of X values that entered into the calculation of the mean of X.* It is clear that the smaller the number of values used to arrive at a mean, the larger will be the standard error, that is, the greater the likelihood that the mean is incorrect.

OTHER COMMON STATISTICAL CONCEPTS AND MEASURES

Two other common statistical techniques, regression and correlation, are widely utilized in the analysis of distributions, and the cartographer is called on more and more to map the resulting data. The cartographer must understand these analytical techniques and be able to utilize them cartographically. The following material is a condensed introduction and reference, not a substitute for a book in statistics.

Regression Analysis

Regression can be used to provide solutions to numerous prediction problems and to summarize relationships between quantitative variables. A retail store chain may wish to predict sales volume for a given store based on its location, or a farmer may wish to predict harvest yields based on climatic factors. In its simplest form, regression analysis in-

cludes at least one independent variable or distribution (e.g., a measure of climate) and one dependent variable or distribution (e.g., yield). A simple model can then be established of the type

$$Y = a + bX$$

where X is the independent variable, Y is the dependent variable, and a and b are constants. One may readily recognize this as the equation for a straight line in a plane coordinate system, in which a is the intercept of the line with the Y axis and b is the slope of the line. Figures 7.8 and 7.9 each illustrate the calculated linear regression equation for the observations of one independent variable X (per capita personal income) and one dependent variable Y (per capita educational expenditure, Fig. 7.8 and number of first-degree graduates, Fig. 7.9), for the areal units in Fig. 7.10.

The solution of a linear regression equation is usually based on the least squares criterion; this means that the regression line is so situated with respect to the graphed observations that the sum of the squares of the differences between observed Y values and those predicted is minimized. (The observed Y values are represented in Figs. 7.8 and 7.9 as points; the predicted values are read on the Y-axis for any given X value by moving vertically from the selected X value to the

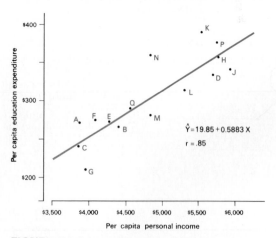

FIGURE 7.8 A scatter diagram with fitted linear regression line. Letters refer to areal units designated in Fig. 7.10 $\hat{Y}$ refers to the predicted values for the dependent variable and r refers to the coefficient of correlation.

*When we are estimating the standard deviation of a total population from a sample, the standard deviation is calculated with $N - 1$ instead of N.

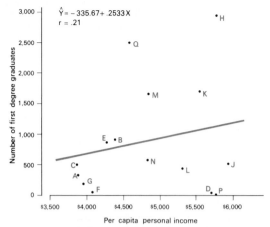

FIGURE 7.9 *A scatter diagram with fitted linear regression line. Letters refer to areal units designated in Fig. 7.10. Ŷ refers to the predicted values for the dependent variable and r refers to the coefficient of correlation.*

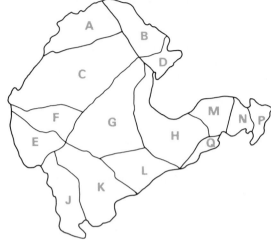

FIGURE 7.10 *A map of a fictitious island showing the areal units to which the data in Table 7.5 refer.*

plotted regression line and then horizontally to the Y-axis to obtain the predicted value.) When this least squares criterion is employed, the correct values for the constants a and b are given by the following formula:

$$b = \frac{\sum_{i=1}^{n} (X_i - \overline{X})(Y_i - \overline{Y})}{\sum_{i=1}^{n} (X_i - \overline{X})^2}$$

$$a = \overline{Y} - b\overline{X}$$

where

$$\overline{Y} = \frac{\sum_{i=1}^{n} Y_i}{N} \quad \text{and} \quad \overline{X} = \frac{\sum_{i=1}^{n} X_i}{N}$$

Table 7.5 and Figs. 7.8 and 7.9 are given as illustrations of this technique.

Maps are often created of either the observed

TABLE 7.5

Area	Per Capita Personal Income	Per Capita Educational Expenditure	Number of First-Degree Graduates
	Data for a given year for a fictitious island having 15 civil districts (data used for Fig. 7.8 to 7.12)		
A	$3882	$273	330
B	4395	266	910
C	3870	240	500
D	5695	333	40
E	4282	273	870
F	4082	276	70
G	3952	210	240
H	5770	357	2920
J	5938	340	530
K	5550	390	1760
L	5304	314	460
M	4840	280	1670
N	4830	360	580
P	5745	376	0
Q	4570	287	2500

values, the predicted values, or the residual values (the difference between the observed and the predicted), as shown in Figs. 7.11 and 7.12. To prepare such maps, one should be aware of the assumption behind this type of analysis, and the selection of appropriate class intervals for mapping is critical (see Chapter 8).

From a geographic point of view the computed relationship of X and Y does not take spatial position into account unless X or Y are position-related variables (i.e., variables whose measurement inherently contains some element of spatial position, such as latitude or longitude). In practice, most cases of X and Y pairs of observations are not locational in that sense but are simply attached to places. For example, a regression analysis using hay production as an independent variable to predict numbers of cattle could be computed for a

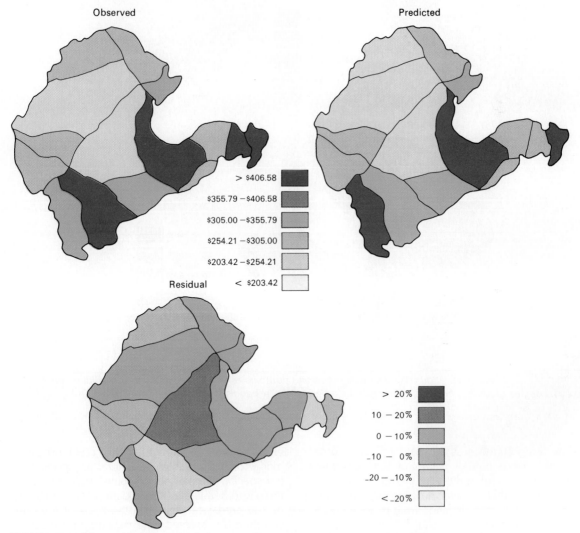

FIGURE 7.11 *Maps of fictitious island showing observed per capita educational expenditure, predicted per capita educational expenditure based on per capita income and the residuals from the regression shown in Fig. 7.8.*

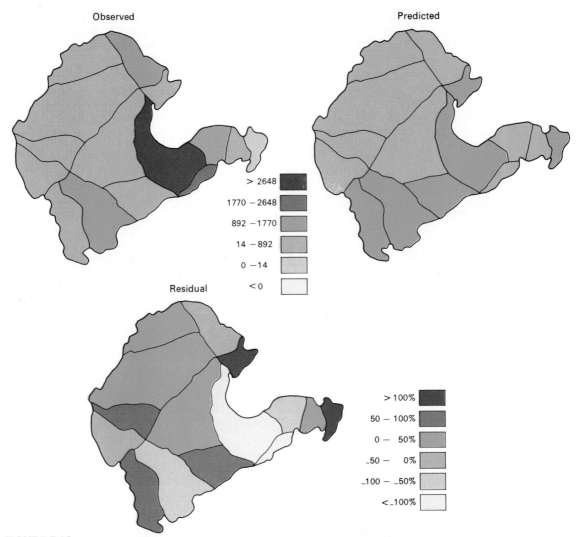

FIGURE 7.12 Maps of a fictitious island showing observed numbers of first-degree graduates, predicted numbers of first-degree graduates based on per capita income and the residuals from the regression shown in Fig. 7.9.

state using counties as the data collection units. In this case a pair of values of the dependent (number of cattle) and the independent (hay production) variables are associated with each county (location); however, location itself is not involved in the computations.

Another assumption underlying regression analysis is that the data are normally distributed. Only on the basis of this assumption can infer-

ences about the strength of the relationship be made for the variables in general as opposed to the specific sample (set of points) used to derive the *a* and *b* values of the equation. The reliability of any prediction made on the basis of the data can then be assessed. The cartographer must be vitally interested in this aspect, since relative reliabilities should generally be indicated when mapping predicted values or residuals. For example, compare the maps in Figs. 7.11 and 7.12. The predicted maps in each figure alone indicate nothing about their reliability; however, compare the corresponding scatter diagrams in Figs. 7.8 and 7.9. The quality of the prediction mapped in Fig. 7.11, as revealed by Fig. 7.8, is much greater than the predicted map in Fig. 7.12. One way to indicate reliability is to map residuals as well as predicted values as shown in Figs. 7.11 and 7.12. It is only the residual map that tells us whether the predicted map is of much or little value.

We have discussed only the simplest case of regression analysis. Cartographers often must map the results of multiple linear regressions (i.e., regressions that include more than one independent variable) and/or curvilinear regressions (i.e., regressions that involve equations of curved instead of straight lines). Occasionally, ordinal or interval information is incorporated into regression analyses. Although the computations become more involved or modified, the mapping problem and requirement remain relatively the same: map the predicted, observed, and/or residual values, and indicate their reliability.

Correlation Analysis

One important indicator of the strength of the linear relationship between two sets of data, Y and X, is the correlation, r, between Y and X. The value of r can vary between -1 and $+1$, where $r = 1$ indicates that an increase in X is associated with a corresponding increase in Y, $r = -1$ indicates that an increase in X is associated with a corresponding decrease in Y, and $r = 0$ indicates the absence of a predictable relationship, that is knowledge of the X value gives no predictive information about Y. The formula for the coefficient of correlation is:

$$r = \frac{\sum\limits_{i=1}^{n} (X_i - \overline{X})(Y_i - \overline{Y})}{\sqrt{\sum\limits_{i=1}^{n} (X_i - \overline{X})^2 \cdot \sum\limits_{i=1}^{n} (Y_i - \overline{Y})^2}}$$

The value r itself is a summary measure relating to an entire set of paired observations; therefore the cartographer cannot map an r = 0.5, for example. It is possible that a separate value of r might be calculated for sets of data (e.g., number of calls to the automobile club for aid in

starting the car and temperature) for a given location. If similar values for r were computed at a series of locations, this set of r values could be mapped.

Perhaps of more importance to the cartographer is the concept of correlation when it is applied to neighboring values in one data set. The appropriate name is *autocorrelation,* and a series of coefficients of autocorrelation can be computed for a single spatially distributed set of data values.

The position of the data in the spatial array is used to determine which observations are paired for calculation of the autocorrelation coefficient (see Fig. 7.13). The formula for calculation can be identical to the above formula for correlation. Each point value is paired with another point value a certain distance (lag) and direction from it (see Fig. 7.14). The number of possible lags is numerous, and a value for r can be calculated for each lag. If the lags are systematically scaled in distance and direction, a plot of r values against lags can be obtained (Fig. 7.15) that is the autocorrelation function for the distribution being analyzed. The importance of this concept for cartographers will become more apparent in the following

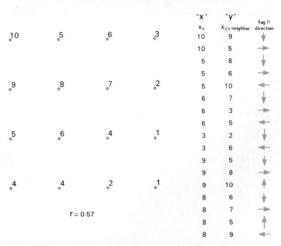

FIGURE 7.13 *The calculation of an autocorrelation coefficient. The value for r, the lag 1 autocorrelation coefficient for the data array at the left, is obtained by defining X and Y as shown in the right-hand columns above. The equation for r is given in the text.*

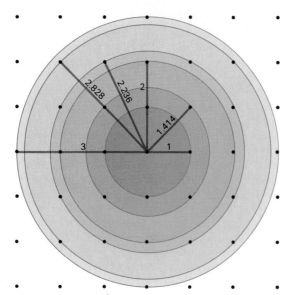

FIGURE 7.14 For the square array of data values indicated by points above, various lags can be defined as illustrated.

chapters, when cartographic generalization and portrayal of surface volumes is discussed.

Finally, the coefficient of correlation described and defined above is perhaps the most common correlation measure for interval or ratio data. Again, however, there are other measures of cor-

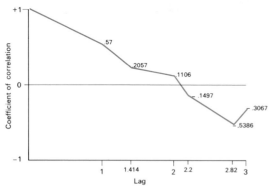

FIGURE 7.15 A graph of approximation of the autocorrelation function for the data array shown in Fig. 7.13 using six different autocorrelation coefficients calculated for increasing lags.

relation more appropriate for nominal or ordinal scale data and for some instances of interval or ratio data. The addition to a map, showing the results for a regression analysis of the value of the coefficient of correlation, is one important way that a cartographer can indicate the reliability of the predicted results being portrayed.

SELECTED REFERENCES

Blalock, H. M., *Social Statistics,* 2 ed., McGraw-Hill, New York, 1972.

Court, A., "The Inter-Neighbor Interval," *Yearbook of the Association of Pacific Coast Geographers, 28,* 180–182 (1966).

Davis, J. C., and M. J. McCullagh (eds.), *Display and Analysis of Spatial Data,* NATO Advanced Study Institute, John Wiley & Sons, New York, 1975.

Draper, N. R., and H. Smith, *Applied Regression Analysis,* John Wiley & Sons, New York, 1966.

Kadmon, N., "Comparing Economic Maps with the Aid of Distribution-Parameter Graphics," *International Yearbook of Cartography, 11,* 278–287 (1971).

Krumbein, W. C., and F. A. Graybill, *Introduction to Statistical Models in Geology,* McGraw-Hill, New York, 1965.

Mendenhall, W., *Introduction to Probability and Statistics,* 4th ed., Duxbury Press, North Scituate, MA, 1975.

Mood, A. M., et al., *Introduction to the Theory of Statistics,* 3rd ed. McGraw-Hill, New York, 1974.

Morrison, J. L., "A Link Between Cartographic Theory and Mapping Practice: the Nearest Neighbor Statistic," *The Geographical Review 60,* 494–510 (1970).

Olson, J. M., "Autocorrelation and Visual Map Complexity," *Annals,* Association of American Geographers, 65, 189–204 (1975).

Taylor, P. J., *Quantitative Methods in Geography: An Introduction to Spatial Analysis,* Houghton Mifflin Company, Boston, 1977.

Thomas, E. N., and D. L. Anderson, "Additional Comments on Weighting Values in Correlation Analysis of Areal Data," *Annals,* Association of American Geographers, 55, 492–505 (1965).

Tomlinson, R. F. (ed.), *Geographical Data Handling,* 2 vols., IGU Commission on Geographical Data Sensing and Handling, Ottawa, 1972.

GENERALIZATION

The earth is so large, relative to diminutive human beings and their ordinary vision, that spatial relationships must be reduced to be comprehensible. As observed earlier, the objective of this essential scientific process is analogous to the employment of magnification to aid in the study of phenomena that are too small to be directly observed. Both reduction and enlargement, as these terms are used here, are means to the same end—comprehensibility; but the consequences of these processes are quite different.

Unmodified magnification (e.g., with a microscope) is accompanied by a general loss of clarity, a reduction in intensity of coloring, and an increased emphasis of the relative visual significance of minor features (i.e., by an increase in the relative visual importance of the specific compared to the general). Reduction, when applied to earth

phenomena, is likewise accompanied by inescapable changes. Distances separating features and widths and lengths of features are reduced in the ratio of the reduction. Adjacent discrete items become more and more crowded, clarity is generally reduced, and there may be an increase in the relative visual importance of the general compared to the specific.

In cartography we are vitally interested in ways in which we can increase the effectiveness of cartographic communication by counteracting the undesirable consequences of reduction. There are a variety of modifications that can, and must, be carried out as a result of reduction; they range from essentially mechanical processes to intellectual exercises. These modifications collectively are called cartographic generalization.

CHAPTER 8
CARTOGRAPHIC GENERALIZATION

Selection

Cartographic generalization is born of the necessity to communicate. It is impossible to portray everything, even at a 1:1 scale, to say nothing of the large reductions required for most maps. In order to portray important aspects of reality, various manipulations of the data that represent information to be mapped are necessary.

A first step, however, is the *selection* of the information to be communicated by the map. Before the cartographer can begin the modifications that we have termed generalization, a selection of available information is made that is consistent with the purpose of the map. In one sense this is generalizing, but selection is not a part of *cartographic* generalization as we define it. Cartographic generalization is the modification of specific data in order to increase the effectiveness of the communication by counteracting the undesirable consequences of reduction. Selection is the intellectual process of deciding which information will be necessary to carry out the purpose of the map successfully. No modification of the information is required in selection; the choice is either to portray roads or not to portray roads, to include or not include major hydrographic features, or to name or not name by lettering all cities over 150,000 population. Selection can be done without regard for map format or scale. Once the cartographer has made this selection, the generalization of each set of data that constitutes the selected information can be accomplished.

The Elements and Controls of Generalization

There are several operations included in generalization; it is convenient to group the manipulations that a cartographer performs on the selected data into four categories of processes termed the *elements* of cartographic generalization, as follows:

> *Simplification:* The determination of the important characteristics of the data, the retention and possible exaggeration of these important characteristics and the elimination of unwanted detail.
>
> *Classification:* The ordering or scaling and grouping of data.
>
> *Symbolization:* the graphic coding of the scaled and/or grouped essential characteristics, comparative significances, and relative positions.

> *Induction:* The application in cartography of the logical process of inference.

These fundamental and complex processes are all consciously accomplished by the cartographer in the practice of mapmaking, but each map provides a different set of requirements. The "mix" of the processes of generalization as they are combined will, therefore, vary from map to map, and the manner in which each will be applied depends on the dictates of the *controls* of cartographic generalization. These controls are:

> *Objective:* The purpose of the map.
>
> *Scale:* The ratio: map/earth.
>
> *Graphic limits:* The capability of the systems employed for the communication, and the perceptual capabilities of the readers of the communication.
>
> *Quality of data:* The reliability and precision of the various kinds of data being mapped.

It is only fair to observe that the separation of cartographic generalization into four categories of processes and four controls is in itself a generalization about generalization. The aim in doing so, like the aim of all generalizing, is to simplify to manageable proportions what is really a very complex intellectual-visual process. To examine this subject in great detail is not possible in a book of this nature, any more than it would be possible to map on a page this size all that could be mapped on a sheet 20 times this size. Generalizations are always needed—even about generalization.

THE ELEMENTS OF GENERALIZATION

The four elements of generalization are not clearly separable in many cartographic situations. Keep in mind that in the subsequent discussions of these categories of processes of generalization, each is being treated individually without regard to the others. This is impossible for the cartographer to do in practice. For example, the process of induction is highly dependent on the way in which the cartographer carries out the other three processes. Although the cartographer has little or no control over induction, through experience, good judgment, and familiarity with the selected data, the cartographer can perform the other three properly, according to the dictates of the controls.

Simplification

When analyzed, the element of simplification includes the detailed examination of the characteristics of the data being mapped and the determination of the data to be retained. The elimination of unwanted details in the data is the most often used form of simplification.

Any map is a reduction, of course, and because items such as line widths, type sizes, and symbol forms must be kept to a visible and legible size, it follows that as reduction occurs in the scale of a map, each map item will occupy a proportionately larger amount of space. Consequently, simplification must be practiced in order to insure legibility and truthful portrayal. Since the reduction of area on a map takes place as the square of the ratio of the difference in linear scales, the amount of information that can be shown per unit area decreases in geometrical progression. Töpfer and Pillewizer have developed a law or principle of simplification that states in general terms the amount of detail that can be shown at different scales.[*] It is called the *radical law* or the *principle of selection* by its authors, and is expressed in its simplest form as

$$n_f = n_a \sqrt{M_a/M_f}$$

where

n_f is the number (n) of items on the newly compiled map (f)

n_a is the number of items on a source map (a)

M_a is the scale denominator of the source map

M_f is the scale denominator of the newly compiled map.

The basic equation of the radical law must be modified depending on the nature of the phenomena being compiled. The modification is made by introducing two constants, C_b and C_z, in the right side of the basic equation.

Constant C_b is called the *constant of symbolic exaggeration* and takes three forms.

$C_{b1} = 1.0$ for normal symbolization, that is, for elements appearing without exaggeration.

[*]F. Töpfer and W. Pillewizer, "The Principles of Selection," (with introduction by D. H. Maling), The Cartographic Journal, 3, 10–16 (1966).

$C_{b2} = \sqrt{M_f/M_a}$ for features of areal extent shown in outline, without exaggeration, such as lakes and islands.

$C_{b3} = \sqrt{M_a/M_f}$ for symbolization involving great exaggeration of the area required on a compiled map, such as a settlement symbol, with its associated name.

Constant C_z is called the *constant of symbolic form* and also takes three forms.

$C_{z1} = 1.0$ for symbols compiled without essential change.

$C_{z2} = (s_a/s_f) \sqrt{M_a/M_f}$ for linear symbols in which the widths of the lines on the source map (s_a) and the newly compiled map (s_f) are the important items in generalization.

$C_{z3} = (f_a/f_f) \sqrt{(M_a/M_f)^2}$ for area symbols in which the areas of the symbols on the source map (f_a) and the newly compiled map (f_f) are the important items in the generalization.

The basic equation, with or without modification by the introduction of the constants, is solely a statement of the number of items that can be expected on the newly compiled map. The value of the radical law is in its expression of the basic relationship between scale and the nature of the mapped phenomena in terms of the number of items that will normally appear on a newly compiled map. When compiling from larger to smaller scales, the number of items that can be shown on the smaller scales will diminish according to the radical law (i.e., in geometric progression).

Although the law specifies, with high probability, how many items or how much detail *can* be retained when working from a larger to a smaller scale, it cannot, of course, specify *which* of the items should be included and which should be discarded in order to characterize (i.e., generalize) a distribution. One cannot, for most thematic maps, make a simplification of the features within a class, such as rivers or cities, on purely objective grounds, such as size. Importance is a subjective quality; a simplification of cities on a map of the United States that included only those of more than 100,000 inhabitants would eliminate many in the western United States that are far more "important" in their region than many of those that would have been included in the more populous eastern United States.

This principle of simplification would appear to have application for cartographers in several important instances. In all cases it provides a guide for the quantity of detail that must be eliminated, given the map scale reduction. Straightforward applications of the principle of simplification would appear appropriate for: (1) point data sets (e.g., towns on a road map), (2) digitized line data (e.g., a file of the coordinate pairs representing the coastline of an island), and (3) area data sets that consist of many small similar areas within the area to be mapped (e.g., lakes in northern Wisconsin).

The question of which individual data element will be retained and which will be eliminated is a more difficult task of simplification. Ideally several solutions exist. The determination of which specific data elements to retain can be deduced from the purpose of the map and the place assigned the particular data distribution in the visual hierarchy specified in the map design. This determination demands that the cartographer be knowledgeable about the data being mapped. In the case of computer-assisted cartographic production, the determination may be made based on ranks assigned to the various data elements after their input to the data file, or on the basis of proximity of position at the scale of the output map, or some other criterion. As more maps are made with computer assistance, the ability to specify such criteria will become more critical. The simplification problem of *which* specific data to retain remains as one of the cartographer's foremost problems.

In computer-assisted cartography it has become possible to reduce all data distributions to essentially a series of discrete points. This process is called "digitizing." Therefore, lines become strings of pairs of coordinates and areas become listed as strings of paired coordinates that define the boundary line of the area. Continuous volumes are represented by dense matrices of Z values or by the isarithmic lines or profile lines that are used to represent the volume. Simplification of these discrete points would then become a straightforward process of elimination of unwanted detail, provided a convenient logically deduced criterion can be specified to indicate which points should be retained.

Initially some simplification schemes advocated the retention of $1/n$th of the points, either randomly or systematically selected; however, more sophisticated schemes employing statistical criteria more coincident with the cartographer's stated purpose are now being used. The cartographer seeks to typify the distribution through simplification by retaining some data elements as they exist and by eliminating others.

Classification

The same goal of expressing the salient character of a distribution is the objective of the second element of generalization, classification, but the methods employed are different. Classification modifies the data elements in its attempt to typify.

Classification is a standard intellectual process of generalization that seeks to sort phenomena into classes in order to bring relative order and simplicity out of the complexity of incomprehensible differences, inconsequential differences, or the unmanageable magnitudes of information. It is difficult to imagine any intellectual understanding, beyond the very elementary, that does not involve classification. We classify without thinking about it; we sort numerical data into averages, above and below average, extremes, and so on, or we condense all the varieties of phenomena into simple classes such as "roads," "rivers," or "coastlines."

Some of the more common processes of classification are: (1) the grouping of similar data elements into categories referred to as range grading; (2) the selection of position and the modification of the data element at that position based on other data elements around the selected position to create a "typical" data element for portrayal on the map at that position; and (3) other forms of typification that may include the exaggeration or actual creation of map data in an attempt to typify.

The manipulations that the cartographer performs during classification are treated in the latter part of this chapter. They include class interval selection and various agglomeration routines. One classification manipulation, called clustering, is appropriately discussed here, since it involves little statistical manipulation and vividly illustrates the distinction between simplification and classification. Cartographically, clustering is necessary whenever numerous discrete items characterize a distribution to be mapped and, at the reduced scale of the intended map, it becomes impossible to portray each item in the distribution individually. Two options are available to the cartographer: one is a method of simplification; the other is a method of classification. Both seek to "typify" the distribution. Figure 8.1 illustrates the two options. In Fig. 8.1A a predetermined number of the discrete items are portrayed on the reduced scale map, and the others are eliminated. This is simplification. In Fig. 8.1B the discrete items are grouped and an average or typical position is designated to represent each group. This is classification. Observe that the results of classification need not be any single existing data element. Methods of classification by agglomeration are varied, and space does not permit a discussion of them.

simplification   A

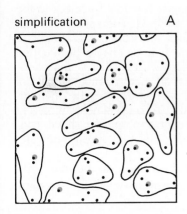

classification   B

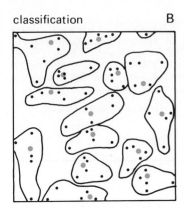

FIGURE 8.1 Examples of the generalization processes. (A) Simplification. (B) Classification.

Symbolization

After simplification and classification have been applied to the data selected for mapping, the cartographer uses the process of symbolization. Symbolization is the graphic coding of the summarizations resulting from classification and the coding of the essential characteristics, comparative significances, and relative positions that result from simplification. This graphic coding makes the generalization desired by the cartographer visible; obviously this process is most critical to the success of any cartographic communication. Judicious simplification and classification can be negated by poor symbolization. On the other hand, good symbolization can enhance the communicability of judicious simplification and classification. Unfortunately, good symbolization can impart an incorrect visual impression of precision and accuracy to poorly simplified or classified data. The cartographer must always strive for honesty here.

All the marks on a map are symbols, from boundary lines to the dots representing cities and from the blue ocean to the hummocks showing marshlands; by using these marks, the cartographer is symbolizing a concept, a series of facts, or the character of a geographical distribution. The very fact of symbolization is generalization, sometimes to a rather small degree, as when we characterize the internal areal administrative hierarchy of townships and counties by a set of boundary symbols, or sometimes to a high degree, as when we symbolize an entire complex urban area with a simple dot.

It is important to observe, furthermore, that the degree of this kind of generalization commonly varies widely within a map. Some symbols, such as the lines of the graticule that indicate latitude and longitude, involve almost no generalization, while on the same map we may include a large, smooth arrow symbolizing the migration of population from one region to another, the arrow taking no note whatever of the many different routes taken by the peoples who moved.

In the total process of symbolization the cartographer must positively assign relative prominence to the features included on the map. Ordinarily this aspect of symbolization is dealt with in the design of the graphic presentation (see Chapter 13), where we are concerned with levels of visual significance and the visual hierarchies of

TABLE 8.1

A cross classification of Data Element Type with Scale of Measurement. An * denotes each potential category of mappable data for the cartographer.

Data Element Type	Scale of Measurement			
	Nominal	**Ordinal**	**Interval**	**Ratio**
Positional	*	*	*	*
Linear	*	*	*	*
Areal	*	—	—	—
Volumetric	—	*	*	*

the communication. On the other hand, in practice, we cannot separate visual and intellectual elements that clearly, and generalization primarily by the processes of simplification and classification during compilation is made easier if some of the basic design decisions have been made prior to beginning the map. The cartographer should have at least an initial idea of the intended visual hierarchy of the data when the initial selection of data is made.

Symbolization is basically a coding process and involves relating the data elements to the marks that a cartographer can use on a map. One aid in doing this is to classify the data elements into a two-dimensional classification based on (1) existent data type, and (2) scale of measurement. In Chapter 5 potentially mappable phenomena were classified into positional, linear, areal, and volumetric data. Also, it was shown that these phenomena may be measured on one or more of four levels of measurement: nominal, ordinal, interval, and ratio. Thus 16 potential categories (see Table 8.1) are possible for the classification of data elements ready to be symbolized.

Second, the cartographer must categorize the visual variables available for the symbolization of the map. It is commonly agreed that, by themselves, the visual variables listed in Chapter 5 can connote on a map only two levels of the scales of measurement of phenomena: nominal (qualitative), or ordinal or higher scales (quantitative). On the other hand, when meaning is assigned to the visual variables through a legend, all four scales of measurement can be communicated. A useful categorization of the visual variables is given in Table 8.2. The first visual variable, relative position in the visual field, is excluded from Table 8.2 because, for the cartographer, this visual variable is

fixed by the geographic ordering of data. The cartographer's only flexibility with respect to relative position is in the selection of the bounds of the visual display (enabling the optimum placement of the most important item within the display) and, to a lesser extent, in the positioning of lettering.

The coding process of symbolization must relate the visual variables listed in Table 8.2 to the phenomena to be mapped. The appropriate link can be made through the scale of measurement. Remember that the phenomena to be mapped can be categorized in Table 8.1 as the cartographer wishes to map them relative to scale of measurement and data type, or as the phenomena themselves exist. Through simplification and classification the cartographer may have consciously generalized a set of data from one category to another. For example, we are all aware that a city is an area on the earth, but the cartographer may have consciously elected to portray

TABLE 8.2

Cross Classification Showing Categories of Marks that a Cartographer Can Use as Symbols on a Map. ("Yes" denotes usable category, "?" denotes a possible category, "no" denotes an impossible category.)

Visual Variable	Scale of Measurement	
	Nominal (Qualitative)	**Ordinal, Interval, or Ratio (Quantitative)**
Shape	yes	?
Size	no	yes
Color (hue)	yes	no
Value	no	yes
Pattern	yes	no
Direction	?	yes

cities as points on the map, thus moving that data set from where it originally was categorized in Table 8.1 to another category. Generally, moving down the scales of measurement from ratio to nominal is simplification; moving from area to line to point is classification. Both, of course, constitute generalization.

The 12 categories denoted by asterisks in Table 8.1 are important to the cartographer. The four remaining categories pose an interesting question. Can we combine areal and volumetric data for cartographic purposes? Strictly speaking, on a two-dimensional map one cannot have a ratio-scaled areal data set without it being a volumetric data set. Likewise, one cannot have a nominally scaled volumetric data set that is not a nominally scaled areal data set. Therefore, it is relatively easy to collapse the 16 possible categories of data to be mapped to 12 categories for cartographic purposes.

If it were possible to link uniquely the available categories of marks shown in Table 8.2 to the 12 data categories in Table 8.1, the process of symbolization could be easily systematized. However, this is not possible, and multiple links are necessary; this means that there is no single cartographic language. It is apparent that the cartographer wields formidable power with the generalization process of symbolization. Because of the importance of this topic, the next three chapters are devoted to the symbolization processes.

Induction

Induction or inductive generalization is a specialized element in the cartographic process. It refers to the results of the logical application of cartographic methods. It applies particularly to the kinds of mapmaking where the cartographer is performing spatial operations that allow the end product to communicate more than just the mapped data. For example, if we have the average January temperatures for a series of stations, we can, by suitable "logical contouring," construct a set of isotherms. The set of isotherms has resulted from making logical inferences about the probable occurrence of average January temperatures in the areas between the data points. Any logical extension of data founded on accepted associations is inductive, such as the mapping of a classification

system (e.g., climates or soils), where data are actually available only for a relatively few points.

The induction that can result from a mapmaking effort may be unknown to the cartographer. The map user, with personal knowledge, may further extend the map information already arrived at inductively. In that sense, the cartographer has little or no control over inductive generalization. The capability of a map for induction suggests that one should not overload the map with information, since that may actually reduce the possibility of induction and thus reduce the effectiveness of the map. The potential for induction would appear to be greater when the map is kept simple and clear. Through induction, map users often uncover hypotheses that cannot be generated in any other manner. This is one of the primary utilities of a map.

THE OPERATION OF THE CONTROLS OF GENERALIZATION

The processes or elements of generalization, mainly simplification, classification, and symbolization, are what the cartographer does; the controls of generalization constitute the factors that influence how each process is performed. It is impossible for a textbook to do more than suggest the operations of the controls, because there is no limit to the number of possible combinations of objectives, scales, graphic limitations, and qualities of data. Nevertheless, a few observations can be made to suggest their functioning.

Objective

The purpose of a map clearly has a great deal to do with its design. The kind of audience to which it is aimed—geographically sophisticated or ignorant, children or adults, and so on—is one basic factor. How it is to be used is another. Is it to be studied with no time limit as might be the case with an atlas map, or is it to be used as an illustration to be shown briefly on a screen during a lecture? Is it primarily to communicate a theme or is it to present evidence to support a thesis? Before beginning, the cartographer must answer these questions; otherwise there will be no basis for making the many decisions that are a part of the generalizing process.

Scale

Obviously the scale of the finished map-to-be is a fundamental factor in the kind and degree of generalization that will be employed. As a general rule, of course, the smaller the scale, the greater will be the degree of generalization required. At large scales, such as in plans and topographic maps, most of the generalization is classification and symbolization, since scale is not a serious constraint.* In thematic mapping, the situation is quite different. When the cartographer has complete freedom to select the scale of a thematic map, that decision may be greatly influenced by the generalization needed to fit the objective. There may well be, as has been suggested, an optimum scale for every objective in thematic cartography.† The influences of scale in cartography are so ubiquitous and pervasive, however, that any such simplification of what is a most complex problem must be applied with care.

Probably the most important task of the cartographer when generalizing the various categories of geographical data within the context of scale is to keep the various degrees of generalization reasonably "in tune" with one another and with the chosen scale. At each scale there is a range of generalization that will "fit" the scale, that is, the treatment will be neither too detailed nor too general: the difficulty comes in the attempt to maintain a balance among categories such as coasts, boundaries, or roads. The combinations of scale, data, and objective are legion; in practice, it is probably wise to leave it as a matter of "good judgment."

The cartographer should remember that when working at very small scales there is likely to be a great scale range within the map. For example, on the conventional Mercator's projection the values of S (see Chapter 4) given in Table 8.3 show the relative exaggeration of areas at the given latitudes. This means that at latitude 60° on Mercator's projection, there is four times as much map area available within which to represent phenomena as there is at the equator. The tendency is

*J. S. Keates, "Symbols and Meaning in Topographic Maps," International Yearbook of Cartography, 12, 168–181 (1972).
†O. M. Miller and Robert J. Voskuil, "Thematic Map Generalization," Geographical Review, 54, 13–19 (1964).

TABLE 8.3

Values of S in Mercator's Projection	
Latitude	S
0°	1.0
20°	1.1
40°	1.7
50°	2.4
60°	4.0

to use all the space available, so there is commonly a great difference in the amount of generalization. Whether one *should* be consistent in the degree of simplification and other factors or not depends on the objective of the map. Whether one *can* be consistent in the degree of simplification and other factors is an equally pertinent question in this age of computer-assisted simplification and symbolization. Algorithms to enable computer-assisted simplification rarely take into account the effects of projection as mentioned above, and most computer-assisted symbolization (plotting) does not compensate for areal or angular distortion that results from projection.

Graphic Limits

The visual variables utilized in symbolization each have two types of limitations, physical and psychological. These limitations act as important controls on the entire generalization process. The physical limitations are imposed on the visual variables by the equipment, materials, and skill available to the mapmaker. The psychological limitations are due to the map user's inabilities to interpret the modulations of the visual variables. Both types of limitations are extremely important in controlling the amount and degree of generalization that a cartographer can successfully use. Sometimes one or the other type predominates.

The physical limitations include factors such as the overall maximum format size that can be utilized, the available line width sizes, lettering styles and sizes, color screens, preprinted symbols, dimensionally stable film or plastics, specialized symbol templates and, of course, the draftsman's or machine's ability to utilize these factors. To date, in computer-assisted cartographic production the graphic limits in terms of the availability and variety of materials and options have been more limiting than in manual cartography, even though the consistency of the application by computer assistance generally exceeds that of manual rendering. Computer-assisted production has extended the possible limitations on accuracy and precision beyond that which is humanly possible in some areas. For example, the repeatability and consistency of computer-assisted products can be measured in thousandths of a millimeter, a capability that manual production cannot attain. It is safe to say that in the areas of accuracy and precision, computer-assisted production has reduced the graphic limitations to such a degree that the psychological aspects are predominant, provided, of course, that the map reader is a human and not a machine!

The psychological aspects of graphic limits are a function of the abilities of the map-reading audience, and these have been shown to vary consistently from one symbol type to another. It can be demonstrated that two graduated circles of slightly different radii separated by more than 10 cm on a map cannot be visually judged correctly by humans (see Fig. 8.2), even though it is quite possible to plot the circles accurately. Similarly, the visual threshold levels for the judgment of tonal values are coarser than our physical abilities to print them. These limits obviously affect the processes or elements of generalization. In many cases they specify the number of classes we may employ, or the minimum distance between points that a computer-assisted simplification algorithm can reasonably use. Such psychologically based limits also determine whether increased clarification, thought to be attained by using color as opposed to black and white, is possible.

Quality of Data

It is readily apparent that the more reliable and precise the data to be mapped are, the more detail is potentially available for presentation. In cartographic generalization the converse is unusually important; that is, the cartographer must sometimes go to considerable pains in order not to let the map give an impression of an accuracy greater than the source material warrants.

Just as basic as the quality of the data is to proper generalization is the scholarly competence and intellectual honesty of the cartographer. A thorough knowledge of the data, which, of course, includes the area being mapped, is indispensable. A Swedish cartographer, Gösta Lundquist, once observed that it always seems easier to generalize faraway places; he stated a familiar and revealing reaction when he pointed out, "I always find . . . that other peoples' maps are extremely good—except for their treatment of Sweden." Intellectual honesty is particularly important in cartography, because a well-designed map has about it an authoritative appearance of truth and exactness. The cartographer must, therefore, take unusual pains to ensure that the data are correct and that their presentation on the map does not convey a greater impression of completeness and reliability than is warranted. For example, there is no theoretical limit to the number of contours that can be interpolated from a set of spot elevations; yet if the set is small, a large number of contours would give an impression of detail and accuracy quite out of line with the quality of the data.

In commenting on the attributes necessary in a mapmaker, a distinguished cartographer and geographer, John K. Wright, expressed it as follows.*

Fundamental among these qualities is scientific integrity: devotion to the truth and a will to record it as accurately as possible. The strength of this devotion varies with the individual. Not all cartographers are above attempting to

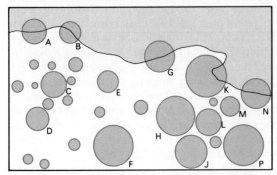

FIGURE 8.2 *Most map readers will visually judge circle C to be larger or equal in size to circle L. This appears to be true because of the distance separating the two circles and because of the differing environments of each. Actually, circle L is larger than circle C.*

*From "Map Makers Are Human . . .," The Geographical Review, 32, 527–544 (1942).

make their maps seem more accurate than they actually are by drawing rivers, coasts, form lines, and so on, with an intricacy of detail derived largely from the imagination. This may be done to cover up the use of inadequate source materials or, what is worse, to mask carelessness in the use of adequate sources. Indifference to the truth may also show itself in failure to counteract, where it would be feasible and desirable to do so, the exaggerated impression of accuracy often due to the clean-cut appearance of a map.

One of the most difficult tasks for the cartographer is to convey to the map reader a clear indication of the quality of the data employed in the map. When writing or speaking, words such as "almost," "nearly," and "approximately" can be included to indicate the desired degree of precision of the subject matter. It is not easy to do this with map data.

There are several ways in which the cartographer may proceed. One is to include in the legend a statement, when necessary, concerning the accuracy of any item. Another and more common method, on larger-scale maps, is to include a reliability diagram, which shows the relative accuracy of various parts of the map.

It is also good practice to include in the legend, if warranted, terms such as "position approximate," "generalized roads," or "selected railroads," so that an idea of the completeness and accuracy may be given to the reader.

Computer-assisted cartography has only compounded the problem of accuracy and completeness. As the number of machine usable cartographic data bases increases in number and availability, the chances for misuse increase. It is highly incumbent on any data base creator that information indicating the quality of the data base be included to aid the cartographer in avoiding misuse. When using an existing map source for data, the cartographer can abide by the rule "Always compile from larger scale to smaller scale." When a machine-readable data base is used, unless information as to the input scale has been included, the cartographer has no way of knowing whether the above rule is being followed or whether an enlargement of a previous generalization is being done.

COMMON SIMPLIFICATION AND CLASSIFICATION DATA MANIPULATIONS

In this section we elaborate on the more common manipulations that a cartographer may have to perform on data during the simplification and classification aspects of generalization. In theory, the manipulations involved in simplification and classification are easily distinguished. In practice, however, a total data manipulation routine may contain elements of both, making it impossible to specify the order in which simplification or classification is performed.

Data may be present after the selection process in any or all of the following forms: (1) tabular values, (2) maps or photographically recorded images, (3) textual or verbal accounts, (4) computer-readable strings of coordinates, or (5) machine-stored picture elements. The actual manipulations employed by the cartographer will vary, depending on the initial data form. We will examine some of the more common manipulations for each data form for both simplification and classification.

Simplification Manipulations

The simplification manipulations performed on tabular values, maps, or photographically recorded images and textual material tend to be very subjective. They mainly take the form of omissions or deletions and, without strict rules, it is improbable that any two cartographers would manually perform exactly the same simplification manipulations for a given map. For example, in manually compiling a map of the North American continent on a Lambert Equal Area base at a scale of 1:20,000,000, certain islands in the Caribbean and in the Canadian Arctic are often omitted, as are lakes in the interior of the continent. Likewise, certain bays or inlets along the coastlines are deleted. Great subjectiveness is here employed by the cartographer-compiler, and the necessity for the cartographer-compiler to be highly skilled and knowledgeable in these processes vitally affects the utility and quality of the map.

Point Elimination. The map in Fig. 8.3 illustrates this process in simplifying the outline of Portugal. The points selected on the left map have been retained and connected by straight line segments on the map at the right. Certainly another

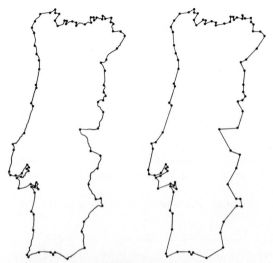

FIGURE 8.3 *The simplification of the outline of Portugal by point elimination. The points indicated on the map to the left were retained on the map to the right where they were connected with straight line segments. All points not selected on the map to the left were eliminated in the production of the map to the right. (Courtesy of the American Congress on Surveying and Mapping.)*

cartographer might choose other points for retention. In this manually systematic elimination of points, the manipulation consists of discarding the points that are unimportant, and a "feel" for this elimination process probably only comes after considerable experience on the part of the cartographer-compiler. The radical law has only marginal utility for this subjective process except in the case of place names or of discrete point objects to be mapped, as in Fig. 8.1. For line features and areas (including islands) the features usually vary considerably in size. Therefore size often overrules the use of the radical law in the decision to eliminate a feature.

In computer-assisted cartography these simplification manipulations are less overtly subjective. Subjectivity is undoubtedly involved in the creation of the computer file but, once the data are in machine-readable form, repeatable simplification manipulations can be performed. Since the amount of data in machine-readable form is increasing rapidly, the aspiring cartographer must pay close attention to the appropriate simplification manipulations that can be performed on

machine-readable data files. Two types of computer-readable data storage currently predominate: (1) data stored as strings of coordinates (mostly the result of digitizing), and (2) data stored as picture elements (remotely sensed or scanned).

Two kinds of operations can be identified in the simplification of data stored as strings of coordinates: (1) point simplification, and (2) feature simplification. Point simplification is the simplification of a string of coordinate points defining a line or the outline of an area. Such simplification takes place by eliminating all but a chosen few points that are deemed more important for the retention of the character of the line being mapped. The feature kind of simplification occurs when many small items of the same class are present in an area. Then, certain of the items are retained and others are omitted in order to retain the essential character while eliminating detail.

Phenomena exist as positional, linear, areal, or volumetric data. When discussing data stored as strings of coordinates, each of these categories can occur in each of the two cases, point and feature simplification; however, if the features themselves are points, there need be no difference in the simplification manipulations between the two cases.

Figure 8.3 illustrates the point case for line data. In Fig. 8.3 the map of Portugal illustrates that simplification of lines results from the retention of some points and the elimination of others. The points denoted by dots on both maps of Fig. 8.3 are exactly the same. The points from the left map are connected by straight line segments on the map on the right.

Figures 8.4 and 8.5 illustrate the following two simplification manipulations: (1) the systematic retention of every nth point, and (2) random retention of 1/nth of the points. In the systematic retention of every nth point, the cartographer creates a new data file by taking the first coordinate point in a data string and every nth point thereafter. The new file is then plotted. The value of n is rather subjectively determined. The larger the value of n, the more will be the simplification. In the random retention of 1/nth of the points, the new file is created by randomly selecting 1/nth of the points. Again, n is usually subjectively determined, and the larger the value of n, the greater will be the simplification (see Fig. 8.6). The sys-

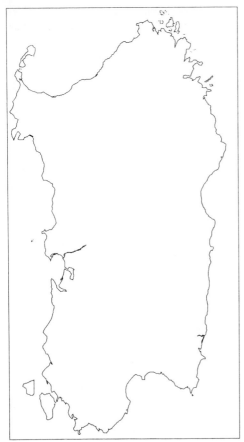

FIGURE 8.4 *This computer-simplified map of Sardinia was produced by systematically retaining every 15th point in the original data file. (Data and algorithm courtesy of D. Brophy.)*

tematic retention and random retention of points are only two of many available computer-assisted simplification procedures. Other criteria can be established to retain or eliminate points in a data file and therefore simplify the data.

Feature Elimination. Simplification by feature elimination is illustrated in Fig. 8.7, which shows forest cover areas of Sheboygan County, Wisconsin. In feature elimination simplification, either a feature is shown in its entirety or it is omitted. In Fig. 8.7 the smaller areas on the left map have been eliminated from the map on the right. Obviously, it is possible to combine point and feature

elimination simplification by (1) eliminating features, and (2) point-simplifying the outlines of the uneliminated features.

This feature elimination routine, if done manually, will be inconsistent. In machine processing the criteria may be size, proximity, or a combination of both. If either strictly size or proximity is used, the simplification routine requires only one parameter. The cartographer may specify the minimum size for retention based on output scale and line width. Likewise, the line width at the output scale would be a factor in proximity determination.

Feature elimination routines can be done

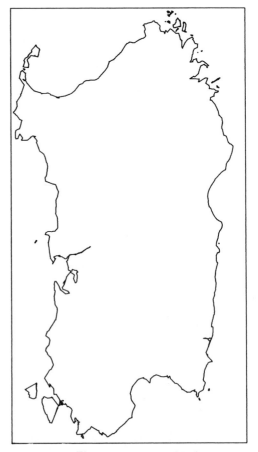

FIGURE 8.5 *This computer-simplified map of Sardinia was produced by randomly selecting 1/15 of the points in the original data file. (Data and algorithm courtesy of D. Brophy.)*

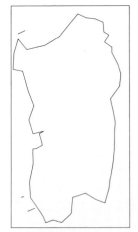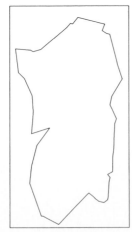

FIGURE 8.6 *The map of Sardinia on the left results from the systematic retention of every 200th point, and the map on the right results from the random selection of 1/200 of the points in the original data file. (Data and algorithm courtesy of D. Brophy.)*

easily by machine if relative importance rankings have been specified for each input record to the data file. Simplification is then accomplished merely by calling for all streams of rank 5 or above, or all roads of rank 2 or above, and so forth. If a tabulation is available of numbers of fea-

tures by rank for a distribution, then the radical law can be used as a guide to help the cartographer call for the correct rank of features for output in accordance with the desired level of simplification.

Simplification can be applied along either of two dimensions: (1) a reduction in the scale dimension (see Fig. 8.8), that is, a change from some original scale to a smaller scale, or (2) at some constant scale dimension (see Fig. 8.9), that is, a detailed representation versus a simplified representation at the same scale. Figures 8.3 and 8.7 are also illustrations of this latter type. Along either of these two dimensions, the process of simplification can be applied to each data category both in point and feature elimination cases. In practice, generalization usually calls for an application of the simplification process primarily along the dimension of scale reduction and secondarily along the dimension of constant scale. For stored data strings, simplification also usually involves both point and feature elimination.

Simplification accompanied by scale reduction is a rather straightforward concept to understand. As the scale of the map is reduced, the physical space available for detail is reduced; therefore, simplification of the details is mandatory. Simplification applied at a constant scale is more subtle and perhaps more important in estab-

FIGURE 8.7 *Simplification by feature elimination. Areas on the left map are either shown in their entirety or completely eliminated in the feature-simplified map on the right. (Courtesy of American Congress on Surveying and Mapping.)*

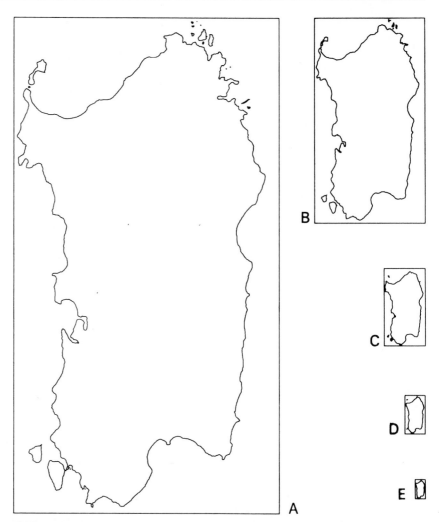

FIGURE 8.8 *Simplification accompanied by scale reduction. Since the scale is reduced in Figs. 8.8A to 8.8E, an increasing number of points in the outline of Sardinia must be eliminated. (Courtesy of American Congress on Surveying and Mapping.)*

lishing a good map design. For example, if the outline of Sardinia as shown in the figures has been selected merely to tell the reader that Sardinia is the area of interest, a rather simplified outline, such as Fig. 8.9D, is appropriate. Presumably the thematic data to be portrayed about Sardinia are the primary message of such a map. On the other hand, suppose the primary purpose of the map deals with ports or coastal shipping. In this case the detail on the delineation of the coastline is important, and an outline like that shown in Fig. 8.9A is more appropriate. Usually the cartographer simplifies along both dimensions simultaneously; that is, the cartographer compiles from a larger-scale map or data base and selects the detail appropriate for the map design.

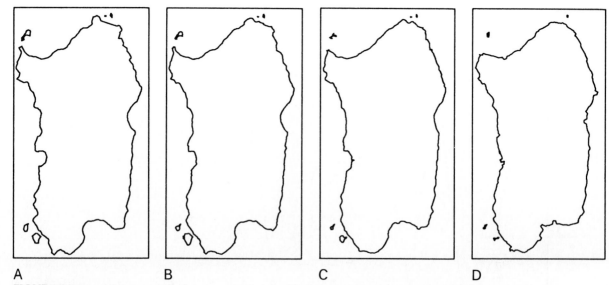

A   B   C   D

FIGURE 8.9 Simplification applied at a constant scale. The cartographer must select from these maps, depending on the importance of the details of the outline of Sardinia to the map message. (Courtesy of American Congress on Surveying and Mapping.)

Smoothing Operators. For data stored as picture elements the simplification manipulations can consist of smoothing techniques, called operators, including surface-fitting techniques, and/or enhancement procedures. Figure 8.10 illustrates the simplification of area/volume data using a smoothing operator. Each picture element or position is compared separately with respect to its neighbors. After evaluation, the picture element may be modified to conform more closely to its neighbors. It is important that the modification does not destroy the individual picture element or combine it with another picture element; it must merely modify its value.

MOVING AVERAGES. Figure 8.11 represents the commonly used smoothing technique of a moving average. As is evident, moving averages can be applied either to linear data (e.g., rivers or terrain profiles) or to mathematically continuous

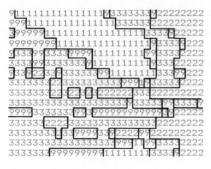

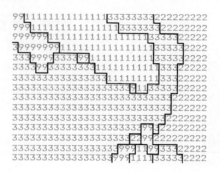

A   B

FIGURE 8.10 A smoothing operator has been applied to previously classed picture elements in Fig. 8.10A to obtain Fig. 8.10B. (Courtesy of S. Friedman.)

LINEAR DATA

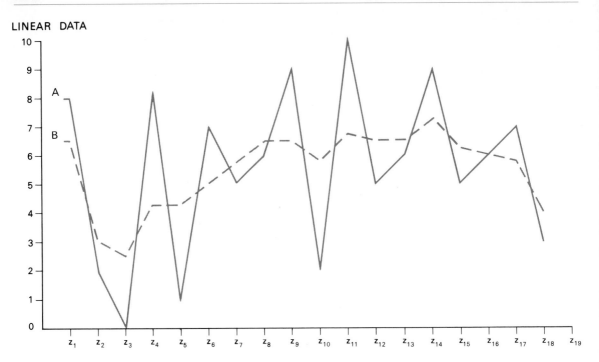

Smoothing operator: for **B**

$$\hat{z}_1 = .75z_1 + .25z_2$$
$$\hat{z}_i = .25z_{i-1} + .5z_i + .25z_{i+1}$$
$$\hat{z}_n = .25z_{n-1} + .75z_n$$

AREA DATA

A – Original data array

0	6	7	0	9	5	3
0	0	10	2	9	1	6
6	3	6	5	1	10	3
8	1	10	7	0	10	2
9	3	4	4	6	1	10
8	6	0	2	7	1	7
7	6	5	5	4	1	1

B – Smoothed data array

0.75	4.875	5.39375	3.3125	6.4375	5.3125	3.485
1.3125	2.375	6.8125	3.9375	6.5625	3.375	5.125
4.625	4.0625	5.375	5.3125	3.25	7.000	4.000
6.8125	3.5625	7.0625	5.75	2.75	7.0625	4.0625
7.6875	4.375	4.0625	3.625	5.00	3.1875	7.500
7.4375	5.625	2.1875	3.1875	5.00	2.8125	5.50
6.9025	5.75	4.6875	4.500	3.625	2.125	1.745

Smoothing operator: for **B**

for z_{ij}

$$\hat{z}_{ij} = .0625z_{i-1,j-1} + .0625z_{i,j-1} + .0625z_{i+1,j-1} + .0625z_{i-1,j} + .5z_{ij} + .0625z_{i+1,j} + .0625z_{i-1,j+1} + .0625z_{i,j+1} + .0625z_{i+1,j+1}$$

for corners (appropriately rotated)

$$\hat{z}_{11} = .6825z_{11} + .125z_{21} + .125z_{12} + .0625z_{22}$$

for edges (appropriately rotated)

$$z_{12} = .125z_{11} + .5625z_{12} + .125z_{13} + .0625z_{21} + .0625z_{22} + .0625z_{23}$$

FIGURE 8.11 Upper: the use of a three-term moving average to smooth linear data. Lower: the use of a nine-term two-dimensional moving average to smooth area data.

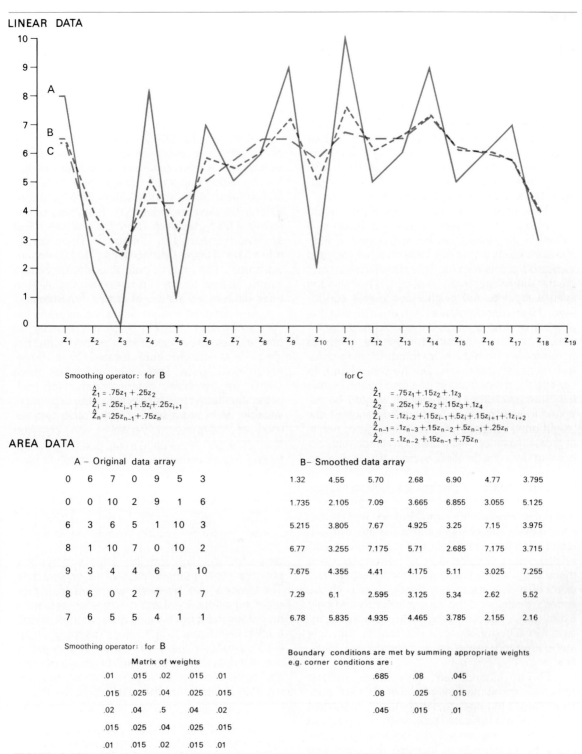

LINEAR DATA

Smoothing operator: for **B**

$$\hat{z}_1 = .75z_1 + .25z_2$$
$$\hat{z}_i = .25z_{i-1} + .5z_i + .25z_{i+1}$$
$$\hat{z}_n = .25z_{n-1} + .75z_n$$

for **C**

$$\hat{z}_1 = .75z_1 + .15z_2 + .1z_3$$
$$\hat{z}_2 = .25z_1 + .5z_2 + .15z_3 + .1z_4$$
$$\hat{z}_i = .1z_{i-2} + .15z_{i-1} + .5z_i + .15z_{i+1} + .1z_{i+2}$$
$$\hat{z}_{n-1} = .1z_{n-3} + .15z_{n-2} + .5z_{n-1} + .25z_n$$
$$\hat{z}_n = .1z_{n-2} + .15z_{n-1} + .75z_n$$

AREA DATA

A – Original data array

0	6	7	0	9	5	3
0	0	10	2	9	1	6
6	3	6	5	1	10	3
8	1	10	7	0	10	2
9	3	4	4	6	1	10
8	6	0	2	7	1	7
7	6	5	5	4	1	1

B – Smoothed data array

1.32	4.55	5.70	2.68	6.90	4.77	3.795
1.735	2.105	7.09	3.665	6.855	3.055	5.125
5.215	3.805	7.67	4.925	3.25	7.15	3.975
6.77	3.255	7.175	5.71	2.685	7.175	3.715
7.675	4.355	4.41	4.175	5.11	3.025	7.255
7.29	6.1	2.595	3.125	5.34	2.62	5.52
6.78	5.835	4.935	4.465	3.785	2.155	2.16

Smoothing operator: for **B**

Matrix of weights

.01	.015	.02	.015	.01
.015	.025	.04	.025	.015
.02	.04	.5	.04	.02
.015	.025	.04	.025	.015
.01	.015	.02	.015	.01

Boundary conditions are met by summing appropriate weights
e.g. corner conditions are:

.685	.08	.045
.08	.025	.015
.045	.015	.01

FIGURE 8.12 Upper: the use of a five-term moving average to smooth linear data. Lower: the use of a twenty-five term two-dimensional moving average to smooth area data.

volume data. The weights and the number of neighboring points considered determine the effect of the application of a moving average. For example, as the number of neighboring values averaged increases, the amount of simplification taking place is increased (see Fig. 8.12). Furthermore, the more evenly apportioned the various weights assigned to the neighboring values are, the greater will be the simplification (see Fig. 8.13). If large weights are assigned only to close neighbors, the amount of simplification is reduced.

Theoretically the selection of weights should be made on the amount of autocorrelation suspected in the distribution being mapped. If high autocorrelation is present, large weights assigned to nearby points result in little simplification, and greater simplification can be obtained by assigning even weights to the neighboring points considered. High autocorrelation also means that the number of nearby points considered must be increased (i.e., consideration must be given to points farther from the point in question) in order to increase simplification. On the other hand, to capture the major features of a distribution having low autocorrelation, high weights should be assigned to nearby values, and consideration of few neighboring values is recommended. Since points at greater distances have little correlation with the point under consideration, there is no justification for considering many neighboring points when low autocorrelation is present. It is best to use few points and let the assignment of weights determine the amount of simplification that is accomplished. The cartographer is making the tacit assumption that the distribution is highly autocorrelated when simplification is performed using many neighboring points and even weights. The effect of such a selection is simply to smooth the surface greatly, and the cartographer should only make such a selection if that is the appropriate intent. Greatly smoothing a distribution with low autocorrelation results in a meaningless simplification.

Photo enhancement procedures utilizing remotely sensed machine-readable data are similar to smoothing operators in that each individual data element is evaluated separately. The purpose is not necessarily smoothing, however; in some cases the purpose is to increase the difference (enhancement) between neighboring values.

Regardless of the purpose, such a procedure is one of simplification, since it treats each data element individually, manipulating it with respect to neighboring values, and retaining it as an individual data element after modification (see Fig. 8.14).

SURFACE FITTING. One final group of simplification manipulations is termed surface fitting techniques. In this category are several techniques that the cartographer can use to approximate lines and surfaces or to estimate values for logical contouring (leading to induction). Examples of the line-type fitting techniques are the regression line analyses mentioned in Chapter 7. These constitute a kind of linear "surface fitting" in two dimensions only. For area/volume data the procedure can be likened to fitting a regression equation in three dimensions instead of in two dimensions.

The simplest surface fitting technique is the "plane of best fit" (see Fig. 8.15). It is necessary to have at least three noncollinear points (points not lying in a straight line) located in a three-dimensional space. Preferably, more than three points are available. One coordinate of each point, say the z value, is selected as the dependent variable, as in regression, and the other two are used as independent variables. An equation $Z = aX + bY + c$ is fitted to the points by the following matrix operation.[*] Most computers have software routines to compute the

$$\begin{bmatrix} N & \Sigma X & \Sigma Y \\ \Sigma X & \Sigma X^2 & \Sigma XY \\ \Sigma Y & \Sigma XY & \Sigma Y^2 \end{bmatrix} \begin{bmatrix} c \\ a \\ b \end{bmatrix} = \begin{bmatrix} \Sigma Z \\ \Sigma XZ \\ \Sigma YZ \end{bmatrix}$$

values for "c," "a," and "b." The resulting equation represents a simplification of the original data. The process can be easily extended to higher-order equations by using machine computation, and widespread use is being made of these simplification techniques in subsurface mapping. Figure 8.15 illustrates a "plane of best fit" as a simplification of a data distribution. Each fit gives a percentage explanation, in the regression sense, that provides an indication of the reliability of the fit.

*For students unfamiliar with matrix manipulations, a good reference text is Frank Ayres, Jr., Theory and Problems of Matrices, Schaum's Outline Series, Schaum Publishing Company, New York, 1962.

LINEAR DATA

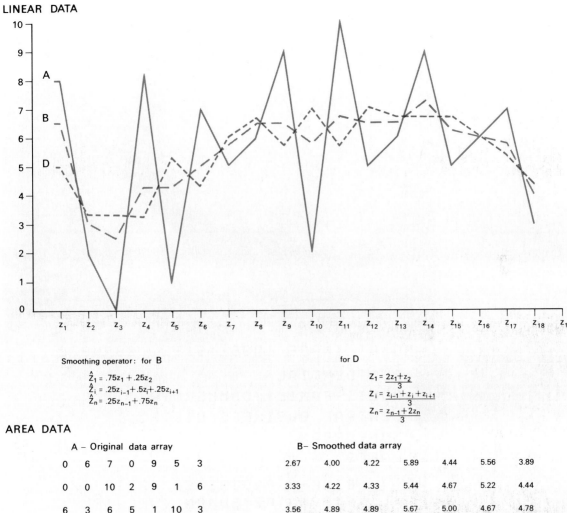

Smoothing operator: for B

$$\hat{z}_1 = .75z_1 + .25z_2$$
$$\hat{z}_i = .25z_{i-1} + .5z_i + .25z_{i+1}$$
$$\hat{z}_n = .25z_{n-1} + .75z_n$$

for D

$$z_1 = \frac{2z_1 + z_2}{3}$$
$$z_i = \frac{z_{i-1} + z_i + z_{i+1}}{3}$$
$$z_n = \frac{z_{n-1} + 2z_n}{3}$$

AREA DATA

A – Original data array

0	6	7	0	9	5	3
0	0	10	2	9	1	6
6	3	6	5	1	10	3
8	1	10	7	0	10	2
9	3	4	4	6	1	10
8	6	0	2	7	1	7
7	6	5	5	4	1	1

B– Smoothed data array

2.67	4.00	4.22	5.89	4.44	5.56	3.89
3.33	4.22	4.33	5.44	4.67	5.22	4.44
3.56	4.89	4.89	5.67	5.00	4.67	4.78
5.89	5.78	4.78	4.78	4.89	4.78	5.67
6.67	5.44	4.11	4.44	4.22	4.89	5.56
7.00	5.33	3.89	4.11	3.44	4.22	4.33
6.89	5.56	4.44	4.11	3.33	3.00	2.33

Smoothing operator: for B

Matrix of weights

.11	.11	.11
.11	.12	.11
.11	.11	.11

Boundary conditions are met by summing appropriate weights e.g. corner conditions are:

.45	.22
.22	.11

FIGURE 8.13 Upper: the use of a three-term equally-weighted moving average to smooth linear data. Lower: the use of a nine-term two-dimensional equally-weighted moving average to smooth area data. Equal-weighting tends to increase the amount of simplification. Compare this figure to Figs. 8.11 and 8.12.

MADISON, WISCONSIN

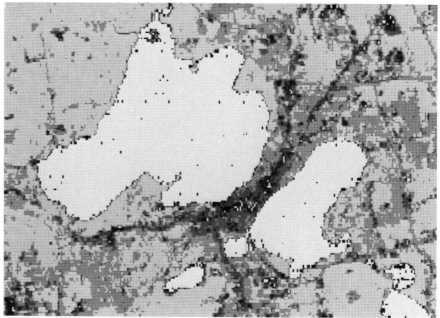

LANDSAT IMAGE 1036-16152   28AUG72

 RESIDENTIAL
MULTI-FAMILY/COMMERCIAL
CENTRAL BUSINESS DISTRICT
NON-URBAN
LAKES

BINARY PROCESSED RATIO 5/6
STEVEN FRIEDMAN   JPL/IPL

FIGURE 8.14 Map of Madison, Wisconsin obtained by examining the changes in corresponding pixels in images taken in different bands of the electromagnetic spectrum. The procedure is one of simplification, since the individual data elements are separately evaluated. (Courtesy of S. Friedman, Earth Resources Applications: Image Processing Laboratory, Jet Propulsion Laboratory.)

A second group of similar techniques employs trigonometric series expansions to approximate a linear profile in two dimensions or a surface in three dimensions.* The mathematics are more involved, but the concepts are similar and the computer routines are readily available.

*John W. Harbaugh and D. F. Merriam, Computer Applications in Stratigraphic Analysis, *John Wiley and Sons, Inc., New York, 1968, pp. 113–155.*

Classification Manipulations

Two kinds of classification routines are important for cartographers: (1) the agglomeration of units, and (2) the selection of class limits. Both may be applied to point, line, and area/volume data. Ancillary problems include the allocation of unit areas to a type or class; for example, if an area is 40% cropland, 30% forest, and 30% water covered, is it classed as cropland or as mixed, since no category covers over 50% of the area?

FIGURE 8.15 *The shaded plane, with respect to the surface shown, is situated so as to minimize the sum of the squares of the deviations between points on the surface and corresponding points on the plane.*

Agglomeration of Data Units. The agglomeration of like units takes place most often by the utilization of point clustering techniques applied to tables of positional locations or to data points in storage. The agglomeration of lines is rare, and the agglomeration of areas/volumes is part of one type of dasymetric mapping. The automation of point clustering techniques is possible and at least one software system can agglomerate irregular areas as in dasymetric mapping.

The agglomeration of points has already been illustrated in Fig. 8.1. Such agglomerations can be done manually or with the aid of computers. In every case the cartographer must subjectively list some criteria that enable the agglomeration procedure to be applied. One such criterion is the specification of a starting point for the agglomeration; a second criterion is the direction of movement followed during the agglomerating procedure. Different results are obtained by starting at different points or by specifying different directions of movement.* The selection of those criteria is slightly more critical in computer-assisted cartography than in manual cartography, since the cartographer retains more subjective control in the latter case.

*Barbara Gimla Shortridge, "Some Aspects of Error in Dot Mapping," unpublished M. A. thesis, University of Kansas-Lawrence, 1965.

The agglomeration of lines is rather uncommon. One example would be the combining of all air passengers using all airlines between any pair of cities. For example, the flow of passengers from Chicago to New York can be individually shown by airline X, airline Y, and airline Z, or these may be agglomerated into one flow line. Likewise, the distributary streams in a river delta may be agglomerated into a few lines in order to convey the essence of the distribution. Depending on the data, this classification manipulation of agglomeration may be performed either manually or with computer assistance.

The agglomeration of areas is a very important manipulation in cartography. The agglomeration of areas can take place on nominally scaled data. Areas have certain characteristics; and Fig.

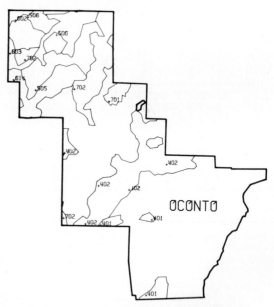

FIGURE 8.16 *Map prepared by the successive overlay of nominal area regions. This computer-produced map of Oconto County, Wisconsin consists of the intersection of the following digital data files: (A) County boundaries, (B) Major water bodies, (C) Soil regions with very severe erosion hazard, and (D) Soil regions and soil associations. The numbers refer to soil region and soil association combinations occurring in regions with very severe erosion hazard. (Courtesy of P. Van Demark.)*

8.16 illustrates an agglomeration of areas based on a specified set of criteria. The numbers associated with the areas designated in Fig. 8.16 indicate regions where particular soils occur with various erosion hazard potentials. This computer-produced map utilizes the common manually applied overlay technique to form the regions. The agglomeration of nominally scaled areal data is dependent primarily on the size of the area unit to be agglomerated in relation to map scale. For example, there might exist two areas of cropland separated by a tree-lined stream, as shown in Fig. 8.17. Consider the area to be broken into 15 small unit areas; at a comparatively large scale, one could map all areas, depending on the percentage of cropland in each small unit area (see Fig. 8.17A), whereas at a comparatively small scale, the area might be mapped as a single unit area consisting entirely of cropland (see Fig. 8.17B).

The problems of agglomerating quantitative as opposed to qualitative area data are only somewhat similar. The agglomeration of areas using ordinal, interval, or ratio scaled data is a form of dasymetric mapping. The agglomeration of the volume data referring to an area is subject to more sophisticated classification manipulations. Foremost among these is class limit selection.

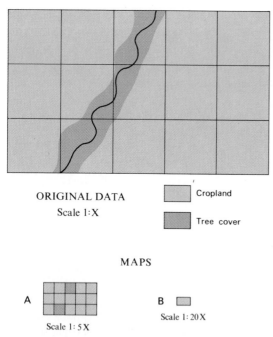

ORIGINAL DATA
Scale 1:X

Cropland

Tree cover

MAPS

A   Scale 1:5X

B   Scale 1:20X

FIGURE 8.17 The ORIGINAL DATA is mapped at a scale of 1:X. (A) represents a smaller-scale representation of the original data. (B) represents the agglomeration of these units for an even smaller-scale representation.

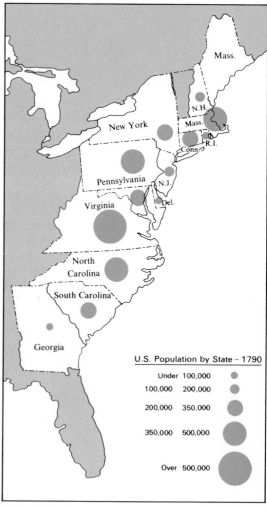

U.S. Population by State – 1790

Under 100,000
100,000   200,000
200,000   350,000
350,000   500,000
Over 500,000

FIGURE 8.18 Range graded point symbols. On this map of the United States in 1790 a standard-size circle is used to represent a specified range of population.

Class Intervals. The agglomeration of volumes at points or lines or over areas usually involves the grouping of like data elements into classes. This is appropriately called range grading. Figure 8.18 illustrates range grading of point volumes; Fig. 8.19 illustrates range grading of line volumes. When a cartographer range grades values that refer to areas on an ordinal, interval, or ratio scale of measurement, the cartographer is mapping according to a choroplethic mapping method (see Chapter 11).

The problem of the selection of the optimal class intervals for a data distribution has been of concern to cartographers for many years, and computer assistance has not lessened the problem. When volumes existing as points or lines

FIGURE 8.19 Range graded line symbols. On this map of immigrants from Europe in 1900, lines of standardized width are used to represent a specified range of numbers of immigrants.

or over areas are mapped by range grading, the cartographer must select and apply a system of generalizing the array of data. Class intervals are the numerical categories of such a system, usually thought of as being bounded by class limits such as 0–2, 2–4, 4–8, and 8–16.

Prior to the specification of the class limits, the cartographer must decide on the number of classes for the map. The number of classes is also a generalizing process. Obviously, fewer classes can be equated with greater generalization, and minimal generalization usually implies more classes. However, much detail is not necessarily good, since excessive detail can easily draw the reader's attention away from more important aspects of the communication.

The graphic limits control on generalization usually specifies the maximum number of classes that a cartographer can successfully employ. The ordinary human eye is *capable* of distinguishing only a limited number of value steps, perhaps five at the most between black and white (and usually less between any other color and white); together with simultaneous contrast, this places severe limitations on the number of classes that can be effectively portrayed. To be sure, we can add pattern (repetitive structure of marks) to the value representation, but even this does not make it possible to increase the number greatly.

The data quality control on generalization along with the desired simplification helps the cartographer to establish the minimum number of classes that can be effectively used. If the data are of good quality, there can be more detail included (i.e., more classes) than if the data are of lesser quality.

A final factor to be considered is the significance of various parts of the range being portrayed. For example, we may be very concerned with the relative changes from place to place, and this would lead us to use a smaller interval in the lower section of the range, since a change from 2 to 4 is the same relative change (100%) as a change from 50 to 100. On the other hand, the interest may be in the absolute values of the higher part of the range; then an opposite choice would be made.

Since the controls of generalization place stringent limits on the number of classes a cartographer may use, the determination of the

number of classes is not as much a problem to the cartographer as the determination of the class limits.

Kinds of Class Interval Series. There are many different kinds of class intervals, and the cartographer needs to analyze the immediate communication problem carefully when choosing one.

As Mackay has pointed out, class intervals are the mesh sizes of cartography, with the chosen limits forming the screen wires.[*] We must choose the mesh size wisely so that the "size sorting" of the distributional data will be most effectively accomplished. Among the more important of the cartographer's concerns in this connection is an analysis of the problem as related to generalization. As mentioned above, there is a direct relation between the number of classes and the amount of information that will be portrayed.

There is an infinite number of kinds of series that can be employed by the cartographer in the setting of class limits. It is almost impossible logically to classify the various series according to their mathematical characteristics or the procedures used to derive them, but we can separate them into three major groups.

One is the equal steps or constant series, which employs some kind of equal division of the data or the geographical area. There are several possibilities in this class.

A second class of series includes those in which the interval becomes systematically smaller toward either the upper or lower end of the scale. Generally, as has been pointed out, the tendency is to put the greater detail at the lower end because, in that section of the range, a small absolute difference is a large relative difference. There seems to be a considerably greater fascination for relative differences than for absolute differences; many distributions are highly skewed; furthermore, in cartography, there is usually much more room for such detail in the regions of sparse occurrences! There is a great variety of series that increase or decrease toward one end of the range.

A third class is the irregular or variable series.

J. Ross Mackay, "An Analysis of Isopleth and Choropleth Class Intervals," Economic Geography, 31, 71–81 (1955).

This kind of series is employed when the cartographer wishes (1) to call attention to various internal characteristics of the distribution (such as some values that may be significant in relation to other analyses), (2) to minimize certain error aspects, or (3) to highlight certain elements of the data range that would not be properly dealt with were a constant or a regular ascending or decending series being employed. These kinds of series are often chosen with the aid of graphic devices, such as the clinographic curve, the frequency curve, or the cumulative curve.*

Before the general availability of computers, the analysis of the consequences of employing various kinds of class interval series was inordinately time consuming. Now, however, we can quickly obtain measures that state the degree of total error (deviation from class means), its distribution among classes, and the ratio of between-class variation to the total variation for any number of classes desired.†

CONSTANT SERIES OR EQUAL STEPS.

There are four major types of constant series that cartographers frequently use for range grading data (see Table 8.4 and Fig. 8.20). The best known and perhaps the simplest is the series of *equal steps based on the data range*. We obtain this kind of constant series merely by dividing the range between the high and low values by the desired number of classes to obtain the common difference. For convenience, the common difference is usually a round number. This common difference is successively added to each class limit, beginning with the lowest value, to obtain the next higher class limit. Major advantages of this series are its simplicity of execution and the comparability of the classes, especially in mapping with

*Jenks and Coulson have devised a method of testing various class intervals series (as applied to a given distribution) against one another in a fashion analogous to the determination of the coefficient of variability. See George F. Jenks and Michael R. Coulson, "Class Intervals for Statistical Maps," International Yearbook of Cartography, 3, 119—134 (1963).

† George F. Jenks, Fred C. Caspall, and Donald L. Williams, "The Error Factor in Statistical Mapping," Abstracts of Papers Presented at 64th Annual Meeting of the Association of American Geographers, Washington, D.C., 1968, p. 45.

isarithms (e.g., contours). If we are mapping a distribution made up of a perfectly rectangular distribution of data observations, there is no difference between this series and the third and fourth series discussed below. A major drawback to this scheme is that the distribution of data observations is not taken into account. Thus a normal distribution or a highly skewed distribution could result in many data observations falling in a few classes and few or no observations in other classes. In fact, with highly skewed distributions, classes with no observations often result from this method.

A second type of constant series can be formed by use of the *parameters of a normal distribution*. If we derive the mean and standard deviation, these may be utilized to set class limits such as the mean plus and minus one standard

TABLE 8.4

Four Class Limits determined by five differing methods utilizing the data given in Table 7.2

I. Equal steps based on the data range
   Range 0 to 100%

Class 1	0 to 25
Class 2	25 to 50
Class 3	50 to 75
Class 4	75 to 100

II. Parameters of a normal distribution
   $\bar{Z}$ = Mean = 34.8   s. = Standard deviation = 26.6

Class 1	$< -1s.$ 0 to 8.2
Class 2	$-1s.$ to $\bar{z}$ 8.2 to 34.8
Class 3	$\bar{z}$ to $+1s.$ 34.8 to 61.4
Class 4	$> +1s.$ 61.4 to 100

Nested means:
First-order mean = 34.8   Second-order means = 13.8 and 58.2

Class 1	0 to 13.8
Class 2	13.8 to 34.8
Class 3	34.8 to 58.2
Class 4	58.2 to 100

III. Quartiles

	rank	data values
Class 1	72 to 55	0 to 9.1
Class 2	54 to 37	15.2 to 33.6
Class 3	36 to 19	33.7 to 52.2
Class 4	18 to 1	55.0 to 100

IV. Areal equal steps
   Each class covers approximately ¼ of area of state.

	data values
Class 1	0 to 15.2
Class 2	16.9 to 32.8
Class 3	33.6 to 49.6
Class 4	52.2 to 100

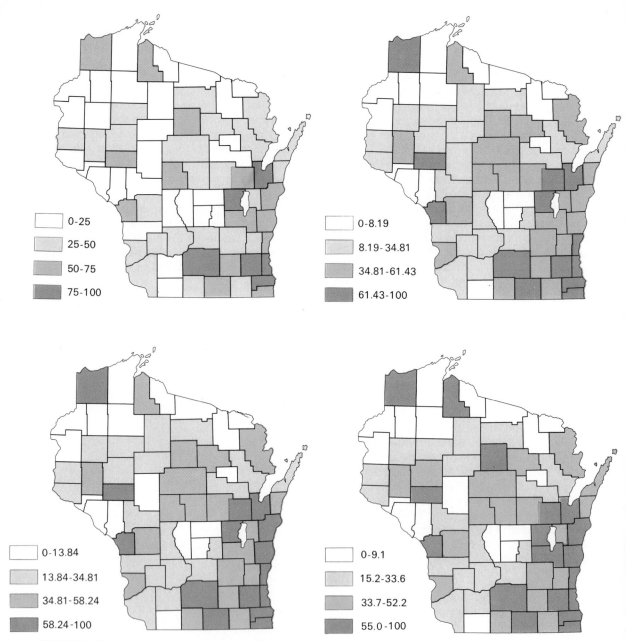

FIGURE 8.20 *Five examples of constant series class intervals applied to the data in Table 7.2. The class limits are defined in Table 8.4.*

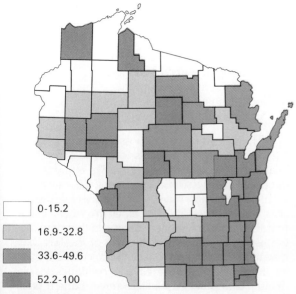

0-15.2

16.9-32.8

33.6-49.6

52.2-100

FIGURE 8.20 (*Continued*)

deviation, from one standard deviation to two standard deviations (above or below the mean), and so on (see Table 8.4). The more nearly normal the distribution, the more useful is this type for graphically portraying the areas and magnitudes of departures from the average.

If the data are normally distributed, the cartographer may employ fractional parts of standard deviations in an effort to create equal areas on the maps in each class or equal numbers of observations in each class. In essence the mean and standard deviations allow the cartographer to specify equal steps based on the data range, or equal numbers of observations per class or equal geographical areas (provided all observations refer to areas of equal size) for normal distributions. If applied to data that are not normally distributed, the use of the mean and standard deviation tends to result in classes with no data observations and to combine many of the observations into one or two classes.

A variation of the use of the mean for class limit determination is the method of "nested means."[*] In this method the overall mean of the

*Morton W. Scripter, "Nested-Means Map Classes for Statistical Maps," Annals, Association of American Geographers, Vol. 60 (1970), pp. 385–393.

data distribution initially divides the distribution into two parts. A mean is again calculated for each part, and each part is divided into two additional parts by this "second-order" mean. Major advantages of this method are that no classes without data observations can be constructed and that at any level the distribution of the data can be said to be in equilibrium, since the absolute sum of the deviations of the class members from a mean dividing a distribution or distribution part into two classes is always equal for both classes divided by that mean. A major disadvantage is that only 2^n classes can be calculated, where n is the order of mean being calculated. Thus 3, 5, 6, 7, or 9, and so forth, classes are impossible with this method.

A third kind of constant series involves the employment of quantiles; a quantile is a division of the number of observations in the data array into equal parts; there may be any division, but the commonly used ones are quartiles (four), quintiles (five), sextiles (six), and so on, up to deciles (ten) or even centiles (one hundred). Quantiles are determined by arraying the observed values in the order of their magnitude, from the lowest to the highest. If we wished to obtain quartile values for interval or ratio data, for example, we proceed by counting one-fourth of the number of observations from the bottom of the ordered array to determine the first quartile value, and so on (see Table 8.4). Quantiles are useful when the mapping is based on unit areas, especially in choropleth mapping but, if there are major differences in the sizes of the unit areas, much of the value of the quantile series is lost. If all the areas are of equal size, this method gives results equivalent to the fourth type of constant series discussed below. This method can be useful for ordinally scaled data since, in ordinal scaling, the data are ranked and, since this method depends only on the number of observations and not their value, it can be applied easily to ordinally scaled data.

A fourth type of constant series is what might be called *equal area steps* or, in a sense, "geographical quantiles." In this kind of series the area of the map is divided into equal regions, the number depending on the cartographer's choice. The determination of the class limits for this kind of series can be accomplished by employing a cumulative frequency graph, as described below. It cannot be done unless the areas of the unit regions are known. Furthermore, it has no applicability to range graded point or line volumes. Finally, if the data are from a rectangular distribution and all unit areas have equal areas, there is no difference between this method and the equal step based on the data range or the quantile methods. An added advantage of this method for cartographic purposes is that it specifies the class limits so that each class occupies an equal portion of the map area (see Table 8.4).

SYSTEMICALLY UNEQUAL STEPPED CLASS LIMITS. This type of class limit determination scheme applies only to data measured on an interval or ratio scale.

Two groups of series with unequal steps are (1) arithmetic series in which each class is separated from the next by a stated numerical difference (not constant), and (2) geometric series in which each class is separated by a stated numerical ratio. The general form of the equation to produce class limits of either type of series is

$$L + B_1X + B_2X + \ldots + B_nX = H$$

where L = the lowest value
H = the highest value
B_n = the value of the nth
    term in the progression.

It is only necessary then to obtain B_n by some method and then solve the equation for X for any given values of L and H to establish the class limits. For *arithmetic progressions* the quantity B_n is obtained by

$$B_n = a + [(n - 1)d]$$

where a = the value of the first term
n = the number of the term
    being determined (the first,
    second, etc.)
d = the stated difference.

In *geometric progressions* the quantity B_n is obtained by

$$B_n = gr^{n-1}$$

where g = the value of the first nonzero term
n = the number of the term
    being determined (the first,
    second, etc.)
r = the stated ratio.

Both arithmetic and geometric series can take any of the following six forms, depending on the values of d and r.

1. Increasing at a constant rate.
2. Increasing at an increasing rate.
3. Increasing at a decreasing rate.
4. Decreasing at a constant rate.
5. Decreasing at an increasing rate.
6. Decreasing at a decreasing rate.

There is an infinity of possibilities using arithmetic and geometric progressions. The difficulty is in assessing when it is appropriate to apply which particular progression. Because of this difficulty, such progressions have seen diminished use in class limit determination schemes. No strictly theoretical grounds have been advanced that demand the use of a progression; in spite of the fact that machine computation makes the calculations easier than ever before, their use continues to diminish. Computer capabilities have made it possible for more complex statistical criteria to be used in class limit determination in place of progressions (see next section below). Appendix E provides examples of all six types of arithmetic and geometric progressions and a step-by-step account of their derivation.

IRREGULAR STEPPED CLASS LIMITS. The irregular stepped interval techniques can be subdivided into two classes: graphic techniques and iterative techniques.

GRAPHIC TECHNIQUES

It must be emphasized at the outset that graphic approaches to the selection of class intervals may not be suitable for two reasons. On the one hand, the limits may be either ill defined or so oddly arranged in the series that it would be difficult for a reader to use them; and on the other hand, the arrangement of the values in numerical order (as is necessary in the preparation of the graphs) clearly destroys their geographical order or association. It is quite possible that some startling characteristic of a curve on a graph, such as a marked peak, depression, or flexure, may have no geographical significance whatever, since the data from which it derives may be widely and unsystematically dispersed. Three common graphic techniques employed by cartographers to aid them in selecting class limits are the frequency curve, the clinographic curve, and the cumulative frequency curve.

THE FREQUENCY CURVE

A frequency graph is prepared by arraying the Z values of the data on the X-axis of a graph plotted against the frequency of their occurrence on the Y-axis. Frequency in this sense means the number of times they occur; in the case of volumes distributed over areas, if the areas of the units are not widely different, the data collection units may simply be counted. If they are greatly different we should employ a cumulative frequency graph. As an illustration of a frequency graph, Fig. 8.21 has been prepared from the data in Table 7.2.

Generally, the low points on a frequency diagram are thought to be the most useful as class limits because they tend to enclose larger groups of similar values. Often, however, there may be few such low points. Frequency graphs are ordinarily quite useful to illustrate the numerical characteristics of the distribution, but they do not always provide clearly desirable class limits. They are relatively simple to construct, and most large computer installations have software that will quickly prepare a frequency curve or histogram of your data set. If low points on the frequency curve exist they are most useful for range grading point and line volume data. This method can be used with ordinal, interval, or ratio data.

THE CLINOGRAPHIC CURVE

A clinographic curve is prepared by arraying the Z values of the data in numerical order on the Y-axis plotted against the cumulative areas to which they

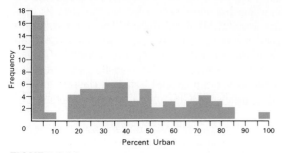

FIGURE 8.21 *A frequency graph of the data in Table 7.2.*

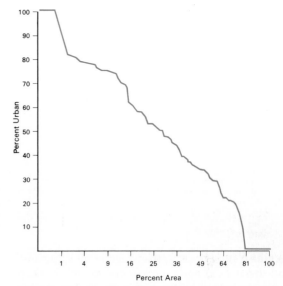

FIGURE 8.22 *A clinographic curve of the data in Table 7.2.*

refer on the X-axis. The Y-axis is scaled arithmetically (evenly spaced steps), while the X-axis is scaled, in percent of total area, with a square root scale from 0 to 100%. If data for enumeration units are being used we may simply use the numbers of unit areas if they are of about equal sizes; otherwise we should employ the areas of the units. Figure 8.22 is a clinographic curve prepared from the data in Table 7.2. If point or line volumes are being used each occurrence receives equal weight on the X-axis.

The critical points in a clinographic curve are the points where the slope of the curve changes. Ordinates at these flexures indicate the Z values separating regions of different gradient, provided, of course, that the data adjacently plotted on the graph are, in fact, reasonably adjacent geographically. It should be possible to obtain plots of clinographic curves computed and drawn with machine assistance at most large computer installations. The clinographic curve would appear to have significance for volumes distributed over areas, since it is possible to take area into consideration.

THE CUMULATIVE FREQUENCY CURVE

A cumulative frequency curve is prepared by arraying the Z values of the data in numerical order

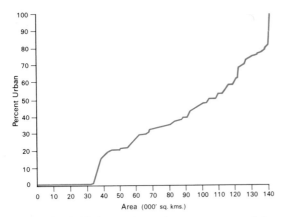

FIGURE 8.23 A cumulative frequency curve of the data in Table 7.2.

on the Y-axis plotted against the cumulative area on the X-axis, as described in Chapter 7. Both axes are scaled arithmetically, in contrast to the clinographic curve. Properly, the Z value of each successive unit area value is plotted against the sum of its area and all preceding unit areas with lower Z values. Again, for equal size enumeration units or for point or line volumes, each numerically higher Z value is equally spaced on the X-axis. The graph will end on the right side with an X value equal to the total area included in the geographical area being mapped or the total number of observations for point and line volumes. A curve so plotted must rise continuously until it reaches the last plotted point. Figure 8.23 is a cumulative frequency graph of the data in Table 7.2.

For class limits that are likely to be significant in terms of a surface volume, the critical points are the tops and bottoms of "escarpments" on the curve, since these tend to segregate areas of different Z values. Probably more important is the fact that the cumulative frequency graph may be used to determine other values having cartographic significance. As described in Chapter 7, we may determine the geographical mean, the geographical median, or other functions of Z values over areas such as "geographical quantiles" or equal area steps from a cumulative frequency curve.

ITERATIVE TECHNIQUES

Iterative techniques are too complex to calculate by hand; they require machine assistance. Nor-

mally, some logical statistical criterion is specified, and the computer iterates a solution so that the data are grouped so as to most closely meet the specified criterion. Jenks first introduced class limit determination schemes employing iterative techniques that have statistical criteria based firmly in cartographic theory.[*]

Two statistical criteria that can be met through machine iteration have been specified more recently by Jenks.[†] One, the goodness of variance fit (GVF), is useful when the cartographer wishes to minimize the squared deviations about the class *means*. The criteria to be satisfied is to maximize the quantity, GVF, where

$$\text{GVF} = \frac{\substack{\text{sum of squared deviations} \\ \text{between classes}}}{\substack{\text{Total sum of squared deviations} \\ \text{from the array mean}}}$$

The total sum of squared deviations from the array mean, SDAM, minus the sum of squared deviations between classes is equal to the sum of squared deviations from the class means, SDCM.

$$\text{SDAM} = \sum_{i=1}^{n} (Z_i - \bar{Z})^2$$

where $\bar{Z}$ = the array mean

$$\text{SDCM} = \sum_{j=1}^{k} \sum_{ij=1j}^{nj} (Z_{ij} - \bar{Z}_j)^2$$

where $\bar{Z}_j$ = the mean of the jth class
where j = class number
i = observation number within class j

Utilizing this criterion, the cartographer must first specify an arbitrary grouping of the numerically arrayed data. The mean of each designated class is computed and the sum of the squared deviations between each observation and its class mean (SDCM) is calculated. The next step consists of moving observations from one class to another in an effort to reduce the sum of SDCM and thereby increase the GVF statistic. After observations have been moved between classes, the new class means, SDCM, and GVF are computed. This

---

[*]*George F. Jenks, "The Data Model Concept in Statistical Mapping," International Yearbook of Cartography, 7, 186–190 (1967).*
[†]*George F. Jenks, personal communication.*

process is repeated until the GVF can no longer be increased.

A second statistical criterion that can be utilized in conjunction with the class *medians* is to maximize the goodness of absolute deviation fit, GADF.

$$\text{GADF} = 1 - \frac{\text{sum of absolute deviations about class medians}}{\text{sum of absolute deviations about array median}}$$

The sum of the absolute deviations about the class medians, ADCM, and the sum of the absolute deviations about the array median, ADAM, are given by the following formulas.

$$\text{ADAM} = \sum_{i=1}^{n} |Z_i - Z_m|$$

where Z_m = array median Z-value

$$\text{ADCM} = \sum_{j=1}^{k} \sum_{ij=1j}^{nj} |Z_{ij} - Z_{mj}|$$

where Z_{mj} = median Z-value of class j

The procedure is similar to the one stated earlier for the GVF criterion. The cartographer must first specify an arbitrary grouping of the numerically arrayed data; then the median for each class and the ADCM value is computed. Observations are moved from class to class in an attempt to minimize ADCM and thereby increase GADF. The process is repeated until the GADF can no longer be increased.

The major advantage of these techniques is that the cartographer can attempt to maximize the homogenity of each class and to maximize the heterogeneity between classes as defined by the statistical criterion employed. This is a worthy theoretical goal in range grading. It is only inappropriate when theory specifies otherwise.

These iterative techniques have come into widespread use by cartographers only since machine computation has been available. Additional iterative techniques could be specified by stating different statistical criteria. Jenks' criteria mentioned above appear to allow the cartographer to approach a theoretical goal in the range grading of data. Because of this, they should receive the serious attention of all cartographers.

SELECTED REFERENCES

Arnberger, E., "Problems of an International Standardization of a Means of Communication through Cartographic Symbols," *International Yearbook of Cartography, 14,* 19–35 (1974).

Chang, K-T, "Data Differentiation and Cartographic Symbolization," *The Canadian Cartographer, 13* 60–68 (1976).

Cox, C. W., "Adaptation-Level Theory as an Aid to the Understanding of Map Perception," *Proceedings of the American Congress on Surveying and Mapping, 33rd Annual Meeting,* Washington D.C., pp. 334–359 (1973).

Douglas, D. H., and T. K. Peucker, "Algorithms for the Reduction of the Number of Points Required to Represent a Digitized Line or Its Caricature," *The Canadian Cartographer, 10,* 112–122 (1973).

Jenks, G. F., "Conceptual and Perceptual Error in Thematic Mapping," *Technical Papers from the 30th Annual Meeting, American Congress on Surveying and Mapping,* Washington D.C., pp. 174–188 (1970).

McCleary, G. F., Jr., "Beyond Simple Psychophysics: Approaches to the Understanding of Map Perception," *Technical Papers from the 30th Annual Meeting, American Congress on Surveying and Mapping,* Washington D.C., pp. 189–209 (1970).

Monmonier, M. S., "Class Intervals to Enhance the Visual Correlation of Choroplethic Maps," *The Canadian Cartographer, 12,* 161–178 (1975).

Monmonier, M. S., "Pre-Aggregation of Small Area Units: A Method for Improving Communications in Statistical Mapping," *Proceedings of the American Congress on Surveying and Mapping, 35th Annual Meeting,* Washington D.C., pp. 260–269 (1975).

Morrison, J. L., "A Theoretical Framework for Cartographic Generalization with Emphasis on the Process of Symbolization," *International Yearbook of Cartography, 14,* 115–127 (1974).

Olson, J., "The Effects of Class Interval Systems on Choropleth Map Correlation," *The Canadian Cartographer, 9,* 44–49 (1972).

Olson, J., "Class Interval Systems on Maps of Observed Correlated Distributions," *The Canadian Cartographer, 9,* 122–131 (1972).

Robinson, A. H., "An International Standard Symbolism for Thematic Maps:. Approaches and Problems," *International Yearbook of Cartography, 13,* 19–26 (1973).

Robinson, A. H., and L. Caroe, "On the Analysis and Comparison of Statistical Surfaces," *Quantitative Geography Part 1,* Northwestern University Studies in Geography, No. 13, Evanston, 252–275, (1967).

Salichtchev, K. A., "History and Contemporary Development of Cartographic Generalization," *International Yearbook of Cartography, 16,* 158–172 (1976).

Steward, H. J., "Cartographic Generalization Some Concepts and Explanation," *Cartographica,* Monograph No. 10, York University, Toronto, Canada, (1974).

SYMBOLIZATION: QUALITATIVE

This chapter and Chapters 10 and 11 deal with the fundamental elements involved in the symbolization of the vast array of data that can be communicated cartographically. In order to introduce this material properly, a few general remarks are made here concerning distribution, statistical, qualitative, and quantitative maps. The remainder of this chapter will deal with the symbolization of qualitative positional, linear, and areal data.

All maps are "distribution maps" in that it is impossible to represent relative geographical location without showing the distribution of something. On the other hand, the term "distribution map" is widely employed in a much more restricted sense. A large group of maps, mostly thematic, employs graphic symbols in a surprising variety of ways to show the general or detailed characteristics of a single class of geographical phenomena; this is in marked contrast to the general class of maps, which portray the locational characteristics all at once of a variety of past or present geographical data.

Many distribution maps deal with statistics derived from censuses, and the like, and are called "statistical maps." Because many geographical phenomena can be differentiated without being obviously numerical and without being subjected to ordinary statistical manipulations, it is not meaningful to use the term "statistical map" except when it is restricted to the portrayal of data ordinarily included in the term "statistics."

Maps showing distributions in numerical form are one of the cartographer's stocks in trade. They are capable of surprising variety and can be used to present almost any kind of data. Few maps can be made that do not in some way present quantitative information, even if the amounts involved result from so simple an operation as the ordinal differentiation of symbols for cities of different size for an atlas map. Column after column of numbers in tabular form frequently are excessively forbidding, and the statistical map usually can present the important geographical characteristics of the material in a far more understandable, interesting, and efficient manner. Tabular quantitative materials of various kinds, from governmental censuses, to the reports of industrial concerns, to the results of our own surveys, exist in staggering

CHAPTER 9
SYMBOLIZATION: QUALITATIVE

variety. With such a wealth of material, it is to be expected that the cartographer would find that a large percentage of effort is devoted to preparing this type of map. Many of the distributions mapped are derived parameters, that is, generalizations of numerical arrays of the sort suggested in Chapters 7 and 8, and the cartographer must proceed warily lest he or she stumble intellectually. It is necessary for the student to appreciate that figures *can* easily lie cartographically if they are improperly presented to the map reader. As one author has observed, "One of the trickiest ways to misrepresent statistical data is by means of a map."*

Hasty evaluation of data, or the selection of data to support conclusions unwarranted in the first place, is thoroughly unscholarly and results in intellectually questionable maps. Their production is reprehensible, since they may be, and unfortunately sometimes are, used by their authors or others to support the very conclusions from which they were drawn in the first place. The exhortation, frequently implied or stated in this book, "to prepare the map so that it communicates what is intended," is not contradictory to the preceding; it is merely a matter of integrity.

Another source of dangerous error in cartography is the map that provides an impression of precision greater than is justified by the quality and quantity of the data used in its preparation. To keep from making this serious mistake, the cartographer is sometimes forced to invent special symbols such as dashed lines or patterns of question marks; these may detract from the aesthetic quality of the map, but they will serve the more important purpose of preventing a map reader unfamiliar with the subject matter from falling into the common trap of "believing everything he or she sees."

Distribution maps employ a great variety of symbolism and to try to classify with precision all the variations in the geographical concepts that can be symbolized is all but impossible; the multiplicity of categories would reach astronomic numbers. Consequently, we cannot illustrate in a practical manner the variety of ways a given

symbol may be used. Instead, here and in Chapters 10 and 11 we will illustrate the means by which the various fundamental classes of geographical data (positional, linear, areal and volumetric) can be and usually are represented.

The examples of distribution maps that appear in this textbook by no means complete the list, since the numbers of possibilities of combination and presentation are almost infinite. Instead of attempting to give directions for the preparation of this vast array of map types, attention is focused here on the basic and most frequently used classes and on the principles involved in their preparation. Other illustrations of their application in specific subject-matter situations may be found in great variety in books and periodicals.*

Distribution Maps

All maps on which data are differentiated nominally (and many in which part of the objective is to show data scaled ordinally, or even according to interval and ratio categories) are clearly *qualitative* in that one of their main functions is to display the relative location of different kinds of phenomena. These may range from topographic maps to maps of classes of precipitation (rain, snow, hail, drizzle, for example), and from categories of employment to predominant races. Their fundamental objective is to present the geographical locations of such categories and, consequently, adequate locational information is usually of vital importance to such maps. Opposed to qualitative maps are those commonly called *quantitative*. These are the maps in which the data are arrayed according to ratio, interval, and ordinal scales; they range from amounts of precipitation to thicknesses of rock formations, and from levels of employment to ratios of racial composition. They naturally cannot escape being somewhat qualitative, since they must display a kind of phenomenon, but their major objective is not the nominal differentiation of the data but the presentation of spatial variations in amounts. This class is much more complex, since the possibilities for manipulation of the data are much greater. Base data re-

---

Darrell Huff, How to Lie with Statistics, *W. W. Norton and Company, New York, 1954.*

A very useful reference, profusely illustrated and documented, is F. J. Monkhouse and H. R. Wilkinson, Maps and Diagrams, *3rd edition, University Paperbacks, Methuen and Co. Ltd., London, 1971.*

quirements are always fundamental but are usually not quite so great for this class of maps; it is not unusual for the major focus of interest to be on the numerical variations within the distribution of the phenomenon being mapped instead of on the precise location of details.

Qualitative Distribution Maps. In principle, qualitative distribution maps, which have as their main objective the communication of nominal categories of data, are relatively simple from the viewpoint of symbolism. Differences can be represented merely by changing the appearance of the symbol.

Colored or uncolored lines can be dotted, dashed, or varied in other ways; point symbols can be dots, stars, anchors, triangles, and so on; area symbols can employ variations in color or pattern.

The selection of symbols for nominal differentiation also poses the problem common to all distribution maps, of symbolizing without employing value contrast. Clarity and visibility are the result of contrast and, of the various kinds of contrast, color value (degree of darkness) is visually the most important. Value changes are generally inappropriate, however, in nominal symbolization on maps because of the universal tendency to assign quantitative meaning to value differences. Generally a darker symbol looks more "important" to the map reader. Of course, such emphasis could be used to advantage if the cartographer were desirous of drawing attention to one or more of the various nominal or nominal-ordinal categories being mapped. The use of hues complicates the problem even further and is described at some length in Chapter 14.

Additionally, on both qualitative and quantitative distribution maps, most of the base data is portrayed on a nominal scale. The neat lines, graticules, coastlines, rivers, area outlines, and the like, are most often portrayed by qualitative or nominal scale symbols when they are used as base data.

Quantitative Distribution Maps. Quantitative distributions employ a variety of numerical measures, some of which were considered in a preceding chapter. The majority are based on absolute numbers or on simple summaries such as averages. The numerical data can refer to positional, linear, areal, or volumetric quantities and can be represented by marks that can be classed as point, line, and area symbols.

When approached from the viewpoint of the frequency of use, the cartographic symbolization of quantitative distributions becomes relatively simple, because only a few classes of symbols are appropriately used on the great majority of maps. Generally, to show variations from place to place of the amounts in a single class of data, it is usually desirable to employ only one basic form of graphic device and then vary it in terms of size, frequency, or visual value to portray the differences in quantity from place to place. To be sure, we can combine several forms of quantitative symbolism on a single map, but this should be approached warily, since it is easy to make the communication too complex for the map reader.

Chapters 10 and 11 concern the more common techniques used to symbolize variations in quantities of data. The remainder of this chapter will focus on the symbolization of nominally scaled positional, linear, and areal data. In a departure from most texts on cartography, a separate discussion on symbolizing the unique distribution of landform will not appear in this textbook. Instead, landform symbolization techniques will be discussed where they appropriately fall in the symbolization classification used in this and the following chapters. A brief section on the history of landform representation does appear at the end of this chapter.

MAPPING QUALITATIVE POSITIONAL DATA

Mapping nominally scaled positional data employs symbols that may be classed as (1) pictorial, (2) associative, or (3) geometric. Several symbol dimensions (visual variables) connote nominal differentiation and are therefore commonly used in the mapping of qualitative positional data. These include shape, direction, color (hue), and pattern.

Pictorial Symbols. Figure 9.1 illustrates the use of pictorial symbols for the representation of nominally scaled positional data. These symbols could also employ differing hues to aid in the communication. A pictorial symbol may be intricate, as

FIGURE 9.1 *Stylized nominally scaled pictorial point symbols dealing with transportation. (From R. Modley, Handbook of Pictorial Symbols, Dover Publications, Inc., New York, 1976.)*

shown in Fig. 9.2, or stylized, as in Fig. 9.1. For maximum effectiveness in communicating the map "message," pictorial symbols should communicate without the necessity for a legend but, in practice, a legend is usually provided. Maps of nominally scaled positional data that employ pictorial symbols often err in two important respects: (1) the symbols are not easily distinguished one from another, and (2) too many symbols of roughly equivalent size are used. In an effort to illustrate that the data are nominally scaled, a cartographer may not vary the size of the various symbols, since doing so might connote a higher scale of measurement. By not varying the sizes of the different symbols, the cartographer may fail to communicate significant patterns in a nominal distribu-

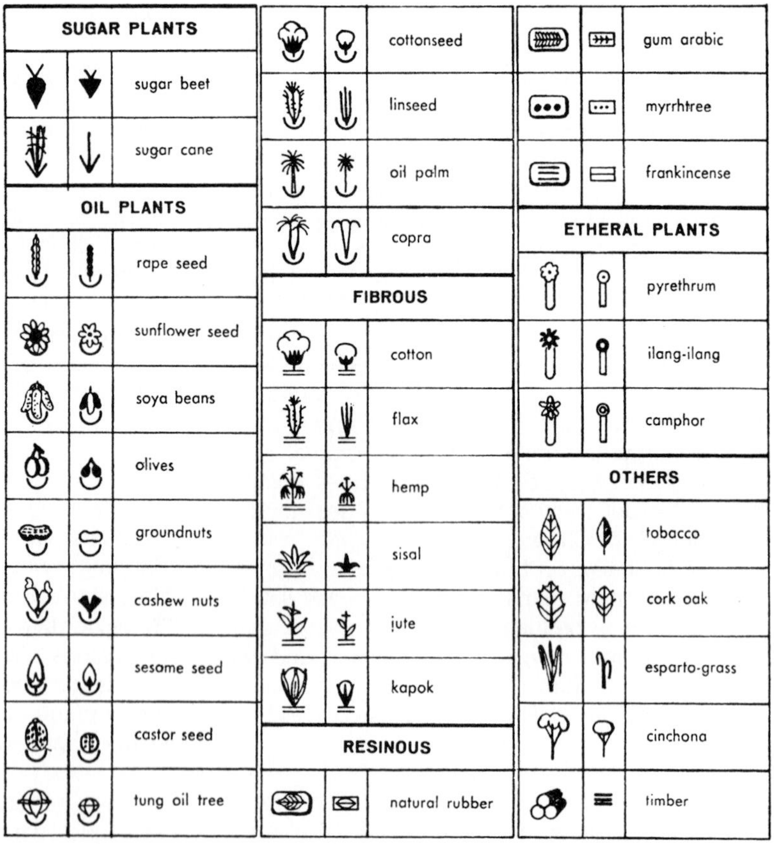

SUGAR PLANTS					cottonseed			gum arabic
		sugar beet			linseed			myrrhtree
		sugar cane			oil palm			frankincense
OIL PLANTS					copra	**ETHERAL PLANTS**		
		rape seed	**FIBROUS**					pyrethrum
		sunflower seed			cotton			ilang-ilang
		soya beans			flax			camphor
		olives			hemp	**OTHERS**		
		groundnuts			sisal			tobacco
		cashew nuts			jute			cork oak
		sesame seed			kapok			esparto-grass
		castor seed	**RESINOUS**					cinchona
		tung oil tree			natural rubber			timber

FIGURE 9.2 *Intricate nominally scaled pictorial point symbols representing crops. (From L. Ratajski, "The Methodological Basis of the Standardization of Signs on Economic Maps," International Yearbook of Cartography, Vol. 11, p. 151, 1971.)*

tion. The trend toward the use on maps of "framed" pictorial symbols of equivalent size especially promotes this error (see Fig. 9.3). The situation is easily avoided by using drastically varying shapes or using differing hues where possible. Different orientations of the same symbol are sometimes useful as well (see Fig. 9.4).

There should be no excuse for a cartographer

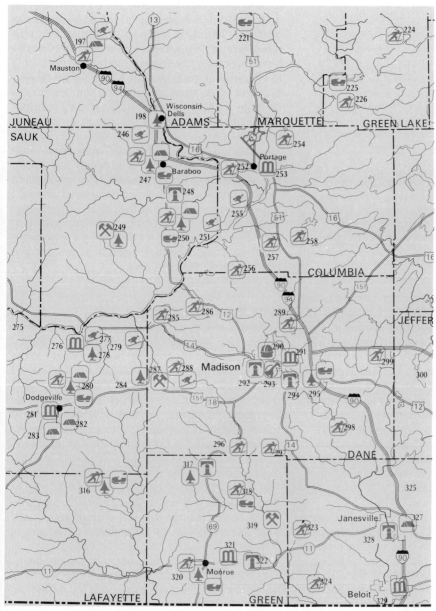

FIGURE 9.3 Framed nominally scaled pictorial symbols on a map promoting winter activities in a portion of the state of Wisconsin. The legend of the map lists 14 symbols. (From a class project, University of Wisconsin-Madison.)

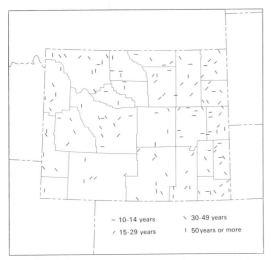

FIGURE 9.4 *Nominally scaled symbols are used to indicate four classes of climatic stations with nonrecording gauges in Wyoming. The classes are delimited on a higher scale of measurement, but the depiction itself is nominal until the legend information is used.*

to err by using pictorial symbols that are so similar that they are hard to distinguish. Yet examples do occur. Note the pictorial symbols shown in Fig. 9.5. Such pictorial symbolization schemes could ultimately lead to a dictionary of standardized symbols that would "handcuff" the cartographer's freedom for design and thus, in many instances, greatly retard the speed of communication of the map information.

The principal concern when employing pictorial representation for nominally scaled positional data is to use shape, hue and, to a lesser extent, orientation and direction to make the symbols easily distinctive and identifiable.

Associative Symbols. Figure 9.6 illustrates some associative symbols for nominally scaled positional data. Symbols in this class employ a combination of geometric and pictorial characteristics to produce easily identifiable symbols. As with pictorial symbols, the most powerful dimensions are shape and hue. A legend is highly recommended for a map that employs associative symbols, since they may be quite diagrammatic compared to pictorial symbols.

Geometric Symbols. The most common geometric symbols used to show nominally scaled positional data are circles, triangles, squares, diamonds, and stars; the dimension of shape obviously is the most important factor. Hue can be used as a secondary dimension to differentiate the data, especially when some of the data classes are related. A legend is *required* on a map that employs geometric symbols. The major problems to be overcome in utilizing geometric symbols are (1) to ensure that each symbol is sufficiently distinctive, and (2) to make sure that the symbols do not connote higher than nominally scaled information. Figure 9.4 illustrates the use of a geometric symbol.

The categories (pictorial, associative, and geometric) can be thought of as occupying successive positions on a continuum of symbolization ranging from the analogic or mimetic at one end to the purely arbitrary at the other. The cartographer must be aware of where the selected symbolization lies on the continuum, its associated problems, and the appropriate corresponding need for providing a legend.

MAPPING QUALITATIVE LINEAR DATA

A surprising amount of qualitative linear information appears on most maps, but often neither the cartographer nor the map reader is fully aware of it. The information includes the outline itself, which defines the area of concern, the graticule definition, coastlines, political boundaries, rivers, roads, and so on; often the cartographer's design ingenuity is taxed to the limit by the range of nominally scaled linear features. To provide sufficient contrast is often difficult, and many maps are less than satisfactory because of shortcomings in the qualitative line work. When symbolizing qualitative linear information the cartographer must work primarily with the character or the type of line. True, line width may also be varied, but only to the extent that this does not connote ordinality. This usually demands that varying line widths of the same type can be used only for widely divergent features.

Line type can be varied in many ways. Figure 9.7 illustrates lines of differing type that can be used for the symbolization of qualitative data. De-

⊕	BEAVER	▦	OLD FORT
⊕	MUSKRAT	▣	BATTLEFIELD
⊕	MINK	▦	BARRACKS
⊕	OTTER	▦	MILL POND
⊕	HERON	⌂	OLD MILL
⊕	CRANE	⌂	BLACKSMITH SHOP
⊕	SWAN	✝	PIONEER CHURCH
⊕	LOON	▦	ONE-ROOM SCHOOL
⊕	EAGLE	▦	GENERAL STORE
⊕	OSPREY	▦	TAVERN, SALOON
⊕	FALCON	▦	OUTSTANDING BUILDING
⊕	RED-TAILED HAWK	▪	COUNTY SEAT
⊕	GOSHAWK	⌂	HISTORICAL HOME
⊕	GREAT HORNED OWL	▦	NATIVE HANDICRAFT (CRAFTSMAN'S SHOP)
⊕	RUFFED GROUSE	✚	OLD CEMETERY
⊕	PHEASANT	▦	GHOST TOWN
⊕	HUNGARIAN PARTRIDGE	▦	HISTORICAL FOLK LORE
⊕	WOODCOCK	▦	SUGAR BUSH
⊕	GEESE	▦	UNUSUAL CROP
⊕	DUCK	▦	UNUSUAL FARM STRUCTURE
⊕	QUAIL	▦	LOG HOME -BARN—UNOCCUPIED
⊕	STURGEON	▦	FRAME STRUCTURE PRIOR TO 1860
⊕	WALLEYE		ASSOCIATED WITH FAMOUS PERSON OR EVENT
⊕	(NORTHERN) PIKE	▦	STONE STRUCTURE PRIOR TO 1860 ETC.
⊕	CATFISH	◑	OUTSTANDING FARMERS MARKET
⊕	BASS	●	OUTSTANDING SOIL CONSERVATION PROJECT,

FIGURE 9.5 Detailed pictorial point symbols. (From the legend of a map entitled Landscape Resource Inventory, Department of Resource Development, State of Wisconsin, May, 1964.)

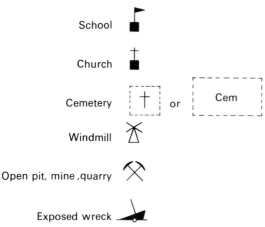

School

Church

Cemetery ... or ... Cem

Windmill

Open pit, mine ,quarry

Exposed wreck

FIGURE 9.6 *Associative symbols for nominally scaled positional data used by the United States Geological Survey on topographic maps.*

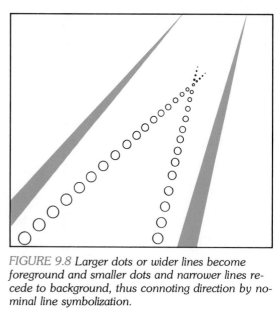

FIGURE 9.8 *Larger dots or wider lines become foreground and smaller dots and narrower lines recede to background, thus connoting direction by nominal line symbolization.*

pending on the cartographic situation at hand, the cartographer may wish either to select lines of varying type that will appear essentially equal in importance, or to use qualitative line symbolization to establish relative visual importance by the selection of line type.

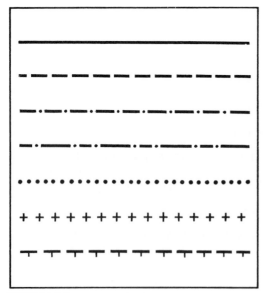

FIGURE 9.7 *Examples of differing line types that are useful for the symbolization of qualitative linear data.*

Several variables can affect the appearance of line symbols. These factors are: (1) size, (2) continuity, (3) brightness contrast, (4) closure, (5) complexity, and (6) compactness.*

Size variation in qualitative symbolization has at least two consequences. First, the larger of two lines of the same color will appear more prominent. Second, direction can be symbolized in two ways by varying size along a single line. For example, in the dotted lines in Fig. 9.8, the larger dots in the foreground trailing to the smaller dots cause the smaller dots to recede into the background; this connotes direction. Also, varying the width of the solid line in Fig. 9.9, as is often done in symbolizing rivers, creates the effect of the direction of flow of a stream (it is presumed that in humid areas a river becomes larger as it flows from its source to its mouth).

Continuity is an important quality for the line symbolization of nominal data. It is illustrated in Fig. 9.7. Continuity can vary from a solid line to a series of essentially point symbols that depend on

*Karen Pearson, "The Relative Visual Importance of Selected Line Symbols," unpublished Master's thesis, Department of Geography, University of Wisconsin, Madison, 1971.

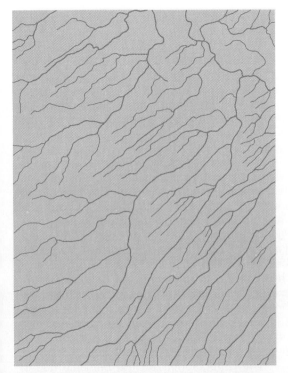

FIGURE 9.9 Example of a stream network where direction of flow is connoted by increased width of lines. (From Fundamentals of Physical Geography by Trewartha, et al. Copyright © 1977 McGraw-Hill Book Co. Used with permission of McGraw-Hill Book Company.)

that a dotted line is judged less important than a solid or dashed line of equivalent width.

Brightness contrast affects the perception of line symbols by the use of color value. Lines of equivalent width but of contrasting brightness may be perceived as varying in importance. The darker tones usually connote "more" importance.

Visual closure occurs among groups of similar marks; therefore a series of dashes or dots will be viewed as a whole. The more closure a symbol has, the more visually prominent it is likely to be (see Fig. 9.7); hence, a solid line should be the most prominent.

Complexity is a term with many meanings. In this context it refers to the sequence of marks that compose the line symbol. All things being equal, the more complex a line, the more visually significant it will be. Figure 9.10 provides one example of increasing line complexity.

Compactness, another visual variable in qualitative line symbolization, refers to the regularity of the symbol. The more compact linear symbol will possess greater visual weight, and a solid line is the most compact. It may only be possible to grade other levels of compactness by referring each line independently to a solid line.

These six variables allow the cartographer to employ lines of differing types that have either equal visual importance or that form a hierarchy of visual importance consistent with the objective

proximity to establish closure and create a line or figure. Linear features occurring on the earth (e.g., roads, railroads, rivers, coastlines) tend to be shown with unbroken or solid lines. Imaginary features (e.g., political boundaries, the graticule, or coordinate system lines) may be shown with broken lines. This is only convention, and the convention is not strong except for perhaps roads and railroads. Often lines are broken to make room for other data. It would be relatively easy to find a broken coastline or power transmission line on a map both of which are continuous linear features that occur on the earth. Likewise, the graticule may be shown by a solid line, especially over water; the division noted above is probably only true for a majority of maps. The relative perceptual merits of continuity are less well known, but there is some evidence to support the statement

solid line	————————————————
dashed long	—— —— —— —— ——
dashed short	— — — — — — — —
dot dash long	· ——— · ——— · ——— · ——— ·
dot dash short	· —— · —— · —— · —— · —— · —
dot	····················
crosses	+ + + + + + + + + + + + + + + +

X+ interspersed X+X+X+X+X+X+X+X+X+X+X+X+X+X+X+X

FIGURE 9.10 Increasing complexity in the marks that compose a line symbol. The solid line is the simplest and the + and X line is the most complex.

of the map. Practice is the best teacher in the selection of which characteristic to vary in nominal scale line symbolization. The cartographer, being aware of the visual effects resulting from the above characteristics of line symbols, should be in a position to employ lines suitably for any cartographic situation.

MAPPING QUALITATIVE AREAL DATA

For mapping a set of data that is measured on a nominal scale and that refers to areas, the cartographer may use either pattern or hue for differentiation. Forests, grasslands, national areas, culture regions—all the ways in which the character of one geographical region can be distinguished from another—are accomplished by utilizing distinctive area symbols.

Mapping such data is relatively straightforward, but some problems are involved. Frequently, nominal categories are not geographically exclusive; that is, two or more categories often occur in the same area, such as botanical species, races, or land use characteristics. Accordingly, if the cartographer does not want to generalize sufficiently to remove the mixture, symbolization that somehow represents the mixture or overlap must be employed. This may be done in a number of ways, as suggested in Fig. 9.11, none of which is suited to all circumstances. If the cartographer

is not careful, this practice can lead to incredibly complicated maps.

If color is being employed, it is possible to choose hues that give the impression of mixture. For example, a red and a blue appear purple when superimposed, and that hue looks like a mixture. On the other hand, a hue produced by mixing blue and yellow produces green, and that hue does not look like a mixture of its components.

Occasionally a cartographer may generalize a geographic volume such that only nominal information will appear on the map. For example, air masses or kinds of precipitation (snow versus rain) may be shown only differentiated by kind, even though in reality an air mass occupies a volume (see Fig. 9.12). Since no quantitative measure is being portrayed, all that is being communicated is the location of the phenomenon, or nominal information. This is identical to the reference map symbolization of a dot for a city or a line for a river. Conversely, any time a quantity associated with an area is portrayed, a volume is present, and it requires quantitative symbolization.

The cartographer can use primarily hue or pattern for qualitative area symbolization. Care must be exercised so as not to connote ordinal relationships. It is well known that darker and lighter tones and, to a lesser extent, color intensities do connote ordinality. Darker tones usually refer to "more" and lighter tones to "less" of a given phe-

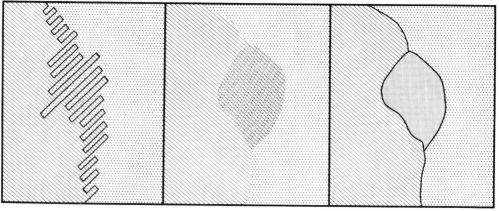

| INTERDIGITATION | OVERLAP OF SYMBOLS | AREA OF MIXTURE SYMBOLIZED |

FIGURE 9.11 Several methods of showing geographical mixture or overlap with area symbols.

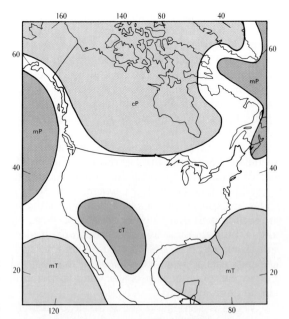

FIGURE 9.12 Portrayal of North American air masses and their source regions. Although data have quantitative characteristics, the intent of this illustration is simply to portray location of air masses. This can be accomplished by using nominal areal symbolization. (From Fundamentals of Physical Geography by Trewartha, et al. Copyright © 1977 McGraw-Hill Book Co. Used with permission of McGraw-Hill Book Company.)

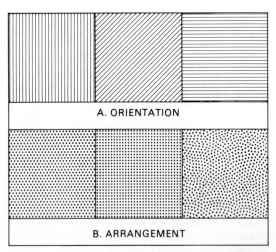

FIGURE 9.13 (A) A single area pattern has been oriented at three different angles to produce three visually different nominal area patterns. (B) The arrangement of the dots that compose the area pattern is varied, again producing three visually different nominal area patterns.

nomenon. Likewise higher intensities usually connote "more" and weaker intensities connote "less." Only hue is adequate in portraying nominal data, since hue does not connote ordinality. Blue is not considered "more or less" than red, or green, and so on.

When employing patterns, the cartographer may differentiate by varying both orientation and arrangement. Figure 9.13 shows differentiation in pattern orientation and pattern arrangement for nominal area data. Care must be exercised not to create a visually complex display that causes excessive eye movement for the reader (see Fig. 14.18).

Certain disciplines have created nominal area data pattern conventions for mapping purposes. Geology has specified the use of particular patterns for certain rock types. The cartographer has

STRUCTURE CONTROLLED LANDFORMS		LANDFORMS OF DEPOSITION	
	sandstone	blocks	
	quartzite	gravel, shingles	
	limestone	sand	
	dolomite	silt, mud	
	marl	clay	
	extrusive rocks	salt	
	crystalline rocks	gypsum	
	unconsolidated rocks	coral detritus	

FIGURE 9.14 Some standardized symbols for indicating lithologic data as suggested by the International Geographical Union Commission on Applied Geomorphology.

no right to violate these conventions. Figure 9.14 illustrates some of the patterns used by geologists for nominal area differentiation.*

Sometimes it is necessary to map nominally a distribution through two levels of its classification. This is referred to as subdivisional organization (see Chapter 12). Preference could be to use hue to distinguish the major categories and to differentiate subclasses within categories by the use of pattern. Hue tends easily to show the major breakdown of the distribution into its first-order classes. Pattern is more subtle and thus lends itself more readily to differentiation at the second order of classification, where the map reader is searching for more detailed information. It is easier to juxtapose patterns that are similar but that still retain their separate identities than it is hues.

MAPPING THE LANDFORM

One distribution that the cartographer is frequently called on to map is the land surface form. This distribution is unique in that it is visible in reality to all viewers, it is relatively unchanging, and it affects human life in many ways, ranging from people's mobility to the weather. Various aspects of the land surface form may be represented by point, line, or area symbols or by combinations of them. Because of its importance to humans, this unique distribution deserves special treatment by cartographers. A thorough understanding of the mapping of the land surface form therefore rightly begins with a brief history of its mapping. Some of the methods used are equally applicable to the mapping of other phenomena, and the mechanics of these techniques will be discussed in the appropriate sections of Chapters 10 and 11. The rich and sometimes controversial history of the mapping of the land surface form is in itself an interesting story for cartographers.

The representations of the X, Y, and Z dimensions of the land surface and their derivatives has, no doubt, always been of special concern to cartographers, but the earliest maps, and even those

of the Middle Ages, showed little of it. This was probably because of the paucity of knowledge about landforms. To be sure, mountains may have been shown as piles of crags, and ranges appeared as "so many sugar loaves;" but until precise elevational and positional data became available on which the cartographer could base the representation, the landform could not be well represented.

It should be noted that, although this section is entitled "Mapping the Landform," the principles of landform representation apply equally to the solid surface beneath the waters of the earth. There are special problems of data gathering in connection with mapping underwater surfaces, of course, but otherwise there is no basic difference. The discussion here will be focused on the landform, since its appearance is relatively familiar and the graphic potentialities of the various methods can be easily grasped.

There is something about the three-dimensional surface that intrigues cartographers and sets it a little apart from other cartographic symbolization. First, it requires more understanding and skill. Moreover, the land surface is a continuous phenomenon; that is, all portions of the solid earth necessarily have a three-dimensional form and, as soon as the land is represented, *all* of it must be represented, at least by implication. Also, the landform is the one major phenomenon with which the cartographer works that exists as a graphic impression in the minds of most map readers, and the map reader is therefore likely to be consciously or unconsciously critical of its graphic representation on the map.

Because of the relative importance to human beings of minor landforms, the representation of landforms together with other data has always been a great problem to the cartographer. If the surface is shown in sufficient detail to satisfy the local significance of individual landforms, then the problem arises of how to present the other map data. On the other hand, if the cartographer shows with relative thoroughness the nonlandform data, which may be more important to the specific objectives of the map, maybe only a mere suggestion of the land surface can be shown, an expedient not likely to please the mapmaker or the map reader. Furthermore, the development of aviation during this century has made the effective

*IGU Commission on Applied Geomorphology, Subcommission on Geomorphological Mapping, Project of the Unified Key to the Detailed Geomorphological Map of the World, *Folio Geographica Series Geographica-Physica*, Vol. II, Krakow, 1968.

and precise representation of the terrain more and more important. The pilot must be able to recognize the area beneath the plane, and the passenger has naturally become more interested in the general nature of that surface, since it can now be seen. The height above the terrain provides a reduction mindful of a map.

Perhaps an even more difficult task has been the depicting of landforms on smaller-scale maps. Small-scale landform representation is a major problem for atlas maps, wall maps, and other general-purpose reference maps, as well as for those special-purpose maps in which regional terrain is an important element of base data. The smaller scale requires considerable generalization of the landform, which is no simple task, as well as the balancing of the surface representation with the other map data, so that neither overshadows the other. No less a creative task is the represen-

tation, in bolder strokes, of the landforms for wall maps, so that important elements such as major regional slopes, elevations, or degrees of dissection are clearly visible from a distance (see Fig. 9.15). The specialized techniques for scientific terrain appreciation and analysis are reserved for the geographer and the geomorphologist.

For many years to come the representation of the landform on maps will be an interesting and challenging problem, since it is unlikely that convention, tradition, or the paralysis of standardization will take any great hold on this aspect of cartographic symbolization. This will probably be particularly true of terrain representation on special-purpose and thematic maps; each such attempt will be a new challenge, since in each case it must be fitted to the special, overall objective of the map.

The Historical Background

The story of the development of landform representation is a recital of the search for methods suitable to a variety of purposes and scales. On large-scale maps the desirable symbolization is one that both appears natural and is capable of exact measurement of elements of the land surface such as slope, altitude, volume, and shape. The major problem arises from the fact that, generally, the most effective visual technique is the least commensurable, whereas the most commensurable is the least effective visually. One of the major decisions of every survey organization has been in what manner to balance these opposing conditions. Although it is somewhat early as yet to judge, there is some indication that advances in color printing have enabled the cartographer to reach a relatively effective combination of techniques, without undue sacrifice of either desirable end. The newer, shaded relief, contour maps of the U.S. Geological Survey are a case in point (see Fig. 1.3).

The earlier representations of terrain on smaller-scale maps were concerned mainly with undulations of some magnitude and consisted of crude stylized drawings of hills and mountains as they might be seen from the side, such as those depicted in Fig. 9.16. The perspective-like, oblique, or bird's-eye view became more sophisticated in the period from the fifteenth to the eighteenth centuries, when the delineation of terrain

FIGURE 9.15 A much reduced monochrome print of a portion of a modern color wall map emphasizing surface. The detailed terrain is derived from photographing a carefully made, three-dimensional model. The map is reproduced by complex color printing analogous to process color. (Map by Wenschow, courtesy Denoyer-Geppert Company.)

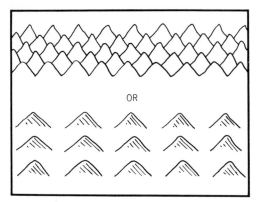

FIGURE 9.16 *Crude, early symbolization of hills and mountains.*

developed along with the landscape painting of the period. The eighteenth century was the period when the great topographic surveys of Europe were initiated and, for the first time, mapmakers had factual data with which to work, albeit the first attempts were crude indeed. Delineation of the form of the land shifted from an employment of the oblique view to the plan view, perhaps because of the availability for the first time of extensive planimetric data. Symbols with which to present the form of the land without planimetric displacement were developed.

Large-Scale Representation. In 1799 the systematic use of a linear symbol called the *hachure* was developed by Johann Georg Lehmann, an Austrian army officer. As advocated by Lehmann, each individual hachure, rather like a vector symbol, is a line of varying width that follows the direction of greatest slope: that is, its planimetric position on the map would be normal to the positions of contours. By varying the widths of the hachures according to the angle of the slopes on which they lie, the magnitudes of the slopes (or steepness) may be indicated. The sense, or the direction of down (or up), is, however, not shown. When many hachures are drawn closely together, they collectively show the forms of the surface configuration. This turned out to be particularly useful on the then recently initiated, large-scale, topographic military maps and, for nearly a century, the hachure was widely employed.

Many varieties of hachuring were employed.

Instead of varying the width of the line according to the slope, some used uniform-width strokes and then increased their density with increased slope. Hachuring provides an illusion of shading, and the light source can be assumed to be orthogonal to the map (Fig. 9.17) or from one side (Fig. 9.18). On small-scale maps or maps of poorly known areas, hachuring easily degenerated into "hairy caterpillars," an example of which is shown in Fig. 9.19. These worms are not yet extinct.

From the earliest use in the sixteenth century of lines of equal depth of water (isobaths), Dutch, and later French, engineers and cartographers introduced the use of these kinds of lines to portray the underwater and dry land configuration. By the early nineteenth century the contour was widely employed, but it was not until relatively late in the nineteenth century that contours became a common method of depicting the terrain on survey maps. One development, which grew out of the use of contours on large-scale maps, was their extreme generalization on small-scale maps, resulting in the familiar "relief map," together with

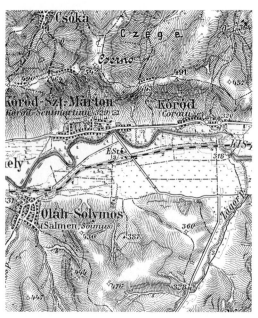

FIGURE 9.17 *A section of a hachured topographic map with vertical illumination. (From the Austria-Hungary, 1:75,000 topographic map series.)*

FIGURE 9.18 *A section of the Dufour map of Switzerland in which obliquely lighted hachure shading is employed. (From Sheet 19 (1858), Switzerland, 1:100,000.)*

"layer coloring" according to altitude. The colors, of course, serve as area symbols between isarithms.

After the development of lithography in the early part of the nineteenth century, it became possible to produce easily continuous tonal variation or shading. It was not until late in the last century that shading as a type of area symbol was widely utilized for the representation of the terrain, the shading applied being a function of the slope.

By the beginning of the twentieth century, the basic methods of presenting terrain on large-scale topographic maps (contouring, hachuring, and shading) had been attempted, and the essential incompatibility between commensurability on the one hand and visual effectiveness on the other was readily apparent. For the last several decades the problem has been one of how to combine the techniques to achieve both ends. The newer topographic maps are the most effective yet produced.

Small-scale Representation. The representation of the land surface at large scales is concerned essentially with the three major elements

of configuration: the slope, the height, and the shape of the surface formed by the combinations of elevations and gradients. The various methods outlined above, and their combinations and derivatives, seem to provide the answer, more or less, for the problem at large scales; but the representation of land surface at smaller scales is quite another matter. Here the generalization required is so great that only the higher orders of form, elevation, and slope may be presented. As knowledge of the land surface of the earth has grown, so also has the need for a variety of methods of presenting effectively that surface at smaller scales.

No one method can satisfy all the small-scale requirements for land surface delineation. Consequently, with the growth of reproduction techniques and drawing media, the variety of the ways of depicting the land surface is steadily increasing, and it may be expected to continue to do so for some time. Some of the older methods are beginning to be discarded because of their obvious deficiencies.

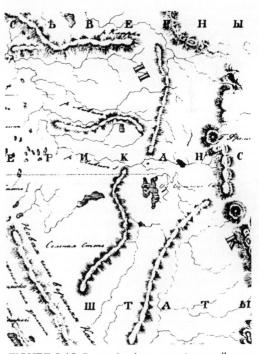

FIGURE 9.19 *Genus hachure, species woolly worm. (From an old Russian atlas of western North America.)*

FIGURE 9.20 *A small perspective map made with a combination of line and shading on Coquille board. Reproduced by line cut. (From A. K. Lobeck,* Things Maps Don't Tell Us, *The Macmillan Company, New York, 1956.)*

Layer coloring, with or without spectrally organized gradations between selected and generalized contours, is one of the earliest devised techniques. It came into wide use during the latter part of the nineteenth century, and it has been more commonly employed than any other technique, mainly because of its relative simplicity. It leaves much to be desired. The character of the surface is presented only by implications of elevation; the generalized contours show little except regional elevations, which are ordinarily not very significant; and the problems of color gradation and multiple printing plates are sometimes difficult. Hachures were quickly employed at small scales after their introduction but, as was observed, they decline easily to the woolly worm. Shading can be very effective at any scale, but it requires unusual skill and, in monochrome, cannot be easily combined with other map information. Without expensive multiple-color plates, this technique, when employed on a map with much other data,

often becomes little more than an uneven background tone, which serves largely to reduce the visibility of the other map data.

In lieu of simple layer tinting, hachuring, and shading at small scales, many other techniques have been tried in more recent years to gain realistic effects. One of the more effective devices is that of drawing the terrain pictorially, as a kind of bird's-eye view with a slight perspective, as in Figs. 9.20 and 9.21. This sort of delineation requires a knowledge of landforms, considerable practice, and at least some manual skills. The cartographer who draws terrain must (1) know what is to be drawn, and (2) have the skill to interpret graphically the three-dimensional relationships to be conveyed on the map. The first requires training in landform analysis, geomorphology, and structural geology; the second requires more training in the visual arts than most cartographers receive. There is practically no end to the combinations of perspective viewpoint, coloring, highlighting, shadowing, and line drawing that can be employed.

Instead of attempting to obtain realistic visual communication of the configuration of the land

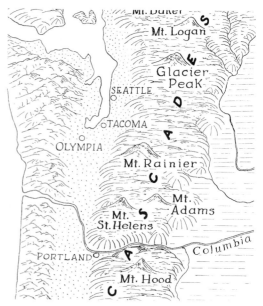

FIGURE 9.21 *A portion of a small perspective map made entirely with line work. (From A. K. Lobeck,* Things Maps Don't Tell Us, *The Macmillan Company, New York, 1956.)*

LAND-SURFACE FORMS

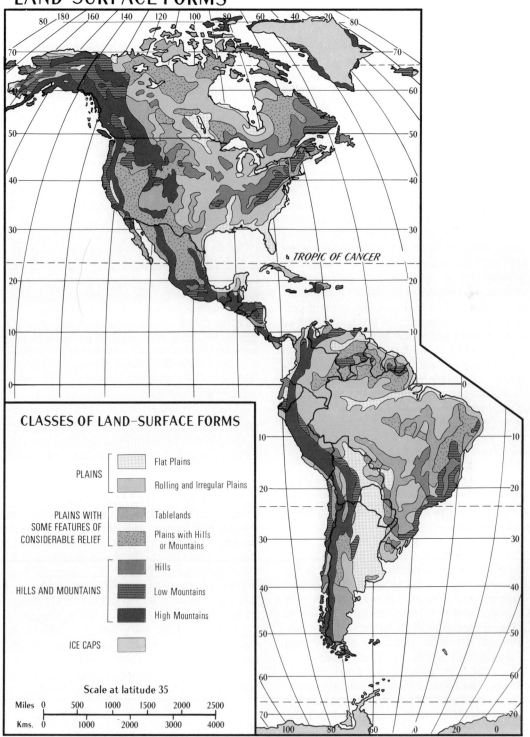

CLASSES OF LAND-SURFACE FORMS

PLAINS
- Flat Plains
- Rolling and Irregular Plains

PLAINS WITH SOME FEATURES OF CONSIDERABLE RELIEF
- Tablelands
- Plains with Hills or Mountains

HILLS AND MOUNTAINS
- Hills
- Low Mountains
- High Mountains

ICE CAPS

Scale at latitude 35

Miles 0   500   1000   1500   2000   2500
Kms. 0   1000   2000   3000   4000

TROPIC OF CANCER

FIGURE 9.22 A nominally scaled portrayal of classes of land-surface form. The data used for this map was ratio scale data, but the cartographer chose to generalize it to nominal scale for mapping. (From Fundamentals of Physical Geography by Trewartha, et al. Copyright © 1977 McGraw-Hill Book Co. Used with permission of McGraw-Hill Book Company.)

surface, the essential characteristics of the forms may be differentiated according to some scaling system. Using nominal scaling for presentation, this may take the form of classes such as flat plains, tablelands, and low mountains (see Fig. 9.22). These categories obviously must be defined by variables measured on a higher scale of measurement than a nominal scale; however, the cartographer elects to communicate the differences on a nominal scale. Areas so outlined may then be shown with area symbols, usually colors. The widespread use of this system is relatively new.

Many other systems of landform representation for special purposes have been tried or suggested. Most of them are relatively complex and intellectually involved. Their use is limited to the professional geographer and geomorphologist, whose knowledge of landforms is sufficient to interpret them. The methods that present certain aspects of the geometry of the land, such as the average-slope technique or the relative-relief technique, would fall into this category.

A thorough understanding of the small-scale possibilities for delineating terrain requires familiarity with the fundamentals of the basic methods (contours, shading, and hachures) used to show terrain on topographic maps.

Qualitative Representation of the Land Surface Form

The majority of the methods mentioned in the preceding section represent quantitative symbolization problems; however, a few require qualitative symbolization.

Landform Diagram. One of the more distinctive contributions of American cartography is the pictorial map in which the terrain is represented schematically on an ordinary map base. Although this type of map is not limited to cartographers in the United States, it reached its highest development through the efforts of A. K. Lobeck, Erwin Raisz, Guy-Harold Smith, and a few others.

Schematic maps in which the pictorial treatment of the landform is treated more systematically are usually called physiographic diagrams or landform maps. The former attempts to relate the forms to their origin. The physiographic diagrams of A. K. Lobeck are of this type. In these, by varying darkness and textures, the major structu-

FIGURE 9.23 *A portion of the small-scale Physiographic Diagram of the United States, by A. K. Lobeck. Note the schematic treatment of the surface. (Courtesy of the Geographical Press, formerly at Columbia University Press, now at C. S. Hammond and Company, Inc.)*

ral and rock-type differences that have expression in the surface forms are suggested. They do not have a particularly realistic appearance, and their common name, "physiographic diagram," is appropriate (see Fig. 9.23).

Landform or land-type maps are those in which more emphasis is placed on the character of the surface forms, with less attention to their genesis. This type of map is exemplified by the maps of Erwin Raisz, who developed a set of schematic symbols with which to represent various classes of the varieties of landforms and land types (see Fig. 9.24).* There is, of course, no sharp distinction between the physiographic diagram and the landform map. All possible combinations of attention may be paid to the underlying structures, rock types, and geomorphic processes.

Whatever the emphasis may be on such maps, the landforms are positioned without per-

*See E. Raisz, "The Physiographic Method of Representing Scenery on Maps," The Geographical Review, 21, 297–304 (1931); E. Raisz, General Cartography, 2nd ed., McGraw-Hill, New York, 1948, pp. 120–121.

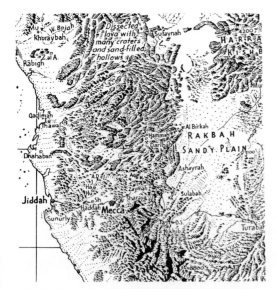

FIGURE 9.24 A portion of a small-scale landform map. Compare this with Fig. 9.23. Note the inclusion of descriptive terms. (From E. Raisz, map entitled Landforms of Arabia 1:3,600,000.)

spective, but the symbols themselves are derived from their oblique appearance. All physiographic or landform maps have one major defect in common: the side view of a landform having a vertical dimension requires horizontal space, and on a map horizontal space is reserved for planimetric position. For example, if a single mountain is drawn on a map as seen from the side or in perspective, the peak or base and most or all of the profile will be in the wrong place planimetrically. This is well illustrated in Fig. 9.23. This fundamental defect in physiographic diagrams and landform maps has, of course, been recognized by the cartographers who draw them, but it has, properly, been justified on the ground that the realistic appearance obtained more than outweighs the disadvantages of planimetric displacement. This is especially true on small-scale drawings in which the planimetric displacement is not bothersome to the reader and only occasionally causes the cartographer concern, for example, when some feature of significance is "behind" a higher area. At larger scales, however, the conflict of the perspective view and the consequent error of planimetric position make it desirable to adopt other techniques.

Terrain Unit or Descriptive Landform Category Method. The terrain unit method employs descriptive terms that range from the simple "mountains," "hills," or "plains" designations to complex, structural, topographic descriptions such as "maturely dissected hill land, developed on gently tilted sediments." This is essentially a modified dasymetric system, and the lines bounding the area symbols have no meaning other than being zones of change from one kind of area to another. This method of presenting landforms has been found useful in textbooks and in regional descriptions for a variety of purposes, ranging from military terrain analysis to regional planning. Its basic limitations are the regional knowledge of the maker and the geographical competence of the map reader.

In many respects it is difficult to separate land surface form representation techniques into qualitative or quantitative categories. The division in this textbook is based on the scale of measurement of the information that the map reader can decode from the map. Obviously, land surface forms possess many qualitative and quantitative characteristics; however, in the preceding categories of representation, only qualitative or descriptive characteristics are involved. In the next chapters methods for mapping quantitative characteristics will be discussed under point, line, and area symbolization.

SELECTED REFERENCES

Baldock, E. D., "Cartographic Relief Portrayal," *International Yearbook of Cartography, 11,* 75–78 (1971).

Curran, J. P., "Cartographic Relief Portrayal," *The Cartographer, 14,* 28–37 (1967).

Head, C. G., "Land-Water Differentiation in Black and White Cartography," *The Canadian Cartographer, 9,* 25–38 (1972).

Irwin, D., "The Historical Development of Terrain Representation in American Cartography," *International Yearbook of Cartography, 16,* 70–83 (1976).

Lobeck, A. K., *Block Diagrams,* 2nd ed., Emerson-Trussell Book Company, Amherst, MA, 1958.

Robinson, A. H., "Scaling Non-numerical Map Symbols," *Technical Papers from the 30th Annual Meeting, American Congress on Surveying and Mapping,* Washington D.C. pp. 210–216 (1970).

Sherman, J. C., "Terrain Representation and Map Function," *International Yearbook of Cartography, 4,* 20–23 (1964).

Sinnhuber, K. A., "The Representation of Disputed Political Boundaries in General Atlases," *The Cartographic Journal, 1,* 20–28 (1964).

Toth, T. G., "Terrain Representation—Past and Present—at the National Geographic Society," *Proceedings of the Fall Convention, American Congress on Surveying and Mapping* Lake Buena Vista, FL, pp. 9–31 (1973).

There is no limit to the variety of phenomena that a cartographer might be called on to map. Each phenomenon exists and can be measured on some measurement scale; however, the cartographer must decide on which scale of measurement to map it. Likewise, each phenomenon exists as positional, linear, areal, or volumetric data. The cartographer must decide how it is to be mapped.

The cartographer can generalize ratio data to interval, ordinal, or nominal scales because these scales of measurement are nested. Similarly, volumetric data can occur at points, along lines, or over areas. However, a cartographer sometimes maps positional, linear, or areal data by symbols other than point, line or area symbols, respectively. Two examples illustrate this. A road is usually thought of as a linear phenomenon, but, at a very large scale and for some purposes, the fact that its surface character is different from its surroundings may be the important matter as, for example, in the study of variations in albedo. Also, a city is frequently thought of as a geographical point phenomenon, but a given array of cities in a region can be considered as distinguishing that area from another with a different frequency of city incidence. The fundamental point is that we must first determine the attribute that is to be given the emphasis before the symbolization system can be chosen. When approached from this point of view, the cartographic representation of quantitative information is not difficult.

In this chapter and in Chapter 11 we will concentrate on the mapping of quantitative data that occur at points, along lines or over areas. Such data are measured on an ordinal or higher scale of

CHAPTER 10

SYMBOLIZATION: QUANTITATIVE DATA BY POINT SYMBOLS

measurement. The organization of this chapter is mapping with point symbols. Chapter 11 treats mapping with line symbols and area symbols. It does not matter how the phenomena actually exist or are collected. Therefore, the dot map is discussed in Chapter 10 under point symbols, even though it may symbolize data collected over an area. The dasymetric map will be discussed in Chapter 11 under areal symbols, even though it may be portraying point phenomena.

MAPPING QUANTITATIVE DATA WITH POINT SYMBOLS

Quantitative data can be symbolized by point symbols in a great variety of ways. Any real or conceptual quantity that can be thought of as existing in variable amounts from place to place can be symbolized simply by assigning a unit value to some point symbol and then putting the right number of these same-sized symbols in the right places on the map. This technique results in what is generally called a dot map, because simple round dots are the most frequently used symbol; however, any point symbol can be used. The kinds of quantities that can be portrayed in this way can range from slope values, to people, and even to percentages. Differently colored or shaped dots (or other marks) can be used to show geographical mixtures; but, fundamentally, all we are doing is repeating a chosen symbol to portray the geographical frequency.

A second system that utilizes point symbols to summarize the quantities that apply to points, lines, or areas is the system that symbolizes the variations in the quantities by variations in the sizes of a chosen symbol, such as a square or a circle. This is generally called a graduated circle (or star, or square, etc.) map. Details can be added by segmenting each symbol to show proportions, and a variety of classes of data can be included on one map by varying the symbol form.

The Dot Map

The simplest of all the maps that use point symbols is the one wherein the data are presented by varying numbers of uniform dots, each representing the same amount. It is possible, of course, to substitute little drawings of men, or sheep, or cows (or whatever is being represented) for the simple dot. This generally reduces the amount of detail that may be presented, but it is sometimes desirable for rough distributions for maps for children.

The dot map is capable of showing more clearly than any other type the details of locational character. Variations in pattern or arrangement, such as linearity and clustering, become readily apparent. The dot map provides an easily understood visual impression of relative density, readily accepted by the reader and easily interpreted on an ordinal scale, but it does not provide any absolute figures.[*]

A second advantage is the relative ease with which such maps may be made manually. No computation is ordinarily necessary beyond that of determining the number of dots required, which merely necessitates dividing the totals for each enumeration unit by the number decided on as the unit value of each dot.

Although the usual map reader does not believe he or she experiences particular difficulty in interpreting the dot map, recent research indicates that it is not as straightforward a kind of symbolization as many have thought. It seems especially difficult for the untrained reader to estimate relative visual densities on an interval scale with much success, although ordinal decisions appear not to be difficult. The latter is probably the most important function to be performed by the dot map.

It may be an important part of the cartographer's objective to have the reader gain considerable information regarding the various densities, relative as well as absolute, from the map. If so, the reader probably will need help. Most current dot maps stop with simply a legend that tells the reader the magnitude of the unit value of the dot used. It has been suggested that both design changes and reader training are needed to ensure that the reader gains the necessary information from the dot map.[†]

---

[*]*Theoretically, it would be possible to count the dots and then multiply the number by the unit value of each dot to arrive at a total but, in practice, it would be done only under duress.*
[†]*Judy M. Olson, "Experience and the Improvement of Cartographic Communication," The Cartographic Journal, 15 (2), 94–108 (December 1975). Some suggested design changes include adding additional dots in dense areas to aid the reader in value determination.*

Dot maps ordinarily show only one kind of fact, for example, population or hectares of cultivated land but, by using differently colored dots or differently shaped point symbols, it is sometimes possible to include several different distributions on the same map.*

Making the Conventional Dot Map

Three important considerations affect the basic utility of a dot map: (1) the size of the dots, (2) the value assigned to a dot, and (3) the location of the dots.

The data needed for a conventional dot map consists of the enumeration of the number of items to be mapped in unit areas, usually civil divisions used as census statistical units. Rarely is it possible to employ so large a map scale that each single item may be shown by a dot, although theoretically the farmhouses on a topographic map might be termed a kind of dot map; instead, it is usually necessary to assign a number of the phenomena to each dot. This is called the unit value of the dot. The number of dots is obtained merely by dividing the total number of items in each statistical division by the chosen unit value. For example, if a county had a total of 6000 hectares in corn, and a unit value of 25 hectares per dot had been chosen, then 240 dots would be placed in the county to symbolize the area devoted to corn production.

The Size and Value of the Dot. If the visual impression conveyed by a dot map is to be realistic, the size of the dot and the unit value assigned must be carefully chosen. The five dot maps shown here have been prepared from the same data; only the size or number of dots used has been changed. The maps show areas of potato production in Wisconsin.

If the dots are too small, as in Fig. 10.1, the

FIGURE 10.1 A dot map in which the dots are so small that an unrevealing map is produced. Each dot represents 16.2 hectares in potatoes in 1947.

distribution will appear sparse and insignificant, and patterns will not be visible. If the dots are too large, they will coalesce too much in the darker areas, as in Fig. 10.2, and give an overall impression of excessive density that is equally erroneous. It appears in Fig. 10.2 that there is little room for anything else in the region. Furthermore, when dots are gross, they dominate the base data and generally result in an ugly map.

Equally important is the selection of the unit value of the dot; naturally, the two problems (size and value of the dot) are inseparable. The total number of dots should neither be so large that the map gives a greater impression of accuracy than is warranted nor should the total be so small that the distribution lacks any pattern or character. These unfortunate possibilities are illustrated in Figs. 10.3 and 10.4.

The selection of unit value and size of dots should be made so that in the denser areas (of a dense distribution) the dots will just coalesce to form a dark area. Figure 10.5 is constructed from the same data as the preceding examples in this section, but with a dot size and unit value more wisely chosen. Of course, if the distribution is

---

Suggested reader training consists of practice in the estimation of dot densities and an explanation of the tendency for underestimation of dot densities. Work will continue in this area as the cartographer seeks to improve the communication effectiveness of this mapping method.
A good analysis of the conceptual variations in dot mapping is given by Richard E. Dahlberg, "Towards the Improvement of the Dot Map," International Yearbook of Cartography, 7, 157–167, (1967).

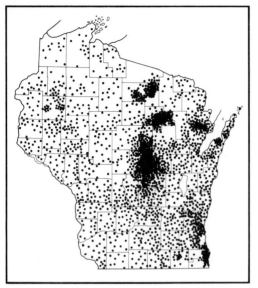

FIGURE 10.2 *A dot map in which the dots are so large that an excessively "heavy" map is produced giving an erroneous impression of excessive potato production. The same data and number of dots are used as in Fig. 10.1.*

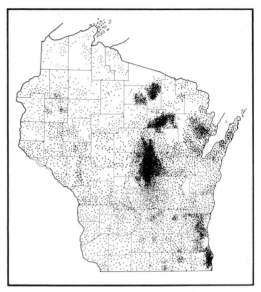

FIGURE 10.4 *A dot map in which the unit value assigned to one dot is too small. Many dots must be drawn causing the production of an excessively detailed map. The dots are the same size as those in Fig. 10.3. Each dot in this example represents 6.07 hectgares.*

FIGURE 10.3 *A dot map in which the unit value assigned to one dot is too large. Few dots can be drawn resulting in a barren map revealing little pattern. Each dot in this example represents 60.7 hectares.*

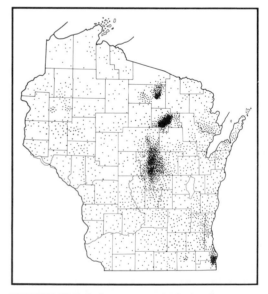

FIGURE 10.5 *A dot map in which the dot size and dot value have been more wisely chosen than in the preceding examples. Each dot in this example represents 16.2 hectares.*

sparse everywhere, then even the relatively dense areas should not appear dark.

On relatively large-scale dot maps showing land-use phenomena, it is possible to relate the dot size to the scale of the map by making the dot cover the scaled actual area. For example, we might make a map showing the area planted in wheat with a unit value of 500 hectares and a dot size that covers exactly 500 hectares at the scale of the map. We should not follow such a procedure blindly, however, since the elusive quality, "relative importance," of areal phenomena is often dependent on factors other than strict area relationship.

Professor J. Ross Mackay developed an ingenious nomograph to assist in determining the desirable dot size and unit value.* This graph, with metric additions shown in Fig. 10.6, requires a knowledge of the sizes of dots that can be made by various kinds of pens. This information is pre-

*J. Ross Mackay, "Dotting the Dot Map," Surveying and Mapping, 9, 3–10 (1949).

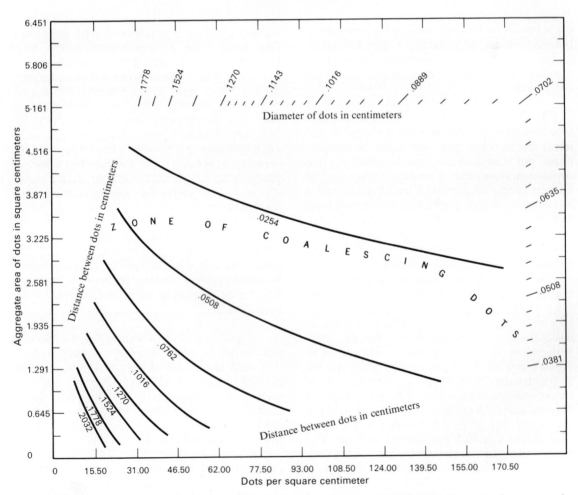

FIGURE 10.6 A nomograph showing the relationship between dot size and dot density. Nomograph courtesy of J. Ross Mackay and Surveying and Mapping with metric notations added by author.)

TABLE 10.1

Dot Diameters in Inches and Millimeters of Various Pens												
Barch-Payzant			Leroy			Pelican-Graphos Type				Wrico		
	Diameter			Diameter		R		O			Diameter	
Number	in.	mm	Number	in.	mm	mm	in.	mm	in.	Number	in.	mm
8	0.012	0.305	00	0.013	0.330	0.4	0.016	0.2	0.008	7,7A	0.018	0.457
7	0.018	0.457	0	0.017	0.432	0.5	0.020	0.3	0.012	6,6A	0.025	0.635
6	0.025	0.635	1	0.021	0.533	0.6	0.024	0.4	0.016	5,5A	0.027	0.686
5	0.036	0.914	2	0.026	0.660	0.7	0.028	0.5	0.020	4,4A	0.036	0.914
4	0.046	1.168	3	0.035	0.889	0.8	0.031	0.8	0.031	3,3A	0.048	1.219
3	0.059	1.499	4	0.043	1.092	1.0	0.039	1.0	0.039	2,2A	0.062	1.575
2	0.073	1.854	5	0.055	1.397	1.25	0.049	1.25	0.049			
1	0.086	2.184	6	0.067	1.702	1.50	0.059	1.60	0.063			
			7	0.083	2.108	1.75	0.069	2.00	0.079			
						2.00	0.079	2.50	0.098			

sented in Table 10.1. By varying the relationship between dot diameter and unit value, the cartographer can settle on a compromise between the two that will best present the characteristics of the distribution.

The Use of the Nomograph. The nomograph may be used in several ways, but perhaps the easiest is first to select three unit areas on the proposed dot map that are representative of (1) a dense area, (2) an area of average density, and (3) an area of sparse density. A tentative unit value can then be selected and divided into the totals for each of the three statistical divisons. The map area of one of the divisions can be estimated in terms of square centimeters, and the number of dots per square centimeter can be deduced.

As in the previous example, assume that a county contains 6000 hectares of corn; a unit value of 25 hectares per dot is chosen, which results in 240 dots to be placed within the boundaries of the division; assume further that the statistical division on the map covers 4.0 cm². This would mean that the dots would be placed on the map with a density of 60 dots per square centimeter. An ordinate at 60 is erected from the X-axis of the nomograph. A radial line from the origin of the nomograph to a given dot diameter on the upper right-circular scale will intersect the ordinate. The location of the intersection on the interior scale will show the average distance between the dots if they were evenly spaced. The height of the intersect on the Y-axis will indicate what proportion of the area will be black if that dot diameter and the

number of dots per square centimeter were used. Also shown is the "zone of coalescing dots," at or beyond which dots will fall on one another.

If the initial trial seems unsatisfactory for each of the three type areas selected for experimentation, either the unit value or the dot size, or both, may be changed. We can enter the graph with any of several assumptions and determine the derivatives. It is good practice while using the nomograph actually to dot the areas (on a piece of plastic or tracing paper) in order to see the results and to help visualize the consequence of other combinations.

The cartographer should remember that the visual relationships of black to white ratios and the complications introduced by the pattern of dots make it difficult, if not impossible, for a dot map to be visually perfect. The best approach is by experiment after narrowing the choices by use of the nomograph.

Locating the Dots. Theoretically, the ideal dot map would be one with a large enough scale so that each single unit of the point data could be precisely located. Ordinarily, dot maps are small scale but, sometimes, if the data are sparse enough, a unit value of one (e.g., paper mills) can still be used. It is usually necessary, however, to make the unit value of the dot greater than one, and the problem then arises of locating the one point symbol that represents several differently located units (see classification Fig. 8.1B).

When generalizing, it is often helpful to consider the group of several items as having a kind of center of gravity and to place the symbol at that point. Usually, all the cartographer knows from the original data is the total number of dots to place within a unit area. Consequently, every available source of information must be used to assist in the placement of the dots as reasonably as possible; such sources may be topographic maps and other distribution maps that correlate well with the map being prepared. The quality of the finished map depends largely on the ability of the cartographer to generalize the distribution, that is, to bring all the pertinent evidence to bear on the problem of where to put the dots.

Considerable detail can be introduced into the map if it is prepared on the basis of the smallest civil divisions and is then greatly reduced. Placing the dots with reference to minor civil divisions can be done easily by using translucent material for the map and putting under it a map that shows the minor civil divisions to serve as a guide. Only the larger administrative units need ordinarily be shown as base data on the finished map. Special care must be exercised not to leave the guiding boundary areas relatively free of dots; if this is done they will show up markedly in the final map as white zones. We should also refrain from inadvertently producing wormlike lines of dots, unwanted clusters, or "tweed" patterns when placing the dots. Such regularity can easily occur and is quite noticeable by contrast with its amorphous surroundings. Machine-produced dot maps usually avoid these problems, but they suffer equally from a different set of possible "pitfalls."

A dot map in which the requisite numbers of dots are evenly spread over the unit areas, like a patterned area symbol (although numerically correct), could better use some symbol system other than repetitive dots. Simple computer routines often err by spacing the dots too evenly. The dot map is best for distributions that are tightly clustered and dense in one area and sparse in other areas. Phenomena that exhibit uniform distribution characteristics are not well mapped by the dot map technique.

With increasing research more and more experimental dot maps can be expected to appear. The utility of the conventional dot map has assured that research devoted to improving the dot map will continue. The cartographer is well advised to warn the map reading audience by appropriate statements in the map legend if the conventional dot map technique has been altered in any manner.

The Graduated Symbol Map

Graduated symbols (i.e., symbols that are differently sized) are widely employed; the most common is the graduated circle. The variation in size is employed to symbolize either amounts at specific locations or totals that refer to enumeration units. Thus, they are useful (1) when point data exist in close proximity but are large in aggregate number, such as the population of a city, (2) for symbolizing totals of quantities such as tonnage, costs, and traffic counts, or (3) for representing the aggregate amounts that refer to relatively large territories. In the latter instance, the territory is considered merely as a location, even though it obviously has areal extent. The area of a statistical unit is naturally a two-dimensional geographical quantity; however, when data referring to it are aggregated and symbolized by a point symbol (in combination with other statistical units), the areal unit has, in a sense, been reduced to only a locational or point value for mapping purposes.*

The graduated circle is one of the oldest of the quantitative point symbols used for statistical representation. Near the beginning of the nineteenth century, it was used in graphs that illustrated the then new census materials, and its first appearance on maps was in the 1830's. Since that time, it has been near the top of any list of quantitative point symbols in the frequency of its use; its ease of construction makes it likely that it will continue to be widely utilized.

The Scaling of Graduated Symbols

When the graduated circle was first employed to symbolize data measured on an interval or ratio scale, the variations in the actual areas of the symbols were made uniformly comparable to the numbers they represented. This has become known as the square root method of symbol scaling. Other methods of symbol scaling are also

*This is an example of the inconsistency that results when we try to match up point data and point symbols, and linear data and line symbols.

employed in cartographic symbolization. They include psychological scaling and range grading.

Proportional Area Scaling or the Square Root Method. If two statistics have values to be represented that are in a 1:2 ratio (i.e., the magnitude of the second is twice that of the first) the second circle is constructed so that its area is twice that of the first. Since the area of a circle is πr^2, and since π is constant, the method of construction is simply to extract the square roots of the data and then construct the circles with radii or diameters proportional to the square roots. The unit radius value, by which the square roots are made into appropriate plotting dimensions, may be any desirable unit, such as so many millimeters or hundredths of an inch; it is selected so that the largest circle will not be "too large" and the smallest circle will not be "too small." As long as the square roots are all divided by the same unit radius value, the areas of all the circles will be in linear proportion to the sizes of the numbers they represent (see Fig. 10.7).

Range-Graded Scaling Method. If the cartographer does not wish to show relative sizes of each individual statistical quantity, the data can be classified or range graded. The cartographer would show all individuals in one class by a "stan-

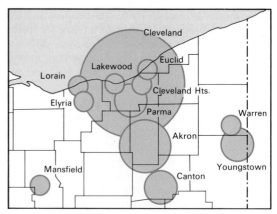

FIGURE 10.8 *The population of some cities in northeastern Ohio. Symbols are range-graded to denote the population of the cities. (See Table 10.2.)*

dardized" symbol constructed to the size of the midpoint of that class (see Fig. 10.8). This process is referred to as the range grading of graduated symbols. In effect, the data are divided into class intervals, and one symbol size is assigned to each class.

Psychological Scaling Method. Extensive research in the psychophysical aspects of cartographic symbols has demonstrated that the perceptual response to differences between symbol areas is not a linear function; instead, the ordinary observer will *underestimate* the sizes of the larger symbols in relation to the smaller ones. The evidence is particularly strong that underestimation occurs between circle symbol areas. For example, to use the previous illustration, if the magnitudes of two quantities were in a 1:2 ratio and the actual areas of two circles representing them were in the same ratio, then a map reader would think the second was significantly less than twice the size of the first. When we make the areas of the circles strictly proportional to the numbers they represent, we have, therefore, in effect reduced the visually apparent sizes of the larger circles in relation to the smaller or, to look at it another way, we have increased the visual significance of the smaller circles relative to the larger. This is unfortunate, since the primary purpose of making such maps is to symbolize quantities so that the map reader will

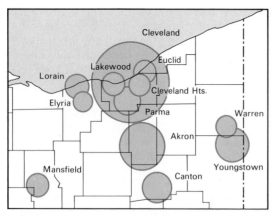

FIGURE 10.7 *The population of some cities in northeastern Ohio. Symbols are proportionally scaled so that the areas of the symbols are in the same ratio as the population numbers that they represent. (See Table 10.2.)*

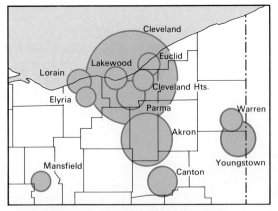

FIGURE 10.9 *The population of some cities in northeastern Ohio. Symbols are psychologically scaled so that the areas of the symbols visually connote the correct ratios between the population numbers that they represent. (See Table 10.2.)*

plotting values and need only be converted to map dimensions by dividing them by some convenient unit value. The maps shown in Figs. 10.7, 10.8, and 10.9 have been constructed from the data in Table 10.2, using a unit radius value chosen so that the smallest circles have the same unit radius value.

The table in Appendix G provides the antilogarithms of $n \times 0.57$ for use in the psychological scaling of graduated circles. Consequently, one only has to enter the table with the raw data, and the corresponding tabular values will provide the plotting values, which need only be divided by a suitably sized unit radius.

Another aid in compensating for the underestimation is to use strategically selected "anchoring stimuli" in the legend that accompanies the map. If several circles of a similar size occur it is appropriate to use a legend circle in their general range. A solitary legend symbol is not often suitable, since the perceptual problem is one of underestimating differences between circles. Therefore, a legend design of at least three nonnested

obtain a realistic impression of the distribution being mapped.

We may compensate for underestimation by scaling the symbols to reflect the amount of expected underestimation (see Fig. 10.9). Instead of extracting the square roots of the data and making the radii of the circles relative to the square roots, the procedure is to (1) determine the logarithms of the data, (2) multiply the logarithms by 0.57 or some other value appropriate for the symbol being used, (3) determine the antilogarithms, and (4) divide the antilogarithms by the chosen unit value for the radii of the circles.* Figure 10.10 shows the difference in the sizes of two sets of circles constructed in the two ways.

An example will demonstrate the procedure. Table 10.2 gives the data for the circle maps in Figs. 10.7, 10.8, and 10.9. The numbered columns of Table 10.2 show the calculations needed to determine the numbers that, when divided by the chosen unit radius value, will give the plotting radii of the circles. The antilogs provide directly the

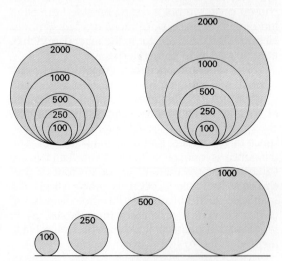

FIGURE 10.10 *Two nested sets of graduated circles prepared from the same data show the difference between circles scaled, left, in linear relation to area (i.e., according to square roots) and right, psychologically scaled to compensate for underestimation. A legend or key designed as in the lower series (the same set of circles as the first four in right above) is probably easier than a nested set for the map reader to use.*

This procedure was devised by Professor James J. Flannery for graduated circles. Note that the square root of log n = log n/2 or log n × 0.5. Multiplying the logarithms by 0.57 instead of 0.5 serves proportionately to increase somewhat the relative sizes of the larger circles so that they appear in proper relation to the smaller ones.

TABLE 10.2

| | | | | | Drafting radii in cm (before reduction) | | |
| | | | | | Fig 10.7 | Fig 10.8 | Fig 10.9 |
	(1) 1970 Population (n)	**(2)** log n	**(3)** Antilog of (log n × 0.5)	**(4)** Antilog of (log n × 0.57)	462.28 = 1.0 cm Column 3 ÷ 462.88	Range Graded	1000 = 1.0 cm Column 4 ÷ 1000
Akron	275,425	5.44000	524.81	1261.23	1.135	1.25	1.26
Canton	110,053	5.04160	331.74	747.68	0.718	0.80	0.75
Cleveland	750,879	5.87557	866.53	2233.96	1.874	2.50	2.23
Cleveland Heights	60,767	4.78367	246.51	532.96	0.533	0.50	0.53
Elyria	53,427	4.72776	231.14	495.25	0.500	0.50	0.50
Euclid	71,552	4.85462	267.49	584.97	0.579	0.50	0.58
Lakewood	70,173	4.84617	264.90	578.52	0.573	0.50	0.58
Lorain	78,185	4.89312	279.62	615.29	0.605	0.50	0.62
Mansfield	55,047	4.74073	234.62	503.75	0.508	0.50	0.50
Parma	100,216	5.00094	316.57	708.82	0.685	0.80	0.71
Warren	63,494	4.80273	251.98	546.46	0.545	0.50	0.55
Youngstown	140,909	5.14894	375.38	860.79	0.812	0.80	0.86

1970 Populations of cities in northeastern Ohio with more than 50,000 inhabitants.

circles (see Fig. 10.10) is advisable for every graduated circle map.

The problem of underestimation is less severe with range-graded circles as long as each chosen circle symbol is distinguishable from all of the others. In the case of range-graded symbols, all symbols should appear in the legend, and they may be nested to conserve space.

The difference between range-graded point symbols and square root or psychologically scaled symbols has its counterparts in the symbolization of volumes by line or area symbols. For example, choroplethic maps (see Chapter 11) are in essence range graded; isarithmic maps are not. The difference is one of uniquely symbolizing every value on a continuum, as in square root or psychologically scaled point symbols and the isarithmic method, compared to the collapsing of values along the continuum into classes, as in range-graded point symbols and the choroplethic method.

Specification of Symbol Size or Unit Radius

Another problem to be solved when employing graduated symbols, and especially with graduated circles, is the selection of the unit radius. The cartographer may find that a desirable unit radius value for the map as a whole may cause the symbols in one part to fall largely on top of one another if their centers are placed approximately at the locations of the data they are to symbolize. Two of several ways to approach this problem are shown in Fig. 10.11. The two small maps are taken from the data in Table 10.2.

The selection of the unit radius value with which to scale the circles is important and should be done with the aid of some preliminary experimentation. The ideal value is one that will provide a map that is neither "too full" nor "too empty." For example, Figs. 10.12 and 10.13 show the area of land available for crops in some counties. When a small unit radius is chosen, the circles are too small

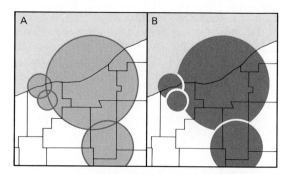

FIGURE 10.11 Two ways of solving the problem of overlapping circles. (A) The circles are allowed to overlap, since they are "transparent." (B) Smaller circles are "above" larger circles.

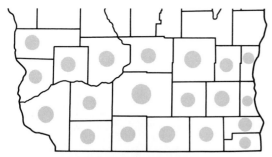

FIGURE 10.12 *Area available for crops by counties. The unit radius is too small.*

to show much differentiation, as illustrated in Fig. 10.12, and the impression is given that there is practically no cultivated land in those counties. When a large unit radius is chosen, the representation does not reveal much, and the impression is given that practically all the land is cultivated, as in Fig. 10.13.

Squares, Cubes, Spheres, and Other Point Symbols

Instead of graduating circles, we may employ almost any other geometric or pictorial figure. The circle is the easiest to construct and scale, but the requirements of design may make some other figures desirable.

The square is relatively simple to use as a symbol to represent magnitudes, since all that is necessary is to obtain the square roots of the data and then scale the sides accordingly. This will

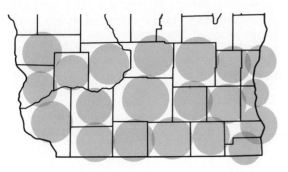

FIGURE 10.13 *Same data as in Fig. 10.12, but the unit radius is too large.*

grade the areas of the squares in linear proportion to the numbers they represent. There is clear evidence that the phenomenon of underestimation affects the map reader's impression of circles so scaled, but the evidence is less impressive with respect to squares. Range-graded squares are an alternative, but they appear to be used much less frequently than other point symbolization schemes.

Apparent Volume Symbols. It is not uncommon that a cartographer is faced with a range of data so large that with interval or ratio scaling, both ends of the range cannot be effectively shown by graduated circles or squares. If the symbols are large enough to be differentiated clearly in the lower end of the scale, then those at the upper end will overshadow everything else and ruin the communication. Many attempts have been made to surmount this apparent difficulty by employing a symbol form drawn to simulate a volume instead of an area, such as pictorial cubes or spheres, as shown in Figs. 10.14 and 10.15.* This has been done by scaling the symbols according to the cube roots of the data, that is, by making the sides of the front of a perspective cube or the circle bounding a sphere proportional to the roots. If the symbols were actually three dimensional, their volumes would be in strict linear proportion to the numbers they represented. The cube roots of a series cover a much smaller range than the square roots, and this, of course, is what makes it *ostensibly* possible to portray graphically the larger range of data.

Although spherelike symbols may be very graphic and visually pleasing, especially when enhanced by good execution and design, they do not serve the purpose of effectively portraying the numbers that they are supposed to represent. Several recent studies have clearly shown that map readers evaluate the spheres not on their volume comparison, but on the basis of the map areas covered by them, that is, as if they were

*The difficulty is more apparent than real. If one segment of a distribution does actually make all else pale by comparison, then that is the fact and it should be so portrayed. The primary responsibility of the cartographer is to prepare the map so that the visual impressions are correct.

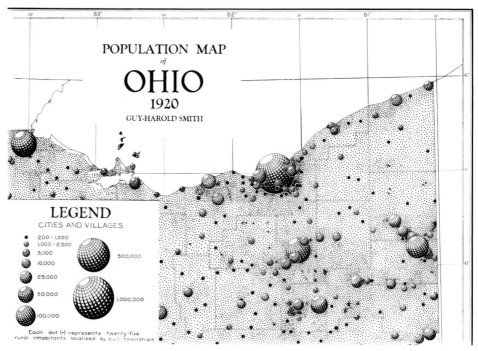

FIGURE 10.14 *A portion of a population map of Ohio using spheres to depict volume data.* (*Courtesy of G.-H. Smith and The Geographical Review, published by the American Geographical Society of New York, 1920.*)

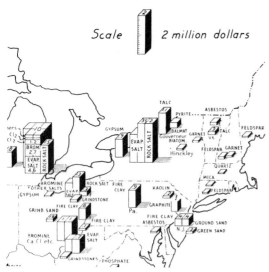

FIGURE 10.15 *A map on which cubelike symbols, called block piles, are used to depict volume data. (From a map by E. Raisz, in Mining and Metallurgy, AIME.)*

graduated circles, including the characteristic underestimation. To what extent cubelike symbols suffer a similar perceptual fate is not yet known.

When graduated pictorial symbols such as soldiers, stars, and animals are employed, care must be taken to scale them properly. Their heights or diameters could be made proportional (as outlined above) to the data to be represented by a system that ensures that the *areas* covered by the symbols are visually related to the magnitudes they represent.* A safer course of action would be to employ these symbols as range-graded symbols. This is especially recommended for pictorial symbols whose areas are almost impossible to evaluate.

There has not been enough research yet to determine to what extent underestimation affects the perception of these more complex symbol types. Determination of relative sizes of some geometric symbols has been studied by Robert L. Williams. See Statistical Symbols for Maps: Their Design and Relative Values, Yale University Map Laboratory, New Haven, 1956.

Segmented Graduated Symbols

Cartographers have attempted to use parts of circles such as pie wedges or other symbols when precise locations are desired but large amounts of overlap would occur if an entire symbol was used (see Fig. 10.16). Such point symbols may be range graded or they may be scaled on an interval or ratio basis. With this type of symbol it is often possible to present several categories of data simultaneously, including nominally scaled data, by showing several symbols at one location and differentiating them by color or pattern (see Fig. 10.17).

Another way of symbolizing quantitative data is by segmenting the point symbol. Instead of using part of the symbol, the entire symbol is used, but it is segmented to illustrate the proportional parts that constitute the data. For example, the ethnic makeup of an urban population, or the value added by several categories of manufacturing, can be portrayed this way. The most common way of doing this is to employ graduated circles and subdivide them as a pie is cut. There is no generic term for this kind of symbol, but the subdivided circles are often called "pie charts."

Any relation of one or more parts to the whole can be shown visually by the pie chart. For example, Fig. 10.18 shows the total amount of farmland in each county and, at the same time, shows what percentage of that total is available for

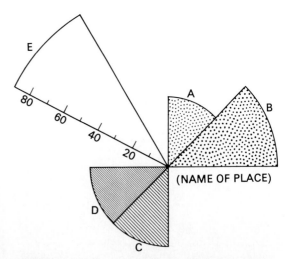

FIGURE 10.17 Examples of several ways to employ circle segments. A and B in the upper right quadrant can indicate related characteristics (e.g., passenger and freight traffic), C and D in the lower left quadrant can indicate different nominal categories, and E, with interval scaling, can indicate yet another category.

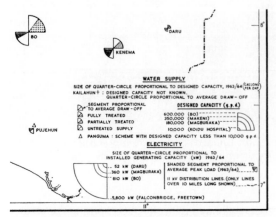

FIGURE 10.16 The legend and a small portion of a map illustrating the amount of information that can be coded in segmented, graduated point symbols. (From J. I. Clarke, Sierra Leone in Maps, University of London Press, Ltd., London, 1966.)

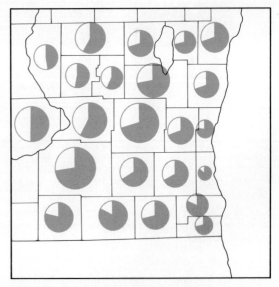

FIGURE 10.18 Land in farms and the percentage of that land available for crops by county in a portion of Wisconsin. The circles represent the land in farms. The percentage available for crops has been shaded in on each circle.

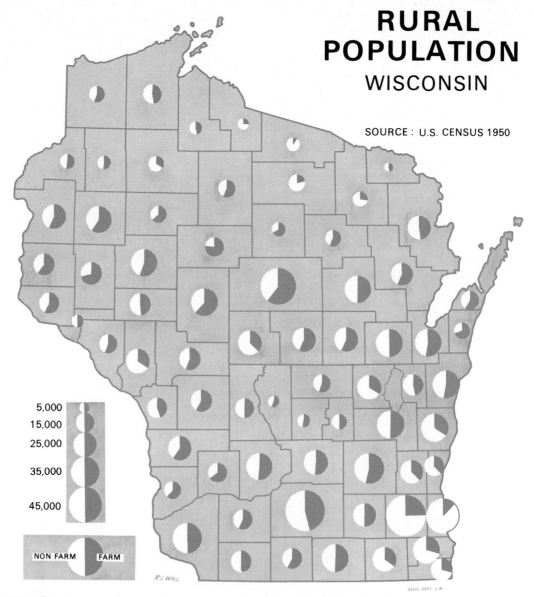

RURAL POPULATION
WISCONSIN

SOURCE : U.S. CENSUS 1950

5,000
15,000
25,000
35,000
45,000

NON FARM   FARM

R.L WALL

GEOG. DEPT. U.W

FIGURE 10.19 Pie chart map showing the proportion of rural farm (shaded) and nonfarm populations in some counties in Wisconsin. The sizes of the circles are scaled in proportion to the total rural population. (From a map by R. L. Wall.)

crops. The procedure merely requires that the percentage be determined and that, by using a "percentage protractor," the various values be marked off on each circle.

It is important that the subdivision of each circle begin at the same point; otherwise, the reader will have difficulty in comparing the values. Also important is the selection of the portion to be shaded or colored.

If only two classes of data are being represented by a pie chart, and if they are of equal importance to the purpose being served by the map, we may use a background with a neutral tone so that the visual contrasts of the two segments of each circle with the background are approximately equal. Figure 10.19 has been constructed in this manner.

Many other, more complex kinds of segmentation can be employed. For example, a kind of map conceptually similar to the pie chart can be constructed by employing one or more concentric circles to represent a portion of the total. The diameters of these interior circles are scaled according to the original data in the same manner as the circle representing the total. The visual impression is not very efficient, however, since percentages are not easily read. Naturally, graduated squares, rectangles, and other figures may be segmented in various ways to show parts of a whole, (see Fig. 10.20).

Directional and Time Series Point Symbols

A variety of more complex cartographic presentations can be accomplished by combining graphs and maps so that variations in directional compo-

nents or changes through time can be given the added quality of relative geographical location. For example, the upper diagram in Fig. 10.21 shows, for each of the stations indicated, the amount of darkness, daytime cloud, and sunshine throughout the year. The lower diagram shows, among other things, the directional component and the percent frequency of winds for August in the same area.

There is almost no limit to the amount of information that we can somehow encode in these kinds of point symbols. Of paramount impor-

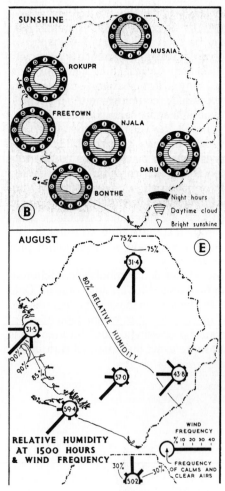

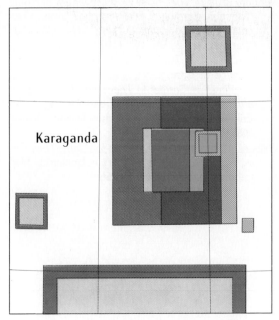

FIGURE 10.20 The use of a complexly divided graduated point symbol (also with overlap). The reader may find it difficult to estimate the various proportions of the symbols correctly.

FIGURE 10.21 Cartographic representation of proportions of directional and sky condition data. (From J. I. Clarke, Sierra Leone in Maps, University of London Press, Ltd., London, 1966.)

tance, however, is the question of how much of the information will actually be available to the reader. As long as the reader can obtain a significant geographical element (i.e., an appreciation of similarities and differences from place to place) from the map, then the map is serving a purpose. If that part of the information is not readily obtainable, then the map is definitely a failure and the data would probably be made more easily available in tabular form.

Readability of Quantitative Data Symbolized by Point Symbols

The detailed readability of most point symbols has not been extensively studied. Therefore the precise degree to which a given symbolization scheme may be successful in communicating information to a map reading audience is unknown. Most such research suggests that significant portions of even simple symbolization schemes may not be successfully communicated sometimes. One may assume that the more complex the symbolization scheme, the smaller the percentage of the information actually communicated to the map reader is likely to be. It follows that cartographers should avoid complex symbolization schemes. Information encoded with intricate symbolization schemes must be expertly scrutinized to be obtained from the map, and cartometric measurements from maps are time consuming and exacting. Such information is much better provided in tabular form.

The use of detailed symbolization schemes should also be avoided for another reason. Data in machine-readable form make maps less needed as storage devices for the spatial tabulation of data. Computer processing of such information is quicker and more exact than manual cartometric

procedures. It should be noted that many mechanical machine-driven plotters have the capability to execute any of the point symbolization schemes discussed in this chapter. Alternative scaling methods that require additional calculations are more accessible using machine-driven plotters than when the calculations must be performed manually. Therefore there is no longer any need for the encoding of large amounts of data into complex symbolization schemes. Machines can produce, in less time, several informative maps that employ simple symbolization schemes.

SELECTED REFERENCES

Cox, C. W., "Anchor Effects and the Estimation of Graduated Circles and Squares," *The American Cartographer, 3,* 65–74 (1976).

Crawford, P. V., "The Perception of Graduated Squares as Cartographic Symbols," *The Cartographic Journal, 10,* 85–88 (1973).

Dahlberg, R. E., "Towards the Improvement of the Dot Map," *International Yearbook of Cartography, 7,* 157–166 (1967).

Flannery, J. J., "The Relative Effectiveness of Some Common Graduated Point Symbols in the Presentation of Quantitative Data," *The Canadian Cartographer, 8,* 96–109 (1971).

Morrison, J. L., "Towards a Functional Definition of the Science of Cartography with Emphasis on Map Reading," in *Studies in Theoretical Cartography,* a *Festschrift* in honor of Professor Dr.-Ing.h.c.Dr. Erik Arnberger, Geographisches Institut, Universität Wien, 247–266 (1977).

Williams, R. L., *Statistical Symbols for Maps: Their Design and Relative Values,* Yale University Map Laboratory, New Haven, 1956.

This chapter explains the more common symbolizations of quantitative data by line and area symbols. The symbolization schemes presented are some of the most important methods that cartographers have developed. As aids to understanding our environment, these methods play significant roles in our ability to survive decently on our planet and perhaps ultimately to survive on other planets. Perhaps nothing the cartographer does for humanity is as uniquely "cartographic"

CHAPTER 11

SYMBOLIZATION: QUANTITATIVE DATA BY LINE AND AREA SYMBOLS

as some of the symbolization schemes presented in this chapter. The chapter has two major divisions; symbolization by lines and symbolization by area symbols.

THE STATISTICAL SURFACE

Since it is possible to assume a statistical surface from either positional or volumetric data recorded for areas and to portray that surface by line or area symbols, the concept of a statistical surface must be clearly understood. If the magnitudes of ratios assigned to unit areas or the sample values of continuous distributions (e.g., temperatures at weather stations) are conceived to have a relative vertical dimension, the sample values or the unit area values can be visualized as forming a three-dimensional surface.

For example, Fig. 11.1 shows an array of numerical data that can either be a set of ratios, such as population densities, or a set of observations, such as temperatures at weather stations, each arbitrarily located at the center of a unit area. Figure

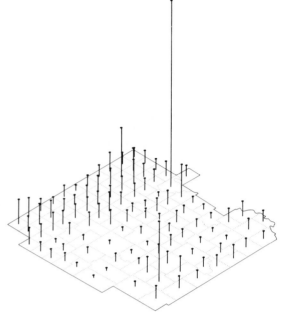

FIGURE 11.2 *Elevated points, the relative height of which is proportional to the number in that unit area in Fig. 11.1. (Courtesy G. F. Jenks and Annals, Association of American Geographers.)*

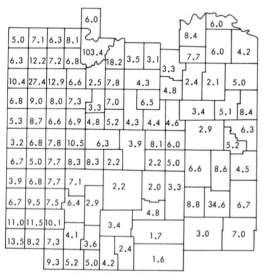

FIGURE 11.1 *An array of statistical data in unit areas. The numbers are rural population densities in some minor civil divisions in Kansas, but the numbers could be sample values of some other continuous distribution. (Courtesy of G. F. Jenks and Annals, Association of American Geographers.)*

11.2 shows the erection of a column in each unit area; the length of the column is proportional to the number in that unit area in Fig. 11.1. In Fig. 11.3 the relative magnitudes of the values in the unit areas are emphasized by erecting over each area a prism that has a base shaped like the unit area and a height that is scaled as the length of the line in Fig. 11.2. In a very real sense, this is a three-dimensional form of the familiar two-dimensional histogram. Figure 11.4, on the other hand, illustrates another way of showing the same distribution, but here the emphasis is put on the magnitudes and directions of the gradients. Both Figs. 11.3 and 11.4 can be thought of as the representation of a statistical surface consisting of a series of numbers each of which has an X, Y, and Z characteristic. The X and Y values refer to the horizontal, or planimetric, locations, and the Z values refer to the assumed relative heights above some horizontal datum, such as the plane of the map.

Figure 11.2 is a symbolization of point data

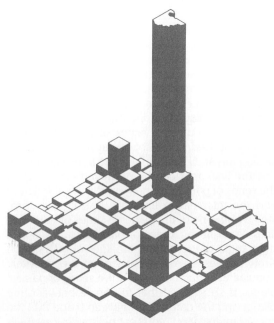

FIGURE 11.3 *A perspective view of the statistical surface produced by erecting prisms over each unit area proportional in height to the numbers shown in Fig. 11.1. (Courtesy G. F. Jenks and Annals, Association of American Geographers.)*

These kinds of quantitative data can be mapped in numerous ways, depending largely on whether the cartographer is more concerned with giving emphasis to the directional aspects at places or to the overall spatial interrelation of the magnitudes. For the former (e.g., wind velocities), we commonly employ lines of some kind that show by their orientation and character the direction and magnitude of the potential or actual flow; for the latter (e.g., atmospheric pressure), we usually conceive of the total mass of data as forming a volume distribution and then portray the continuous statistical surface of that volume by some means. The best-known example of this general class of quantitative distribution, which can be represented in a variety of ways, is the surface configuration of the land. This can be thought of as an aggregation of an infinite number of related linear magnitudes and gradients existing at points. These can be shown directly by a type of gradient or vector line symbol called a *hachure,* or indirectly by a set of lines called *isarithms* (contours). These

by line symbols, while Fig. 11.4 is another symbolization by line symbols of point data conceived as forming a statistical surface. Figure 11.3 is a symbolization of a statistical surface by varying heights of area units. The concept of a statistical surface is important, because its visualization allows for the derivation of a great deal of additional information. For example, gradient is an important concept, and there are numerous kinds of gradients, such as the slope of the land, and variations from place to place of air temperature or atmospheric pressure; and many other more complex gradients derive from concepts such as population or economic potential. Such quantities may be thought of as spatial vector quantities in that at each point of the distribution on the map there is a direction, a magnitude, and a sense. To use the example of the slope of the land surface, there is a degree or percent of slope, a direction of maximum value in relation to the compass (such as NE-SW), and a sense in that the potential energy flow is in one of the directions.

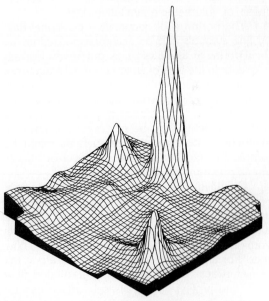

FIGURE 11.4 *A perspective view of a smoothed statistical surface produced by assuming gradients among all the numbers shown in Fig. 11.1. The elevated points in Fig. 11.2 would all just touch the surface shown here. (Courtesy G. F. Jenks and Annals, Association of American Geographers.)*

linear magnitudes may also be classed and mapped as areas of uniformity. Such a map is called dasymetric and will be discussed under the symbolization of quantitative data by area symbols below.

The statistical surface is one of the most important concepts in cartography, and the methods for mapping a statistical surface must be thoroughly known to cartographers. Simply, a statistical surface can be assumed to exist for any distribution that can be conceived of as being continuous and measured on an ordinal, interval, or ratio scale of measurement.

MAPPING QUANTITATIVE DATA USING LINE SYMBOLS

The use of line symbols to symbolize quantitative data can be divided into the general classes of pictorial lines, associative lines, and geometric lines similar to the division of point symbols. However, the distinction is less rigid and useful. Line symbols can be used to portray either positional, linear, or volumetric quantitative data.

The portrayal of quantitative positional data by line symbols can be accomplished either by their portrayal as flows or as gradients. For example, both the flow of emigrants to Chicago from the rural southern United States and the number of passengers flying from Chicago to New York represent flows measured at positions on the earth. Gradients may be portrayed by lines by assuming a statistical surface such as might be formed by elevations above sea level or atmospheric pressures measured at positions.

Quantitative linear data and volumetric data from areas can also be portrayed by the use of line symbols. For example, population density can be symbolized by the isoplethic technique, or imports from one nation by another can be shown by flow lines. For linear data, tonnage shipped along rail or waterways can be shown by graduated flow lines, and profiles of a statistical surface can be shown by line vectors.

The organization of our discussion of the symbolization of quantitative data by line symbols will first consider the important portrayal of the statistical surface with its gradients and will then consider the symbolization of flows.

Mapping the Statistical Surface by Line Symbols

A statistical surface implies a base datum and a distribution of Z values on an ordinal, interval, or ratio scale measured at right angles to that datum. By connecting the Z values, a smoothly undulating statistical surface is formed and the character of its surface is displayed on the map.

If one assumes a series of parallel (usually equally spaced, but not always) planes intersecting this statistical surface, the intersections define the lines that can be portrayed on a map. If the series of parallel planes is parallel to the datum and the intersection lines are orthogonally projected onto one of the planes, a series of isarithmic lines result. If the series of parallel planes is at a right angle to the datum, the lines of intersection show profiles across the surface of the smoothly undulating surface. Finally, if the series of parallel planes is so situated that the angle between the planes and the datum lies between 0 and 90°, the lines of intersection represent oblique traces. Each method is important in cartography, and every cartographer should be aware of the assumptions, merits, and demerits of each method.

Isarithmic Mapping

When the interest in a geographical distribution is focused (1) primarily on the form of the distribution, that is, the organization or arrangement of the magnitudes, and directions (sense) of the myriad gradients that together constitute it, or (2) on the values at points of a truly continuous distribution, such as the land (elevation), air temperature, or pressure, one of the two forms of the isarithmic method is usually used.

In isarithmic mapping, the distribution is clearly conceived of as a volume and, in order visually to comprehend a volume, it is obviously necessary to see the shape of the outside surface enclosing it.

Together the Z values suggest this three-dimensional surface (see Figs. 11.1 to 11.4). From the geometric form of this surface we can infer the extent of the volume resting on the base. In many instances, the actual *volume* itself is less interesting and of less concern than is the form of its surface. This is not always the case; for example, a road engineer is greatly concerned with actual volumes, as in cut-and-fill problems. Nevertheless, the character of a three-dimensional distribution is most clearly mapped by delineating its surface.

The symbolization of real or abstract three-dimensional surfaces is difficult, and more time and effort probably have been devoted to it than to all other problems of symbolization put together. The principles involved in delineating such a surface are best illustrated by beginning with the familiar example of the surface of the land.

If the irregular land surface has been mapped in terms of *planimetry,* that is, the relative horizontal position of all points on the land, it is evident that there exists an infinity of points, each of which has, by reason of its location, an X, Y, and Z coordinate position with respect to the *datum* surface (spheroid) to which the earth's surface has been orthogonally transferred (see Fig. 11.5). By definition, the land surface is at all points either above or below the smooth assumed reference elevation, called a *datum.* If an imaginary surface, parallel to the horizontal datum and a given Z distance from it, is assumed to intersect the irregular land surface, it must do so at all points having that Z value. The *trace,* or the line of intersection, of these two surfaces will be a closed line. When this line, an isarithm, is orthogonally viewed, that is, perpendicularly projected to the map, it shows by its position the X and Y locations of all the points that have the particular Z value it represents.

Figure 11.6 shows, in perspective a hypothetical island, evenly spaced Z levels, and below, an isarithmic map of the distribution of the Z values on the surface of the island. In this case, Z is elevation above the average level of the sea, which is defined as 0; it is therefore a contour map. The lowest or outer isarithm represents the average position of the shoreline. The next isarithm in Fig. 11.6 is the trace of the plane spaced 20 Z units above 0; it is in the same location as the average shoreline would be were the sea 20 Z units higher.

The configuration of a three-dimensional surface is symbolized by the characteristic shapes and patterns of spacings of a set of isarithms, especially when the intersecting Z planes are equidistant from one another. Smooth, steep, gentle, concave, convex, and other simple kinds of gradients and combining forms may be readily visualized from isarithmic maps, as indicated in Fig. 11.7. For example, the bends of contours always point upstream when they lie athwart a valley; they always point down slope when crossing a spur. The angles of slope, that is, the gradients or rates of change of elevation, of the land are shown by the relative spacings of the set of contours provided by equally spaced Z planes. Profiles of the land surface along a traverse, or along a road or railroad, can easily be constructed from a contour map by working backward from the map to the profile. The recognition of detailed topographic forms and often even genetic structural details are readily revealed by the patterns of the contours on topographic maps.

It is not necessary that the Z surface to be rep-

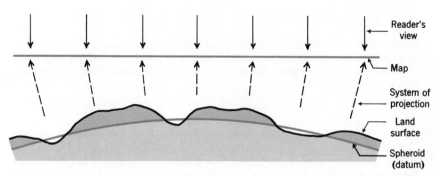

FIGURE 11.5 *The building of an isarithmic map. The positions of elevations are first projected orthogonally to a spheroid. The "map" on the spheroid is then reduced and transformed to a plane, which the reader in turn views orthogonally. Curvature is exaggerated in this illustration.*

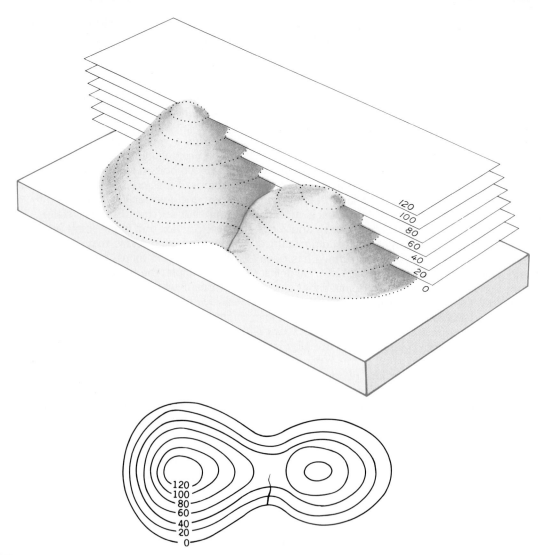

FIGURE 11.6 In the upper diagram, horizontal levels of given Z values are seen passing partway through a hypothetical island. The traces of the intersections of the planes with the island surface are indicated by dotted lines. In the lower drawing, the traces have been mapped orthogonally on the map plane and constitute the representation of the island by means of isarithms.

resented by isarithms be an actual visible, tangible surface, such as the land; the configuration of any three-dimensional surface may be mapped in the same way. For example, the form of a defined pressure surface in the atmosphere, such as the 500 millibar surface, may be mapped by passing Z levels through the atmosphere. The shapes in na-

ture are not visible, but the patterns of the isarithms show the gradients, the troughs, and the ridges of that surface in the same way that contours show the ups and downs of the land. The three-dimensional surface may even be an abstraction. For example, the Z values may consist of some sort of ratio or proportion, such as persons

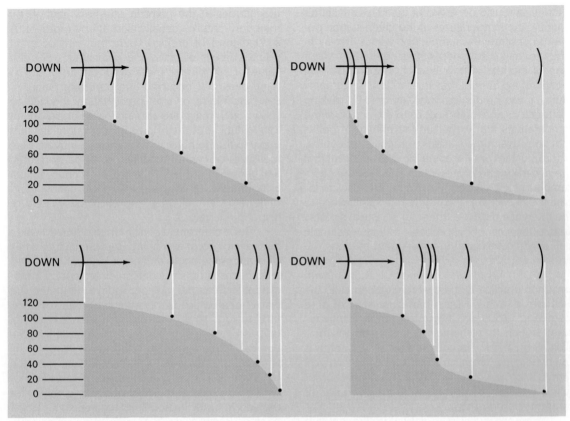

FIGURE 11.7 The isarithmic spacings above and the profiles below illustrate in each diagram, the manner in which the spacings of the isarithms show the nature of the configurations from which they result. All the forms may be described as variations of the general rule that if the isarithms are the traces of equally spaced Z levels, the closer the isarithms are, the greater the gradient will be.

per square kilometer. Anything that varies in magnitude and either actually exists or can be assumed to exist in continuous fashion over area constitutes a statistical surface. Its configuration can be mapped isarithmically.

Inferring the Statistical Surface. In order to map the traces of the intersections of horizontal levels with a statistical surface, it is necessary, as in following the proverbial recipe for rabbit stew, first to "catch" the statistical surface. This is easier said than done.

Only since the development of stereoviewing equipment for air photographs has it been possible to specify in its entirety the infinity of points on the land, each of which has its Z value of rela-

tive elevation. In *all* other cases the statistical surface must be obtained from a limited number of Z values, and the totality of the surface must be *inferred* from these. In a very real sense, the infinity of Z values constitutes a statistical population or universe but, because of practical limitations, only a sample of these points is usually available. Since we cannot know precisely the characteristics of the universe from which the sample has been taken, any extension of the characteristics of the sample to the universe or to a particular part of it constitutes an inference, the validity of which can only be determined provided we have certain basic knowledge about the characteristics of the universe.

An example will clarify this. If air temperature

is simultaneously observed at a series of weather stations, the temperatures at the given station positions constitute the sample of Z values. If it is desired to make a synoptic (instantaneous view) map of the total distribution of temperatures (by means of *isotherms,* that is, isarithms of temperature), it can be done inductively only by making assumptions and inferences as to the nature of the temperatures that existed at the infinity of points that lie between the stations, for which, of course, actual Z values are not available. It is apparent that the accuracy and representativeness of the given sample values, which the cartographer uses to locate the traces, are of considerable significance to the inference drawn of the total statistical surface. Various aspects of the probable validity of sample values are considered in subsequent sections.

Kinds of Isarithmic Mapping. It has been common practice in the past to give names to the isarithms employed for a particular kind of phenomenon. Thus we have isotherms (temperature), isobars (barometric pressure), isohyets (precipitation), and so on; the number of such terms is overwhelming. Such proliferation of the technical terminology serves no very useful purpose (except when necessary to distinguish among two sets of isarithms on one map).

There are two classes of Z values that differ in terms of the precision with which they can specify a statistical surface: (1) actual or derived values that can occur at points; and (2) actual or derived values that cannot occur at points.

Actual values that can exist at points are exemplified by data such as elevation above or depth below sea level, a given actual temperature, the actual depth of precipitation, and thickness of a rock stratum. These kinds of values do exist at points. Only errors in observation or in the specification of the XY positions of the observation points can affect the validity of the sample values.

Derived values that can exist at points are of two kinds. One kind consists of averages or measures of dispersion, such as means, medians, standard deviations, and other sorts of statistics derived from a time series of observations made at a point. We can calculate a mean monthly temperature, an average retail sales figure, or an average land value for some particular place; the resulting numbers, although representative of

magnitudes at the point in question, cannot, by their very nature, actually exist at any moment. A second kind of derived value that can occur at points consists of ratios and percentages of point values. Examples of these are the ratio of dry to rainy days that occurred at a particular place, or the percentage of total precipitation that fell as snow. Such ratios are also incapable of existing at any instant, but they do represent quantities that apply to the point for which they are derived. Like measures of central tendency or dispersion, they are generally subject to more error than simple actual values. If they are rigorously defined and uniformly derived, they approach the validity of actual point values.

Quite different in concept is the derived value that cannot occur at a point. Representative of this class are percentages and other kinds of ratios that include area in their definition directly or by implication, such as persons per square kilometer, the ratio of beef cattle to total cattle, or the ratio of cropland to total land in farms. With such a quantity, only an average value for a unit area can be derived. Consequently, although it is perfectly legitimate to assume a statistical surface specified by these kinds of quantities, such a surface is dependent only on a series of average values for unit areas. Since each unit area represents a larger or smaller aggregate of XY points, no single point can have such a value. Nevertheless, in order to symbolize the undulations of the statistical surface by isarithms, it is necessary to assume the existence of such Z values at specific points.

Absolute values that cannot occur at points but, instead, refer to areas, such as the populations of minor civil divisions, number of cows in counties, or potato production in states should not be assumed to constitute a statistical surface because their magnitude is affected by area; for example, 1000 persons in 20 km^2 is the same as 500 persons in 10 km^2. Unless such total values have been related in some manner to the area of the unit of collection, conceptually no statistical surface should be assumed to exist.

Because of the fundamental differences in the concepts of the two classes of statistical data, it is conventional to make a distinction among the isarithms employed to display their form. Different names are applied to them, and considerable confusion in definition and spelling exists in the literature. The following terminology seems to be consistent with modern usage.

An *isarithm* is any trace of the intersection of a horizontal plane with a statistical surface. It is thus the generic term; it may also be called an *isoline* or *isogram*. Isarithms showing the distribution of actual or derived quantities that can occur at points are called *isometric lines*. In contradistinction isarithms that display the configurations of statistical surfaces that are based on quantities that cannot exist at points, and that are therefore likely to be subject to a somewhat larger inherent error of position, are commonly called *isopleths*.

An observation is in order here. Much less confusion will result if the student will transfer the distinctions just made from the symbols to the surfaces they delineate. An isarithm is defined simply as the trace made by the intersection of a horizontal Z-level plane with *any* three-dimensional surface. Whatever the nature of the surface may be, the function of the trace as a cartographic symbol is the same. Consequently, the "difference" between an isopleth and an isometric line is really an attempt to distinguish between the type of data and, as a consequence, the precision of the symbolization scheme; it is not a distinction between kinds of cartographic symbols.

Isarithmic maps serve two purposes: (1) they may provide a total view of the configuration of the statistical surface, for example, the form of the land, and (2) they may serve to portray the location of a series of quantities, for example, elevations at points. The ability with which they perform these functions is dependent on the validity and reality of the surface. For example, we can safely obtain values of elevations by interpolation from a contour map, within a certain margin of error. On many isarithmic maps this cannot be done so readily; this is especially true of large-scale isoplethic maps. In such maps the isopleths serve more as "form lines," to delineate the general character of the surface, and less as a method of portraying specific values at points.

Elements of Isarithmic Mapping. When we are called on to prepare an isarithmic map, whether we are working with an isarithmic or isoplethic surface, we follow much the same procedure. In each case the numbers on our map, constituting the series of Z values, are spaced some distance apart; whether they can or cannot actually occur at points, we must, by assumption, inference, and estimate, produce isarithms that represent a continuous statistical surface. Three elements are present that may be called the elements of isarithmic mapping: (1) the location of the control points, (2) the gradient assumed, and (3) the number of control points.

LOCATION OF THE CONTROL POINTS. The location of each Z value of the assumed statistical surface is called the *control point*. The XY positions and Z values of these points constitute the statistical evidence (called "the control" in cartography) from which the character of the statistical surface is assumed and from which the locations of the isarithms on that surface are inferred.

The problem of choosing a location for the control point is not difficult when the statistical surface being mapped is based on actual or derived quantities that occur at points, as in isometric mapping. The location of each observation is then the location of the control point. This is not so, however, when the mapping distributions are derived from ratios or percentages that involve area in their definitions, as in isoplethic mapping [e.g., the density of population (persons per unit of area)]. These kinds of quantities are derived from two sets of data based on unit areas. The resulting numbers, on which the ultimate locations of the isopleths depend, refer to the whole areal unit employed, and each is "spread" over the entire area of the unit. Therefore, there can be no points at which the values used in plotting the isopleths exist. Nevertheless, in order to use the isarithmic method, it is necessary to assume positions for control points so that the isarithmic lines may be positioned.

When the distribution is uniform over an area of regular shape, the control point may be chosen as the center. If the distribution within the unit area is known to be uneven, the control point is shifted toward the concentration. Center of area may be considered as the balance point of an area having an even distribution of values without any unevenness of the distribution taken into account. The center of gravity takes into account any variation of the distribution. Figure 11.8 illustrates the concept. The four diagrams of statistical divisions show possible locations of the center of gravity

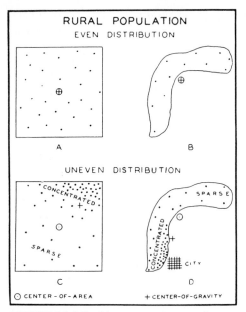

FIGURE 11.8 *Problems in positioning the control points arise because either the center of area or the center of gravity may be used. These centers may not always coincide, or be representative of the distribution being mapped. (From J. R. Mackay, courtesy of Economic Geography.)*

and the center of area for uniform and variable distributions. They also illustrate the problem of locating the control point in regularly and irregularly shaped divisions. In Fig. 11.8A, which is square, the center of gravity and center of area would, of course, coincide at the intersection of the diagonals. Because the distribution in Fig. 11.8B is even, the two centers also coincide, but they lie outside the irregularly shaped area at some point that is more representative of the whole division than any point within it. The distributions are uneven in Figs. 11.8C and 11.8D, so the center of gravity is displaced away from the center of area. In each case, however, the center of gravity is probably the more reasonable location for the control point. In isoplethic mapping the cartographer has a great responsibility, since the control point location is assigned by the map author. The effects of poor control point location will be seen in the section on error in isarithmic mapping.

INTERPOLATION. The selection of the isarithmic method follows on the assumption that the distribution being mapped is continuous and smoothly undulating (i.e., is composed of slopes). The surface of the land is visible either by stereophotogrammetric methods or in the field, and the infinity of Z values can, therefore, be directly or indirectly employed as a guide for the delineation of the form of the land surface by the isarithms. In contrast, the basic characteristics of the configuration of most other statistical surfaces are known only imperfectly, and an assumption is necessary as to the kinds of gradients that exist between the Z values at control points.

A cross section of a statistical surface that shows a profile along the top is an ordinary graph in which Z is the ordinate and XY is the abscissa. Consequently, as is illustrated in Fig. 11.9, if two Z values at different XY positions are represented, then the gradient between them may be represented by the straight line a or by some other gradient, such as the dotted lines b or c. It is entirely possible that c may represent the true slope; but unless evidence is available to indicate that Z varies in a curvilinear relation with change in XY, such an assumption is obviously more complex than the gradient shown by a. In science, whenever several hypotheses can fit a set of data, the simplest is chosen. As previously pointed out, the slopes of the actual land surface are known, in most cases, to bear certain kinds of curvilinear relationships to XY positions of the land. On the other hand, for most other statistical data distributed over the earth, this is not known. Consequently, in the majority of manually drafted

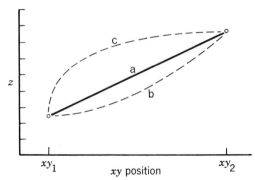

FIGURE 11.9 *Three possible gradients which can be assumed between two control points.*

isarithmic maps, a linear gradient is assumed in the construction of the isarithms. There are theories in some cases, such as that population density tends to be curvilinear (for example, *b* in Fig. 11.9).

Interpolation is the name for the process of estimating the magnitude of intermediate values in a series, such as the *Z* values along the lines shown in Fig. 11.9. The control points of a statistical surface to be mapped by isarithms constitute the series; when no evidence exists to indicate a non-linear gradient between control points, interpolation becomes merely a matter of estimating linear distances on the map in proportion to the difference between the control point values. For example, Fig. 11.10 is a map on which are located *Z* values at *XY* positions *a*, *b*, *c*, and *d*. If the position of the isarithm with a value of 20 is desired, it will lie $^3/_{10}$ of the distance from *a* to *b*, $^3/_7$ of the distance from *a* to *c*, and $^3/_4$ of the distance from *a* to *d*. Lacking any other data, the dotted line representing the 20-isarithm would be drawn as a smooth line through these three interpolated positions.

Figure 11.11 illustrates one of the common problems that arises when linearly interpolating manually among control points located in a rectangular pattern. Many census data are based on the rectangular minor civil divisions of the United States and Canada, and many other surveys use more or less rectangular subdivisions. When control points are arranged rectangularly, it is usual that alternative choices will arise concerning the location of an isarithm when one pair of diagonally opposite *Z* values forming the corners of the

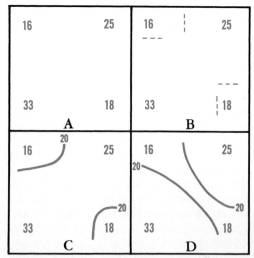

FIGURE 11.11 (A) The Z values are arranged rectangularly. (B) The positions of the isarithms of value 20 between adjacent pairs have been interpolated. (C,D) The two ways the isarithm of 20 could be drawn through these points. (Redrawn, courtesy of J. R. Mackay and The Professional Geographer.)

rectangle is above and the other pair below the value of the isarithm to be drawn.

A careful examination of other relevant information may help to indicate which choice of the two alternatives is better. If this is not possible, averaging of the interpolated values at the intersection of diagonals will usually provide a value that will remove the element of choice (see Fig. 11.12). We are forced, in the absence of other data, to assume the validity of the average. This is often used in computer contouring routines. In

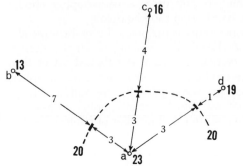

FIGURE 11.10 Linear interpolation between control points.

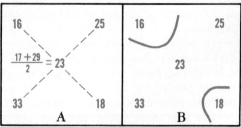

FIGURE 11.12 If the average of the interpolated values at the center is assumed to be correct, the isarithm of 20 would be drawn as in B. (Redrawn, courtesy of J. R. Mackay and The Professional Geographer.)

isoplethic mapping, if the cartographer has control over the shapes of the unit areas, the problem of alternative choice can be prevented by designing the pattern of unit areas so that the control points have a triangular pattern.

In recent years several programs for computer interpolation of Z at XY positions in isarithmic mapping have been developed. All common output devices can produce computer-interpolated and devised isarithms. As the use of computer interpolation increases, the complexion of the interpolation problem also changes.

Most often in today's computer routines, interpolation is done in two stages: (1) a primary interpolation, not necessarily linear, from the irregular original data points to a square or rectangular matrix of interpolated values, and (2) a secondary, usually linear, interpolation from the interpolated matrix of values to the positioning of the isarithmic lines. This two-stage interpolation can only be avoided when the original data set exists as a square or rectangular matrix of Z values or when the complete set of original data is approximated by a surface approximation equation. As we have already indicated in Chapter 7, surface approximations are also used for generalization and can result in greatly simplified surface forms, especially when the original data are irregularly spaced.

As in all isarithmic mapping (except mapping of the land surface), a fundamental assumption is required concerning the configuration of the statistical surface. This is made necessary by the use of sample points and, irrespective of whether the interpolation is done by machine or by hand, the character of the surface away from the points must be inferred. The relation among the spacing of the control points, the interpolation models, the number of control points, and the fundamental character of the "total" statistical surface is obviously complex.

Because of cost (time) constraints, manual interpolation is almost solely linear. Machine interpolation is not necessarily linear. Many different mathematical procedures can be used for first-stage machine interpolation. For example, Fig. 11.13 shows the same initial data set using four common first-stage machine interpolation methods. The choice among outputs can only be made with regard to the variation assumed by the cartographer to be present on the surface. Since the variations of only a few phenomena are actu-

ally known, cartographers face a major problem in selecting a first-stage interpolation method. For example, which map in Fig. 11.13 most adequately portrays the data? In an actual mapping situation, there is presently no way to answer that question.

NUMBER OF CONTROL POINTS. The final element of isarithmic mapping is the number of control points. There are several important concerns regarding the number of control points of which the cartographer must be aware. Initially, one of two conditions presents itself to the cartographer: (1) the data to be mapped must be collected by the cartographer, or (2) the cartographer is presented with a fixed set of data and there is no way to enlarge the number of data observations and no way to improve their accuracy. Unfortunately, the second case predominates.

In general, the larger the number of control points, the greater can be the detail of the surface portrayal. Since the three elements of isarithmic mapping do interact, however, a large increase in the number of control points is not always a blessing. There is a point of diminishing returns beyond which the increased improvement in surface portrayal is not worth the extra effort to process the additional control points. When this point of diminishing returns is reached depends on all the controls of generalization and the other two elements of isarithmic mapping. Certainly more control points that tend only to increase the clustering of the control points are not necessarily beneficial for surface portrayal, since this is more likely to lead to uneven generalization. Likewise, the purpose of the map may require a generalization level that requires only a minimum number of points. In any case, a cartographer should always avoid using too few points.

Control can be maintained over the three elements of isarithmic mapping when the cartographer gathers the data. A supplied set of data presents a different problem for the cartographer, and there are some guidelines for isarithmic mapping that will permit a cartographer to assess a fixed data scatter and number of control points.*

* *Joel L. Morrison,* Method-Produced Error in Isarithmic Mapping, *Technical Monograph No. CA-5, Cartography Division, American Congress on Surveying and Mapping, March 1971.*

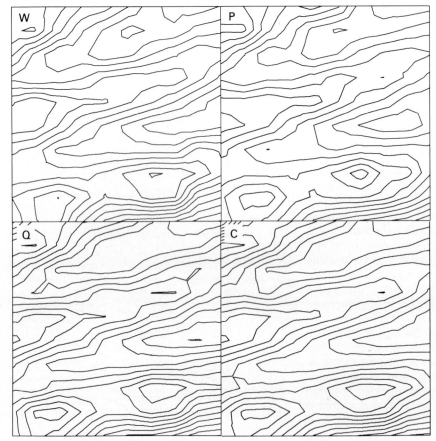

FIGURE 11.13 *Four first-stage interpolation models applied to the same data set. The models are: W, weighted by distance squared; P, least squares plane of best fit; Q, least squares quadratic of best fit; and C, least squares cubic of best fit.*

Error in Isarithmic Mapping. Error in maps is an elusive thing and depends a great deal on the various definitions we employ. It ranges from the generalized "error" in a coastline to the "error" involved in ascribing a mean value to an entire unit area. Such a complex subject can be touched on only lightly in a textbook of this sort and, in many cases, only by implication. The process of isarithmic mapping, however, is subject to a considerable variety of errors stemming from the three elements of isarithmic mapping and from the quality of the data and the selection of the isarithmic interval.

For isometric mapping the errors arising from the three elements in isarithmic mapping are straightforward but, for isoplethic mapping, the lo-cation of the control points, since the cartographer establishes their position during the mapping process, is subject to more error than in isometric mapping. The quality of the data obviously affects both isometric and isoplethic mapping, as does the selection of the isarithmic interval. The selection of an interval is fundamentally a generalization (classification) process, and the control quality-of-data serves as a control on generalization. Each of these factors will be discussed briefly.

METHOD-PRODUCED ERROR. Errors rising from the elements of the isarithmic mapping procedure are collectively called method-produced errors. The number of control points, their locations, and the selected interpolation model interact to affect the accuracy of any isarithmic map. Figure 11.14 illustrates the effect of increasing the number of control points. Figure 11.15 illustrates the effect of improving the location of the control points; Fig. 11.13 illustrated the results obtained from employing different interpolation methods. Obviously, when three elements are used to the best of a cartographer's ability, the reliability of the isarithmic portrayal will be enhanced.

Figure 11.16 illustrates the hypothetical relationship between number of points and accuracy. Accuracy is usually very slight when few control points are present; then a rather dramatic increase in accuracy occurs as the number of control points increases. The diminishing returns mentioned earlier result in the leveling off of the curve as more and more control points are added. The cartographer should strive to balance the accuracy level desired against the cost of obtaining or using more control points.

Figure 11.17 illustrates the hypothetical relationship between location of control points and accuracy. From this figure it is obvious that clustered control point scatters (near the left hand edge of Fig. 11.17) are to be avoided and that equispacing (2.0 on abscissa) is not as desirable as is "not quite equispacing" (to the immediate left of 2.0 on the abscissa). The best results are usually obtained by a scatter of control point locations that is more evenly spaced than random.*

In isoplethic mapping the cartographer has more control over the number and location of control points, because the data exist or are collected for areas. The sizes and shapes of these areas affect the locations of the isopleths. Elongated unit areas can create, in some instances, strong gradients transverse to the orientation of the elongation; this in turn will induce the isopleths to lie in the direction of elongation. Figure 11.18 illustrates this phenomenon. In Fig. 11.18A a hypothetical series of unit areas is shown together with isopleths located with an interval of two. In Fig. 11.18B a smaller series of elongated minor civil divisions has been formed by aggrega-

---

*Joel L. Morrison, "Observed Statistical Trends in Various Interpolation Algorithms Useful for First-Stage Interpolation," The Canadian Cartographer, 11 (2), 142–159, (1974).

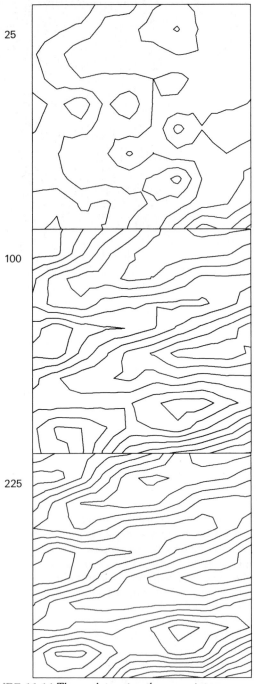

25

100

225

FIGURE 11.14 Three plots using the same interpolation model but with successively more initial data. The top plot utilizes 25 points, the middle plot utilizes 100 points, and the bottom plot utilizes 225 points.

FIGURE 11.15 *Six plots illustrating the effect of sample scatter. Plots 1, 2 and 3 are produced from random or more clustered than random scatters, especially plot 1, which is highly clustered. Plots 4, 5 and 6 are produced from scatters that are more systematic than random, especially plot 5, which is a square lattice of points.*

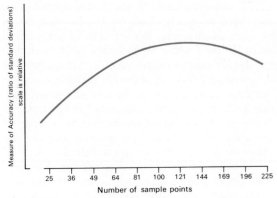

FIGURE 11.16 *The characteristic curve of the hypothetical relationship between number of points and accuracy of the isarithmic map. The Y-axis scale is relative.*

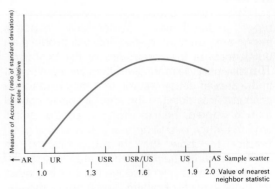

FIGURE 11.17 *The characteristic curve of the hypothetical relationship between sample point scatter and accuracy of the isarithmic map. The Y-axis is relative.*

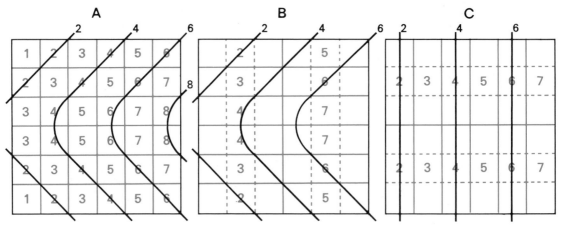

FIGURE 11.18 Diagram illustrating how elongated unit areas can sometimes affect the positions of isopleths. See text for explanation.

tion from exactly the same distribution of data as shown in Fig. 11.18A. The plotted isarithms show the same pattern. But when this process is followed with a different orientation of the elongated unit areas in Fig. 11.18C, the isopleths are straightened and "pulled" into a different pattern.

There is not much the cartographer can do to compensate for this sort of effect, except to look for additional data that will either reinforce the pattern in an area of elongated unit areas or, conversely, lead to a modification of the possible artificial effect created by the elongation. Sometimes, by estimating densities of parts (discussed in Chapter 7), the cartographer can counterbalance this trend.

When the control points are unevenly arranged so that there is a greater density of points in one area than in another, this will produce an unevenness in the consistency of the treatment. This may provide more detail than we would want in the area of dense control, or it may lead to the converse, less detail in the area of sparse control than is wanted; hence, clustered control points tend to reduce overall accuracy. It is always good practice to show the control points on an isarithmic map to help the reader judge the quality of the map. This may be done by showing either the data points for isometric line maps or the unit areas (or control points) for isopleth maps.

With the advent of computer-assisted car-

tography the possibilities for using various nonlinear interpolation methods became practical for the first time. As illustrated in Fig. 11.13, cartographers can only specify with confidence an interpolation method when a theory exists about the variation present in the phenomenon being portrayed. Where there is no such theory, the selection of the interpolation method must be done in conjunction with the number and location of control points and the generalization level desired for the given map. An assessment of the method-produced error is difficult.

QUALITY OF DATA. As illustrated in Fig. 11.19, any error in the Z value at a control point can have as much effect on the location of an isarithm as changing the location of the control point. If the Z value of a control point is correct but the XY position is incorrect, a displacement of the isarithm will result. If the XY position is correct but the Z value is in error, the same thing will happen if a linear gradient is assumed. Clearly both the positions of the control points and the validity of the value at the control point have considerable effect on the accuracy of the statistical surface.

Several kinds of factors affect the validity of the control point value and, hence, the certainty with which the cartographer can locate the isarithm. Whenever there is a question concerning the accuracy of the Z value, this doubt is automati-

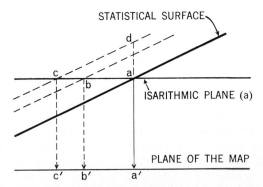

FIGURE 11.19 If a' is the XY location of a control point with Z value a, and if the isarithmic plane has the value a, then a' will be the orthogonal position on the map of the isarithm. If the Z value a is incorrect and really should be d, then the a isarithmic plane would intersect the surface at c and the isarithm on the map would be located at c'. If the a value were correct by the XY position of a should be at b, the b' would be the map position of the isarithm.

cally projected to the map; it there becomes transformed to a question as to the XY position of the isarithm. There is, therefore, always a zone on the map within which any isarithm may be located, depending on the certainty of the Z values, and the width of this zone depends on the amount of error in the Z value, assuming that the control point (the XY position) is correct.

There are three kinds of errors that commonly affect the reliability of the Z values from which an isarithmic map is made: (1) *observational error*, (2) *sampling error*, and (3) *bias or persistent error*. A fourth, *conceptual error*, is related not so much to error as it is to the validity of the concept being presented by the map.

Observational error refers to the method used to obtain the Z values. If they are derived by means of instruments operated or read by human beings, there is usually some inaccuracy, both in the instruments and in their reading by the observers. Observational error is not limited to instruments, however. A considerable amount of statistical data is based on various kinds of estimates, such as a farmer's estimate of the extent of cropland, the yield per acre, and the extent of soil

erosion. Most statistical data are subject to observational error of one sort or another.

Sampling error is of several kinds. The most obvious is that associated with any map that purports to represent the distribution of an entire class of data, only a sample from which is actually known. Any map of mean climatological values, which is not specifically for a particular period, is, in effect, attempting to describe a total average situation (or statistical population) from only one relatively small time sample. Many kinds of data in censuses are collected by a sampling procedure. In this case, the statistical population exists at one time, but the cost of ascertaining it in its entirety is too great; thus, a sample is taken.

Bias or persistent error may be of many kinds. Instruments may consistently record too low or too high; the majority of weather stations may be in valleys or on hilltops; people show preferences for certain numbers when estimating or counting; and so on. Bias is difficult to ascertain but may have considerable effect on Z values.

Conceptual error in the Z value may be illustrated by the use of mean values computed from a series of observations with which to make a map of mean monthly temperatures. The mean may not describe the actual fact very well, since the dispersion of the values around the mean may be large. For example, the standard deviation of mean December temperatures in the central United States is around 4°F. There is therefore a high probability that approximately one-third of the time an individual December mean will be more than 4° above or below the average of the December means. A cartographer constructing a map that purports to show by isarithms the distributions of mean December temperatures will, in effect, be forced to locate each isarithm near the center of a map zone or band represented horizontally by several degrees of temperature gradient. Furthermore, it is necessary to realize that if individual year-to-year maps had been made, approximately one-third of the time the isarithm would lie outside the zone delineated by an average of 8° of temperature gradient. On even a small-scale map, such a zone would be of noticeable dimensions. It follows that minute isarithmic wiggles and sharp curves would be an affectation of accuracy not supported by the nature of the data.

It is important to take into account the various kinds of possible error in the validity of the Z values. In many cases, some of the kinds of sampling error have already been ascertained and need only be obtained from the sources, for example, the U.S. Census of Agriculture. Standard deviations and standard errors of the mean are often available or are not difficult to compute. Simple logic and common sense will often provide enough of an answer for highly generalized maps. The total effect of all the possible sources of error and inconsistency from one part of an isarithmic map to another can function toward only one end, that is, to *smooth out the isarithms.*

CLASS INTERVALS IN ISARITHMIC MAPPING. Relative gradients on isarithmic maps are easily judged only when there is a constant interval; the general rule that the closer the isarithms, the steeper the gradient, holds only under such circumstances. Therefore, if primary interest is in the configuration of the statistical surface, the interval should be constant. By far, most isarithmic maps employ equal-spaced intervals.

The recognition of form on the statistical surface from the patterns made by the isarithms, such as concave or convex slopes, hills or valleys, escarpments, and the like, is difficult, if not impossible, when irregular intervals are used. The nature of slopes is easily revealed by the spacings of lines from an equal steps interval in the one case and almost completely hidden by the use of an unequal interval in the other. In some cases, when interest is concentrated in one portion of the range of values, an increasing or decreasing interval can be justified. This requires much more mental effort on the part of the map reader when trying to infer the form of the surface, but it does allow the cartographer to concentrate detail in one portion of the range. This should only be applied rarely and with sufficient notice to the map reader.

An equal steps or constant interval is ordinarily employed on contour maps of the land and isometric maps of other phenomena. On the other hand, Mackay has found that an equal steps interval is the exception in isopleth mapping.*

---

*J. Ross Mackay, *"Isopleth Class Intervals: A Consideration in their Selection,"* Canadian Geographer, 7, 42–45 (1963).

Summary of the Isarithmic Method

Perhaps no cartographic method is as useful or often requested for surface portrayal as the isarithmic method. First, for the practiced reader, the configuration of the surface can be very graphic when equal-step intervals are used. Second, for extracting data from maps, detailed cartometric methods can be used (especially where accuracy permits, for example, on topographic maps). Third, many other cartographic methods are derived from isarithmic portrayals, among them profiles and oblique traces (to be discussed below), hachuring, shading or shaded relief, and some dasymetric maps.

The variation between the two forms of isarithmic map, the isometric and the isoplethic, is not visually evident. The distinction only refers to the form of the original data used for the map: isometric maps result from positional data, and isoplethic maps result from areal data. Since the error characteristics of isarithmic maps have received more attention than the comparable characteristics of other cartographic methods, the cartographer has more information to enable the production of effective communication with isarithmic maps than with most other methods.

The Isarithmic Method Applied to Land Surface Form: Contours

Representation of the form of the land by means of equal-interval contours is the most commensurable system yet devised. As described above, contours are isarithms that are the traces obtained by passing parallel "planes" through the three-dimensional land surface and projecting these traces orthogonally to the plane of the map. A contour is, therefore, a line of equal elevation above some assumed starting elevation (see Fig. 11.5). The assumed horizontal surface of zero elevation is called the *datum,* which is the surface of a particular earth spheroid (a regular geometric figure) projected beneath the land. It is apparent from Fig. 11.5 that the spheroid surface is not a flat surface, but is curved in every direction. It is the problem of the mapper to establish the horizontal position on, and the vertical elevation above, this surface of a large number of points on the land. When enough positions are known and the curved datum surface has been transformed into a plane surface by means of a projection system, the map may then be made. The map reader *sees* the represented land surface orthogonally.

The representation of form by contours is an artificial system in that it has few counterparts in nature and, therefore, it is not in the normal experience of the average person. Figure 11.20 shows the usual way we see form—by the interplay of light and dark—and, for comparison, a contour representation of the identical form.

Contours on a topographic map are remarkably expressive symbols if they have been correctly located and if the interval between them is constant and relatively small. All the error sources outlined above still operate on contour maps, ex-

cept, of course, the isoplethic problem of control point location specification.

We must remember that not all contour maps are of the same order of accuracy. Before the acceptance of the air photograph as a device from which to derive contours, the lines were drawn in the field with the aid of a scattering of "spot heights" or elevations. Consequently, they were often not precisely located.

It is apparent that much of the utility of contours depends on their spacing, and the choice of a contour interval is not an easy task. Because the portrayal of the relative slopes (gradients) of the landform is one of the major objectives of contour maps, contour intervals are almost always equal-

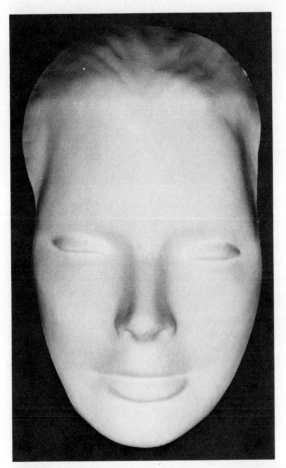
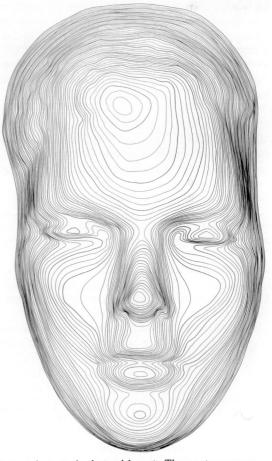

FIGURE 11.20 A plaster model and contours with a 1 mm interval precisely derived from it. The contours were obtained by photogrammetric methods. (Courtesy G. Fremlin.)

step progressions. As Fig. 11.7 illustrated, equal-spaced contours convey profile types. When an uneven contour interval is employed, slopes are difficult to calculate, and misleading impressions can be gotten if strict attention is not focused on the contour interval.

As the interval is increased, the amount of surface detail lost between the contours becomes correspondingly greater. If, because of a lack of data or scale, the contour interval must be excessive, other methods of presentation, such as shading, are likely to be preferable. In some cases, lack of data requires the use of *form lines* in place of contours. These are discontinuous lines which, by their arrangement, suggest slopes and the general configuration, but from which precise information regarding slopes and elevations are not to be read.

Although contours do not present quite so clear a visual picture of the surface as shading, the immense amount of information that may be obtained by careful and experienced interpretation makes the contour by far the most useful device for presenting the land on topographic maps.

Profiles. Another method of portraying the surface shell of an assumed statistical surface is to use profiles. A profile trace results from the intersection of a plane perpendicular to the XY datum and the statistical surface. Strictly speaking, a single profile trace does not constitute a map. A series of profile traces placed in relative position to one another, however, can enable the practiced observer to visualize a surface quite well (see Fig. 11.21).

Cartographers have paid scant attention to using series of profile traces for displaying the statistical surface, and there has been little investigation of the perception of profiles.

The construction of a profile begins with an isarithmic map. The important steps are illustrated in Fig. 11.22. The line along which a profile is desired is marked on the isarithmic map. The end points of the profile and the intersection of each isarithmic line with the profile line are transferred to a blank sheet of paper. A set of parallel ruled lines, spaced at a distance that reflects a suitable vertical exaggeration, is numbered to include the isarithmic values of each line crossing the profile on the map. The appropriate altitudes on the par-

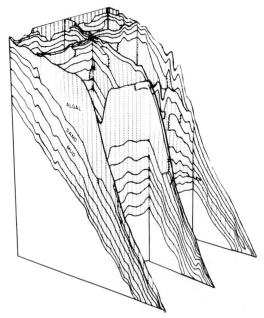

FIGURE 11.21 *A series of profiles situated at right angles to each other gives a clear picture of the surface and its underlying strata. (Courtesy of John Wiley and Sons and J. W. Harbaugh and D. F. Merriam.)*

allel lines are marked, depending on the isarithmic line value, and these altitudes are connected by a smooth line on the ruled parallel line set.

Figure 11.21 represents a series of profiles constructed in this manner along perpendicular directions. These surface profiles can also form the basis for the construction of an isometric block diagram. This technique can also be very instructive in helping a reader to learn to "visualize" an isarithmic map.

Oblique Traces. Oblique traces result from the intersection of a series of planes with the base datum at some angle $\theta°$, where $0° < \theta° < 90°$. These traces may be graphically portrayed either in planimetrically correct position or in one or two point perspective.

Planimetrically Correct Oblique Traces. A vertical profile across the surface of the ground, when viewed from the side and placed on a map,

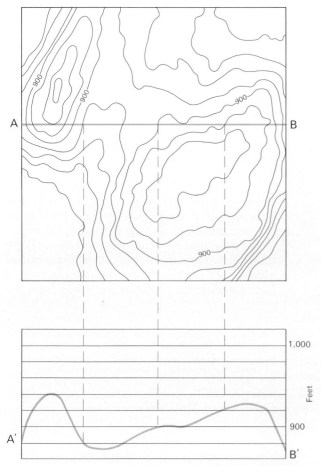

FIGURE 11.22 *The construction of a profile from an isarithmic map. Line AB is drawn on the map. Intersections of the isarithms with line AB are perpendicularly projected on the appropriate line of the set of parallel ruled lines below. The resulting points are connected by a smooth line resulting in a profile A'B'.*

utilizes planimetric position in two dimensions, yet the top and base of the profile are actually in the same planimetric position on the earth. It is impossible, therefore, to utilize planimetric position to express vertical dimension without producing planimetric displacement. In order to overcome this fundamental difficulty, it is sometimes desirable to substitute for the usual vertical or perspective profile across the land a line that gives a similar appearance but is not out of place planimetrically. This kind of cartographic legerdemain may be accomplished by mapping the traces of the intersections of the land surface

with a series of parallel inclined planes (Fig. 11.23).

When a single plane parallel to the base datum is passed through the land surface, the mapped trace of the intersection is a contour. If the plane is rotated on a horizontal axis, *ab* (see Fig. 11.24) and the successive traces plotted, and still viewed from above, the traces produced run the gamut from the contour trace, *r*, to a straight line, *t*. The trace produced by the inclined plane, *s*, when viewed from directly above, has much the same appearance as would a conventional vertical profile if it were being viewed in perspective at an

FIGURE 11.23 A much reduced portion of a planimetrically correct terrain drawing of the Camp Hale, Colorado area. The map was drawn at a scale of 1:50,000. (Drawn by N. J. W. Thrower, courtesy of The Geographical Review.)

trace is a vertical profile, and its orthogonal representation is a straight line. When θ is greater than zero but less than 90° (s in Fig. 11.24), the trace will be that of the intersection of the ground with an inclined plane. A series of such traces on parallel inclined planes produces an *appearance* of the third dimension while still retaining correct planimetry (Fig. 11.25).

The construction of an inclined trace is not difficult. It can be done either manually or with computer assistance.*

PERSPECTIVE TRACES. One of the first cartographic methods to be programmed to take advantage of computer assistance was the calculation and automatic plotting of perspective traces. Commonly used computer programs allow (1) the rotation, θ, about the vertical axis of the statistical surface, (2) changes in the viewing distance, *d* (i.e., the distance of the viewer's eye from the

oblique angle as in a landform diagram. If the angle of inclination from horizontal is designated as θ, then when θ is zero, the trace of the ground with the plane orthogonally represented on a map is the conventional contour. When θ is 90°, the

The interested student is referred to the appropriate references for the detailed construction procedures. Arthur H. Robinson and Norman J. W. Thrower, "A New Method of Terrain Representation," The Geographical Review, 47 507–520 (1957). Pinhas Yoeli, "Computer-Aided Relief Presentation By Traces of Inclined Planes," The American Cartographer, 3, 75–85 (1976).

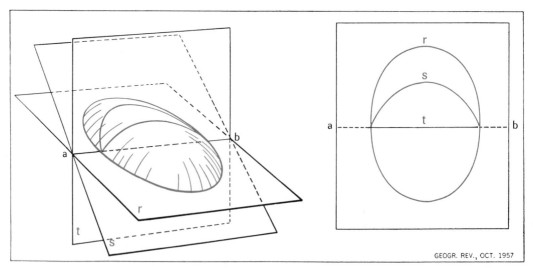

GEOGR. REV., OCT. 1957

FIGURE 11.24 A comparison of the appearance on a map (right) of three traces produced by passing three differently oriented planes (r, horizontal, contour; s, inclined; and t, vertical, profile) through a single landform. The drawing at left shows the relationships in perspective. (Courtesy of The Geographical Review.)

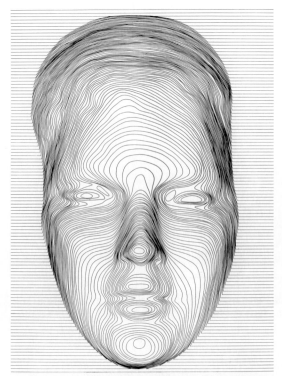

FIGURE 11.25 The model illustrated in Fig. 11.20 is shown here mapped with inclined traces from parallel planes angled 63 degrees upward toward the rear. (Courtesy G. Fremlin.)

corner of the map), and (3) changes in the viewing elevation, ϕ (i.e., the angle at which the viewer is above the horizon) to be specified by the cartographer (see Fig. 11.26). Computer-drawn perspective traces can be drawn along two directions perpendicular to each other to form "fishnets" that tend to give a realistic impression of surface form (see Fig. 11.27) or along either the X or Y direction only. Either one-point or two-point perspective can be employed (see Fig. 11.28). The ease of production and the resulting realistic impression insure this method a secure place in the cartographer's "bag of tricks" for the future.

FLOW LINES

A second major category of line symbolization of volumetric data is flow lines. Quantitative flows may be shown by vectors or lines running from position to position, along a route, from area to

area, or any combination of these. The cartographer has primarily two different methods in this category with which to symbolize quantities. One of these methods has been used for the portrayal of the statistical surface.

Inferring the Statistical Surface by Hachures. When the slopes of a statistical surface are considered as linear values existing at points, the surface may be portrayed by a line symbol called a hachure. The hachure (see Figs. 9.17 and 9.18) was widely employed during the nineteenth century to show the surface configuration, but its use has largely died out. The hachure is a line symbol drawn down the slope, and it is varied in width or spacing in relation to the slope of the land on which it lies. The original form of hachures was based on the assumption of a vertical source of illumination. Therefore, the steeper the surface, the darker the representation. Accordingly, some of the light would be reflected in the direction of origin (to the reader) in some proportion to the angle of the slope.

Not long after hachuring with vertical lighting was employed on topographic maps, it was discovered that much more realistic modeled effects would be attained by varying the line widths to give the effect of oblique lighting.

The major difficulty experienced with hachures is that, although slope is their basis, it cannot practically be measured from the map, regardless of the precision underlying the representation.* Flat areas, whatever their location, appear the same, and only rivers and streams or spot heights strategically placed make it possible for the reader to tell valleys from uplands. Another difficulty of hachuring is that its effectiveness, when printed in one color, is greatly dependent on the darkness of the ink. Thus, a considerable problem is created; as darker inks are used to make the terrain more effective, the other map detail becomes correspondingly obscured.

It is interesting to note that precise, effective hachuring depended on a considerable knowledge of the terrain. In practice, contours were often drawn on the field sheets of a survey, and the hachures were drawn in the office from the contours. Thus, the original French survey for the nineteenth-century 1:80,000 map had contoured field sheets but was published only in the hachured form!

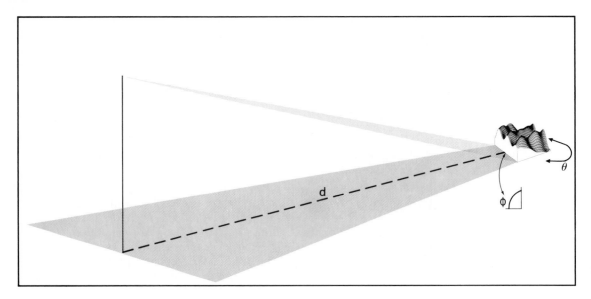

FIGURE 11.26 *Three principal variables that the cartographer must specify when using computer assistance to plot a series of perspective traces. θ is the angle through which the block is rotated in the XY plane, φ is the viewing elevation, and d is the viewing distance.*

Naturally, the hachurelike symbol need not be restricted to showing the gradients on the land surface; it can be employed to show the directions and magnitudes of any sort of gradients, from temperature to population potential. It has not been used very often in this way, however, since cartographers and geographers generally have thought that the gradients were adequately portrayed by the arrangements of sets of isarithms, which provide considerably more information than just gradients. It is reasonable to suppose that this attitude will continue, not only because the isarithm is extremely efficient as a symbol, but because it can be more easily automated than the hachurelike symbol. Nevertheless, if emphasis on the gradients of a distribution is paramount, the hachurelike symbol can be employed. As noted above, an isarithmic map must be constructed first.

Other Flow Lines. The principal method of symbolizing quantitative data by line symbol does not show a continuous distribution. The symbols that portray particular "connections" or "routes" belong in this group. For example, the flow of

petroleum from the Middle East to Europe, of vacationers from the urban northeast of the United States to Florida, or the draft capacity of a canal system all occur along specific routes, either in the actual sense with the canals or in the more general sense with the commodity or population flows. No statistical surface is assumed.

The techniques used in the construction of flow maps are easily observed from well-made examples. Interval or ratio scaling is most often accomplished by merely graduating the thickness of the lines in proportion to the values by using a convenient unit width. Range grading is also possible. The unit width is selected so that division of the data by the unit width results in a number that represents a map dimension in inches or millimeters, in the same fashion that a unit value is used to determine the drafting sizes of graduated circles. For example, if we were representing the flow of freight traffic between two cities, we could use a unit value of 1 mm for each 1 million metric tons; therefore, if the traffic total were 4.5 million metric tons, a line 4.5 mm wide would be shown between the two cities. There appears to be no tendency toward underestimation in map readers

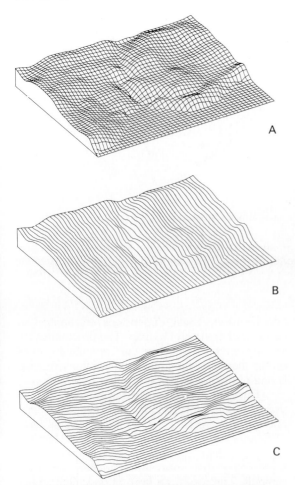

although the varying thickness and angle with which "tributaries" enter frequently show flow adequately. Arrows may be placed along the lines, as in Fig. 11.31, to show the direction or even to show relative movement in opposite directions. Lines can increase or decrease in width as the values change but, between tributaries, a line should maintain a uniform width. Tributaries should, of course, enter smoothly in order to enhance the visual concept of movement.

In some instances, the range of the data is so large that a unit width value capable of allowing interval or ratio differentiation among the small lines would render the large ones much too large. It is, consequently, sometimes necessary to symbolize the smaller lines in some way, such as by dots or dashes (see Fig. 11.31).

Instead of using a system of symbolization that depends on variable width, we can employ

FIGURE 11.27 Three computer-drawn sets of perspective traces: (A) "Fishnet" where traces are drawn in both the X and Y directions. (B) Traces drawn only parallel to the X direction. (C) Traces drawn only parallel to the Y direction.

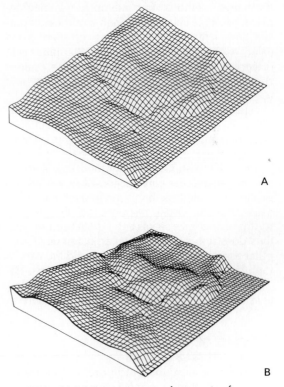

who view flow lines of variable width. Therefore, psychological scaling does not differ from numerical scaling. Lines may be shown as smooth curves, as in Fig. 11.29, or as angular lines, as in Fig. 11.30.

Movement along an actual route may be represented or, as in "origin and destination" maps the terminal points may simply be connected by straight lines. Arrowheads at the ends of lines are often used to show the directions of movement,

FIGURE 11.28 Two computer-drawn sets of perspective traces: (A) Using one-point perspective. (B) Using two-point perspective.

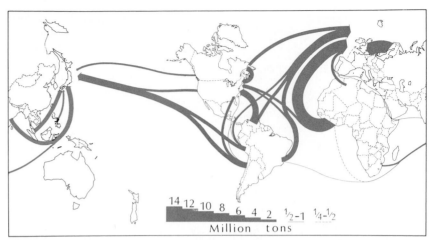

FIGURE 11.29 A portion of a flow-line map showing the movements of iron ore. [Map by G. B. Lewis. From G. Manners, "Transport Costs, Freight Rates and the Changing Economic Geography of Iron Ore," Geography, 52, 260–279 (1967).]

alternative methods. For example, we may replace the single line with a variable number of parallel lines, letting each of the lines represent a unit value, as in dot mapping. Another alternative is to categorize the data and symbolize the various classes with symbols that vary in pattern and value.

Range-graded line symbolization is also possible. As with all range-graded symbolization, a finite number of symbol sizes, each representing an interval part of the range of the data, are used. Figure 8.19 illustrates range-graded flow lines.

In general, the construction of flow lines is a relatively simple task that is made even easier because human visual response to the varying widths of flow lines seems to be quite accurate and thus does not require scaling adjustments. Computer assistance is also available to aid the cartographer in the preparation of flow line maps. Because of their excellent perceptual qualities, flow lines are highly recommended where appropriate.

MAPPING QUANTITATIVE DATA USING AREAL SYMBOLS

The use of areal symbols to portray quantitative data in cartography is relatively straightforward. The symbol dimensions of most utility are color-value and pattern-texture. Color-intensity may

also be used. These dimensions are applied to designated areas on the map. The procedures for arriving at meaningful lines that separate the areas on the map (i.e., the areal extent of the symbols), is one of the cartographer's major problems.

The data may be positional or areal, but the use of areal symbolizations in conjunction with any quantitative data implies a statistical surface. The cartographer can map positional data by clustering the data into homogeneous groups and symbolizing the areas thus created by areal symbols. The result is a form of dasymetric mapping. The cartographer can also map areal data by applying areal symbols to the various map areas, resulting in either a simple choropleth or a dasymetric map. Finally, the cartographer may wish to display positional or areal data constituting a statistical surface subjectively by continuous tonal shadings.

MAPPING THE STATISTICAL SURFACE BY AREAL SYMBOLS

When confronted with quantitative areal data that, by definition, constitute a statistical surface, the cartographer must conceptually decide whether to employ line symbols or area symbols. Essentially three techniques are available: the isoplethic technique, which employs line symbols, as described earlier in this chapter, or either the simple choro-

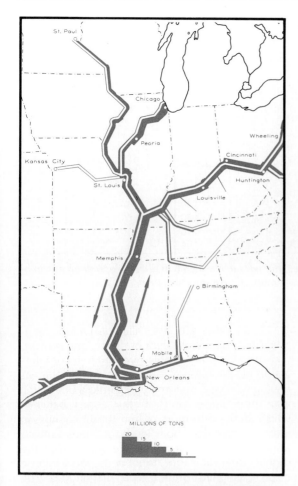

FIGURE 11.30 A portion of a flow map showing 1949 tonnage of barge and raft traffic in the United States. The legend has been moved. Note how direction of movement is indicated. (From E. L. Ullman, American Commodity Flow, University of Washington Press, Seattle, 1957.)

one of the two choroplethic techniques is appropriate. As noted earlier, when the emphasis in the mapping is on the gradients among the Z values, their magnitudes, and their directions, the isoplethic technique is appropriate.

The simple choropleth, the dasymetric, and the isoplethic techniques are shown comparatively in Fig. 11.32. The diagnostic characteristics (i.e., the basic differences in appearance) among them are clearly evident. In the simple choropleth map the locations of the class limits employed coincide with the boundaries of the data collection unit areas. In the dasymetric map the map positions of the class limits are independent of the boundaries of the enumeration districts, but the lines separating one class from another have no numerical value, as is also true in the simple choropleth map. In the isarithmic map the positions of the lines showing the isarithmic class interval employed are independent of the unit areas; they are lines along which the value is assumed to be constant.

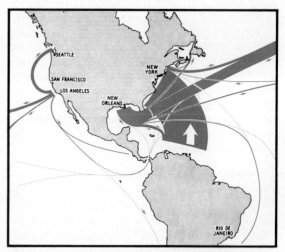

FIGURE 11.31 A portion of a quantitative flow map symbolizing 1947 tanker traffic of the United States. Since the thickest line represents well over 20 million tons, small values of less than 200,000 tons cannot be adequately shown by proportional thickness. Consequently, they are symbolized by line character (dots or dashes). (Map drawn by R. P. Hinkle. From E. L. Ullman, American Commodity Flow, University of Washington Press, Seattle, 1957.)

pleth or dasymetric techniques, which employ area symbols. The choice among these methods depends on whether the interest is in (1) as a distribution where the focus of interest is on the specific quantitative values at particular places, or (2) as a distribution where the focus of interest is on the geographical organization of the magnitudes and directions of the gradients (i.e., in its surface configuration). When the emphasis in the mapping is on the locational representation of the Z values,

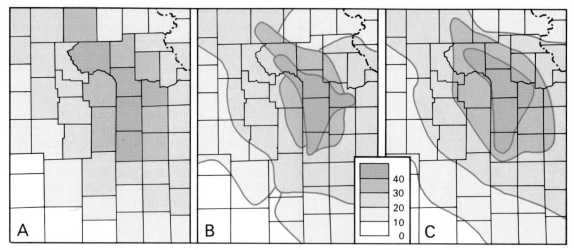

FIGURE 11.32 *Examples of three ways a set of Z values that refer to enumeration districts, or unit areas can be mapped: (A) A simple choropleth map. (B) A dasymetric map. (C) An isoplethic map.*

In contrast, the cartographer must employ a form of dasymetric mapping, as explained below, when confronted with quantitative point data to be symbolized by areal symbols.* Since, by definition, *every* set of quantitative data, either referring to areas or to points that is to be symbolized by areal symbols constitutes a statistical surface, there is only one *basic* method for representing a surface by areal symbols in a highly commensurable manner. It is called the choroplethic method. Other less commensurable methods for symbolizing surfaces by areal symbolization will be discussed at the end of this chapter.

CHOROPLETHIC MAPPING

The choroplethic method consists of two highly commensurable techniques for the portrayal of a statistical surface by areal symbols. These techniques can best be described as follows. In one, called simple choropleth mapping (from Gk. *choros,* place, and *plethos,* magnitude), the primary objective is to symbolize the magnitudes of the statistics as they occur within the boundaries of the unit areas, such as counties, states, or other kinds of enumeration districts.

* George F. McCleary, The Dasymetric Method in Thematic Cartography, *unpublished Ph.D. dissertation, Department of Geography, University of Wisconsin, Madison, WI, 1969.*

In the other subclass, called *dasymetric* mapping (from Gk. *dasys,* thick or dense, and *metron,* measure) the primary objectives are to focus interest on (1) the location and *Z* magnitudes of areas having relative *Z* uniformity, regardless of the unit area boundaries, and (2) the zones between which there occur more or less abrupt changes in these magnitudes. Both kinds of maps employ area symbols and, since they are quantitative, the darker visual value (gray tone) or the more intense color is assigned to the greater magnitudes. The simple choropleth map depends entirely on data collected by areal unit, but the dasymetric map can be created from either data that initially refer to areas or to points.

The Simple Choropleth Map. The simple choropleth map is, in effect, a spatially arranged presentation of statistics that are tied to enumeration districts on the ground. Tabular statistics are very convenient for many purposes, but frequently we want to refer to the array geographically instead of alphabetically and to be able to compare magnitudes in various places easily. This kind of map is made simply by symbolizing in some manner the quantitative magnitude that applies to the enumeration district or some other unit area. It is a range-graded symbolization scheme.

The simple choropleth map requires the least

analysis on the part of the cartographer. This technique may be used for many kinds of ratios and, when the statistical units are small, it provides sufficient variability in the map (if there is variability among the data) to take on some of the visual character of a continuous distribution. Ordinarily, however, it is employed when the problems of doing any other sort of map make it out of the question. For example, when mapping at a large scale, the cartographer may not know enough about the details of the distribution, or at a small scale, the regional variations may be too inconsistent to attempt an isoplethic or dasymetric map. In effect, the simple choropleth map presents only the spatial organization of the statistical data, with no effort being made to insert any inferences into the presentation.

Absolute numbers alone are not ordinarily presented in a simple choropleth map. For many classes of data, the absolute quantities of a phenomena (i.e., persons, automobiles, farms, and vacationers) are naturally functions of the areal size of the enumeration district. When we are concerned with the geographical distribution, we are normally concerned with comparisons of ratios involving area (density) or ratios that are independent of area (percentages or proportions), that is, when the effect of variations in the size of the enumeration districts has been removed.

The quantity mapped is therefore almost always some kind of average that is assumed to refer to the whole of a unit area, and the unit areas are range-graded in representation. Where class limits call for a change of symbolization, it occurs only at the unit area boundaries in a simple choropleth map. Since the boundaries are usually quite unrelated to the variations in the phenomenon being mapped, this adds a further measure of geo-

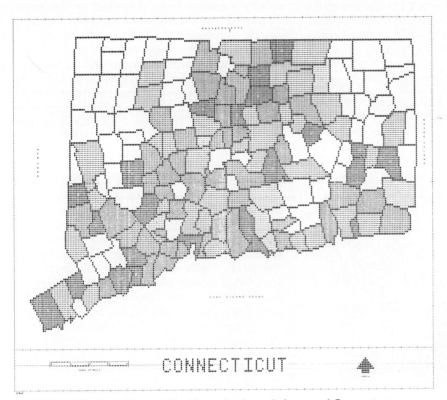

FIGURE 11.33 A machine-produced simple choropleth map of Connecticut showing the average sale price per house by towns. (Map prepared by the Laboratory for Computer Graphics, Harvard University.)

graphical obscurity to that supplied by the practice of dealing entirely with averages.

Choropleth mapping is "safe" in that we simply symbolize the quantities where they are, but it is the least informative of the commensurable methods of mapping quantitative areal data. The production of relatively coarse simple choropleth maps is easily automated. Because of the speed and relatively low cost with which the computer can handle the data and direct the line printer, this kind of map is important as a first "glance" at a distribution. Figure 11.33 is an example of a simple choropleth map produced on a line printer.

The difficulties in completely automatically producing choropleth maps of high visual quality are related to computer hardware and software developments. High visual quality choropleth maps can now be obtained by (1) photomechanical enhancement of computer line drawing output (see Fig. 11.34), or (2) computer output on microfilm plotters (see Fig. CI.12). As use of the latter output mode becomes more widespread, the cartographer should finally be able to obtain good quality automatic production of simple choropleth maps.

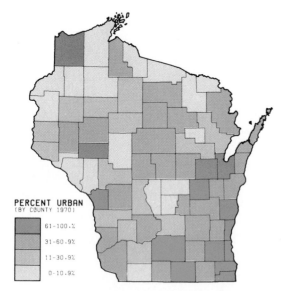

PERCENT URBAN
(BY COUNTY 1970)

61-100.%

31-60.9%

11-30.9%

0-10.9%

FIGURE 11.34 A map produced by photomechanically enhancing a computer line drawing obtained from a CALCOMP 1163 drum plotter. (Courtesy University of Wisconsin Cartographic Laboratory.)

The Dasymetric Map. The nature of and knowledge about the phenomenon being mapped, or the inefficiency and variation of the unit areas employed to derive the data values, may make it desirable to employ the dasymetric system of representation for areal quantitative data. Some kinds of phenomena do not extend, even when abstractly conceived, across certain kinds of areas. Neither do they have distributional characteristics like those related to the sloping-surface-concept associated with a continuously changing set of values. For example, agricultural population density in a region of extreme variation in character of agricultural land is likely to show relatively uniform high density in favorable areas together with sparse occupance in unfavorable areas. Changes would be rapid in zones that are not likely to coincide with the boundaries of the data collection unit areas.

The dasymetric map, although made from exactly the same initial data used for the simple choropleth map, assumes the existence of areas of relative homogeneity, which are also assumed to be separated from one another by zones of rapid change. By subdivision of the original statistical unit areas, additional detail may be added to the presentation. These additions are made on the basis of whatever knowledge the cartographer has about the data. This may be founded on field knowledge or any other kinds of data the cartographer may obtain.

The arithmetic involved in reconciling the numerical assumptions with the original unit area data has been explained and illustrated in Chapter 7. It should be emphasized that there is nothing in the data itself that will indicate to the cartographer the kinds of subdivision that should be made or where the zones of rapid change occur. Instead, this kind of information must come from other knowledge concerning the spatial relationships among the data and other geographical distributions. It may be observed in this connection that it is very much more difficult to program the dasymetric process for automation than it is the simple choropleth, since it must always involve additional information, including associated distributions. The phenomena found in association with the distribution being mapped and on which the cartographer must base the rearrangement of the basic unit area data are of two kinds, (1) limiting variables, and (2) related variables.

LIMITING VARIABLES. These are variables that set an absolute upper limit on the quantity of the mapped phenomenon that can occur in an area. For example, suppose we were mapping percent of cropland and the data showed that county A had 60% of its total area as cropland. If 15% of the area of the county were devoted to urban land uses, it is obvious that that area could not be cropland. Exclusion of the urban area allocates all the cropland to the remaining sections of the county.

Suppose, furthermore, that a second limiting variable in county A were woodland, the areal percentage of which was known for minor civil divisions such as townships. If a township had 60% of its area in woodland, then the maximum cropland it could have would be 40%. Other areas of the county would, therefore, need to have a much higher percent of cropland to make possible the original statistic of 60% for the county as a whole.

The diagrams in Fig. 11.35 show how a knowledge of the geographical occurrence of limiting variables enables the cartographer to add considerable geographic detail in dasymetric mapping to the simple choropleth representation. In Fig. 11.35A all that is known is that the county in question consists of 55% cropland. In Figs.

11.35B and 11.35C the limiting variables have been mapped: these consist of urban land use, consisting of 15% of the area of the county and an area of 50 to 70% woodland (occupying 42% of the area of the county). Using the formula for estimating densities of fractional parts and starting with urban land use, if we assume there is no cropland in the urban area, we may determine that the county outside the urban area consists of 65% cropland. That section, as shown in Fig. 11.35B, is almost half occupied by an area of 50 to 70% woodland. Assuming that the most cropland that could occur in the woodland area would be 50%, we may next determine that the area between the woodland zone and the urban area is approximately 80% cropland. A choropleth map would have found the county in only one class; approached dysametrically, there are three classes represented.

RELATED VARIABLES. The functioning of related variables in the dasymetric mapping process is much more complex than that of limiting variables. Related variables are those geographical phenomena that show predictable variations in spatial association with the phenomenon being mapped, but that cannot be employed in a limiting sense. To continue with the example of percent of cropland, we may have available a map of surface configuration that shows that the land form varies from level on the one hand to very hilly, steep country on the other. There would normally be a high positive correlation between level land and percent of cropland. Similarly, we could think of several other variables that would assist in predicting variations in the geographical occurrence of percent of cropland, such as types of farming regions and soil characteristics.

Related variables are more difficult to use properly because they cannot be employed in the strict manner possible with limiting variables. They are, however, very useful in helping the cartographer put some geographical "sense" into the bare statistics that are commonly gathered on the basis of enumeration districts.

The dasymetric technique can also be used to portray areas of uniform positional data. The cartographer creates these areas of uniformity by clustering or grouping similar quantities into regions. Many different clustering techniques can be utilized. The equivalent of related and limiting

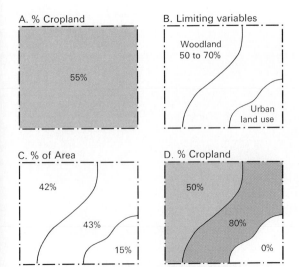

FIGURE 11.35 *How a knowledge of limiting variables can assist in dasymetric mapping. See text for explanation.*

variables may also be used to assist the cartographer in this form of dasymetric mapping.

Elements of Choroplethic Mapping. In choroplethic mapping the elements involved include the size and shape of the unit areas, the number of classes, and the method of class limit determination. Only the latter two are important for dasymetric maps created from positional data.

SIZE AND SHAPE OF UNIT AREAS. The unit areas of a region taken together form a pattern, and this pattern acts as a generalization filter. If the unit areas are large, the spatial variation of the data tends to be reduced or averaged out; if the unit areas are small, variation is preserved. If the unit areas vary greatly in size from one an-

other, variation is preserved in one part of the region and lost in another, which results in uneven generalization, an undesirable effect for a map. Ideally, for the best use of the choropleth method, unit areas should be of relatively equal sizes (preferably small) and of similar shape (see Fig. 11.36).

Since the data collection unit area boundaries must coincide with the discontinuities on the assumed statistical surface when the simple choroplethic technique is used, irregular shapes often cause distracting patterns in the lines on the map. Unit areas of similar shapes avoid this problem. In dasymetric mapping, the shapes of the data collection unit areas are of less concern, since they are much less prominent in the mapping procedure.

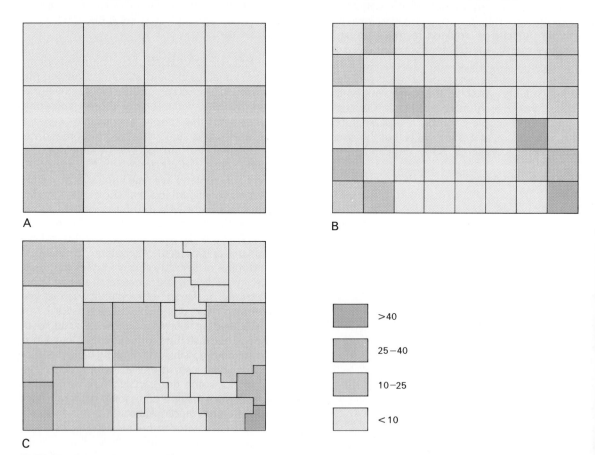

FIGURE 11.36 *A single distribution mapped with three different sets of unit areas employing the same class interval scheme: (A)* Uses large units of equal area. *(B) Uses small units of equal area. (C) Uses irregularly sized unit areas.*

THE NUMBER OF CLASSES. The choroplethic method results in a range-graded symbolization. The number of classes symbolized therefore determines the detail that can be read from the map. The cartographic ideal is to present the maximum number of classes that can easily be read from the map. This depends on (1) whether the map produced is monochromatic or color, and (2) the nature of the distribution (see Fig. 11.37). Perceptually, a reader is limited to a relatively few classes on a monochromatic map. Even when both pattern and value are employed, five to eight classes approach the maximum. When color is used, the number of classes can be increased.

Normally the cartographer, by a process of trial and error, arrives at a convenient balance between the number of readable classes and the complexity of the distribution.

CLASS LIMIT DETERMINATION. In Chapter 7 methods for the determination of class limits for range-gradeede the data such as to present a biased fortunately, it is easy for a cartographer to select class interval determination schemes that will range-grade the data such as to present a biased view or to mask the basic character of a distribu-

tion. Obviously, the potential to enhance a portrayal by the judicious selection of class limits also exists.

Theoretically the cartographer should seek to highlight critical values in the distribution. For example, if a minimum of 40% of a county's population must be below the poverty level to qualify for a certain federal program, then this value might be chosen as one limit. If, however, no critical values are present, the cartographer should then seek to maximize both the homogeneity within each class and the differences between classes. Fortunately, with computer assistance, these criteria stated statistically can be approximated.

Error in Choropleth Mapping. Jenks and Caspall have studied in detail the error characteristics of choropleth maps that result from the number of classes and the class limit selection.* They have proposed that error index numbers be

*George F. Jenks, and Fred C. Caspall, "Error on Choroplethic Maps: Definition, Measurement, Reduction," Annals, Association of American Geographers, 61, 2 (June 1971), 217–244.

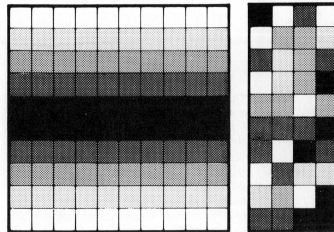
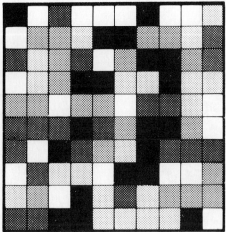

A   **B**

FIGURE 11.37 Both A and B employ five classes. Becauses of the rather simple distribution pattern shown on A, it would be possible to increase the number of classes used. For B the lack of pattern in the distribution would make it even more confusing to the map reader if more classes were used. (Maps courtesy of J. M. Olson.)

placed in the legends of all choropleth maps. Errors can be of three types, depending on whether the map is to be used to portray (1) an overview of the distribution, (2) the geographical positions of tabular values from the distribution, or (3) significant boundary lines for the distribution. According to Jenks and Caspall, the overview error can be calculated as the absolute volumes of prisms lying between the unclassed data in Fig. 11.38 and a classed map model (Fig. 11.39). Similarly, they present statistics for calculating tabular and boundary value indices.

The errors resulting from the size and shape of the unit areas mainly affect the visual qualities of the map; these qualities also depend on the graphic design of the map. The choropleth method assumes that each unit area is homogeneous to the extent that the size and shape of a unit area masks internal variations, and this, of course, adds error to any choropleth map. The dasymetric technique is a way to reduce this type of error.

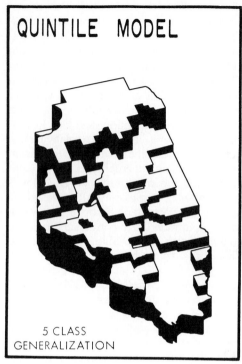

FIGURE 11.39 *A five-classed, three-dimensional map of the statistical distribution shown in Fig. 11.38. The differences between the models in Figs. 11.38 and 11.39 are termed the overview error. (Courtesy of G. Jenks and F. Caspall and the Annals, Association of American Geographers, 1971.)*

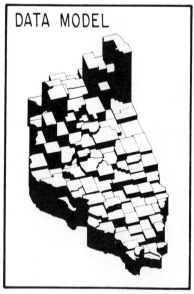

FIGURE 11.38 *A three-dimensional map of a statistical distribution. Each unit area is shown as a prism whose height is proportional to the value of the mapped distribution, and whose base is a scaled representation of the unit area. (Courtesy of G. Jenks and F. Caspall and the Annals, Association of American Geographers, 1971.)*

Other Methods for Symbolizing Quantitative Areal Data

The choroplethic mapping method, with its simple choropleth and dasymetric techniques, represents highly commensurable ways of cartographically portraying quantitative data by areal symbols. Other methods useful for treating the same types of data are less commensurable, but some are much more visual. These methods will be introduced in this final section of the chapter. Most of the examples will deal with representation of the land form. The methods may be applied to any distribution that the cartographer can conceptualize as a statistical surface, provided the data available are detailed enough.

Shading. Shading or, as it is sometimes called, plastic shading, is the gaining of a three-dimensional impression by employing variations in light and dark.* These variations, as in chiaroscuro drawing, are applied according to a number of different systems. In theory the system of shading is based on the principle that the lighting of a three-dimensional form will result in varying amounts of illumination on the varying slopes. The direction from which the light comes makes a significant difference. For example, vertical lighting would cause maximum illumination on the horizontal surfaces and minimum on the vertical surfaces. This would result in the logical pro-

gression, the steeper the darker. Oblique lighting produces more illumination on one side than the other of a form (Fig. 11.40).* In its simplest form, shading attempts to create the impression, appropriately exaggerated, which we might gain from viewing a carefully lighted model of the land, as is illustrated in Fig. 11.41. Of course, the usual shaded map is not quite the same as an area seen from above, since the observer is, in theory, directly above all parts of the map, there being no perspective.

One of the problems facing the cartographer who plans to employ shading is the direction from which the light is to come. A curious phenomenon is that with light from some directions, depressions

---

*The adjective "plastic" as it is used here derives from the German plastik (from Gk. plastikos), which refers to the modeled, three-dimensional effect; it has nothing to do with the modern synthetic materials for which the same term is used.

*In the actual practice of manual shading, it is usually not done systematically. John S. Keates, "Techniques of Relief Presentation," Surveying and Mapping, 21, 459–463 (1961).

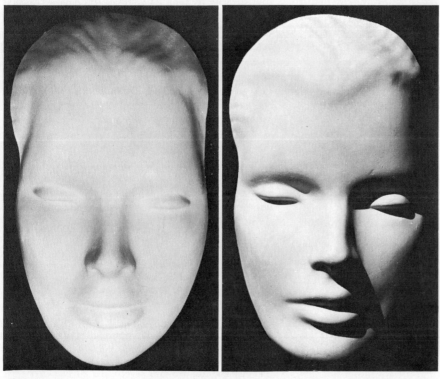

FIGURE 11.40 Two views of the same plaster model, one lighted vertically (left) and the other from the upper left (right). (Courtesy of G. Fremlin.)

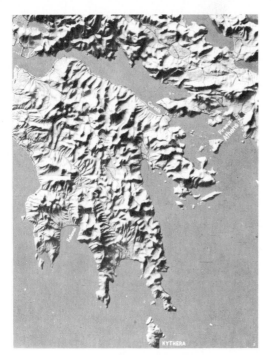

FIGURE 11.41 A much reduced portion of a topographic model.

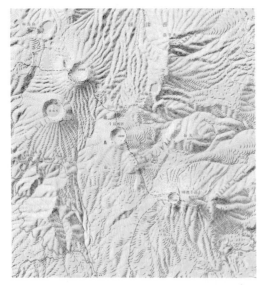

FIGURE 11.42 An example of Professor Tanaka's illuminated contour method. (Courtesy of The Geographical Review.)

and rises will appear reversed. Consequently, the direction must be chosen so that the proper effect will be obtained. When the light comes from an upper direction, elevations generally appear "up" and depressions, "down." In addition to providing the correct impression of relief, the direction of lighting is important in portraying effectively the surface being presented. Many areas have a "grain" or pattern of feature alignment that would not show effectively if the light came from a direction parallel to it. For example, a smooth ridge with a northwest-southeast trend would result in about the same illumination on both sides if the light were to come from the northwest. The cartographer must also select the direction of lighting and apply the darkening and highlighting so that the items of significance will be effectively portrayed.

The utility of shading as a visual vehicle for presenting the configuration of a surface has long been appreciated. It was to be expected that there would be attempts to make it in some way commensurable as well as visually effective. The first such attempt was the hachure. Professor Tanaka of Japan* suggested the illumination of contours systematically so that they provide an impression of shading with oblique lighting as well as being commensurable. Figure 11.42 is an example. A number of other possibilities have been suggested, but none of them has been tried beyond the experimental stage.

Analytical Shading. One of the interesting examples of the role the computer can play in cartography is provided by Professor P. Yoeli's experimentation with what he calls analytical hill shading. K. Brassel has recently expanded the work.† Although not yet put to practical use, the

*Illuminated contours are lines on a neutral gray background. On the light side of a hill they are lighter than the ground; on the dark side they are darker than the ground. K. Tanaka, "The Orthographical Relief Method of Representing Topography on Maps," The Geographical Review, 40, 444–456 (1950).

†For Yoeli's work, see the series of explanatory papers in Kartographische Nachrichten, 15, (1965), Heft 4; 16 (1966), Heft 1; 16 (1966), Heft 3; and 17 (1967), Heft 2. The first two appeared in English translation in Surveying and Mapping, 25, 573–579 (1965), and 26, 253–259 (1966). Also Kurt Brassel, "A Model for Automatic Hill-Shading," The American Cartographer, 1 (1), 15–27 (April 1974).

method is nicely illustrative of the scientific elements in the attempt to portray the configuration of a surface by light and shade.

Nearly 100 years ago, the equations having to do with the amount of light intensities at various points on an irregular surface were worked out. Stated simply, the intensity of light reaching an observer from a point on a three-dimensional surface under illumination by parallel rays is a function of the inclination and orientation of that part of the surface in relation to the directions of the observer's line of sight and the source of light. This remained in the realm of theory, however, because of the immense amount of calculation that would be needed to make use of the equations, as well as the practical problem of producing any graphic result. The computer now makes possible extensive computation and the basing of the degree of light and shade on a strict analysis of the surface configuration as delineated by contours.

In Yoeli's experiment the surface to be shaded is divided into small rectangular segments or facets. From the contours may be calculated the average inclination and orientation of each facet. By then assuming a direction and an inclination of parallel light rays, the resulting relative brightness of each facet may be calculated. All that remains is to darken each facet accordingly and, if the facets are small enough, a strictly analytical shading results.

The elements in Fig. 11.43 show the original contour map, the appropriately darkened facets, and the combination of the two. As Yoeli points out, the visually ideal shading results when the direction of the light can be shifted so as to take advantage of the orientation of the configuration. Even this can be programmed for the computer to a certain degree.

Brassel has attempted to incorporate some additional parameters into Yoeli's analytical hill shading technique. Adjustment of the light source, possible in both the vertical and horizontal directions in response to local relief forms, is illustrated in Fig. 11.44. These adjustments avoid the accenting or obscuring of edges that lie in unfavorable positions with respect to the light source. Brassel also has added an atmospheric perspective modification. Figure 11.45 diminishes contrasts in the lowlands and accents contrasts in the uplands.

Pictorial Representation

Even on ancient maps the major terrain features were represented pictorially in crude fashion. As artistic capabilities have increased, and as knowledge of the character of the earth's surface has expanded, pictorial representation has become increasingly effective. Within the past 50 years, this method of presenting the land features has made great strides.

Many varieties of perspective delineation of the land are in common use today. Most of them stem from the attempts made during the nineteenth century to portray the concepts that were being rapidly developed by the growing science of geology. One such delineation, the block diagram, has today become a standard form of graphic ex-

  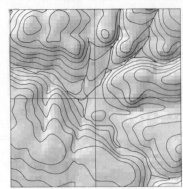

FIGURE 11.43 *A contour map, the shaded facets, and the two combined to illustrate an instance of an analytically shaded map. The direction of light is from the northwest at an inclination of 45 degrees. (Courtesy of P. Yoeli.)*

FIGURE 11.44 *Left map illustrates illumination with constant light source from the northwest; center map illustrates illumination with a constant light source from the southwest; and right map illustrates local adjustment of horizontal light direction with a maximum change of 60 degrees (general light direction from west). (Courtesy of K. Brassel and the American Congress on Surveying and Mapping.)*

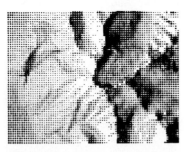

FIGURE 11.45 *Top map illustrates no atmospheric perspective modification; lower map illustrates atmospheric perspective modification where the value contrast has been enhanced in the upper elevations and reduced in the lower elevations. General obscuring of detail occurs in the lower elevations. (Courtesy of K. Brassel and the American Congress on Surveying and Mapping.)*

pression, and its utility has been extended to illustrate land use, land types, and other kinds of earth distributions. Almost any degree of elaborateness may be incorporated in a block diagram, ranging from the successive geologic stages in the development of an area to multiple cross sections of the structure. Uncomplicated block diagrams are not difficult to draw, and even the simplest is remarkably graphic (see Fig. 11.46). As discussed above, perspective plots can be produced by computer-driven machines (see Figs. 11.27 and 11.28).

The block diagram, being obliquely oriented to the reader and thus in perspective, is not an ordinary map in the sense that a map is viewed orthogonally and its scale relationships are systematically arranged on a plane projection. From the block diagram, however, have come other map types that combine the perspective view of undulations of the land and the planimetric (two-dimensional) precision of the map. These are the landform map or physiographic diagram developed primarily in the United States and discussed in Chapter 9.

Perspective drawing also requires quantitative areal data about the distribution being portrayed and, therefore, the perspective drawing differs from the landform map in that the former attempts to show a portion of the earth as if seen in perspective from some distant point, while the landform map puts pictorial symbols on an otherwise conventional map base (Fig. 11.47). On the perspective drawing, the appearance of the whole

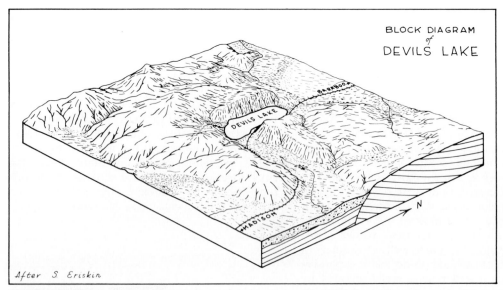

FIGURE 11.46 *A simple block diagram prepared for student field trip use. The natural appearance of the sur-*
face forms on a perspective block makes the concepts easily understandable.

is relative (as in a block diagram of a large section of the earth), and it has become very popular in recent decades. Drawings of this type are usually done on an orthographic projection (for very small scales), or on an "oblique" photograph of a portion of the globe. The terrain is then modeled so that the earth's curvature is simulated and so that the entire drawing provides the impression of a view of the earth as seen from a point far above.

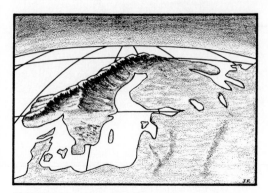

FIGURE 11.47 *Sketch illustrating a perspective*
drawing. Compare with the block diagram shown in
Fig. 11.46 and the land form map in Fig. 9.24.

The student should have clearly in mind why pictorial terrain must be generalized (and exaggerated) to a tremendous degree. Departures of the earth's surface up or down from the spheroid are actually very small, relative to horizontal distances, and they can hardly be shown at all at most medium and small scales. For example, the highest mountain on the earth, Everest, is a bit over 8800 m above sea level. If the terrain of the continent of Asia were represented accurately on a three-dimensional model at a scale of 1:10,000,000 (a map about 1.2 m square), Mount Everest would be less than 1 mm high! Consequently, almost all pictorial terrain representation must greatly exaggerate and simplify the terrain. The problem is even more severe when the distribution is other than terrain, because the heights assigned as Z values are completely arbitrary.

Layer Tinting. Perhaps the most widely used method of presenting land surface information on wall maps, in atlases, and on other "physical" maps is the method variously called layer tinting, hypsometric coloring, or altitude tinting. This is merely the application of different area symbols (hue, pattern, or value) to the areas between the lines (on isarithmic maps). On small-scale maps the simplification of the chosen isarithms must, of necessity, be large. Consequently, the "contour" lines on such maps are not particularly meaningful, and the system degenerates into a presentation of large categories of surface elevation. It should be emphasized that the larger the scale, assuming a reasonable degree of isarithmic simplification, the more useful is the layer system. When the scale has been increased to the point at which the character of the individual isarithms and their relationships become meaningful, the representation graduates to being a most useful map.

Colored layer tints at small scales on terrain maps, when combined with pictorial terrain or shading, nearly satisfy most of the landform requirements of the general map reader. They present the major structure as well as the details of form. On the other hand, to do this is expensive and demands a skill not generally enjoyed by most cartographers. The Wenschow wall maps and the works of Richard Edes Harrison are outstanding examples of such combinations. Their reproduction is necessarily by process color (see Chapter 16) or some equally expensive method, and their reproduction cost effectively removes them from the endeavors of the average cartographer.

Additional Methods. Because of the relative inadequacy of the layer tinting method for showing detail, geographers and cartographers are continually searching for ways to present a more useful representation of the land surface or the statistical surface. Several methods have been suggested: (1) the relative-relief method, and (2) the gradient-value method. Neither method has attained acceptance for general maps that even approaches hachuring, shading, or layer tinting. Their symbolization and presentation is straightforward; area symbols are used to reinforce either isarithms or dasymetric lines. The major problem, inherent in these methods, is the determination of what to present, not how to present it. Consequently, to utilize these methods, the cartographer must be essentially a specialist about the distribution being mapped or else must simply present the work of others.

In both Europe and the United States, the concept of relative relief on surfaces, as opposed to elevation above a datum, has been tried. Rela-

tive, or local, relief is the difference between the highest and lowest elevations in a limited area, for example, a 5' quadrangle. These values are then plotted on a map, and isarithms may be drawn or the distribution may be symbolized in the simple choropleth manner. Area symbols, as in layer tinting, may also be applied. The method is of value when applied to areas of considerable size because basic features and divisions of the distribution are emphasized, but it seems to be unsuited for differentiating important details too small to extend beyond the confines of the unit area chosen for statistical purposes. It is best adapted to relatively small-scale representation.

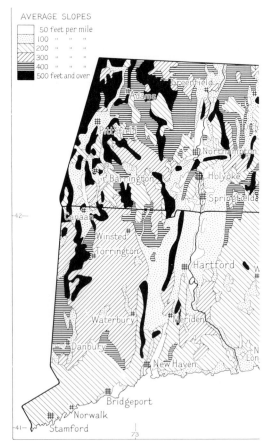

FIGURE 11.48 A portion of a slope-value map of southern New England. The areas of similar slope were outlined on topographic maps by noting areas of consistent contour spacing. (Courtesy of The Geographical Review.)

From the time the hachure became popular, the cartographer and geographer have been concerned with the representation of the slope of the land. Shading and hachuring, although not particularly commensurable, provide a graphic account of slopes on medium- and large-scale maps. The problem of presenting actual slope or gradient values on small-scale maps is not easily solved. One technique is the gradient category method. In this method, areas of similar gradient are outlined and presented by means of area symbols in the dasymetric fashion (see Fig. 11.48). Other slope techniques, such as percent of flat land per unit area, have been tried but, except for specialized teaching or research purposes, they have not been widely used.

SELECTED REFERENCES

Blumenstock, D. I., "The Reliability Factor in the Drawing of Isarithms," *Annals,* Association of American Geographers, *43,* 289–304 (1953).

DeLucia, A., "The Effect of Shaded Relief on Map Information Accessibility," *The Cartographic Journal, 9,* 14–18 (1972).

Dixon, O. M., "Methods and Progress in Choropleth Mapping of Population Density," *The Cartographic Journal, 9,* 19–29 (1972).

Hammond, E. H., "Analysis of Properties in Land Form Geography: An Application to Broad Scale Land Form Mapping," *Annals,* Association of American Geographers, *54,* 11–19 (1964).

Hsu, M-L., and A. H. Robinson, *The Fidelity of Isopleth Maps: An Experimental Study,* University of Minnesota Press, Minneapolis, 1970.

Jenks, G. F., "The Data Model Concept in Statistical Mapping," *International Yearbook of Cartography, 7,* 186–188 (1967).

Jenks, G. F., "Contemporary Statistical Maps—Evidence of Spatial and Graphic Ignorance," *The American Cartographer, 3,* 11–18 (1976).

Kimerling, A. J., "A Cartographic Study of Equal Value Gray Scales for Use With Screened Gray Areas," *The American Cartographer, 2,* 119–127 (1975).

Miller, O. M., and C. H. Summerson, "Slope-Zone Maps," *The Geographical Review, 50,* 194–202 (1960).

Porter, P. W., "Putting the Isopleth in Its Place," *Proceedings of the Minnesota Academy of Science, 15–16,* 372–384 (1957–1958).

Peucker, T. K., "A Theory of the Cartographic Line," *International Yearbook of Cartography, 16,* 134–143 (1976).

Ridd, M. K., "The Proportional Relief Landform Map," *Annals,* Association of American Geographers, *53,* 569–576 (1963).

Robinson, A. H., "The Cartographic Representation of the Statistical Surface," *International Yearbook of Cartography, 1,* 53–61 (1961).

Robinson, A. H., "Mapping the Correspondence of Isarithmic Maps," *Annals,* Association of American Geographers, *52,* 414–425 (1962).

Thrower, N. J. W., "Extended Uses of the Method of Orthogonal Mapping of Traces of Parallel Inclined Planes with a Surface, Especially Terrain," *International Yearbook of Cartography, 3,* 26–35 (1963).

Yoeli, P., "An Experimental Electronic System for Converting Contours into Hill-Shaded Relief," *International Yearbook of Cartography, 11,* 111–114 (1971).

Yoeli, P., "Analytical Hill Shading," *Surveying and Mapping, 25,* 573–579 (1965).

Yoeli, P., "Analytical Hill Shading and Density," *Surveying and Mapping, 26,* 253–259 (1966).

COMPUTER-ASSISTED CARTOGRAPHIC
  HARDWARE SYSTEMS

INPUT
CONTROL
OUTPUT
  Line Printer
  Plotters
  CRT
  COM Units

EFFECTS OF COMPUTER-ASSISTED
  SYSTEMS ON BASIC CARTOGRA-
  PHIC PROCESSES

MAP BASE CONSTRUCTION
COMPILATION
  Base Data
  Thematic Data
GENERALIZATION
LETTERING
PRODUCTION

During the past 15 years the field of cartography has been notably affected by the rapidly increasing use of computers and automated equipment. Computer-assisted cartography is a better term than automated cartography to describe the present state of affairs because, in fact, complete automation of the mapmaking processes is far from being attained. The changes are quite uneven; in some aspects the developments have been profound, but in others relatively little has occurred.

The magnitude of the changes has been sufficient to give rise to a false dichotomy in the eyes of those who view the field from a short-term perspective. This division, between the manual production of cartographic works and the automated or computer-assisted aspects of production, implies that the two are quite different. On the contrary, what has occurred is simply that machines have been introduced to perform some of the necessary time-consuming, exacting, and repetitive operations. It is worth emphasizing that the basic cartographic processes have not been changed but, in many respects, the way these processes are carried out has been modified to take advantage of computer technology. Thus, what has seemed to some to be a division is nothing more than procedural developments resulting from the introduction of useful technology into an existing field.

There are several important consequences of

these changes. One of these is simply that the use of computer-assisted methods requires that cartographers become familiar with the new techniques, since it is unlikely that a cartographer who is expert in some aspect of manual cartography can immediately interact with sophisticated computer technology. In some operations the transition is easier than in others. For example, the process of line digitizing (recording the X-Y positions along a line) is similar to scribing and, in some instances, they are almost identical, since a visible line is created to allow the cartographer to know what part of a given line has been digitized. However, the concomitant entry of necessary auxilliary information into the digital record has no counterpart in the scribing operation and, in fact, the process of digitizing demands that the cartographer learn something about the structures of digital data files.

A second consequence stems from the time lag between technological development and its full-scale implementation. Technology has advanced relatively fast compared to its incorporation in efficient cartographic systems. Although the directors of cartographic units may have long known that many operations can be carried out with the assistance of computer technology, the lack of capital and/or the exact hardware configurations to make an efficient system for a specific cartographic unit may not have been available. In

CHAPTER 12

COMPUTER-ASSISTED
CARTOGRAPHY

some cases this may have created a frustrating situation for those who recognized the potential to perform various operations more efficiently with computer assistance but who were unable to do so because of a skeptical attitude on the part of some administrators regarding the exact hardware configurations necessary to perform specialized tasks. Such skepticism is normal in a rapidly changing field.

A third consequence of this rapid introduction of a new technology has been a predictable misuse of its capabilities. The adoption of a new technology is usually accompanied by instances of persons who are adequately trained in the technology but utilize the technology inappropriately, because they are not trained in the discipline. Jenks has indicated some of the results of these misuses in cartography.* Hopefully this is a short-lived phenomena. As cartographers adopt and learn to work with the technology of computer-assisted cartography, it can be expected that the instances of incorrect application of cartographic processes will decline. Likewise, as people not trained in cartography continue to utilize the technology, they can be expected to sense that they are misusing it and seek more fundamental training in cartography.

What has been transpiring is simply that a relatively new technology is being superceded by an even newer technology, making possible in some instances more precision, more accuracy, and faster speed. All of these are, of course, great advantages to the field of cartography. The total impact of these advantages and of the computer-assisted technology in general has not been fully assessed as yet. The problems of the manual cartographer of the 1960s are not necessarily the problems of the computer-assisted cartographer of the late 1970s. The problems have shifted in emphasis.

Nothing is more illustrative of this shift in emphasis on a broad scale than to point out that in 1960 a cartographer who wished to prepare (by hand) an isarithmic map and a choroplethic map of the same data could produce the choroplethic map in far less time than the isarithmic map. For the isarithmic map, one had laboriously to measure and perform linear interpolation between known data points. With the introduction of computers the isarithmic technique became one of the first methods to be programmed (Fig. 12.1). As a result, cartographers have been producing with ease isarithmic maps with computer assistance for about 15 years, but it has only been recently that high quality choroplethic maps have been made by machines.* (Fig. CI.12)

Other shifts in emphasis have also occurred. For example, the literature of the 1950s and 1960s in cartography shows much concern for class interval determination. Elaborate mathematical-graphic schemes were devised for generalizing a set of data into a set of given classes. With the advent of computer assistance, a whole new set of methods for deriving class intervals, largely iterative, became available to the cartographer. Such schemes allow the cartographer to satisfy significant statistical criteria in class limit selection. As a consequence, the earlier methods are much less important. A further example is provided by the interpolation problem inherent to the preparation of isarithmic maps. Before computer assistance, the cartographer could employ only one or two methods of interpolation to fit the isarithms to the set of data points. In contrast, machine interpolation allows the application of highly complex mathematical interpolation models. The problem today is *which* interpolation method to use or *what* statistical criteria does the cartographer wish to satisfy with a selected class interval scheme.

A final example of a shift in emphasis involves projections. Until the use of computers, the manual construction of projections was dealt with elaborately in books on cartography. Today, with commonly available software, any projection that can be specified by a set of equations or table of values can be produced by computer-plotter methods in a very short period at very little cost. In addition, oblique projections and interrupted projections are now obtainable with relative ease. Therefore, the construction of projections is now of relatively minor concern to most who have available computer assistance. The net effect of

---

*G. F. Jenks, "Contemporary Statistical Maps—Evidence of Spatial and Graphic Ignorance," The American Cartographer 3, 11–19 (1976).

* M. A. Meyer, F. R. Broome and R. H. Schweitzer, Jr., "Color Statistical Mapping by the U.S. Bureau of the Census," The American Cartographer, 2, 100–117 (1975).

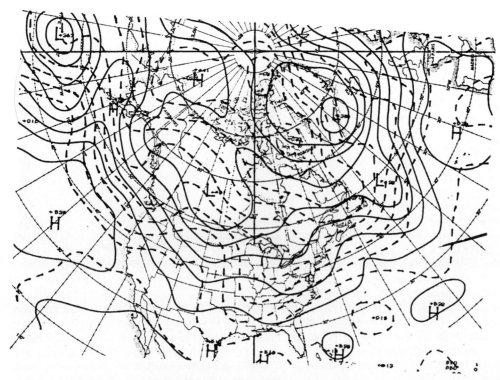

FIGURE 12.1 *A chart representing the 850 mb constant pressure surface as observed at 7 p.m. EST, 01 March 1961. Produced on the U.S. Weather Bureau's Dataplotter. The Dataplotter began service in December 1960. (Courtesy of Association for Computing Machinery.)*

this new technology has been to give the cartographer more opportunity to be responsive to meeting the objectives of a map by allowing more time to consider the problems of design, since the actual execution often can be more efficiently completed by machines.

This chapter is organized to show how automation is affecting the basic processes in the field of cartography, to provide an indication of how thoroughly computer assistance has been integrated into cartography, and to show how the problems that the cartographer faces today have changed from those that the cartographer faced, say, 15 years ago. A section on computer-assisting hardware configurations will be presented first. Specific hardware configurations will not be discussed or recommended; it is not our purpose to engage in publicity for private firms. The body of the chapter will discuss four major areas of cartographic processing: base construction, compila-

tion, generalization, and production. For the use of the reader, a glossary of basic terms related to computer-assisted cartography appears at the end of this chapter.

One final introductory comment concerns the fact that most of the computer assistance in cartography to date has been digital. Some cartographers are investigating analog systems for the production of cartographic products. This chapter will deal primarily with digital cartography and perhaps understate the developments that are current or on the near horizon in analog work.

COMPUTER-ASSISTED CARTOGRAPHIC HARDWARE SYSTEMS

The main hardware components of a computer-assisted cartographic system can be discussed under input, control, and output.

Input

Input essentially involves the compilation devices that allow the cartographer to put the data necessary for a map into machine-readable form. At present there are two principal methods for capturing data from existing sources, for example maps and photographs. The first method employs computer equipment such as manually operated line-following digitizers (Fig. 12.2); the second employs machines called scanners (Fig. 12.3).* These two types of information-capturing equipment are fundamentally different, and both are being applied to cartography; today, the line-following digitizing operation probably has more utility for thematic applications.

Scanners tend to be used in connection with photographic images or for the sensing of primary data, such as the earth itself, but they are not limited to that. First, a scanner resolves an analog map or photo of an area of the earth into a set of discrete picture elements called pixels. The resolution capability of the scanner determines the pixel size. Naturally, the scale of the map or photo would therefore determine the overall scale of the information being input into machine-readable data formats.

The scanner is, in effect, an electronic eye that moves essentially as one would read a page

*The equipment pictured in Fig. 12.3 can also operate as a high-speed plotter for output in addition to scanning input.

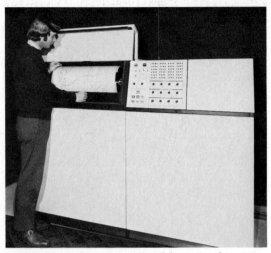

FIGURE 12.3 A raster scanner/plotter used in cartographic production. (U.S. Army Photograph.)

of text, from left to right and top to bottom. As in some remote sensing, it does this very rapidly, and the resulting digital record is in terms of a series of numbers each representing a gray level of a pixel. There is no differentiation of classes of features that are put into the data file in the initial scanning. Therefore, this initial scanned data is of little direct utility to the cartographer until it has been processed into categories of information. The ease with which this can be accomplished depends on the categories of information present. After categorization, the data can be processed cartographically and an output image (photograph) or, in some cases, a line plot is produced. Certainly the use of scanners to digitize a color separation plate for an existing map is one way of obtaining that information in machine-usable form and the speed with which it can be done is certainly far faster than employing the hand-operated line digitizers.

The second type of compilation equipment, or analog to digital converter system, is the line-following digitizer. Predominantly these are manually operated machines. The drafting tablelike equipment includes a movable cross-hair (see Fig. 12.2). The position of the cross-hair is sensed by the equipment; as the operator moves the cross-hair over the map or photograph, the trace of the cross-hair is translated into arbitrary X-Y positions and these are automatically recorded, usually on punch cards or magnetic tape.

FIGURE 12.2 The Bank Datagrid digitizing hardware. (Photo courtesy of Bendix Aerospace Systems Division, Ann Arbor, Michigan.)

In thematic cartography most of the data capture has been through this type of digitizing procedure. It is quite easy to compile or trace a single distribution from an existing map or photograph. Since most manually run, line-following digitizers have keyboards for the entry of auxilliary information into the file, it is possible to create a sophisticated file that contains, for example, a transportation network, the areas covered by forest, or the location of geodetic control points for an area. The operation can become tedious for the operator, but it has been shown that the utility of the resulting file is well worth the tedious manual effort.

Once a distribution has been entered into a file with requisite accuracy and resolution, it is possible to convert this information to any scale or projection that a cartographer wishes and, therefore, any given distribution need be digitized only once. This is in marked contrast to manual compilation where, for example, each time a map of South America is wanted, one must recompile the thematic information, if the scale or projection has been changed.

A more recent type of digitizing method is accomplished by automatic line-following equipment. A sensor whose X-Y position is known is "locked" onto a line, and it follows the line automatically to the edge of the sheet. This is presently just becoming available, and the test results suggest that it may ultimately relieve the digitizer operator of that tedious task. However, the amount of editing (addition of header information, identification of lines, etc.) that needs to take place is not insignificant. Certainly for manual line-following digitizing, editing time is a major expenditure sometimes exceeding the actual time spent digitizing.

In summary, with today's technology, any distribution can be transformed into machine-readable form.

Control

The second category of automated components are the control units. These units receive the input data and process it. Such processing includes manipulation of the data such as the computation of means, medians, or standard deviations, generalization such as simplification or classification routines, and composition of the individual data sets to check for overlap of features, and the like.

FIGURE 12.4 THE PDP 11/34 system: A "minicomputer." (Copyright © 1976, Digital Equipment Corporation. All rights reserved. Reproduced with the permission of Digital Equipment Corporation.)

Finally, the control unit can write a data file that will drive one of the computer-assisted output devices.

A control unit can be anything from a dedicated minicomputer, for example, a small capacity machine whose single (dedicated) function might be to receive one type of data, process it in one manner, and prepare it for output from a single device (Fig. 12.4), to a very large CPU (central processing unit) (Fig. 12.5), which operates in a time-sharing environment, handling many varied tasks from a variety of input devices seemingly simultaneously. The important operation required of the control unit is the ability to read either the digitized or the scanned data and to manipulate the data into a form that the cartographer needs as output. At this stage the cartographer must have all of the normal procedures translated into machine instructions. This means that the operations must be set forth in a logical order in a finite number of steps. A finite set of logical instructions constitutes a *software* program. The software programs available to the cartographer must be able to carry out all the different processes that the cartographer might wish. Separate software routines presently are required to process scanned data as opposed to line digitized data.

It is necessary at times for there to be interaction between the cartographer and the control unit. Interaction with the control unit is vital for the efficient use of computer-assisted hardware by the

FIGURE 12.5 *The IBM 370/168 system. (Photo courtesy of the IBM Corporation.)*

cartographer. Such a control unit may consist of a teletype or typewriter or a graphics display unit (Fig. 12.6). Most of these units can also double as output devices (these will be discussed below). Using an interactive graphic unit in a mode that allows the cartographer to interact (make changes) with the control unit, the cartographer is able to observe how the final product will look. Manipulation of the different variables that are being processed can be done in real time (i.e., immediately) and, in essence, the cartographer can choose a preferred display before arranging for hard copy output (separate map in hand).

Equally important is that at the control stage in the processing the cartographer is able to interact with the data file, which allows for the updating or editing of the data file. Certainly a transportation network digitized on a line-following digitizer can be expected to change through time. The amount and timing of the information may not be sufficient to warrant going back to the line-following digitizing table and redigitizing the

entire transportation network; instead, all that is required is a simple update of the existing file, and this can be accomplished in a supplementary operation.

FIGURE 12.6 *A control unit consisting of a keyboard similar to a typewriter. (Photo taken in the University of Wisconsin Cartographic Laboratory.)*

Sufficient capacity exists today in control units to solve most cartographic problems. What is lacking is sufficiently versatile software to enable the cartographer to perform all the manipulations on the various data files that are necessary. The improvement of cartographic software is steadily occurring.

Output

The final component of a computer-assisted cartographic system is the output device. There are several output devices. Four will be discussed here: line printers, coordinate plotters, CRT's (cathode ray tubes), and COM units (computer output on microfilm).

Line printer. A rather crude map can be produced on an ordinary computer line printer (Fig. 12.7). Cartographic output from a line printer is relatively familiar, partially because the SYMAP software program from the Harvard Laboratory for Computer Graphics, which initially relied heavily on the line printer, has received widespread use. This type of hard copy map can be improved by the use of special chains that employ symbols different from the common alphanumeric symbols.* (see, for example, Fig. 12.8.) Today,

*A chain is the string of characters capable of being printed by a line printer. It is analogous to the keys of a typewriter or a type head on an IBM Selectric typewriter.

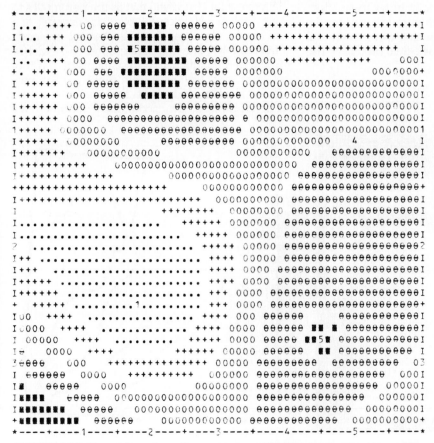

FIGURE 12.7 Output from the software program SYMAP. Such output is readily obtainable because of the widespread availability of this software package, but it is crude. Reduction of the printer output does enhance the visual appeal; this illustration has been reduced from an original that measured 30 cm².

FIGURE 12.8 *Examples of output using special chains on line printers. (Courtesy of K. Brassel.)*

line printer output of conventional cartographic products is justified in only two instances: (1) to produce a list (e.g., all place names that belong within a geographic area or an index to an atlas), and (2) to produce an interim rough map for the cartographer to use in planning or research, but that will not be published as a final product.* The

* *Any output from a line printer is too coarse for most cartographic purposes and is not recommended.*

production of special effects employing special chains, (e.g., Brassel's work) is an exception (see Fig. 12.8).

Plotters. The second major class of output devices are plotters, of which there are two major types. One is called a drum plotter; here the paper is affixed to a drum that can be rotated (Fig. 12.9). The plotting pen is held stationary in the direction of the drum's movement, but it can be moved back and forth at right angles to the drum's movement. Since the plotting paper moves with the drum, it is thus possible to plot anywhere within the limits of these movements. The drums are ordinarily either 11 or 30 in. wide; the length of the paper for an individual plot may be considerably longer. Most drum plotters are what is called "incremental," meaning that as the pen moves to draw a diagonal across the page, for example at a 30°-angle, the line must be stepped or incremented from one end to the other (Fig. 12.10). Some of the plotters have resolutions of 1/100 of an inch, and diagonals from these plotters produce "steplike" increments that are quite noticeable. The more advanced plotters not only have

FIGURE 12.9 *The CALCOMP 936 drum plotter. (Picture courtesy of California Computer Products, Inc.)*

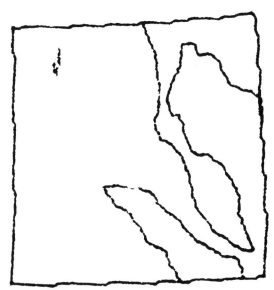

FIGURE 12.10 *An enlargement of a map section produced on an incremental plotter. The "steps" are readily apparent, especially along the bounding graticule lines.*

FIGURE 12.11 The Kongsberg KS5000 flatbed plotter. This is a high-quality, high-speed drafting table of use to cartographers. (Picture courtesy of Kongsberg Systems, Inc.)

higher resolutions, which makes these increments less noticeable, but they also have the ability to move in more than just the horizontal and vertical dimensions without incrementing; therefore, much smoother lines can be obtained.

The second class of plotter is the flatbed plotter (Fig. 12.11). This plotter is usually a large horizontal tabletop where the material that is receiving the plot is positioned. The pen or stylus can be moved in any direction in a horizontal plane following the instructions written on tape or coming on-line from the control unit. Plotters are essentially line work production units, and they can draft line work more consistently, precisely, and accurately than a human draftsman.

Plotters can scribe, or they can draft in a number of different colored inks. They can also draft on a number of different materials, plastics, paper, or film that is exposed by a point light source being driven by instructions from a control unit. The advantage of plotters for the cartographer is that very precise line work can be produced. Plotters cannot efficiently fill in areas that are to receive a given tint, tone, or value.

CRT. A third major type of output unit is the cathode ray tube or CRT (Fig. 12.12). These de-

vices are essentially like a television picture tube, and there are two basic types. One type is called a refresh tube. This tube is essentially a normal TV screen that produces the entire image at regular intervals. Although it is possible to add to an image on this screen, it is not possible to erase selectively. For final output this does not seem to be a problem but, when using a CRT with a control unit for manipulating cartographic data or for designing cartographic products, the inability to erase selectively is a major drawback.

The second type of CRT is the storage tube device. The storage tube displays information as directed by the control unit and continues to display this information as long as it is desired. Portions of the information can be deleted and additions can be made; in fact, it is a composition machine where one can move things around and arrange them as desired.

One problem with CRTs in general is that they do not provide hard copy outputs. The image appears on the screen and is then lost unless the screen is photographed or videotaped.

COM units. A final class of output devices useful for cartographic purposes that has received fairly little attention to date is that of a microfilm unit. Examples are COM (computer output

FIGURE 12.12 The Tektronix 4081 system. This CRT has the dual capabilities of operation as both a refresh tube and a storage tube device. (Picture courtesy of Tektronix, Inc.)

FIGURE 12.13 *The CALCOMP 1675 computer output on microfilm (COM) system. (Picture courtesy of California Computer Products, Inc.)*

on microfilm) units such as the CALCOMP 1675 or the FR-80 system (Fig. 12.13). In these units a light beam is directed by on-line or off-line instructions sent from a control unit or peripheral device to trace or shade an area as directed by the car-

tographer. Film sizes have normally been in the 35-, 70-, or 105-mm range. However, the resolution obtained is so great with these units that it is possible to enlarge these films to normal map size without sacrificing quality (Fig. 12.14).

The potential of microfilm units in cartography seems to be enormous. One only need consider the products of the U.S. Census Bureau in the form of choroplethic maps (see Fig. CI. 12), or their urban atlas series, to realize the capability of the COM unit.

In summary, it seems clear that for cartographic purposes, line printers are being superseded by other output units. Currently the plotter is the most trusted and reliable workhorse for the output of cartographic products; the CRT is probably more valuable in the creation and design of products than in their output. The full potential is as yet undetermined for microfilm units, but it would seem that this is an area of great potential.

Therefore, in a complete automated system for handling digital cartographic data, we have seen that an input device in the form of a scanner or a digitizer is necessary. In addition, a control

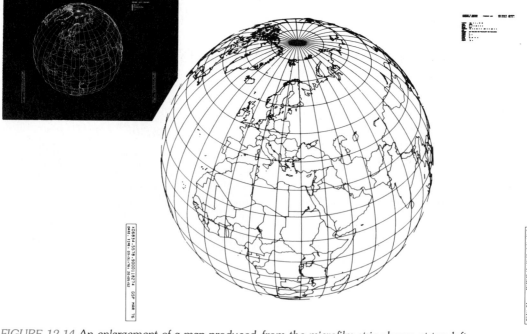

FIGURE 12.14 *An enlargement of a map produced from the microfilm strip shown at top left.*

unit by means of which manipulation and design composition are possible is required and, finally, an output device either in the form of a line printer, a plotter, a CRT, or a microfilm unit is necessary. Only in the past few years have complete sets of equipment such as these become available at one place. An increasing number of such installations can be predicted for the future.

Each component can operate either on-line or off-line, depending on the specific hardware. In general, in a configuration where a large CPU is available in a time-sharing mode, and many noncartographic uses are required of the CPU, most cartographic work is off-line. Digitizing on-line to a large time-sharing CPU is costly and risky, and so is on-line plotting. In a hardware configuration where the CPU is dedicated to cartographic

operations, the work can be on-line. As the use of dedicated minicomputers increases, in all probability most computer-assisted cartographic systems will be on-line to a dedicated minicomputer.

In this cursory discussion we have avoided discussing analog or nondigital computer-assisted cartographic operations, because analog computers as yet are not as well known as digital computers.

One analog computer useful for cartographic purposes is called an optical processing unit. Figure 12.15 illustrates one possible hardware configuration. It consists of a laser beam with a series of mirrors and filters, input position, and final recording of the information. The use of such units for cartographic purposes is relatively new, and very few cartographic products have been pro-

FIGURE 12.15 An optical bench configuration that can be used for the analog processing of cartographic products. (Photo courtesy of E. Wingert.)

duced on such equipment. The full potential of nondigital automatic applications to cartography is as yet unknown.

EFFECTS OF COMPUTER-ASSISTED SYSTEMS ON BASIC CARTOGRAPHIC PROCESSES

The remainder of this chapter will focus on five basic areas of cartographic processing. An attempt will be made to show how the processes of cartography have been affected by the introduction of computer assistance.

Map Base Construction

The first of these areas is map base construction; map base construction is the projection and coordinate systems on which the compilation will take place, or the use of primary control that will aid the compiler in locating the requisite information. In this area of base construction, primarily in conjunction with projections and coordinate systems, the impact of computer assistance has been large. The result has been a greatly increased flexibility in the use of different projections and coordinate systems, with different orientations and origins.

This has produced a change in the type of information that a cartography student must now know about projections.

In the manual construction of the map base for compilation purposes, the cartographer did have to work with the projectional equations and did have to know how to plot any projection selected for use. On the other hand, any deviation in the form of the projection (e.g., a different orientation or origin) required considerably more calculation and in itself constituted a barrier to the utilization of these different capabilities. Often such unique orientations were simply not used because of the effort required to produce them.

With computer assistance any projectional or coordinate system that can be specified by a series of equations (usually two) is capable of being automated, and all the graticule lines can be drawn by machine from any origin or unconventional orientation (Fig. 12.16). Therefore the only limitations that a cartographer now has in map base construction is in a personal understanding of projections and the size limitations of the automated equipment. The limitation or constraint of dealing with complex mathematical equations has been eliminated, as has the tedious drafting of curved

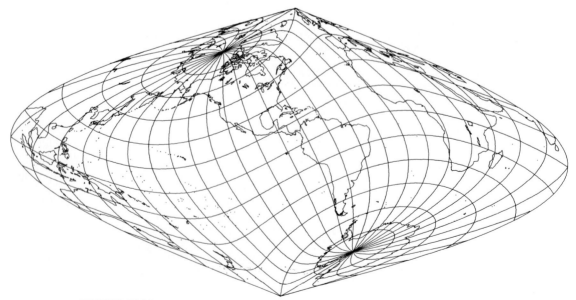

FIGURE 12.16 An unconventional orientation of the sinusoidal map projection showing the world.

lines that do not have a constant rate of curvature.

The cartographer need concentrate only on the design aspects and the more theoretical aspects of what use the base is being prepared for. For example, it is literally possible for the cartographer to make the map fit the format that is available. For special effects the cartographer can easily produce elongated, sheared, or unconventionally rotated projections and overlay them with coordinate systems. The cartographer must be thoroughly schooled in the use of projections and their properties, and must be able to select the projection possessing the property that is needed and that best fits the required format. The formidable constraints of construction have been removed.

From a theoretical point of view, by being able to specify the equations for positions, and by having the machine plot the points after making the calculations from the equations, the cartographer is now able easily to obtain any desired projection with any arrangement of its distortion. Thus a greater responsibility for the cartographer in addition to greater flexibility is present with computer assistance.

The study of projections and coordinate systems in cartography must concentrate on their properties and their uses and on how to employ a projection to fit the characteristics of the distributions that are being mapped. This greater flexibility and responsibility on the part of the cartographer is already in practice in the cartographic installations that have even a modest amount of computer hardware. For the future, because of the readily available software in this particular area, it can only be predicted that cartographers will be able to produce any projection for which the equations can be specified at any scale and in any orientation to fit the prescribed format. It will be possible to do this with more precision than is possible with a manual construction technique.

Compilation

Compilation can be divided into two aspects for the purposes of this discussion: first, the compilation of the base data for the map, and second, the compilation of the thematic data to be represented on the map. It is best to treat these two types of data separately in any discussion of computer assistance and to provide new definitions for these terms. The major reason for this is that computer assistance is most useful for repetitive tasks; therefore, if we define base data as the data that will be used over and over again and thematic data as the data that will be used a limited number of times, the computer-assisted compilation of each type will differ considerably. For example, the outline of the United States and the individual states within it, the basic hydrography of the United States, and perhaps the basic highway system of the United States constitute base data. However, other data are thematic data such as income per capita by minor civil division in the United States. These data are subject to rapid change, and the number of times that any single data set might be utilized would be small. Maybe such data should not be kept in a readily available file structure.

It is important for the cartographer to have the distinction between base data and thematic data in mind, because it is entirely possible that the most efficient map to be produced is one that is produced using all digital files for the base data and a manually constructed overlay for the thematic data. This is especially true where the thematic data are of relatively questionable reliability or of a rapidly changing nature.

Base Data. The primary considerations in the compilation of base data for use in computer-assisted cartography are: (1) scale and accuracy of input, and (2) scale and resolution of potential output. To qualify as base data, a data file will be expected to (1) be used regularly, and (2) change in basic detail infrequently relative to frequency of use. These criteria suggest that editing other than initial editing in setting up the base data file and updating will be minor considerations.

A base data file would be encoded at the largest scale that it will ever be utilized. This is simply application in computer operations of the general rule that compilation is always done from larger to smaller scale. Obviously a cartographic installation must determine its production limits (requirements) and then create a data file at the largest scale it ever hopes to create cartographic products. Accuracy should be consistent with the selected scale of input.

It may well be that a cartographic installation has production requirements at a series of scales so varied that base data in file form will have to exist at two or more scales. For example, if the cartographic unit has responsibility for regional, state and local mapping at scales ranging from 1:50,000 to 1:500,000, and also responsibility for continental and world products at scales from 1:10,000,000 to 1:200,000,000, the same data file for the state outlines of the United States may not serve both needs. In such instances cost considerations alone might dictate that two or more series of base data files for the United States be available for use. The cost of handling, storing, manipulating, and trying to plot a base file input at 1:50,000 for output at 1:20,000,000 is too great. Most of the cost of such an attempt would be for reading, manipulating, and attempting to plot stored data points whose location at output scale are closer together than the resolution of the output hardware. Similarly, if data files exist at 1:50,000 but the requirements are only for 1:500,000 or larger, it is too costly to generalize the data set each time output is desired. Thus a reduction and generalization to 1:500,000 for subsequent storage is required. These considerations are most applicable to string coordinate digitized data.

Scanned data as base data are entirely different. An essentially photographic base for cartographic products can be stored as scanned data. On the other hand, it is probably not wise to attempt to store base data such as highway networks or river systems as scanned data. It is only advisable to do so if the data are going to be used repeatedly at one scale and projection base combination. Perhaps land areas versus water areas as base data could be stored in scanned form, but little has been done with this as yet simply because the utility of scanned data is less than the utility of string coordinate data for most basic cartographic manipulations (e.g., projectional change) and for different levels of generalization of various phenomena. As we learn more about optical processing, perhaps neither scanned data files nor digital coordinate string data files will be optimal for cartographic purposes.

Already base data sets are available for the world's coastlines and political boundaries input at a scale of approximately 1:12,000,000 for smaller-scale use: also available are the U.S. county boundaries, and sections of the world at 1:3,000,000 showing basic hydrography, coast-

lines, and political units. Additionally, elevations for the land surface and sea bottom are available by one degree quadrilaterals (geographical coordinate system) for the entire world and at considerably larger scale for the North American and European continents. One can expect these data files to be augmented regionally and locally with larger-scale bases for states, regional planning distributions, and metropolitan areas. It is unlikely that these more localized files will be compatible with the more comprehensive files, but standardization of data file structure will be a part of every cartographer's future; the first step in this direction is being attempted by the National Cartographic Information Center at the USGS in Reston, Virginia.

Thematic Data. The compilation with computer assistance of thematic data is a more involved and less cost-saving step. Certainly, as the data collected by censuses become regularly used for mapping purposes, some algorithms for tying census data to base data will be used. This is not an insurmountable problem. Some census data may well become essentially base data for the intercensual time periods.

Other particular data (e.g., property boundaries, taxation records and maps) present other problems. For these data the scale and base data are usually fixed. The major problem becomes one of keeping the thematic data current by almost daily updating. Once an entire system is in data file form, the problems are usually minor. Technology can make this an easy way to proceed after the large amount of capital investment required to initiate the system becomes available. As more governmental bodies expend the necessary one-time capital investment, and begin to reap the vast rewards of computer-assisted record and map keeping, others are likely to follow quickly. For example, it is possible today to sit before a CRT and call up the part of a cadastral map of interest and to obtain the relevant data available about the individual land parcels.

In thematic data another problem is transforming it from one projection or coordinate system to another. It has become more and more evident that for data bases referring to large areas, the most useful storage system is latitude and longitude. To convert some thematic data distribution from its existing coordinates to coordinates that a machine can process and that are compatible with the base data file often is more expensive than to produce a manual overlay for use with a machine-produced base. As thematic maps (whose purpose is visual communication and not necessarily extreme cartometric accuracy) increase in importance, this will be even more true. It simply is not prudent to do the conversions for a thematic data set that is to be used only one time.

Generalization

Prior to computer assistance, generalization was very subjectively done by cartographers. Often an ability to "feel" the correct generalization was obtained by practicing mapmakers, and a good mapmaker's rendition was indisputable. With the advent of computer-assisted generalization, all operations had to be reduced to a finite number of steps that could be executed in a logical sequence. Machines can consistently follow any programmed generalizing procedure. The degree of consistency produced by machines is something humans find very difficult to do.

All generalization schemes are being programmed. Some have defied programming longer than others, but any often used cartographic method is being programmed today. Several generalization schemes were discussed in Chapters 7 and 8, and the results of the algorithms were illustrated. There are essentially two necessary forms for each generalization scheme, and these correspond to the two basic forms of data storage. Since much of the detail of the more important schemes has already been suggested, this discussion will concentrate on the changes wrought by computer assistance and will only allude to the desired ultimate system.

The changes are profound simply because never before have generalization schemes been (1) so rigidly defined, or (2) so consistently applied. For the first time generalization routines are being precisely defined. The true nature of some of the previously accepted methods is becoming more obvious, and it has enabled a better understanding of cartographic techniques and a more justifiable map product. A consistent application of a technique is mainly a by-product of computer assistance. Most cartographers have attempted to be consistent where consistency was required;

however, because of their inability to obtain detailed knowledge of each distribution to be mapped, variations in consistency were necessarily a part of every map. When a machine performs the manipulations according to a program, obviously consistency of treatment relative to input will be assured. Sometimes in cartography inconsistency is a virtue, and now the cartographer must plan desired inconsistencies.

The final point of this section indicates that it is desirable to have available manual overrides for all computer-assisted generalization. This adds increased complexity to the software programming, but allows for greater flexibility for the cartographer. Ideally, a cartographer should be able to watch the production of each segment so that as problems arise it is possible to intercede, or if a chance to be subjectively inconsistent will add to the map's communicative effectiveness, the production can be interrupted and the desired effect created. It is also possible that data from various data files may conflict, and the cartographer must be able to override the system manually in such cases.

As computer hardware becomes more sophisticated, concern for smaller resolution will decrease except for machine reading. Resolution of output for human use must be tempered with the human ability to read maps. If the machine can plot to 1/1000 in., but the eye can see only 1/200 in., why should the data be plotted at the resolution of the machine? This argument extends to abilities to plot detailed gray levels. If 128 gray levels can be sensed, recorded, manipulated, and plotted, but only 9 or 10 can be visually distinguished, why do it? The machine resolution currently available already seems adequate for most cartographic products. Resolution "overkill" becomes a fact when machines produce line width deviations or gray shades not distinguishable by the eye.

Lettering

Typography has to some extent been the stepchild of computer-assisted cartography. Lettering has not been successfully automated. With the use of computer assistance, lettering for routine maps can be accomplished, but limitations in flexibility have made this the area of least progress. Some estimates show that lettering accounts for almost

40% of the map production costs. If this is true, cartography has only succeeded in computer assisting the production of 60% of a map product.

The major problem with computer-assisted lettering is that cartographers have enjoyed extreme flexibility in lettering on maps for centuries. Whereas the dot denoting a city had to be exactly placed, the name for the dot could be placed almost anywhere in the vicinity of the dot. Furthermore, any type style could be utilized and, in hand lettering, certain variations of the various letters could be added for better fit or to create moods or flairs with the name. In contrast, computer-assisted typography is rigid and cold, without character and, although many styles are already possible by machine (see Fig. 12.17), each letter is stiffly drawn and placement tends to be firm and mechanical. Letter spacing and curved names (e.g., hydrographic names curved to fit meandering rivers) are next to impossible to do well without manual override.

If cartographers can take advantage of computer assistance in regard to lettering, it will probably be in the following ways. First, a data base of place names is useful. It is quite useful to be able to have a list of all place names for a given specified earth area. With this computer-generated list the cartographer can select those for portrayal and further select type styles and sizes. The cartographer can even get a first approximation of the number of conflicting names. The computer can generate the "type order" if need be, and drive output devices that provide the stickup lettering for the map.

At a more advanced level, the computer or output device can quickly plot all selected place names. Problems of overlap and illegible names can then be spotted, and the cartographer has essentially a worksheet over which the stickup can be positioned. At the most advanced level of computer assistance, the output device will actually plot the names in position on a piece of film. Two items are important here: first this will only work with manual override capabilities; and second, any cartographer is well advised to preview the entire plot before it is produced in final form. This preview can be accomplished on a fast drum plotter or on a CRT or microfilm plotter.

It is evident that for lettering on maps, manual override of a computer-assisted system is and will be a fact of life for cartographic production. Lettering will probably also be more rigid and mechanical in appearance and more stiff and less flowing in positioning for some time. Ultimately the cartographer may experience greater flexibility in map lettering but, for the near future, the cost savings by utilizing computer assistance will only result in less flexibility as far as map lettering is concerned.

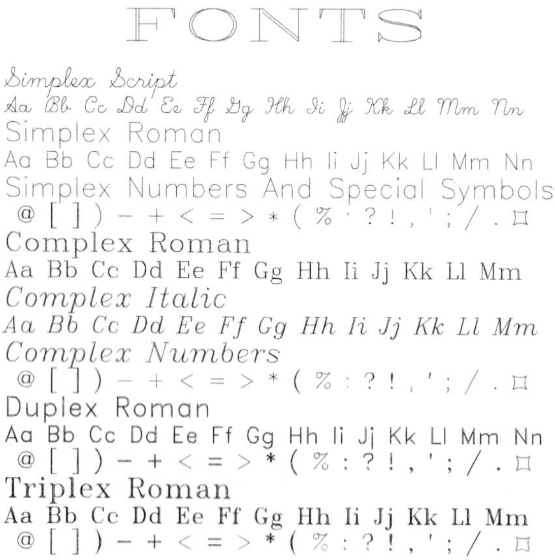

FIGURE 12.17 Examples of lettering fonts currently capable of being used in computer-assisted cartographic production.

Production

The final phase of cartographic processes is the actual production of the map itself. To date no problems exist. The cartographer can receive printable artwork suitable for reproduction from line printers, drum or flatbed plotters, and microfilm plotters. Photographs of CRT displays are also possible. Color separations as well as black and white output can be obtained. Some CRTs can be interfaced with electrostatic plotters to yield hard copy, Xerox-like output of CRT displays (Fig. 12.18). Resolution and quality are sufficient to equal manually drawn products, and the time savings are enormous. Dotted or dashed or other line types can easily be produced on most output devices. Registration can be held.

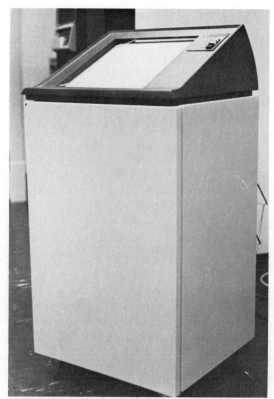

FIGURE 12.18 *An electrostatic printer/plotter. (Photo taken in the University of Wisconsin Cartographic Laboratory.)*

Once the capital investment has been made, the overwhelming savings by using computer assistance is in time and, therefore, cost. For the most part, the cost of operating computer equipment has held constant or actually decreased during the usable life of the equipment. The reverse is always the case for the human draftsman. This fact coupled with the increased speed of the computer equipment makes it certain that computer-assisted cartographic output is extremely important in the practical future of the field.

When one considers that computer-assisted cartography is less than 20 years old, the effects of it on the field are revolutionary. An entirely new technology has been invented and implemented. It has concentrated on doing by machine many operations that were formerly painstakingly done by hand. The increased flexibility and capabilities of the new technology will enable even further ex-tensions of the field which, at present, are only partially conceived.

GLOSSARY

accuracy—the degree to which a measurement is known to approximate a given value.

address—a numeric or symbolic designation of a memory location or peripheral unit where information is stored.

algorithm—a set of rules of procedure for solving a specific problem.

alphanumeric—(1) a character set of letters, integers, punctuation marks and special symbols; (2) term used to denote a combination of letters and numbers.

analog—the representation of a numerical quantity by a physical variable, for example, electric voltage; contrasted with digital.

assembler—a computer program that translates computer instructions written in a source language directly into machine language instructions.

batch processing—a method whereby items are coded, collected in groups, and processed sequentially.

binary—a numbering system based on two. Only the digits 1 and 0 are used.

bit—abbreviation for binary digit.

block—a group of records or words treated as a logical unit of information.

buffer—a storage register or section of memory where information from one part of the computer system may be held temporarily until another part is ready to receive it.

bug—an error in a program or routine; an equipment malfunction.

byte—a group of binary digits handled as a unit.

CAI—computer-assisted/computer-aided instruction.

card image—a binary representation in storage of the hole patterns of a Hollerith card.

chain—(1) the printing part of a line printer: a set of printing heads, like those of a typewriter, that strike an inked ribbon to leave an impression; (2) a synonym for string, as in "a string of coordinates."

character—a basic symbol used to convey information.

COM—computer output on microfilm.

command—the portion of an instruction word that specifies the operation to be performed.

compiler—a computer software program that converts a source language program into object language.

configuration—the specific arrangement of computer hardware and peripherals making up a system.

coordinate—an ordered set of data values that specifies a location; may be absolute or relative.

core—the most accessible information storage unit(s) of a digital computer.

CPU—central processing unit.

CRT—cathode ray tube: a vacuum tube in which a slender beam of high-speed electrons is projected on a fluorescent screen to produce a luminous spot.

CRT display—graphic output utilizing the screen of a cathode ray tube as the viewing element.

cursor—a movable part of an instrument (e.g., a digitizer or CRT) that indicates a position visible to the operator and whose *X-Y* coordinates are perceived by the machine.

data bank or data base—a store of information (usually in digital form) organized such that retrieval can be done on a selective basis.

data structure—the arrangement and interrelation of records in a file.

digital—the representation of a quantity by a number code; contrasted with analog.

digital incremental plotter—a device used for drawing line segments. The length and direction of the line segments are specified by a series of commands in machine-usable form.

digitizer—an instrument that converts graphically represented information into machine-readable form.

digitizing—the process whereby graphic images are converted to digital coordinates.

disc—a rotating magnetic storage device for digital information.

drum plotter—a digital incremental plotter in which the output material is mounted on a rotating drum or cylinder.

feature code—a numeric label attached to a line or point.

file—an organized collection of records.

flatbed plotter—a digital incremental plotter in which the output material is mounted on a fixed or moveable plane.

format—(1) the specific arrangement of data and their descriptors, identifiers, or labels; (2) the arrangement of data in bit, byte, and word form in the CPU.

hard copy—any permanent map, chart, or graphic presentation.

hardware—the physical components of a computer and its peripheral equipment.

Hollerith card—a machine-readable, informational storage medium of special quality paper stock, usually 7 ⅜ × 3 ¼ in. having 80 columns with 12 punch positions per column.

input—information or data, transferred (or to be transferred) from an external storage medium to internal storage of a computer.

instruction—a code that defines an operation to be performed as well as the data or unit of equipment to be used.

interactive mode—a method of operation that allows on-line communication between a person and a machine; commonly used to enter data or to direct the course of a program.

interface—the common boundary between individual components of a system.

label—a group of characters used as a symbol to identify an item of data, an area of memory, a record, or a file.

library routine—a prewritten standard algorithm for use in computer programs.

light pen—a cursor the size of a ballpoint pen; used for pointing to a location on a CRT screen.

line printer—a peripheral computer device like a typewriter that prints one line at a time; a set of alphanumeric characters is available at each position along the line.

machine language—instructions written in machine code that can be obeyed by a computer without further translation.

magnetic tape—ferrous-coated tape; selective polarization of the surface permits the sequential storage of digital data.

memory—an organization of storage units retained primarily for the retrieval of information.

minicomputer—a relatively low-cost CPU with limited core capacity.

node—a point common to two or more line segments.

noise—unwanted variations introduced into data during measurement or display.

object language—a machine language resulting from the output of a compiler; the target language to which a source language program is converted by the assembler or compiler.

off-line—processing of information that is not directly under the control of the CPU.

on-line—processing of information that is directly under the control of the CPU.

output—computer results; for example, answers to mathmetical and statistical problems or graphic plottings.

peripheral—the input/output equipment used to transmit data to and receive data from the CPU.

pixel—a sensed unit of a digital image, usually a very small area; for example, 25, 50, or 100 microns square. The decomposition (and recording) of a viewed surface into pixels results in a digital image matrix.

precision—a statistical measure of repeatability, usually expressed as a variance or standard deviation (root mean square, RMS) of repeated measurements.

program—a series of logical statements or instructions written in source language or object language to specify what operations the machine is to perform.

programming language—a language used to prepare computer programs, for example, FORTRAN, ALGOL, COBOL, PL1, BASIC.

raster—the screen area of luminescent material in a CRT where an image is produced.

record—(1) a collection of related items of data, treated as a unit; (2) one segment of a file; for example, the outline of an area may form a record, and a complete set of such records may form a file.

register—a storage location in a hardware unit.

resolution—a measure of the ability of a device to differentiate a value; high resolution means fine differentiation.

scanner—any device that systematically decomposes a sensed image or scene into pixels and then records some attribute of each pixel.

software—the already established programs and routines used in computer operation.

source language—the language used by a programmer to communicate computational requirements to the compiler for that source language.

string—a consecutive sequence of characters or physical elements.

stylus—a cursor or device in a recording machine used to designate data to be digitized or stored.

subroutine—a set of computer instructions to carry out a predefined computation; usually exists as part of a larger computer program.

throughput—the total computer processing of data from input to output.

time sharing—allocating in turn small divisions of a total time period to two or more functions so that machine time can be used to peak efficiency.

update—the modification of data records as necessary.

word—an ordered set of characters handled as a unit by the computer.

word length—the number of bits in a computer word.

SELECTED REFERENCES

American Congress on Surveying and Mapping, *Proceedings* of International Conference on Automation in Cartography, December 9–12, Reston, VA. 1974.

American Congress on Surveying and Mapping, *Proceedings, Auto-Carto II,* International Symposium on Computer-Assisted Cartography, September 21–25, Reston, Va. 1975.

Boyle, A. R., "The Requirements of an Interactive Display and Edit Facility for Cartography," *The Canadian Cartographer, 13,* 35–59 (1976).

Brassel, K., "A Model for Automatic Hill Shading," *The American Cartographer, 1,* 15–26 (1974).

Brassel, K., "A Survey of Mapping Software," *Proceedings of the American Congress on Surveying and Mapping, 37th Annual Meeting,* Washington, D.C., pp. 509–515 (1977).

British Cartographic Society, *Automated Cartography.* Papers presented at the Annual Symposium of the British Cartographic Society, Southampton, 1973, Special Publication No. 1, British Cartographic Society, London, 1974.

Broome, F. R., "Micrographics: A New Approach to

Cartography at the Census Bureau," *Proceedings of the American Congress on Surveying and Mapping, Fall Convention,* Washington, D.C., pp. 1–14 (1974).

Le Blanc, A. L. (ed.), "Computer Cartography in Canada," *Cartographica,* Monograph No. 9, York University, Toronto, Canada, 1973.

Linders, J. G., "Computer Technology in Cartography," *International Yearbook of Cartography, 13,* 69–80 (1973).

Meyer, M. A., F. R. Broome, and R. H. Schweitzer, Jr., "Color Statistical Mapping by the U.S. Bureau of the Census," *The American Cartographer, 2,* 100–117 (1975).

Peucker, T. K., *Computer Cartography,* Commission on College Geography Resource Paper 17, Asso-ciation of American Geographers, Washington, D.C., 1972.

Peucker, T. K., and N. Chrisman, "Cartographic Data Structures," *The American Cartographer, 2,* (1) 55–69 (1975).

Rhind, D. W., and T. Trewman, "Automatic Cartographic and Urban Data Banks—Some Lessons from the U.K.," *International Yearbook of Cartography, 15,* 143–157 (1975).

Riffe, P. D., "Conventional Map, Temporary Map or Nonmap?," *International Yearbook of Cartography, 10,* 95–103 (1970).

Yoeli, P., "Computer-Aided Relief Presentation by Traces of Inclined Planes," *The American Cartographer, 3,* 75–85 (1976).

Design in the graphic context is both a noun and a verb. As a noun the term refers to the visual qualities of a display with attention paid to the appearance of the individual components and the character of their arrangement. As a verb the term refers to the planning and decision making involved.

Graphic design is a vital part of the cartographic process because effective communication requires that the various marks (lines, tones, colors, patterns, lettering, etc.) be carefully modulated and fitted together. Just as an author—a literary designer—must employ words with due regard for many important structural elements of the written language, such as grammar, syntax, and spelling, in order to produce a first-class written communication, in parallel fashion the cartographer—a map designer—must pay attention to the principles of graphic communication.

Maps are made with the fundamental objective of conveying some geographical information to a reader, and the processes of compilation, symbolization, choice of scale, and projection are focused to that end. Although the substantive aspects of compiling a map can be quite complex, the designing of it as a graphic communication is equally important and complicated. The manner of presentation of the many map components so that together they appear as an integrated whole, devised systematically to fit the objectives, includes a variety of elements. Regardless of the positional accuracy or essential appropriateness of the data, if the map has not been carefully designed, it will be a poor map.

In this chapter we will survey the principles of graphic communication and some of the fundamental elements of map design. Two elements of cartographic design, color (and pattern) and typography, are unusually complex and require individual treatment in the following two chapters.

The Design Process

In the sense that cartographers prepare functional structures, they are like engineers and architects; they try to find solutions either to new problems or new solutions to problems solved another way. The cartographer's objective is to evoke in the mind of a viewer the desired image of the spatial

CHAPTER 13

GRAPHIC DESIGN
IN MAPMAKING

environment appropriate to the intended purpose of the map, whether it be the composite details of a general map or the structural character of a distribution, as in a thematic map. Furthermore, the cartographer must make available the necessary geographical information if the map is to be used for specific operations, such as navigation by automobile or water craft. Because graphic structures involving spatial data are often relatively complex, being made up of a variety of marks with many possibilities for variation, it is likely that any one problem will have a considerable number of possible solutions. Consequently, it is desirable that a cartographer be freely imaginative in the approach to cartographic design.

Imagination is the primary element in the first stage of the designing process. It is here that the mapmaker scans the various possibilities and considers all the ways the problem can be approached and tries to visualize the different solutions. At this stage the cartographer is being the most creative in the design process. The result is a general idea of the approach involving decisions such as the relation of the map to others, the format (i.e., its size and shape), the basic layout, the graphic organization of the components, and so on.

The second stage in the design process involves the development of the specific graphic plan. Here the mapmaker weighs the various alternatives within the limits of the general plan. At this stage decisions are made regarding particular kinds of symbolism, color use, typographical relationships, general line weights, and the like, in terms of how they will fit together graphically. By the time the second stage is completed, all but minor decisions have been made.

The third stage is that of preparing the specific plans and specifications. This is analogous to the architect preparing the detailed drawings to be followed by a builder. It involves (1) the preparation of the compilation worksheet where everything is put in proper planimetric relationship, and (2) the preparation of detailed specifications for the construction of the artwork—all line weights, screen values, ink colors, lettering sizes, and so on. Nothing is left unplanned. It is obvious that preparing detailed specifications requires that the cartographer have a thorough understanding of all the processes involved in map construction.

Relation of Map Design to the Arts

As suggested in Chapter 1, cartography is not an aesthetic art, such as painting, music, fiction, or dance. Generally, the functionalism of cartography together with geographical reality put too many constraints on the cartographer to allow "full freedom of expression." On the other hand, one cannot survey the creations of past cartographers and say that mapmaking is artistically sterile. The variety of delineations and the significant role of decoration show that mapmakers have often been stirred by aesthetic urgings.

From the point of view of graphic design, mapmaking is clearly creative in that there is an almost unlimited number of options for organizing the display involving a combination of intuition and rational choice. Since aim is excellence in communication, some will come closer than others, but some inner subjective sense is not a requirement for good design. Like the study of written composition, the basic elements of graphic composition lend themselves to systematic analysis, and the principles can be learned. This is not meant to imply that every individual can become highly creative merely through study. As has been often repeated, one can be taught a craft (i.e., the mastery of the materials and techniques with which one works), but one cannot be taught the art.

A basic requirement in graphic design is a willingness to think in *visual* terms, uninhibited by prejudices resulting from previous experience. The range of imaginative innovation must, of course, be disciplined to some extent since, like many fields, cartography has developed traditions and conventions; to disregard them completely would inconvenience the user of the maps, which would in itself be proof of poor design.

Three propositions may be advanced as guides to the student of functional map design. They have been suggested by students of aesthetics and visual expression:

1. Beauty or elegance may occur in functional graphic design, but it will be a consequence of good design in a favorable context.
2. Something well designed will not look so; in other words, the design itself should not appear contrived.
3. Simplicity is highly desirable and is a result of excellence. Because simplicity is relative in a

context, it cannot be defined, but it can be recognized,

GRAPHIC PRESENTATION AND MAPS

The designing of a map involves working with a variety of components, ranging from the symbolism treated in earlier chapters to the lettering and map reproduction methods covered in the following chapters. Because cartography is a visual medium, there are various principles of graphic communication involved that are as indispensable to good visual composition as things such as syntax, paragraphing, organization, phrasing, and so on, are to good literary composition. Although the general objectives of communication by means of language and by means of graphic presentation may be similar—clear, unambiguous transfer of information—there are many differences between the two media.

Among the more important of these differences is the seemingly obvious fact that the components of a graphic display are visual stimuli and that people react to graphic presentation in quite different ways than they do to written and spoken communication.

Perception of Graphic Complexes

Through long familiarity the sounds of language and the written words that encode the sounds tend to be "transparent" in that their meanings become known to us without our paying much attention to the actual sounds or the physical appearance of words. This is not true with graphic presentation. The marks are "opaque"; we pay a great deal of attention to them and their arrangement.

A person reading or being spoken to receives the information in serial fashion, that is, the words follow one another in sequence. The reader or listener is thus programmed to receive the thoughts in a definite order. Graphic communication is quite different in that a person receives the visual impressions synoptically (i.e., all at once) instead of in a sequence. Our perception of every mark on a map is simultaneously affected by its location and appearance relative to all the other marks. This means that instead of structuring our communication sequentially, we must deal with it always as a whole. Everything on a map is related

visually to everything else; if we change one thing, we change everything.

When people view a display of any kind, they process it perceptually. Just as when people hear a series of words or other sounds, there is an unconscious attempt to "make sense out of it," so do they—unconsciously—attempt to organize what is seen in a visual field. Perceptually this involves assigning visual meaning and importance to the various marks and to the shapes, sizes, directions, values, colors, and so on. Because we tend to resist visual monotony and ambiguity, we will "organize" the display so that it "makes sense" visually. It is inevitable, therefore, that viewers of graphic displays will see them structurally: some marks are more important than others, some shapes "stand out," some things are crowded, some colors dominate, and so on. If these meaningful relationships among the graphic elements coincide with the cartographer's intentions, effective communication has been established.

Objectives of Map Design

The fundamental objective in designing a map is the communication of some kinds of spatial relationships. As we have seen in the preceding chapters, there are a great many ways to symbolize (i.e., to encode) geographical data, concepts, and relationships. The basic ways of doing this (assigning specific meaning to the various kinds of distinctive marks, to their variabilities, and to their combinations) is, in a sense, the first of two steps in cartographic design. The second step is to arrange the marks in a total composition that will evoke in the viewer the desired perceptual response. The two aspects are not really separable, of course, since one must select the symbolism with the ultimate objective in mind. The comparison with literary composition is apt. In written composition one selects the appropriate words— the symbols of various kinds of meaning—but one also arranges the words in sequence in sentences, paragraphs, and sections to fit the objectives of the communication. Although the two are conceptually distinct, in actual practice, one cannot easily separate them.

In Chapter 1 we saw that the communication objectives of cartography ranged along a continuum from general maps to thematic maps. An individual map usually involves some of each aim, but it is worthwhile to contrast the two as they affect map design. A "pure" general map tries to display a variety of spatial data in such a way that selected individual attributes of each place are displayed. In theory neither any class of phenomena nor any place are more "important" than any other. Although the parallel has some considerable limitations, the pure general map is like a geographical dictionary. In a dictionary no word is emphasized and the relation among the ideas the words encode is not dealt with explicitly, although it is available, so to speak, to the reader if the provided definitions are analyzed.

Thematic cartography, at the other end of the continuum, is primarily concerned with conveying the overall character or structure of a given spatial distribution. It is the structural relationship of one part to another that is important. In that sense it is not a dictionary defining the geographical character of places; it is a kind of graphic essay dealing with the spatial variations and interrelationships of some geographical distribution. The fundamental design problems of general maps and thematic maps are therefore distinct.

In general mapping each spatial attribute to be mapped must be encoded in a unique fashion so that the meaning will not be confused with some other attribute. The graphic character of each mark must be distinct so that confusion will not arise about which mark is which. Unless specifically desired, no mark should appear more important than another, and no group or class of marks should dominate. Obviously, that description refers to a "pure" general map. A "pure" general map will rarely occur, because usually some spatial attributes are thought to be more "important" than others and will, consequently, be given visual emphasis. Individual legibility and graphic contrast are primary aims.

In thematic mapping the marks and symbolic system must be chosen so as to work together graphically to evoke the overall character of a distribution. The possible modulations of marks, the visual variables, are to be employed in related fashion. Although spatial images and graphic integration are primary aims, adequate locational "base data" must also be provided.

It bears repeating that although the fundamental objectives of general and thematic cartography are opposite and lead to distinct graphic treatments, in practice, most maps actually combine the objectives to some degree. It is simply necessary that the cartographer recognize the combination and proceed to design accordingly.

Components of Map Design

A useful way of considering the process of map design is to conceive of it as consisting of a set of primary graphic elements that may be modulated by the cartographer to attain the desired end. In each map design problem the manipulation and treatment of the elements, such as clarity, legibility, visual contrast, and balance, are dictated by the operation of a set of controls. The controls may be thought of as operating influences that, to a greater or lesser degree, establish the conceptual framework for the manipulation of the design elements and provide tolerances and limits.

The controls of map design are the objective of the map, the reality of the area being mapped, the scale, the audience, and the technical limits within which the map must be prepared. Some maps will not involve managing all the elements of map design, but the full set of controls is operative in all cases. The controls interact, of course, and it is often useful to contemplate the set altogether, so to speak, at the outset of a map project. If one can anticipate the kinds of problems likely to be faced, it is often possible to influence the initial conception of the project before it has progressed to the point where some unanticipated constraint has made the graphic design problem unduly demanding.

Controls of Map Design

Brief comments on the character and operation of the individual controls are given below. Of necessity, they are general. The student may find it useful to consider each one in relation to a specific map.

Objective. It is obvious that the communicative purpose for which the map is being made is the essential determinant of its final form. All aspects of symbolization and graphic design must be consciously fitted to the purpose as carefully as possible.

Of particular importance in the operation of objective as a control is the breadth of the purpose. In thematic mapping, the greater the number of identifiable objectives, the more difficult it is to arrive at a graphic design that incorporates solutions to all of them. Because maps are difficult, costly, and slow to make, there is a strong tendency to combine many purposes in one map. Furthermore, there seems to be a kind of fear among some cartographers that a map may be used for some purpose not intended by its maker. Obviously, any knowledgeable map user will not measure distances improperly, or the like, but the nearest map is often the easiest place to look for something, such as a boundary, river, or city. It is a mistake to try to anticipate possible improper uses, and to make allowance for them complicates the graphic design and leads to visual complexity.

Another significant aspect of the objective is that of the total appearance or "look" of the map. Whether the map should appear dark or light, crowded or open, precise or approximate, and the general subjective feeling to be conveyed, such as graceful, bold, traditional, or modern, are important aspects of map design; such affective objectives may be as important as the substantive objectives in many cases.*

Reality. The overall geographical dimensions and character of the distribution or distributions being mapped are given and, except in unusual circumstances, they cannot be much modified by the cartographer; projectional modulation can have some effect, but regardless of what is done, Chile will remain long and narrow; the Boston-New York-Philadelphia-Washington megalopolis area is heavily populated with many comparatively large cities; soil variations are inherently complex; and areas of recent glaciation have enormous numbers of lakes. Each of these kinds of reality, and many more like them, place serious constraints on the graphic design possibilities; they must be anticipated at the outset of the planning stage of a map.

Scale. Scale is prescribed by the format and by its relation to the area being mapped. From a conceptual point of view, scale operates in subtle ways. The smaller the scale, the "further away" one is from the area being mapped, and this intu-

itive feeling should be matched with the graphic design. Although one can say, in general, that the smaller the scale, the smaller the line weights and lettering sizes should be, no specific prescription can be made because other controls are involved. Nevertheless, under ordinary circumstances, a magnifying glass should never be needed to use a map. In addition to line weights and lettering sizes, patterns and perhaps colors may be affected in series maps of different scales. The relation among coarsenesses of pattern and intensities of color and the various scales of related maps should appear to be logical.

Audience. Both general and thematic maps are made for a variety of audiences. These range from young schoolchildren to college students, and from the lover of the outdoors to the technical engineer. Each group may vary from another in terms of its familiarity with graphic and symbolic conventions, its geographical knowledge, and its perceptual limitations in general. For the geographical unsophisticate, the important basic shapes, important names, and the like should be emphasized in the graphic design. A map on a wall being viewed by an audience seated at desks will require a different treatment of graphic character than a map of the same sort in a textbook on the desk. The perceptual limits of the audience may sometimes be a significant element in the design. For example, older people have more difficulty seeing small type and must hold visual material closer to them. An extreme example of this aspect of "graphic" design is provided by the problems of making maps for the partially sighted and totally blind, wherein attention must be paid to things such as tactile discriminability.

The functioning of the audience as a control may be extended to include the conditions under which a map may be used. Legibility requirements for road maps to be used in a moving automobile must be higher than for a map to be viewed in a stable condition. Maps and charts to be used at night in boats, ships, and aircraft must be usable in dim light so that night vision will not be greatly affected by bright illumination.

Technical Limits. How a map is to be constructed and reproduced will affect the graphic design in several ways. Is it to use color or is to be monochrome, scribed or drafted in ink, hand let-

---

*Barbara Bartz Petchenik, "A Verbal Approach to Characterizing the Look of Maps," The American Cartographer, 1 (1), 63–71 (1974).

tered or made with printed type? The list of technical influences on graphic design is very long and must be thoroughly examined before beginning the design process.

GRAPHIC ELEMENTS OF MAP DESIGN

The graphic elements of map design are those attributes of the marks used for representation that either by themselves or in organized array are visually significant in the total graphic presentation that is the map. Only the most important can be considered in this book. They are clarity and legibility, visual contrast, figure-ground, balance, hierarchic structure, color and pattern, and typography. The subjects of color (including value) and pattern and typography are complex and of sufficient importance that they will be treated in separate chapters to follow.

Although space allows only a summary treatment of each of the graphic elements, it should be kept in mind that the abbreviated attention that can be paid to them here is not a proper measure of their importance. It is worth observing that cartography involves the combination of many kinds of elements, and that it is simply not possible to provide instructions "from the ground up" in topics that are fundamental to it, whether it be mathematics, geography, or graphic composition. Nevertheless, the basic principles involved can be treated.

Clarity and Legibility

The transmission of information by means of the coding built into marks of various sorts (lines, letters, tones, etc.) requires that the marks be clear and legible. Although the various graphic elements treated in this chapter have other functions to perform in a graphic composition, one of their basic objectives is to promote clarity and legibility.

Clarity and legibility are broad terms, and many of the techniques and principles considered in other parts of this textbook are important factors in obtaining these qualities in a presentation. Furthermore, a considerable portion of the task of achieving clarity and legibility will have been accomplished if the mapmaker has made sure that the intellectual aspects of the map are not open to doubt or misinterpretation.

If it is assumed that the geographical concepts underlying the purpose and data of a map are clear and correct, then legibility and clarity in the presentation can be obtained by the proper choice of lines, shapes, and colors and by their precise and correct delineation. Lines must be clear, sharp, and uniform; colors, patterns, and shading must be easily distinguishable and properly registered (fitted to one another); and the shapes and other characteristics of the various symbols must not be confusing.

One important element of legibility is size; no matter how nicely a line or symbol may be produced, if it is too small to be seen, it is useless. There is a lower size limit below which an unfamiliar shape or symbol cannot be identified. Although there is some disagreement as to the exact measure of this threshold, a practical application sets this limit as being a size that subtends an angle of about 1 min at the eye. That is, no matter how far away the object may be, it must be at least that size to be identifiable. This limit sets rather an ideal, since it assumes perfect vision and perfect conditions of viewing. Because of the unreasonableness of these assumptions, it is wise for the cartographer to establish the minimum size somewhat higher, and it may be assumed that 2 min is more likely to be a realistic measure for average (not "normal") vision and average viewing conditions. Table 13.1 (based on 2 min) is useful in setting bottom values of visibility. It should be remembered that some map symbols have length as well as width; in such cases (e.g., with lines) the width may be reduced considerably, since the length will enhance the visibility. Similarly, other elements, such as contrasting

TABLE 13.1

	Approximate Minimum Sizes for Legibility		
	Viewing Distance	Size (Width)	
Metric		U.S. Customary	
50 cm	0.3 mm	18 in.	0.01 in.
2 m	1.15 mm	5 ft	0.03 in.
5 m	2.9 mm	10 ft	0.07 in.
10 m	5.8 mm	20 ft	0.14 in.
15 m	8.7 mm	40 ft	0.28 in.
20 m	11.6 mm	60 ft	0.42 in.
25 m	14.5 mm	80 ft	0.56 in.
30 m	17.4 mm	100 ft	0.70 in.

colors or shapes, may increase visibility and legibility, but even though the existence of a symbol on the map is made visible by such devices, if it does not stand at or above the sizes given in Table 13.1, it will not be legible. In other words, it might be seen (visible), but it might not be read or recognized (legible).

A second element regarding legibility is also operative in cartography. Generally, it is easier to recognize something we are familiar with than something that is new to us. Thus, for example, we may see a name in a particular place on a map and, although it may be much too small to read, we can tell from its position and the general shape of the whole word what it is.

Visual Contrast

The fact that symbols, lines, and the other elements of a map are large enough to be seen does not in itself provide clarity and legibility. An additional graphic element, that of visual contrast, is necessary. No element is as important as contrast. Visual contrast is the basis of seeing and, assuming that each component of the map is large enough to be seen, the manner in which a mark differs from its background and the adjacent marks determines its visibility.

Visibility is indispensable, of course, but it must not be assumed that maximum contrast is automatically desirable. In some cases, separate components of the map differ only to a limited degree, and it is desired that the marks indicate this. As will be pointed out later, many degrees of contrast can be incorporated in hierarchic structures. Contrast is achieved by modulating the visual variables (position, shape, size, hue, value, pattern, and direction), as outlined in Chapter 5. There are a few general principles of contrast that can be suggested here.

An observer resents visual ambiguity, and the contrast among marks should always be sufficient so that there is no doubt about it. Furthermore, the critical eye seems to accept moderate and weak graphic distinctions passively and without enthusiasm, so to speak, whereas it relishes greater contrasts. The desirability quality of a display, variously termed "crisp," "clean," and "sharp," seems to be obtained in large degree by the amount of contrast it contains. Figure 13.1, a simple matrix of line widths, illustrates one aspect

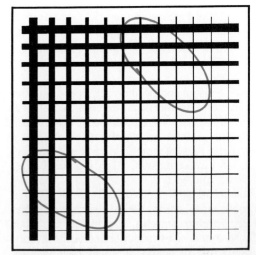

FIGURE 13.1 Size contrast of lines. Uniformity produces unpleasant monotony. The general areas outlined in red appear to be more interesting or "crisp" than where there is less contrast.

of this. The most "interesting" sections are those where there is considerable contrast among the widths. On the other hand, Fig. 13.1 also illustrates that too much contrast can be unpleasant.

The visual variables are additive in that if two marks are varied in two ways (e.g., in both shape and size), the contrast between them will be greater than if only one variable had been modulated. Variation of only one variable may be sufficient to achieve the needed contrast if the mark is simple, but the more complex the mark, the larger the number of elements that must be varied to obtain contrast.

Contrast is a subtle visual element in some ways, and in other ways it is blatant. The character of a line and the way its curves or points are formed may set it completely apart from another line of the same thickness. The thickness of one line in comparison to another may accomplish the same thing. The relative darkness of tonal areas or the differences in hue of colored sections may be made similar or contrasting. The shapes of letters may blend into the background complex of lines and other shapes, or the opposite may be true. If one element of the design is varied as to hue, value, thickness, or shape, then the relationship of all other components will likewise be changed. It requires careful juggling of the lines, shapes, and color characteristics on a kind of trial-and-error basis to arrive at the right combination.

Balance

Designing a visual composition requires a number of preliminary decisions pertaining to balance. These decisions involve problems of layout concerning the general arrangement of the basic shapes of the presentation. The basic shapes may include land-water masses, titles, legend, boxes, color areas, and so on.

Balance in graphic design is the positioning of the various visual components in such a way that their relationship appears logical or, in other words, so that it does not unconsciously or consciously disturb the viewer. In a well-balanced design nothing is too light or too dark, too long or too short, in the wrong place, too close to the edge, or too small or too large. Layout is the process of arriving at proper balance.

Visual balance depends primarily on the relative position and visual importance of the basic parts of a map, and thus it depends on the relation of each item to the optical center of the map and to the other items, and on their visual weights.

The optical center of a map is a point slightly (about 5%) above the center of the bounding shape or the map border (see Fig. 13.2). Size, value, brilliance, contrast and, to some extent, a few other factors influence the weight of a shape. The balancing of the various items about the op-

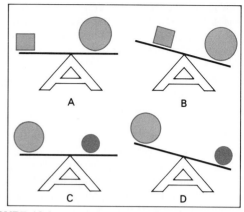

FIGURE 13.3 *Visual balance. A, B, C, and D show relationships of balance. A and B are analogous to a child and an adult on a "teeter-totter"; C and D introduce relative density or visual weight, darker masses being heavier than lighter.*

tical center is akin to the balance of a beam or lever on a fulcrum. This is illustrated in Fig. 13.3, where it can be seen that a visually heavy shape near the fulcrum is balanced by a visually lighter but larger body farther from the balance point. Many other combinations will occur to the reader.

The aim of the cartographer is to balance the visual items so that they "look right" or appear natural for the purpose of the map. The easiest way to accomplish this is to prepare thumbnail sketches of the main shapes and then arrange them in various ways within or around the map frame until a combination is obtained that will present the items in the fashion desired. Figure 13.4 shows some rough sketches of a hypothetical map in which the various shapes (i.e., land, water, title, legend, and shaded area) have been arranged within the border in various ways. Many reference map series utilize the sheet margins for titles, legends, scales, acknowledgements, and the like. These masses must also be appropriately balanced.

The format, that is, the size and shape of the paper or page on which a small-scale map is to be placed, is of considerable importance in the problem of balance and layout. Shapes of land areas vary to a surprising degree on different projections; in many cases the desire for the greatest possible scale within a prescribed format may

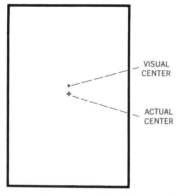

FIGURE 13.2 *The visual as opposed to the actual center of a rectangle. Balancing is accomplished around the visual center.*

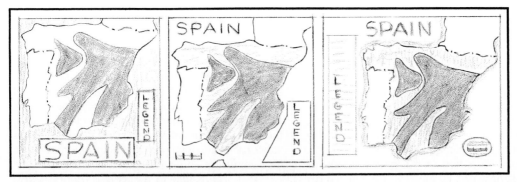

FIGURE 13.4 *Thumbnail sketches of a map made in the preliminary stages to work toward a desirable layout and balance.*

suggest a projection that produces an undesirable fit for the area involved. Likewise, the necessity for fitting various shapes (large legends, complex titles, captions, etc.) around the margins and within the border makes the format a limiting factor of more than ordinary concern. Generally, a rectangle with sides having a proportion of about 3:5 seems to be the most pleasing shape (see Fig. 13.5).

The possibilities of varying balance relationships to suit the objective of the map are legion. The cartographer will do well to analyze every visual presentation, from posters to advertise-

ments, in order to become more versatile and competent in working with this important factor of graphic design.

Figure-Ground

A complex occurrence in human visual perception is the *figure-ground* phenomenon. The eye and the mind, working together, react spontaneously to any array of visual stimuli, whether or not it is familiar. They tend immediately to organize the display into two basically contrasting perceptual impressions: a figure on which the eye settles and sees clearly, and the amorphous, more or less formless ground. The figure is perceived as a coherent shape or form, with clear outlines, that appears in front of or above its surroundings. The figure is not confused with the less distinct, somewhat formless ground that surrounds it. While this spontaneous visual organization is occurring, the previous experience of the observer is also brought into play in terms of mentally processing the probabilities of what the particular or several geographical form relationships may be.

The separation of the visual field into figure and ground is automatic; it is not a conscious operation. It is a natural and fundamental characteristic of visual perception and is therefore a primary graphic element in map design.

The figure-ground phenomenon is especially important in the design of thematic maps.* In such

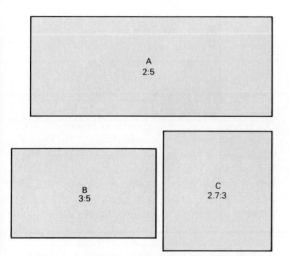

FIGURE 13.5 *Various rectangles. B, with the ratio of its sides 3:5, is considered to be more stable and pleasing than the others.*

*Borden Dent, "Visual Organization and Thematic Map Communication," Annals, Assoc. of American Geographers, 62 (1) 25–38 (1972).

maps the immediate perception of the fundamental structural elements of the map is of primary importance. In both general and thematic maps the reader must be able to tell land from water, and recognize and separate the outline of the city, island, or harbor from its surroundings; in short, the reader should be able to focus immediately on the cartographer's objectives without visually fumbling and groping to find what he or she is supposed to be looking at.

A simple illustration of this phenomenon of perception is provided by Fig. 13.6. In Fig. 13.6A the simple line, wandering across the visual field, merely separates two areas and is quite ambiguous. Assuming that the map reader knows he or she is supposed to see an area separated into familiar shapes, the reader is confused as to which is land and water; unless the reader is unusually familiar with the area, he or she will continue to be perceptually confused. In Figs. 13.6B and 13.6C, areas are clearly defined from one another, but there is still the problem of which is land and which is water. In Fig. 13.6D, by the use of land-based names, familiar boundary symbols that usually occur on land areas, the graticule to "lower" the water, and slight shading to separate

the basic land from the water, the land clearly emerges as the figure. Of course, once we have clarified the display by observing Fig. 13.6D and recognizing the area, Figs. 13.6A, 13.6B, and 13.6C become much easier. It is important to observe, however, that the fundamental perceptual relationships are always there; regardless of how well acquainted a map user may be with an area, the kind of presentation illustrated in Fig. 13.6A is subconsciously perceptually ambiguous and therefore confusing and irritating.

It is impossible to develop in depth in this textbook the variety of visual stimuli that are involved in the promotion of desired figural perception. Like all elements of graphic design, complex interrelations are involved. Only some of the basic principles can be suggested.

Differentiation must be present in order for one area to emerge as the figure. The desired figure area must be visually homogeneous, and the homogeneity of the whole visual field (the entire map) must not be stronger than the desired figure; otherwise the map itself will become the figure and the ground its surroundings. Such differentiation may be promoted in a variety of ways, such as by color, value, and texture.

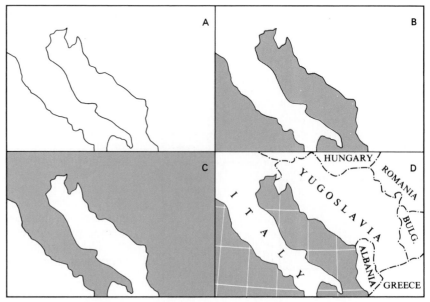

FIGURE 13.6 Four simple sketch maps to illustrate various aspects of the figure-ground relationship. See the text for explanation.

Closed forms, such as islands, entire peninsulas or countries, and other complete entities are more likely to be seen as figures than if they are only partially shown. If all of the peninsula of Italy had been shown in Fig. 13.6A, it would have appeared more easily as figure. In general, the smaller surrounded area tends toward figure.

Familiarity exercises considerable influence in the promotion of figure. We readily fix on a well-known shape. In terms of map design, for example, this suggests that the less well known an area is that is desired as figure, the greater must be the application to it of the factors that promote figure.

Brightness (tonal value) difference promotes the emergence of figure, and all other things being equal, the darker tends toward figure. In graphic composition, all other things are not usually equal, and attention must always be focused on the particular set of visual circumstances. For example, in Fig. 13.6D, the lighter area tends toward figure because of the overriding effect of other influences.

Good contour is a term that means the graphic equivalent of logical and unambiguous. When something appears continuous, symmetrical, or sensible, it will lead toward figure-ground differentiation. Good contour also involves the tendency for the viewer to move toward the simplest "visual perceptual explanation" of graphic phenomena. For example, in D in Fig. 13.6 the graticule appears to be continuous "beneath" the land, thus "raising" the land area above it and helping it to become figure.

The principle of good contour finds many applications in cartography in addition to the common practice in thematic mapping of placing the graticule only on the water in order to help the land become figure. The breaking of a line for lettering or other map components is possible because, given the requisite character of the line, visual logic—good contour—will "continue" the line where it is not actually shown.

Articulation of an area, in the broad sense of involving a complex of internal markings, tends to lead toward the emergence of figure. In Fig. 13.6D there is considerable detail on the land, which is made up of complex lettering and boundaries in contrast to the simplicity of the graticule on the water. This helps to cause the land to emerge as figure. Things such as city symbols, names,

rivers, transportation routes, and relief representation are all components that can contribute to figure.*

Area is important in leading to figure-ground differentiation. Generally, the tendency is for smaller areas to emerge as figure in relation to larger areas. Some recent research has suggested that for thematic maps the figure to ground ratio (total map area minus area of figure divided by area of figure) should be between approximately 1:4 and 1:1.5.† When ratios are larger than 1:4, the ground may overwhelm the figure; when ratios are smaller than 1:1.5, there may be confusion between figure and ground.

In composing a graphic communication, the cartographer must work with the elements of the figure-ground relationship with great care. Some geographical relationships have "bad contour," are large, cannot be closed, and so on; in these cases the cartographer must lean more heavily on other factors, such as brightness or texture. Each composition is a new problem, and standard specifications cannot be made any more than we can say that a well-designed paragraph will contain a given number of words or sentences.

Hierarchic Organization

The communication of spatial phenomena always involves substantive elements of differing significance, and the foregoing sections have dealt with graphic elements of design that can be employed to portray this. In addition, the effective influences of hue, value, and intensity of color, as well as pattern, will be treated in the next chapter. Most communications about reality involve more complex relations, which require a kind of internal graphic organization or structuring. In a sense this is the graphic counterpart of the internal structuring of a written communication wherein basic sections are differentiated and each is, in turn, internally organized. The parallel applies only to the general concept of structuring a communication, since graphic presentation is fundamentally different from language presentation.

*C. Grant Head, "Land-Water Differentiation in Black and White Cartography," The Canadian Cartographer, (9), 25–38 (1972).

† P. V. Crawford, "Optimum Spatial Design for Thematic Maps," The Cartographic Journal, 13 (2), 134–144 (1976).

The substantive spatial phenomena of a map are often of differing importance, both to the general functioning of the map and in relation to one another. For example, the basic structure of a distribution in a thematic map is more important than the base data against which it is displayed; the classes of roads shown on a map may range in importance; or a map may show numerous areal categories, such as types of soil, vegetation, or crops, that may be separated into two groups in which the components of each group are more related than are the two groups. It is the cartographer's objective to separate the meaningful components visually and to portray graphically the significant interrelationships. In other words, the mapmaker must attempt to translate the substantive relationships into graphic expression.

Because this usually involves "levels" of relative importance, we can call it hierarchic organization, and it is among the most important of the graphic elements of map design. It calls for sophisticated application of the visual variables in interrelated fashion. Three kinds of hierarchic organization can be recognized: extensional, subdivisional, and stereogrammic.*

Extensional organization is primarily concerned with the portrayal of networks of lines of varying significance, although it can also apply to point marks. For example, road systems are comprised of several classes of roads; drainage systems involve several orders of streams; and railroad and airline systems have mainlines or routes and feeder components. The central place concept in geography recognizes a hierarchy of settlements ranging from a metropolitan center downward to other cities, towns, villages, and hamlets. A producing center may be connected to consuming centers.

The reader is cautioned that these three terms are here used this way for the first time and therefore do not appear elsewhere in cartographic literature. We are not in favor of jargon, but we do need handles with which to refer to these important elements of graphic design.

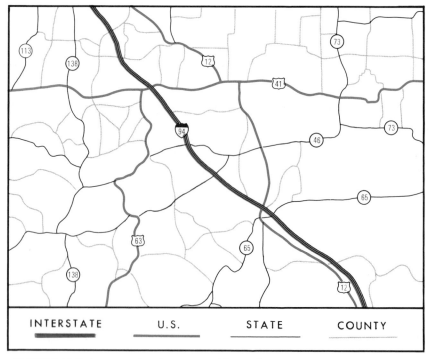

FIGURE 13.7 *An example of extensional hierarchic organization in which a set of roads is graded according to relative importance.*

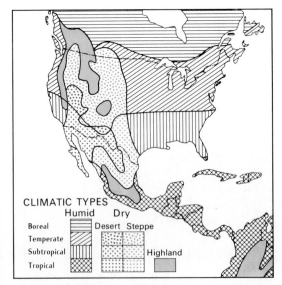

CLIMATIC TYPES

Humid   Dry

Boreal   Desert Steppe
Temperate
Subtropical   Highland
Tropical

FIGURE 13.8 An example of subdivisional hierarchic organization in which the primary division is between humid and dry climates, with a secondary subdivision based on temperature, and a tertiary subdivision based on desert versus steppe.

Extensional organization is usually basically ordinal. Even though interval or ratio scales may be employed in the individual symbolization, the overall graphic objective is to portray relative importance, and the visual variables of size, value, and color are the means most often employed. Figure 13.7 is an example of extensional organization.

Subdivisional organization is employed to portray the internal relationships of a hierarchy that usually involves areas. For example, cities are divided into wards and precincts; the land use categories of agricultural and nonagricultural may each be subdivided into several related uses; and different classes of bedrock or soil may each comprise several associated types. Figure 13.8 is an example of complex subdivisional organization.

Subdivisional presentation is primarily concerned with area symbolism and generally employs the visual variables of color and pattern. The differentiation may be solely nominal or may be made up of a combination of nominal and ordinal scaling. Since at least two levels of organization are involved, each of which incorporates subdivision, the elements of visual contrast and similarity

must be judiciously employed. The primary categories must have distinctive visual homogeneity, and the visual variance, so to speak, between them must be greater than the within-class visual variance of the several subdivisions of each major category.

Stereogrammic organization is basically different from subdivisional structure in that it is concerned with evoking in the map user the impression that the components lie at different visual levels. Such an objective is fairly common in both general and thematic cartography. The purpose is to assist the reader to focus on one set of data and, while doing so, to subordinate the rest of the data.

General maps of all scales portray many categories of data; it is often desirable to prepare them in such a way that a user can separate one category (boundaries, the drainage system, the contours, etc.) from the remainder in order to be able to concentrate on the single element.

Often in thematic cartography, the cartographer wishes to emphasize one portion of the map or a particular relationship on the map. For example, as illustrated in Fig. 13.9, the objective may be to show territories that changed hands in Europe since 1900. In order to accomplish this, it is helpful to make them appear above the background base data so that the eye will focus on these areas and will only incidentally look at the locational base material "beneath" them.

Stereogrammic organization must, of course, draw heavily on the principles involved in figure-ground differentiation, but many related cues to depth perception may be invoked, such as superimposition, progression of size and weight, value progression (all illustrated in Fig. 13.10), as well as differences in hue and saturation, which have some connotations of depth.

Hierarchic organization is sometimes made difficult by the constraints of reality. "Important" linear and areal phenomena may well be smaller than less important items; significant items may unfortunately occur in a faraway corner of the map; and color convention, such as blue water, may prevent a desirable series of hues. In any given design problem any one of the graphic elements of design may be of paramount importance, but solving problems of hierarchic organization are likely often to be the most important and demanding tasks of the map designer.

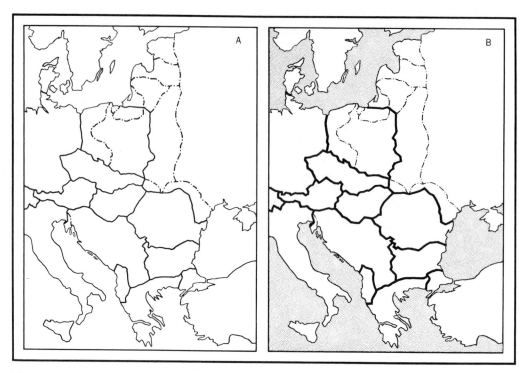

FIGURE 13.9 *An example of stereogrammic hierarchic organization. (A) All elements lie generally in the same visual plane. (B) The land has been made to appear above the water, and the more prominent boundaries have been made to rise above the visual plane of the land. Lines of the graticule on the water only would also tend to make the land appear above the water level.*

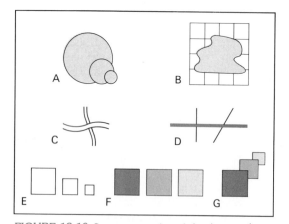

FIGURE 13.10 *Some examples of depth cues that may be useful in stereogrammic organization. A, B, C, and D illustrate various kinds of superimposition, E illustrates a progression of size, and F illustrates a progression of value. The visual variables of depth cues may be used additively, as shown in G.*

PLANNING MAP DESIGN

Thoughtful and effective communication of complex relationships demands careful planning and, when graphic means are employed, the display must be as carefully organized and planned as any other sort of communication. When we plan to write something, we first prepare an outline; it is also necessary to outline a graphic communication.

Each graphic element of the map should be evaluated in combination with the other elements in terms of its probable effect on the reader. To do this requires a full and complete understanding of the objectives of the map to be made. We can scarcely imagine writing an article or planning a lecture without first arriving at a reasonably clear decision concerning (1) the audience to which it is to be presented, and (2) the scope of the subject matter. With these considerations in mind, the writer assesses the relative significance of each item to be included.

We use the term "outline" in its broad sense as a brief characterization of the essential features of a communication. Although notes can be jotted down to keep track of ideas, the outline for a graphic presentation is *graphic.* Some of it can be done by visualizing relationships, but often it requires preliminary sketches for layout decisions, and the roughing in of the major elements. As planning progresses, a representative section of the map can be prepared to evaluate the graphic relationships. Final determinations of colors and values can often be decided at the proofing stage—if there is one—(see Chapter 17), but the basic decisions must be made at the outset.

The Graphic Outline

As an illustration of how a graphic communication may be outlined, Figs. 13.11A, 13.11B, 13.11C, and 13.11D have been prepared. The assumption is made that the planned thematic map is to show two related (hypothetical) distributions in Europe. The fundamental organizational parts of the communication are as follows.

1. The place—Europe.
2. The data—the two distributions to be shown.
3. The position of the data with respect to Europe.
4. The relative position of the two distributions.

Any one of these four elements may be placed at the head of the graphic outline, and the order of the others may be varied in any way the cartographer desires. In Fig. 13.11A the design places the items in the general order of 1-2-3-4; in Fig. 13.11B, 2-3-4-1; in Fig. 13.11C, 3-1-4-2; and in Fig. 13.11D, 4-2-3-1. Other combinations are, of course, possible. It should not be inferred that

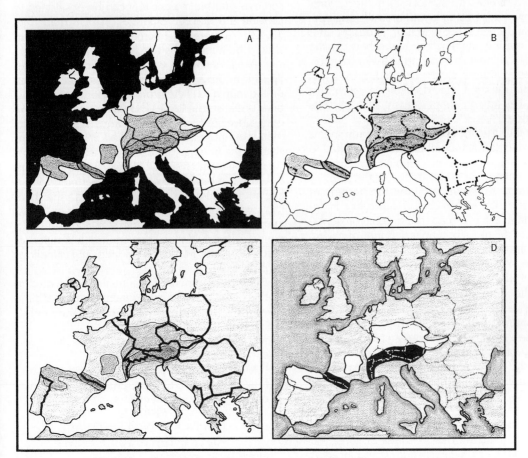

FIGURE 13.11 Examples of variations in the primary visual outline. See text for explanation.

the positioning of items in the graphic outline can be as exact and precise as in a written outline. In the latter, the sequence is reasonably assured, since the reader traditionally starts at one point and proceeds systematically in one direction to the end, whereas in a graphic presentation, the viewer, at least potentially, sees the items all at once. It is up to the cartographic designer to attempt to lead the viewer by applying the various principles involved in the graphic elements of design.

It is appropriate at this point to digress slightly in order to emphasize one of the more difficult complicating factors that a cartographer must face in cartographic design. Any component of a map has, of course, an intellectual connotation as well as a visual meaning in the design sense. It is difficult to remove the former in order to evaluate the latter, but many times it is not only necessary but definitely desirable. Illustrators turn their works upside down; advertising layout artists "rough in" outlines, and even basic lettering, as design units without "spelling anything out"; and because the intellectual connotations cannot always be predicted, cartographers, for their own purposes, will obtain better design if they do likewise, except for obvious, well-known shapes, such as continents and countries.

After the structuring of the major components has been decided on, attention must be shifted to the second stage of graphic outlining. Just as with a written outline, the major items, or the primary outline, are first determined; then the position of the subject matter within each major rubric is decided. In the case of a map, the graphic presentation of the "detail" is primarily a matter of clarity, legibility, and relative contrast of the detail items.

Often, the problems of cartographic design for a series of maps, whether general or thematic, call for the preparation of one or more "trial maps" in order that primary decisions may be made regarding various forms of symbolism, lettering styles, area patterns, consistency of layout, and many other important design elements. Not infrequently, the decision making in such instances includes the opinions of noncartographers. Such trial maps should be made with great care so that the substantive aspects are *as correct as they can be.* Even though such accuracy may confuse the design issues involved, all names should be properly spelled, lines placed in

their correct planimetric position, and so forth. Unfortunately, it appears to be difficult, if not impossible, for those inexperienced in design problems to ignore the substantive aspects of a trial design. The noncartographer, with seeming perversity, is likely to dwell on any planned "inaccuracies" and fail to give the needed attention to the design problems. One experience in such a situation serves clearly to exemplify the fact that any map component has both an intellectual and a visual connotation; but it is an experience the cartographer can well avoid.

Titles, Legends, and Scales

These standard elements serve two major purposes in cartography. Naturally, they have a denotative function in identifying the place, subject matter, symbolization, and so on, but they also serve as graphic masses that can be positioned to provide the graphic organization of a map (Fig. 13.12). Generally, fragmentation of the map should be avoided as much as possible.

A title serves a variety of functions. Sometimes it informs the reader of the subject or area on the map and is therefore as important as a label on a bottle of medicine. But this is not always the case, since some maps are obvious in their subject matter or area and really need no title. In these instances the title may be useful to the designer as a shape to be used to help balance the composition.

It is impossible to generalize as to the form a title should take; it depends entirely on the map, its subject, and its purpose. Suppose, for example, that a map had been made showing the 1975 density of population per square kilometer of arable land in France. The following situations might apply.

1. If the map appeared in a textbook devoted to the general worldwide conditions at that time with respect to the subject matter, then only
FRANCE
would be appropriate because the time and subject would be known.

2. If the map appeared in a study of the current food situation in Europe, and if it were an important piece of evidence for some thesis, then
FRANCE
POPULATION PER SQUARE
KILOMETER ARABLE LAND
would be appropriate.

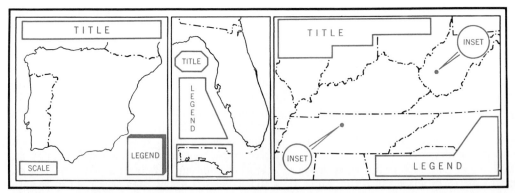

FIGURE 13.12 *Titles, legends, scales, and insets may be arranged in various ways in the graphic organization of a map.*

3. If the map appeared in a publication devoted to the changes in population in France, then
POPULATION PER SQUARE
KILOMETER ARABLE LAND
1975
would be appropriate, since the area would be known but the date would be significant.

Many other combinations could be worked out, but there is no need to belabor the fact that the title must be tailored to the occasion. Similarly, the degree of prominence and visual interest displayed by the title, through the style, size, and the boldness of the lettering must be fitted into the whole design and purpose of the map.

Legends or keys are naturally indispensable to most maps, since they provide the explanation of the various symbols used. It should be a cardinal rule of the cartographer that no symbol that is not self-explanatory should be used on a map unless it is explained in a legend. Furthermore, any symbol explained should appear in the legend *exactly* as it appears on the map, drawn in precisely the same size and manner.

Map legends can be emphasized or subordinated by varying the shape, size, or value relationship. Figure 13.13 illustrates several variations. In the past it was the custom to enclose legends in fancy, ornate outlines, called *cartouches* which, by their intricate workmanship, called attention to their presence. Today it is generally conceded that the contents of the legend are more important than its outline, so the outline, if there is any, is

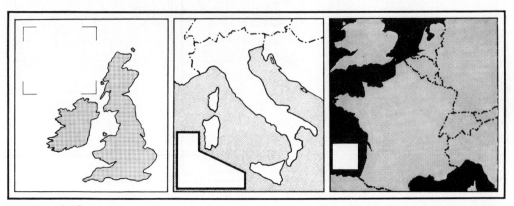

FIGURE 13.13 *Variations in the prominence of map legends. Note the operation of the principles of the figure-ground relationship.*

usually kept simple, and the visual importance is regulated in other ways.

The scale of a map also varies in importance from map to map. On maps showing road or rail lines, air routes, or any other phenomenon or relationship that involves distance concepts, the scale is an important factor in making the map useful. In such cases the statement of scale must be placed in a position of prominence, and it should be designed in such fashion that it can be easily used by the reader.

The method of presenting the scale may vary. For many maps, especially those of larger scale, the Representative Fraction, (RF), is useful because it tells the experienced map reader a great deal about the amount of generalization and selection that probably went into the preparation of the map. It should, however, be remembered that if a map is to be changed in size by reduction, this will not change the printed numbers of the RF. On a map designed for reduction, the RF must be that of the final scale, not the construction scale. A graphic scale is much more common on small-scale maps, not only because it simplifies the user's employment of it, but because an RF in the smaller scales is not so meaningful. Furthermore,

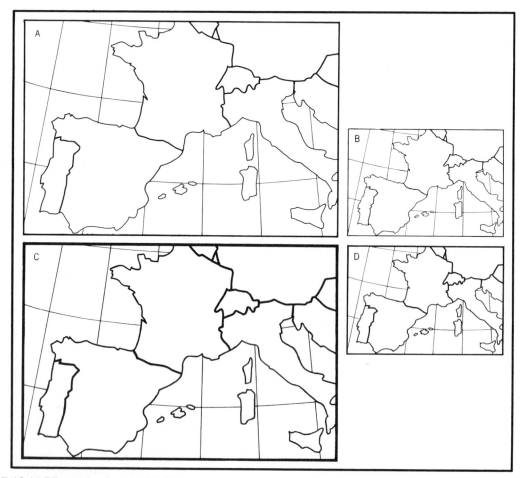

FIGURE 13.14 Effects of reduction. (A) When artwork is designed for proper line contrasts at drawing scale, then (B) reduction will decrease the contrasts too much. (C) When artwork is designed for reduction, then (D) reduction produces proper line contrast relationships.

enlargement or reduction have no effect on a graphic scale.

Effects of Reduction

With the scribing technique, most maps are constructed at the final reproduction scale. Nevertheless, a considerable number of maps are still prepared by pen and ink methods; these are usually drafted at a scale larger than the reproduction scale, that is, for reduction. This is done for a variety of reasons, the most important of which is that it is often impossible to draft with the precision and detail desired at the scale of the final map. Also, reduction frequently "sharpens" the line work of the drawing.

Drafting a map for reduction does not mean merely drawing a map that is well designed at the drafting scale. On the contrary, it requires the anticipation of the finished map and the designing of each item so that when it is reduced and reproduced, it will be "right" for that scale. A map must be designed for reduction as much as for any other purpose.

The greatest problem facing the cartographer in designing for reduction is that involving line-width relationships. In general, a map on which the line relations appear correct at the drafting scale will appear a little "light" if it is reduced. Consequently, the mapmaker must make the map overly "heavy" in order to avoid its appearing too light after reduction. It is also necessary for the cartographer to "overdo" the lettering somewhat, just as it is necessary to make lines and symbols a little too large on the drawing. Figure 13.14 illustrates these relationships.

Maps of a series should appear comparable and, if they are drawn for reduction, they should be drafted for the same amount so that similar lettering, lines, and the like, will appear comparable. This may necessitate changing the scale of base maps, which is troublesome, but it will insure that the line treatment and lettering in the final maps will be uniform.

Specifications for drafting and for reduction are given in terms of linear change, not areal relationships. It is common to speak of a drawing as being "50% up," meaning that it is half again wider or longer than it will be when reproduced. The same map may be referred to as being drafted for one-third reduction; that is, one-third of the

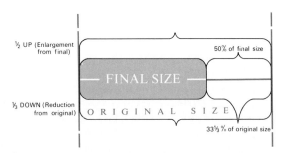

FIGURE 13.15 *Relation of enlargement to reduction.*

linear dimensions will be lost in reduction. Figure 13.15 illustrates the relationships. Since it is common practice in large printing plants to photograph many illustrations at once, it is also desirable, for economy, to make series drawings for a common reduction.

SELECTED REFERENCES

Arnheim, R., *Art and Visual Perception,* University of California Press, Berkeley, CA, 1971.

Arnheim, R., "The Perception of Maps," *The American Cartographer, 3,* 5–10 (1976).

Arvetis, C., "The Cartographer-Designer Relationship, A Designer's View," *Surveying and Mapping, 33,* 193–195. (1973).

Bartholomew, J. C., and I. A. G. Kinniburgh, "The Factor of Awareness," *The Cartographic Journal, 10,* 59–62 (1973).

Dent, B. D., "Visual Organization and Thematic Map Communication," *Annals,* Association of American Geographers, Vol. 62, 1972, pp. 25–38.

Dent, B. D., "Simplifying Thematic Maps through Effective Design: Some Postulates for the Measure of Success," *Proceedings,* Fall Convention, American Congress on Surveying and Mapping, 1973, pp. 243–251.

Fahey, L., "An Approach to Map Design," *Proceedings of the 32d Annual Meeting,* American Congress on Surveying and Mapping, Washington, D.C., pp. 132–142, 1972.

Head, C. G., "Land-Water Differentiation in Black and White Cartography," *The Canadian Cartographer, 9,* 25–38 (1972).

Keates, J. S., *Cartographic Design and Production,*

Longman Group Limited, London, 1973 pp. 10–12, 29–33.

Castner, H. W., and G. McGrath (eds.), "Map Design and the Map User," *Cartographica, 2* (1971).

Petchenik, B. B., "A Verbal Approach to Characterizing the Look of Maps," *The American Cartographer, 1,* 63–71 (1974).

Robinson, A. H., "Map Design," Proceedings, International Symposium on Computer-Assisted Cartography, pp. 9–14, American Congress on Surveying and Mapping, Falls Church, VA, 1977.

Sorrell, P., "Map Design—With the Young in Mind," *The Cartographic Journal, 11,* 82–90 (1974).

Wood, M., "Human Factors in Cartographic Communication," *The Cartographic Journal, 9,* 123–132 (1972).

Wood, M., "Visual Perception and Map Design," *The Cartographic Journal, 5,* 54–64 (1968).

COLOR IN CARTOGRAPHY

Throughout the history of cartography map-makers have paid special attention in their graphic design to the two visual variables, color and pattern. Color especially has played a major role, partly because of its aesthetic character, but more important because of its utility as an aid to clarity. When available, color was used to distinguish among classes and subclasses of phenomena, ranging from the land-water difference to individual countries, cities, and transportation routes. Even a small amount of color can make an enormous difference in the comprehension of a map.

A graphic complex of numerous different black lines, representing boundaries, coasts, rivers, roads, lakes, to say nothing of things such as canals, railroads, contours, and transmission lines, in an extreme combination can convey nothing but utter confusion. Yet, if the various phenomena are carefully color coded (bounded areas tinted to show their extent, rivers and coastlines distinguished from roads and boundaries, etc.), visual order is apparent instead of chaos. Pattern is not so versatile a variable as color because it is largely restricted to representing areas, yet that is a major

CHAPTER 14
COLOR AND PATTERN

element of many maps. Pattern may be used alone (i.e., monochromatically) or it may be employed in conjunction with color.

Color in Cartography

The desirability of color and pattern as visual variables has been so great that considerable ingenuity has been devoted to ways of applying them to maps. Before maps were printed, color and pattern were simply drawn or painted on the manuscript maps. After the fifteenth century, when the printing of maps became common, color still had to be applied mostly by hand to each sheet because color printing was prohibitively expensive and technically difficult. Map tinting became a trade, and many people were so employed by mapmaking establishments. Often templates and other elaborate ways of ensuring that the colors were put in the right places were worked out. Such practices extended even to large-scale topographic series.

The achievement of pattern and tones on printing surfaces was very difficult, especially on woodblocks, but it could be done. Pattern became a standard graphic component of maps, often serving in place of color, especially to set off water from land. With the development of lithography and then photography in the nineteenth century, techniques for obtaining color and pattern were incorporated in the printing process; since that time, an increasing number of maps have utilized them.

Color on a map allows greater detail; it adds visual interest; it increases the design potentialities; and it adds greatly to the possibilities for hierarchic graphic structures. Because color may be easily used to code similarity and dissimilarity between and within classes of phenomena, it is a great aid to clarity. All these advantages are not without their cost, of course. Color is relatively expensive, and it calls for special treatments in map construction and reproduction.

Cartographers have employed color for so long that a good many conventions have developed that are important in considering the design of a map. Furthermore, color arouses aesthetic reactions and connotes concepts (red with warm, blue with cold, green with vegetation, etc.) that may be important in creating an effective graphic communication. Equally important in cartographic design involving color are the physiologically based perceptual aspects, such as sensitivity, visual acuity, and contrast.

In order to discuss color as a primary graphic element in cartography, it is necessary first to examine this perceptual phenomenon from the point of view of its occurrence as a visual sensation.

THE PHENOMENON OF COLOR

The Spectrum

Light is the visual sensation aroused by the stimulation of receptors in the eye by a portion of the electromagnetic spectrum. The electromagnetic spectrum ranges from the very short wavelengths of X rays and gamma rays to the very long wavelengths used by radar and television (Fig. 14.1). Only a very small portion of the spectrum is visible, the wavelengths ranging from approximately 400 to 700 nm.* When the source of illumination emits the full range of visible wavelengths in suitable portions, it is called white light. When white light (e.g., sunlight) is separated into its component wavelengths by the refraction of a prism or of raindrops, and these are reflected to the eye, the separate wavelengths provide the attribute of color called hue (blue, yellow, red, etc.). The shorter wavelengths are the violet-blues, near the 400-nm end of the spectrum, and the longer wavelengths are the reds, near the 700-nm end. The order of these spectral hues is that of the rainbow, and is shown in Fig. CI.4

Pure spectral hues are not often seen except when white light is refracted, as by a gem, a water droplet, and the like, but they provide most of the basic terms we use to identify hues (violet, blue, green, yellow, orange, and red). The order of their wavelengths is of significance in cartography because hues sometimes are employed in what is called a spectral progression.

Common Colors

The myriad colors we see in nature and in all fabricated things are almost always combinations of wavelengths. They occur because illuminated sur-

---

*A nanometer (nm), previously referred to as a millimicron ($m\mu$) is 1 meter times 10^{-9}, or one billionth meter.

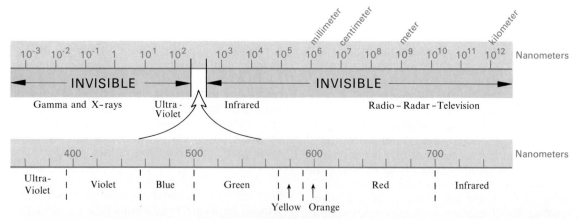

FIGURE 14.1 *A portion of the electromagnetic spectrum with wavelengths shown in nanometers. Only a small portion is visible. One nanometer (nm) = one billionth meter (10^{-9} meter). See Fig. CI.4.*

faces absorb some of the wavelengths of the light that falls on them (incident light) and reflect other wavelengths in different proportions. For example, a living tree leaf absorbs much of the wavelengths of the spectrum except for those in the 500 to 550-nm range, a large proportion of which it reflects. Because of this selective absorption and reflection, it appears green to our eyes. Paper that reflects all the wavelengths of sunlight in about the same proportion will appear white, and a surface that absorbs most of the wavelengths and reflects very little will appear black.

Most ordinary surfaces will reflect at least a small proportion of all the wavelengths, but they will gain their distinctive character from those wavelengths they reflect the most. The concept of colors as combinations of wavelengths is most easily gained by considering the spectral reflectance curve.

The Spectral Reflectance Curve. A spectrophotometer is an instrument that measures the proportions of the various wavelengths of light reflected from a surface. The measurements may be plotted on a graph whose x-axis is wavelength and whose y-axis is percent reflectance. Graphs of several colored surfaces are shown in Fig. 14.2.

In addition to the general color represented by the dominant section of the curve on a graph, such as green or yellow, other characteristics are revealed by the shape of the curve. A rel-

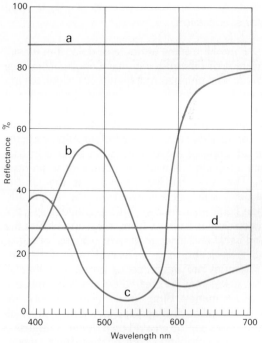

FIGURE 14.2 *Four spectral reflectance curves: a—a white surface; b—a greenish-blue ink; c—a red-purple ink; and d—a dark gray surface. (Curves b and c after Nimeroff.)*

atively flat curve indicates a lack of a dominant spectral hue; if the flat curve is high on the graph it approaches white; if it is low it is a dark gray. The higher the curve, the lighter the color; the more peaked the curve, the more intense or saturated the color will appear.

Dimensions of Color

Color is a complex total response to a variety of physical stimuli plus our perceptual processing of them. Depending on whether one is considering color from the point of view of the stimuli or the response, one arrives at somewhat different definitions of the major components. In both cases, however, the phenomenon of color may be subdivided into three attributes or dimensions. A description of a color is incomplete unless all three are included. The cartographer is, of course, concerned with what the map reader sees', consequently, we will consider the dimensions primarily from the perceptual point of view. The three dimensions are: hue, value, and intensity or chroma. As we will see later, specific terms and definitions are employed in particular systems of color specification.

Hue. Hue is the attribute of color that we associate with differences in wavelength. When we identify something as red, green, or yellow, we are describing its hue. There are an infinite number of hues because wavelengths can be combined in various proportions. The arrangement of hues into an orderly sequence or classification requires the adopting of some point of view or purpose; each adoption will result in a different sequence.

The most familiar sequence is that in the rainbow, which occurs when the wavelengths in white light are refracted. This is called a spectral progression and ranges from violet and blue at the short wave end through greens, yellows, oranges, and reds at the long wave end (see Fig. CI.4). Some hues are described as primary in that all other colors may be created by a suitable mixture of the primaries. Blue, green, and red light are what are called the *additive* primaries, since all other colors can be created on a "white" reflective surface by combining light of these three wavelengths in the required proportions. If the three are combined equally, they form white (Fig. CI.6).

Colors on maps (usually) result from white surfaces to which some pigment has been applied so that when it is illuminated by white light some of the wavelengths are absorbed or subtracted, and what is reflected to the observer are the remaining wavelengths. The *subtractive* primaries are blue (cyan), yellow, and red (magenta), and any color can be created by a suitable mixture of these pigments (Fig. CI.5). This is an ideal, of course, since the pigments are not "pure" and since the paper and the incident light will have some special characteristics of their own.

Value. All colors may be ranked in terms of their lightness or darkness in a given situation. Various terms are used to describe this attribute, the more common being brightness, reflectance, and value. Brightness is simply the attribute of sensation by which an observer is aware of differences in luminance (the effective luminous intensity of a surface per unit area); for example, we may say that one surface is brighter (i.e., lighter) than another. Reflectance is a measurement of the ratio of the incident light reflected from a surface compared to the amount reflected from a reference white surface; for example, we might say that the reflectance of surface A is 30%, meaning that 30% is reflected and 70% absorbed. Reflectance (and transmission) is measured by a photometer, often called a densitometer.

Value is the sensation of lightness or darkness as rated on a gray scale. A gray scale is a progression of regularly spaced gray tones from black to white (Fig. 14.3). Because the appearance of the lightness or darkness of a color is affected by its surroundings, the values of two surfaces may be different, even though their reflectances are the same (Fig. 14.4). Furthermore, perceptual judgments of equal differences in value do not parallel equal physical differences in reflectance. Consequently, as we will see later, an equal-value gray scale is not the same as one in which black-white ratios change linearly.

Intensity. A third attribute of any color has to do with its richness. Some colors are brilliant in hue, such as a strong red, while another example of that same red hue may be quite weak, as would be the case were we to mix the red with a large amount of gray of the same value. Various terms, such as chroma, saturation, and purity, are em-

FIGURE 14.3 *A photographic gray scale in which the steps are equal differences in black/white ratios (i.e., percent reflectance measures). Note the apparent "waviness" caused by induction. Note also that a middle gray does not appear to be halfway between black and white. (From a Kodak Gray Scale, courtesy Eastman Kodak Co.)*

ployed to describe this dimension of color in different systems of color description or specification. In each case they are precisely defined, and this gives the terms somewhat different meanings.

The two other dimensions of color, hue and value, are uncomplicated attributes in the sense that the hues in a chromatic range may be thought of as varying only in hue without paying any attention to the fact that yellow is usually lighter than blue. Similarly, the grays of an achromatic range are uncomplicated by hue. The third dimension—intensity, chroma, saturation purity —whatever it is called, involves scaling the colors according to a combination of hues and grays.

Both the spectral colors (Fig. CI.4) and the colors of the Munsell hue circle (Fig. CI.8) vary in their richness and brilliance, part of this being a consequence of their inherent luminosity. In many circumstances it is convenient and more meaningful to refer to the relative "intensity" of a color,

although the term has no precise definition in color science.

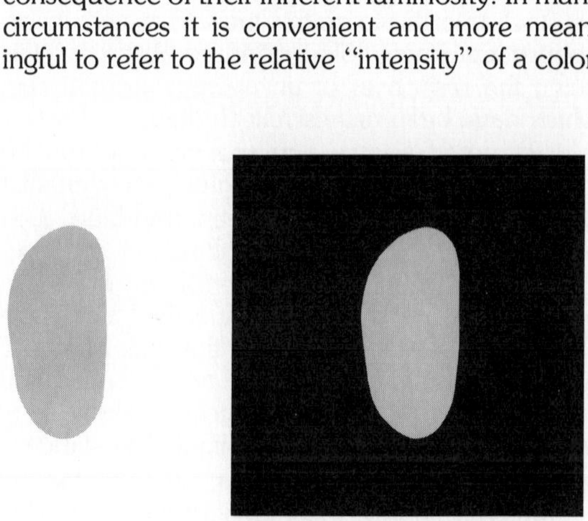

SYSTEMS OF COLOR IDENTIFICATION

Color has intrigued scientists and artists for centuries. Theories as to the nature of light and the physiologic basis for the occurrence of color in the eye-mind perceptual mechanism are very complex, and even today there is less than a full understanding of this mysterious perceptual phenomenon. On the other hand, the practical problem of describing and specifying colors without ambiguity so that they may be recreated, specified in plans and contracts, and studied with rigor has been solved. There are several systems of which two are well-known in the United States One, known as the CIE system, is based on instrumentation and the mathematical analysis of the physical characteristics of light. The other, known as the Munsell system, is based on the human perceptual reaction to light and its colors. This is not the place to go deeply into the theory and application of these two systems, but they will be briefly described. An understanding of them and the basic difference in their approaches to the phenomenon of color will help to impress on the mapmaker the complexity and importance of color in cartography.

The CIE System

The *Commission International de l'Éclairage* (CIE), also known as the International Commission on Illumination (ICI) has developed a widely used system of colorimetry. It allows the precise specification of any color in numerical terms. The procedure is employed by the U.S. National Bureau of Standards to define color, and it is the

FIGURE 14.4 *Value is a perception that cannot be measured instrumentally. The two gray spots have the same reflectance, but they are clearly different in value.*

basis for a standardized color identification system employed in the U.S. Defense Mapping Agency.

There are three prerequisites for an objectively based scheme: the definition of standard illuminants, standard observers, and standard primaries.

Standard Illuminants. It is readily apparent that any surface potentially can reflect only the wavelengths of light contained in the incident light, and the color qualities of any surface will change as the quality of the incident light changes. The CIE early defined several light sources: a tungsten incandescent lamp (A), noon sunlight (B), light from an overcast sky (C); subsequently the CIE added several others. Each standard illuminant can be illustrated by a spectral reflectance curve (Fig. 14.5).

Standard Observer. Since color as we experience it is obviously a human perception, any

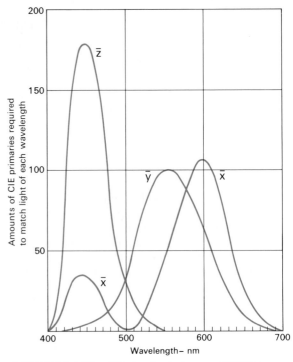

FIGURE 14.6 *The CIE standard observer (1931) is defined by the amounts of three selected primary lights needed to be mixed to match the wavelengths of the spectrum.*

useful system of specification must be based on that fact. In order to do this the CIE system incorporates the concept of a "standard observer," this being the responses of an average normal eye, which have been extensively studied.

As noted above, any spectral hue can be matched by mixing additively lights (not pigments) of three primary hues, red, green, and blue, each of which will consist of a combination of wavelengths. The required amounts of the three primaries needed to match a specified color are called the tristimulus values of that color and are numerically designated by X, Y, and Z. These may be determined by mathematical analysis from measurements of the reflected light. The standard observer in the CIE system is defined by the varying proportions of the three primaries needed to match light of each spectral wavelength. These plot as spectral reflectance curves and are also called tristimulus values. These curves are designated $\bar{x}$, $\bar{y}$, and $\bar{z}$ and are shown in Fig. 14.6. The

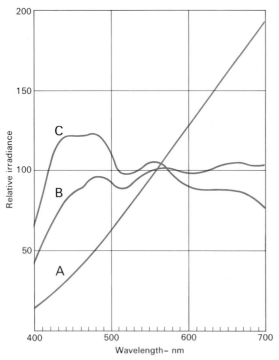

FIGURE 14.5 *Relative spectral irradiance of CIE standard illumination sources: A—incandescent lamp; B—noon sunlight; and C—average daylight from an overcast sky. (From Nimeroff.)*

curve of tristimulus value $\bar{y}$ corresponds to the relative luminosity of the spectrum. Consequently, the tristimulus value Y for a colored surface also expresses its luminous reflectance.

CIE Color Specification

The characterization of a color in the CIE system is made by first obtaining a spectrophotometric analysis of a particular colored surface under specified illumination and then determining by computation the tristimulus values, X, Y, and Z. Tristimulus value X, for example, is in effect the summation of the amounts of the red primary (in combination with the other two primaries, of course) needed to match each wavelength of the color in question. Tristimulus values Y and Z are similarly obtained. The tristimulus values are then converted to chromatic coordinates by obtaining the ratios $x = X/X + Y + Z$ and $y = Y/X + Y + Z$. The analogous ratio $z = Z/X + Y + Z$ is not needed; x and y carry all the information, since all three ratios sum to unity. The values x and y are the fractional amounts of the red and green primaries

needed to match the hue and saturation of the particular color. As we saw earlier, in the CIE system Y equals the total luminous reflectance of the colored surface. The chromatic coordinates x and y are then entered on the CIE chromaticity diagram.

The chromaticity diagram is shown in Fig. 14.7. The smooth red curve shows the collection of points (locus) representing all the hues of the visible spectrum. The positions along this line of selected wavelengths (nanometers) are identified. The sloping straight line at the bottom represents the magentas and purples, which fall between red and blue but do not occur in the spectrum. The use of the chromaticity diagram is as follows:

1. The chromatic coordinates of the illuminant, for example (C), are plotted. This position on the diagram is achromatic (without color) for that illuminant. One can visualize the chromaticity diagram as showing all the spectral hues at full saturation around the inside of the locus, all of which fade in saturation toward C until there is no color at the point representing the illuminant.
2. The intersection of the chromatic coordinates of the color being analyzed is plotted. For example, a green ink might have chromatic coordinates of $x = 0.212$ and $y = 0.348$. These are plotted as F on Fig. 14.7.
3. A straight line from C through F will intersect the locus of the spectrum at G. This intersection designates a wavelength that is called the *dominant wavelength*. In the illustration this is 495 nm.
4. The *luminosity* (luminous reflectance) of the color is given by tristimulus value Y.
5. The ratio CF to CG (percent) defines the *purity* (excitation purity) of the color.

The three values, *dominant wavelength, luminosity,* and *purity,* specify the color in the CIE system.* The *dominant wavelength* is the hue and may be defined as the monochromatic light which, when mixed in suitable proportion with the illuminant light, will match the color in question. The *luminosity* is the lightness or darkness of the color. The *purity* is the degree of saturation

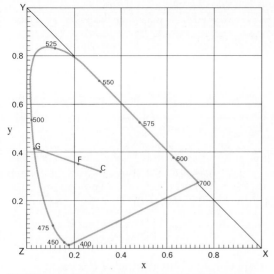

FIGURE 14.7 The CIE chromaticity diagram (also called a Maxwell diagram) shows the primary colors at the apexes. The spectral hues are shown by the smooth curve (locus), and the chromaticities of the mixture of the extremes of the spectrum is shown by the sloping line joining them. All physically realizable chromaticities occur within these limits. See text for explanation of plotted points C, F, and G.

---

*Large-scale charts from which dominant wavelength and purity can be read directly have been prepared to accompany A. C. Hardy, Handbook of Colorimetry, Technology Press, Cambridge, MA, 1936.

ranging from full color (100%) to the achromatic stimulus of the specified illuminant.

It is apparent in Fig. 14.7 that the spacing of the spectral hues along the locus is not even and that the hues vary greatly as to their inherent lightness as indicated by the y chromaticity coordinate. Generally, the higher the lightness ranking of the hues, the more difficult it is to obtain noticeably different tints without changing hue. Conversely, it is easier to do so with the blues and reds at the extremes of the spectrum, which are inherently darker.

The great advantage of the CIE system is that any color may be precisely defined in physical terms. It must be emphasized, however, that it is based on the characteristics of additive light and not on the characteristics of subtractive pigments and our perceptions of them. Pigments are what are put on colored maps and it is people (who are not spectrophotometers) who look at them. Consequently, although the CIE system provides a rigorous method for analyzing and specifying colors that is extremely useful in production and research, it is also necessary for the cartographer to become well acquainted with a system based on human perception. The difference between the two kinds of systems is profound, and it will become apparent after considering the perceptual system. Several such systems have been advanced; the best-known and most widely used system in the United States is the Munsell system.

The Munsell Color System

This system, named for its originator, A. H. Munsell, an American painter, identifies a color in terms of three dimensions: hue, value, and chroma. Each dimension is divided into a sequence of steps arranged so that the steps are equally spaced from a perceptual point of view. Each step in each scale is assigned a designation, and the combinations of these constitute a notation system that refer to the more than 1500 actual samples of the Munsell colors.

The Munsell color system is widely used by industry and government for many purposes that range from matching soil colors and color photogrammetry in survey work to color coding of wires in electronics. It forms the basis for a standardized system of color names worked out by the Inter-Society Color Council and the National Bureau of

Standards (ISCC-NBS). Munsell colors physically exist as painted chips.*

The Hue Circle. Hues are equally spaced from one another in the sense of equal visual steps from one hue to the next. The 100 hues are arranged in a circle so that each hue is perceptually related to the next to it (i.e., adjacent hues are chromatically similar). There are 10 major hues, arranged as shown in Fig. 14.8 and illustrated in Fig. CI.8. The order of the hues will be seen to correspond to the order of the hues in the spectrum with the addition of purples. A clockwise progression of colors in the hue circle is similar in order to a counterclockwise progression around the spectral wavelength locus on the CIE chromaticity diagram.

*Information and catalogs of materials are available from Munsell Color, 2441 No. Calvert S., Baltimore, MD 21218.

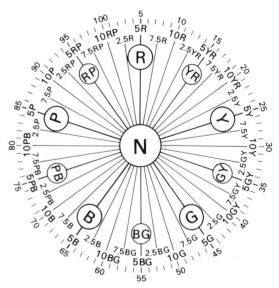

FIGURE 14.8 The Munsell circuit of 100 equally spaced hues. The 10 major hues, 10 steps apart, are named R (red), YR (yellow-red), Y (yellow), GY (green-yellow), G (green), BG (blue-green), B (blue), PB (purple-blue), P (purple), and RP (red-purple). They increase outward in chroma from an achromatic stimulus, N (neutral), at the center. (Courtesy Munsell Color.)

The 10 major hues, consisting of 5 principals and 5 intermediates, may be referred to by their initials (e.g., R = red or BG = blue green). More commonly, each of the spaces between the major hues is divided equally (visually) into four divisions, as shown in Fig. 14.8, and these are designated by a numerical/letter combination, such as 5BG or 10YR. Numerals alone from 1 to 100 can be used.

The Value Scale. All colors are ranked in relation to a range of achromatic grays (neutrals), from black to progressively lighter grays to white; this is called the value scale. Although the value scale is in theory continuous from black (low value) to white, it is usually thought of as consisting of the steps shown in Fig. 14.9. Black is designated as 0/ and white as 10/ on the scale, with equally spaced grays between. A gray midway between black and white would be designated as 5/. Fine distinctions can be made by employing decimals. The spacing on the Munsell value scale is perceptual, and is not the same as a photographic gray scale (Fig. 14.3) in which the spacing is made by equal changes in the physical black-white ratios.

The Chroma Scale. The dimension of chroma in the Munsell system is the degree to which a hue departs from a neutral gray of the same value. It is analogous to the descriptive term intensity and to the concept of purity (saturation) in the CIE system, except that in a chroma range a value ranking is specified and is kept constant. The scales of chroma extend from /0 for a neutral gray in equal steps toward full saturation for that hue and value. Figure CI.9 shows the ranges of value and chroma for complementary Munsell hues 5.0 blue and 5.0 yellow-red.

Munsell Notation and Color Solid

The relationship among the three dimensions of color in the Munsell system can be visualized as forming a three-dimensional color space or solid. We may use the earth as an analogy (Fig. 14.10). The axis of the earth forms the scale of values from black at one pole to white at the other. Halfway between, at the plane of the equator, is value 5/. The hue circle is arranged longitudinally with the highest chroma values theoretically at the

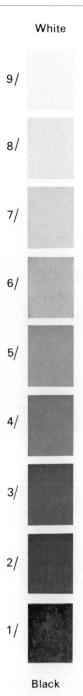

White

9/

8/

7/

6/

5/

4/

3/

2/

1/

Black

FIGURE 14.9 The Munsell value scale from 1/ to 9/ of equally spaced grays (0/ is black and 10/ is white). The range is continuous, but is here shown in steps. (Courtesy Munsell Color.)

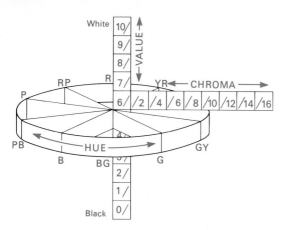

FIGURE 14.10 *The arrangement of the Munsell hue, value, and chroma scales in color space. If the hue circuit at value 5/ and chroma /10 were made to correspond to the equator of the earth, then value would be latitude (0 to 180 degrees), hue would be longitude (0 to 360 degrees), and chroma would increase with perpendicular distance outward from the axis.*

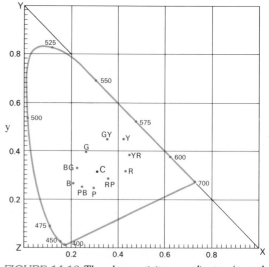

FIGURE 14.11 *The three-dimensional shape of the Munsell color space.*

"equator." In this view "latitude" from one pole to the other represents value and "longitude" represents hue. Chroma would vary along a latitudinal plane from high chromas at the surface inward to the achromatic axis.

Because hues vary in their intrinsic lightness and because of physical limitations of materials, the Munsell color solid is irregular, assuming the shape illustrated in Fig. 14.11. When selected color chips are arranged on vertical planes radiating out from an axis, they form the "color tree" shown as Fig. CI.10.

In Munsell notation the designation of a particular color is given symbolically by H V/C; H is the hue designation, such as 5R; V/ is the value ranking; and /C is the chroma. A strong red might be 5R 5/14, a pale pink 5R 8/6, and a deep red 5R 3/10.

All the Munsell colors have been analyzed by the National Bureau of Standards according to the CIE system, and the relation between the two systems may be appreciated by viewing the plot on the chromaticity diagram (for illuminant C) of the 10 major hues at value level 5/ and chroma level /6, the highest even chroma level at which all the hues exist (Fig. 14.12). Although the hues are

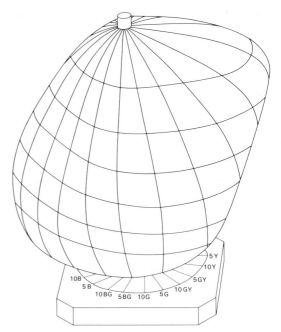

FIGURE 14.12 *The chromaticity coordinates (x and y) of the 10 major Munsell hues at value 5/ and chroma /6 plotted on the CIE chromaticity diagram. (Data from Kelly, Gibson and Nickerson.)*

equidistant visually in the Munsell system, they are irregularly arranged on the chromaticity diagram. Furthermore, all the hues are the same chroma level but, as is shown by the relative distances from *C* to the points on the locus, in the CIE system they have marked differences in purity.

THE EMPLOYMENT OF COLOR IN MAPS

When the cartographer has the opportunity to use color on a map, there are two major aspects to the procedural problem of doing so. The first is the choice of what colors to use, and the second is how to obtain what is wanted in the final printed product. Both involve a careful consideration of a variety of alternatives, both perceptual and technical.

The mechanics of constructing artwork, manipulating film, and preparing the directions for the printer will be dealt with in Chapters 16 and 17. Here we need only observe that the normal procedure involves choosing colors from a chart showing actual printed examples of the various colors that result when various combinations of percentages of inks are overprinted. A portion of such a chart is illustrated in Fig. CI.11.

In the printing process any one ink (any color) can be printed either *solid* (all color) or *screened*, that is, as very small areas of solid color (tiny dots, thin lines or other kinds of marks) separated by uninked areas. The screened printing often employs such small units of ink and no ink that the result simply appears as a tint of that ink color. If desired the marks can be made large enough to be seen, thereby creating a visible pattern. In any case, a solid or screened ink area of a map may be overprinted with another solid or screened ink, thereby creating a third color different from the two colors of the inks being combined. More than just two colors of inks may be used this way and, as we saw earlier, if the three colored inks approximate the subtractive primaries, red, yellow, and blue, a full range of hues may be obtained.

The cartographer works ideally from a chart that approximates as closely as possible the inks, screens, and paper to be used in printing the map. This allows the cartographer to come close to obtaining the effects desired, although there is usually some difference between the specification

and the realization because of variations in ink supply on the press, absorbency of the paper, and the like, during the actual printing. In any case we are concerned here not with the mechanics involved, but with the various aspects of color perception and convention that will guide the cartographer in the effective employment of the visual variable of color.

The Relative Importance of Color Dimensions

The three graphic dimensions of color, hue, value, and intensity, are not equally significant from the perceptual point of view. Although we can suggest a general order of importance, it must be emphasized that in some specific instances, the order may be upset.

Value (relative darkness or lightness) is generally considered to be the most significant dimension, since it is the primary factor in the recognition of graphic differences. The interplay of light and dark, as in shadows and other such contrasts, is basic to clarity, legibility, and the recognition of three-dimensional form. As long as there is value contrast, legibility is assured—in the sense of recognition of differences among map marks.

Whereas value is a critical dimension from the point of view of perceptibility, hue is by far the most interesting dimension. People actively like or dislike hues and find them intriguing; when asked to comment on the appearance of something, they usually mention hue before value or intensity. Different cultures rank hues differently in desirability. We tend to associate hues with various reactions such as warm and cool. Hues are profoundly involved in aesthetic reactions.

Intensity (chroma, purity) seems to be the least significant of the three dimensions of color although, on occasion, especially when misused, it can be extremely important.

It bears repeating that the dimensions do not occur separately. All colored areas on maps may be ranked in all three. Although we may react quite positively to the "redness" of a bright red ink, we must keep in mind that it also needs to be judged as to its value and chroma.

Principles in the Employment of Color

In selecting colors for mapping, the cartographer must be guided by a knowledge of a variety of perceptual reactions. Some of these are physiologically based; that is, the human eye-mind perceptual mechanism functions in certain ways because of the way it is constructed. These primarily set limitations on what we can do. Some of the perceptual patterns are psychologically based, resulting in a variety of connotative and subjective reactions to color. Because of the importance of color in mapping and because of its long use, a variety of conventions have developed. Some are old and some are new, but all are important, if for no other reason than if we violate a convention, it may raise questions in the reader's mind and diminish the effectiveness of the communication.

In the following sections we will examine briefly the perceptual aspects of hue, value, and chroma and, in some cases, we will divide the consideration into those elements that are physiologically based, those that are subjective and connotative, and those that are conventional. It is important to remember two things. One is that there are many aspects of the employment of color that are still mysterious and have not been sufficiently investigated to allow one to formulate valid precepts. The other is that in each situation the perceptual elements will vary in relative importance. The successful employment of color involves arriving at the appropriate mix of the elements.

The Employment of Hue on Maps

The physiologic reactions of one individual to hue are likely to be slightly different from those of anyone else, yet people have much in common. A number of generalizations can be made of significance in cartography having to do with sensitivity to hue, the apparent advance and retreat of hues, their relative luminosity, the relation of hue to visual acuity, and the phenomenon of simultaneous contrast.

Sensitivity. Most people are quite sensitive to differences in hues when the colors are placed side by side against a neutral background. On the other hand, the ability to remember hues and to carry an impression, such as from a legend to a map, appears to be severely restricted, especially when the surroundings of the colors are different. As a general rule, then, when hues are being employed to distinguish one thing from another, they should be made as different as possible within the logic of the hierarchic organization of the map. It

should also be remembered that a significant proportion of the population has some degree of the abnormality called "color blindness," which should emphasize the need for considerable differences in hue.

An aspect of sensitivity, that is important in cartography is the fact that when colored map marks (lines or dots) are very thin or small, it is difficult to distinguish hues, and additional contrast of shape or character is often necessary.

Our sensitivity to hues varies; the eye is "attracted" more to some hues than others. No precise values of relationship are available, but most students rank the hues ordinally in the following order (most sensitive first): red, green, yellow, blue and purple.

Advance and Retreat. One physiological phenomenon that has had a profound effect on cartography is that light rays entering the eye are refracted in inverse relation to wavelength. This means that, theoretically, blue items focus in front of the retina, while red ones focus behind the retina. In fact, the eye accommodates with the result that a red object should appear slightly nearer than a blue. The recognition of this relationship led to the proposal in the later nineteenth century of arranging hues in spectral order to represent elevations; the highest elevations, since they were red, would therefore appear closer to one looking "down" on the map. In fact, the effect is usually slight, and so many other factors are involved (such as the combinations of wavelengths that usually make up a color, as shown in Fig. 14.2) that in most instances the phenomenon of advance and retreat can be ignored. The spectral progression of blues for water, greens for lowlands, through yellows, buffs, and reddish colors at the highest elevations is a strong convention in cartography.

Visual Acuity. Many maps show a wealth of detail by fine lines and point marks on a colored background. Other things being equal, the more monochromatic the background, the easier it is for the eye to resolve the detail. A background color such as yellow, which combines monochromaticity and high luminosity, will display fine black detail well. Background colors that mix a variety of wavelengths, such as brown, would make it more

FIGURE CI.1 Images from the four bands in LANDSAT's multispectral scanner differ in appearance because the amount of energy reflected from features varies with the wavelength. Band 4, the green band, is useful for delineating areas of shallow water. Band 5, the red band, emphasizes cultural features. Bands 6 and 7 sense the near infrared and infrared, respectively, and emphasize vegetation and land-water boundaries. Band 7 provides the best penetration of atmospheric haze. (Photos by NASA.)

GREEN

NEAR INFRA-RED

COLOR COMPOSITE

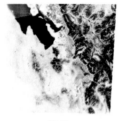

RED

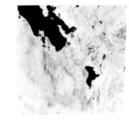

INFRA-RED

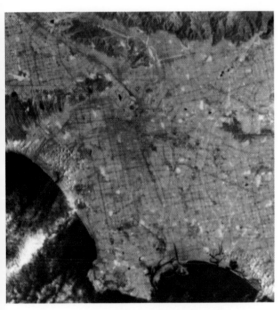

FIGURE CI.2 False color LANDSAT(ERTS) photograph of part of Los Angeles. Tonal variations of bands 5, 6 and 7 have been color-combined to show vegetation as yellow and commercial concentrations as a blue-purple gray.

FIGURE CI.3 Electronic enhancement of the center section of Fig. CI.2 processed from the LANDSAT computer compatible digital tapes. Statistical techniques are used to correlate and assign colors to combinations of the tonal levels of the several spectral bands. Heavy commercial areas show as yellow. [(Photos for Figs. CI.2 and CI.3 by NASA, Hughes Aircraft Company and General Electric Company.) From brochure prepared by John E. Estes and Leslie W. Senger. Copyright 1975 by John Wiley and Sons, Inc., New York.]

FIGURE CI.4   The visible portion of the electromagnetic spectrum. Wavelengths shown in nanometers. Compare with Fig. 14.1.

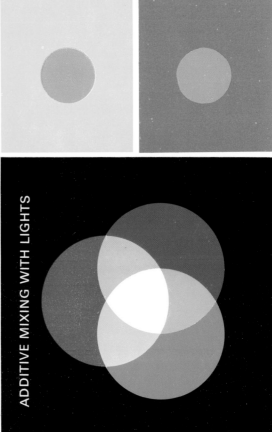

ADDITIVE MIXING WITH LIGHTS

FIGURE CI.6   Additive mixing of light primaries as in electronic displays (e.g., television).

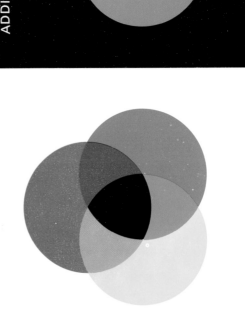

FIGURE CI.7   (Right) Environment affects appearance (simultaneous contrast). The green discs are the same mechanically,

FIGURE CI.5   Subtractive mixing of pigment primaries as in painting or printing.

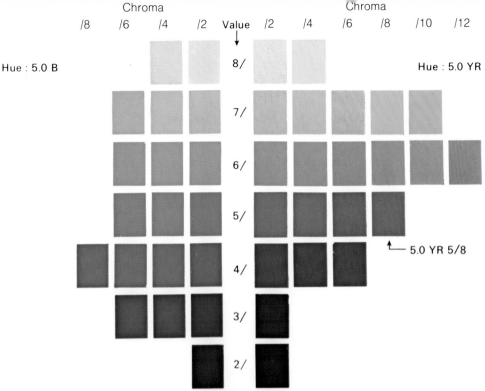

FIGURE CI.9   Value and chroma variations of two complementary Munsell hues: 5.0 blue and 5.0 yellow-red. Note that extreme values are not shown and that chroma differences are shown by double steps.

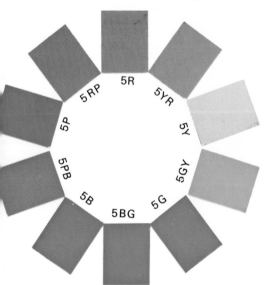

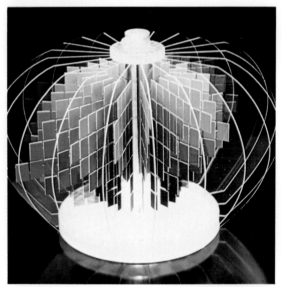

GURE CI.8   The Munsell color circle showing the five ncipal hues, red. yellow, green, blue, and purple and e five intermediates at midvalue and strong chroma.

FIGURE CI.10   Munsell color chips, such as those in Fig. CI.9, arranged in three dimensions. Hue, value, and chroma variation is continuous, of course, and the complete array would take the form of the solid shown in Fig. 14.11.

INTERRELATIONSHIP OF EDUCATIONAL ATTAINMENT AND PER CAPITA INCOME

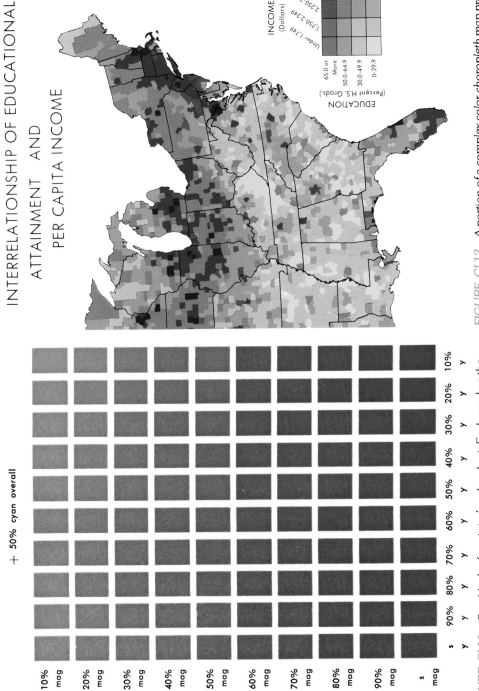

+ 50% cyan overall

FIGURE CI.11 One block of a printer's color chart. Each row has the percentage screen of magenta indicated at the left, and each column has the percentage screen of yellow indicated at the bottom. S stands for solid (i.e., no screening). A 50% screen of cyan has been added overall. Other blocks on the complete chart have other percentages of cyan overall.

FIGURE CI.12 A portion of a complex color choropleth map produced in this case by computer output on a microfilm plotter. This "two-variable" map was made by combining two "single-variable" maps. See M.A. Meyer, F.R. Broome, and R.H. Schweitzer, Jr., "Color Statistical Mapping by the U.S. Bureau of the Census," The American Cartographer, 2 (1975) 100-117. (Courtesy Geography Division, U.S. Bureau of the Census)

difficult to distinguish fine complexity. It is unfortunate that a lightish brown has become the conventional color for contours; they are difficult to follow in complex graphic situations.

Blue is relatively poor as a defining color; as a consequence, an involved blue line coast with a blue tint on the water areas makes it difficult to see the line clearly. Because blue is conventionally so employed, one must use care that value relationships are used to assist as much as possible.

Simultaneous Contrast. Whenever two colors are adjacent, they modify one another, not only in hue, but also in value. When a hue is surrounded by another color, the surrounded color tends to shift in appearance toward the complimentary color of the background. For example, a green on a yellow background will appear bluer than the same green on a blue background (Fig. CI.7). The phenomenon of simultaneous contrast is more significant in cartography in connection with value, but its existence in connection with hue makes it necessary to refrain from using hues that are very similar, since they may be difficult to identify in different parts of a colored map.

There are three subjective aspects of the perception of hue that the cartographer must keep in mind when utilizing color on maps: the phenomenon of individuality of hues, their symbolic connotations, and their affective value.

Individuality of Hues. Some hues appear quite unique and distinct, while others look like combinations of these individual hues. These unique hues are *perceptual* primaries, not to be confused with the three physical additive and subtractive primaries, which can be mixed to produce any other color. These individual hues are blue, green, yellow, red, brown, black, and white. None looks like a combination of any of them. For example, it is well known that a mixture of blue and yellow pigments results in green, but green does not look like a mixture. On the other hand, orange looks like a mixture of red and yellow, and purple looks like a mixture of red and blue. Many such apparent mixtures are possible, such as reddish brown, greenish yellow and, of course all the tints (white plus hue) and the shades (gray plus hue).

The phenomenon of individual hues is important in cartography in two ways. Individual hues should be used to symbolize distinctly different phenomena, and apparent mixtures should be used to portray items that share some of the attributes that are symbolized separately by the individual hues. For example, an area of predominately ethnic group A might be blue, another area mostly ethnic group B might be red, and a third area of mixed A and B might be purple.

Symbolic Connotations. Colors are widely associated with sensibilities and moods. Green is cool, red is warm, buffs are dry, blue is wet, yellows are sunny, and so on. Different cultures associate hues with various meanings, and the diverse symbolisms (white for purity, black for mourning in Western culture, etc.) and significances are very complex.[*] Most such attributes of color are not directly important to the cartographer, but the need to parallel portrayals of things such as temperature, wetness, and so on, with generally associated hues is apparent. Furthermore, if a map is to be attractive and the cartographer wishes the general reaction of viewers to be one of confidence, the color associations would be different than if the objective was to be to suggest the inherent complexity of a distribution.[†]

Affective Value. Some colors are liked more than others. As is the case with symbolic connotations, this apparently varies from one culture to another. Generally, the cartographer is primarily concerned with designing an effective communication, and aesthetics play a minor role. Nevertheless, other things being equal, it is better to employ colors that are liked. Studies of affective value in the United States—which probably reflect Western culture—suggest that blue, red, and green are generally considered "pleasant"; blue is at the top, while orange and yellow are rated significantly lower. The differences do not seem to be large and would depend greatly on the specific character of a color and its strength.

During the long history of cartography, many

---

[*]See Henry Dreyfuss, Symbol Sourcebook, *McGraw-Hill Book Co. New York, 1972, pp. 231–246.*
[†]*The significance of the "look" of maps is discussed in Barbara Bartz Petchenik, "A Verbal Approach to Characterizing the Look of Maps." The American Cartographer, 1 (1), 63–71 (1974).*

color conventions have developed. Some are based on the connotations of hue, some have been contrived in order to standardize symbolism, and some have grown up more or less by accident. An example of a connotative convention is blue water; an example of a devised convention is red for igneous rocks, pinks for metamorphic, yellows for the Tertiary period, and so on; and an example of an accidental convention would be brown contours. The subject of color conventions in cartography is very complex, and space does not allow a full treatment. Some conventions are generally employed, while other, very complex systems have been devised for particular kinds of mapping, such as hydrogeologic or vegetation mapping. An example of a generally followed convention is the spectral progression for hypsometry referred to earlier and discussed in a later section. Some conventions are conventions only in a limited area, such as a country. Some are widespread. Only a few of the generally accepted, broad, conventional uses of hue can be listed here:

1. Blue—water, cool, positive numerical values.
2. Green—vegetation, lowlands, forests.
3. Yellow, tan—dryness, paucity of vegetation, intermediate elevations.
4. Brown—landforms (mountains, hills, etc.), contours.
5. Red—warm, important items (roads, cities, etc.), negative numerical values.

The Employment of Value on Maps

Of the three dimensions of color, value is probably the most important in terms of the fundamental perceptual aspects of map design. Value is basic to clarity and legibility because of the effect of value contrast in definition. The basic precept is that the greater the value contrast, the greater the definition, and the greater the clarity and legibility.

As with the color dimension of hue, the effective employment of value by the cartographer is based on a number of perceptual aspects. Some of these are physiologically based, some are founded on our subjective reactions, and some are established conventions.

Of great importance is the fact that physiologically we are not very sensitive to differences in value, and our recall and recognition of a particular value are limited. When a quantitative geo-

graphical series is being represented by value differences alone, such as on a precipitation map or a choropleth map of economic data, it is wise to limit the symbolization to four or five value steps, not including solid and white. If the hue being used is inherently light, even fewer steps can be employed.

Our limited sensitivity is further complicated by the phenomenon of simultaneous contrast, illustrated earlier by Figs. 14.3 and 14.4. When different tones are used on a map, the value of any one area is affected by the adjacent lighter or darker tones. This phenomenon, called induction (not to be confused with induction in cartographic generalization), may be lessened by edging the tones with black (or very dark) or white lines. Although the use of smooth tones on a map may be aesthetically pleasing, the cartographer must be wary of potential recognition problems. In many instances it is wise to add some visible pattern to the tones to aid their recognition.

Another physiologically based perceptual phenomenon is irradiation, an effect that occurs with extreme values, especially black on white and white on black. Given the same width of line as, for example, in type, the dark on a light background will appear slightly smaller than the reverse (white on a dark background) (Fig. 14.13).

Perhaps the most important of the subjective and connotative aspects is the fact that variation in value conveys an implication of magnitude variation; the darker indicates more of whatever is being symbolized. It makes no difference whether gray or some other hue is being employed; we naturally make the deeper water, the denser population area, and the greater per capita income darker than the areas of lesser magnitude. Some phenomena in which the range from low to high magnitudes is complementary create problems. For example, a high rate of literacy is the same as a low rate of illiteracy. In such instances, the cartographer must judge which aspect is to receive the emphasis.

There is an additional subjective and conventional aspect that sometimes aids the cartographer in making a decision when faced with the kind of problem just mentioned. Where a range of value steps is used to portray a distribution that has qualitative implications, the darker is usually associated with the less favorable. In the example

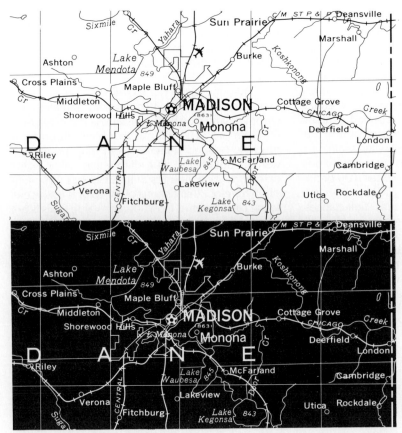

FIGURE 14.13 *The phenomenon of irradiation. The black and reverse (white) lines and lettering are the same dimension, but the reverse appear larger.*

above, we would normally use the darker values for the higher rates of illiteracy (or the lower rates of literacy) on the ground that not being able to read is unfortunate.

One other subjective aspect of value in cartography is important in the overall graphic plan for the symbolism on a map. Extreme values tend to dominate a composition, and the larger the area of such a value, the greater the dominance. What constitutes an extreme value depends on the graphic context, so to speak. White and black are the extremes in the range of values, but maps are usually printed on white or light-colored paper. Ordinarily, then, black would be the extreme, and large areas of black, such as if a considerable area of ocean were made black, would draw the attention. This would be undesirable if,

in fact, it was the land areas that were the regions of main concern. The opposite would occur if a white area were the only high value on an otherwise dark map.

The Equal Value Gray Scale

A cartographer frequently uses values, whether gray or color, to present a graded series of information. The quantitative variations in many geographical phenomena, such as amounts of precipitation, temperature, depth of water, intensity of land use, and so on, are usually depicted by some technique that depends for its effectiveness on a series of classes that are graphically differentiated by value contrasts. It is desirable to make the value contrasts correspond to the differences between the data classes. For example, if we have a range of five equally spaced data classes (e.g., 0–20, 20–40, 40–60, 60–80, 80–100), it would be most appropriate that the visual intervals between the values of gray (or color) also be equal. To do this we need to know the ratios of black to white (i.e., percent reflectances) in a series of grays that will result in a progression of equal visual steps. This problem has received a great deal of study. Recent analysis tends to confirm the equality of spacings of the Munsell value scale, but this is a very complex problem in cartography because of the variety of ways that values may occur in juxtaposition on maps.*

The relation between percent reflectances and their perceptual effects (value) in the Munsell system is shown in Fig. 14.14. Two characteristics of the relationship are noteworthy: differences in percent reflectance in dark grays result in relatively large differences in value; and the opposite is the case at the light ends of the scales. Commercial dot screens with which to obtain tones are prepared in tonal densities of 10 or 20% intervals (see

*A. Jon Kimerling, "A Cartographic Study of Equal Value Gray Scales for Use with Screened Gray Areas," The American Cartographer, 2 (2), 119–127 (1975).

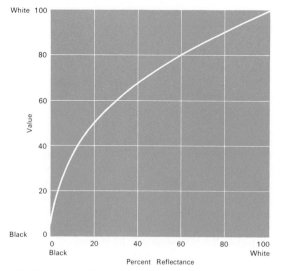

FIGURE 14.14 The relation between percent reflectance and value as specified by the Munsell value scale.

Chapter 17).* This means that the value difference between 10 and 20% screened areas will be considerably smaller than the value difference between 60 and 70% screened areas.

The Employment of Intensity on Maps

Our sensitivity to differences in intensity (or chroma), with hue held constant, is not very great. Furthermore, it is difficult to obtain chroma (as precisely defined) differences in map production without obtaining value differences. Also, value differences of a hue result in differences in intensity. Therefore this dimension of color is less useful to the cartographer than hue and value. This is not to say that intensity is unimportant. Conventionally and subjectively, the greater the intensity, the greater the magnitude implication.

A significant physiologically based aspect of chroma is that the larger the map area of a color, the more intense it will appear. This is important in two ways. Unless the cartographer is very careful in selecting colors from the small samples on a printer's color chart (Fig. C1.11) by keeping in mind the relative map areas on which the colors are to be used, it may turn out that a color that seems modest on the chart will be overwhelming in its intensity on the map. Just as extreme values dominate a composition, so do extreme intensities. This principle (the larger the map area, the more intense the color) is important in a second way. If two similar hues or two values of the same hue are used in the legend and a primary distinction between them is their intensity, then large map areas of the less intense color (in the legend) may be confused with small map areas of the more intense color.

Color Selection in Mapping

In addition to the foregoing considerations of employing hue, value, and intensity, there are some important general considerations that one must keep in mind.

In mapping nominally scaled phenomena with color area symbols, the main problem of the cartographer is to select colors that match the

*The tonal densities are ostensibly percent area inked, which is not quite the same as percent reflectance because of variations in dot size, ink quality, and the like. In any case, they are only approximate.

communication objectives. If there is no ranking involved (i.e., if no area is more important than another), then colors must be selected that carry no magnitude implication. If some of the nominal categories are logically thought to be "more important" than others as, for example, in urban-nonurban or agriculture-forest, then a magnitude implication may be desirable. If many colors are required, it will be difficult to select a set that does not include noticeable differences in value and intensity and, as we have seen, the darker and more intense colors inherently carry a magnitude implication.

A second concern of the cartographer in color selection is to make sure that the colors selected for the map areas do not result in undesirable linear contrasts. A large area of a light color (e.g., yellow) next to one or more colors that are considerably darker will result in the separation line being very prominent when, in fact, it may be of no great consequence. Selection must be based not only on the color dimensions, but also on the relative sizes and locations of the areas being symbolized on the map.

Graded Series. One of the common tasks of the cartographer is to select colors to be used to portray progressive classes on maps in which the phenomena have been ranked on interval or ratio scales. Two avenues are open: (1) to use several hues, and (2) to use progressive values of one hue.

The first is quite common, stemming in part from connotative associations of hues, such as red-warm and blue-cool, and in part from the common convention of portraying successive elevations on relief maps with a spectral progression of hues. On the latter the water is blue, lowlands are green, intermediate elevations are yellow-brown, and high areas are reddish. The spectral progression on relief maps is a convention of long standing and thus is well known through regular appearance on school maps and on atlas maps for the general public. Except for its familiarity, it is graphically illogical with little to recommend it both as to the connotative associations of some of the hues and because the value changes are inconsistent, resulting in marked variations in the perceptibility of other data being shown. A complete spectral progression should not be used for data other than elevation. Figure CI.12 illustrates a complex graded series in which variations in hue in two directions have been overprinted to provide the legend for a "two-variable" map. Sixteen different hues result, and the significance of value in the display is evident.

Several tests have shown that a simple progression of values of one hue is the most efficient in conveying the magnitude message of a graded series; it is even more effective than a two-hue series with strong connotative aspects such as a red to blue series to portray warm to cool temperatures.* When the range of data is too large to be portrayed by distinguishable values of one hue, additional hues must be used. When this is done, the value progression should be systematic from low (dark) at one end to high (light) at the other so as to be consistent with the strong subjective implication of magnitude gradation pointed out earlier. The gradation of intensities should also be kept in mind.

THE EMPLOYMENT OF PATTERN IN CARTOGRAPHY

Pattern is the term used to describe the perceptual impression provided by any visible, more or less systematic repetition of marks that covers an area on a graphic display. Pattern is widely used in cartography for several purposes. Probably the most common use of pattern is as an area symbol to impart a unified quality to a geographical area, such as an area of a given kind of bedrock, climatic regime, or administrative jurisdiction. Another associated use is to break up printings of uniform tones to give the tonal area more character, allowing it to be more easily identified and matched with the legend.

Kinds of Patterns

There are many kinds of patterns in a geographical context. The word has a variety of meanings ranging from a "pattern" of settlements to a "pattern" of temperature distributions. In cartography we are concerned with pattern as a graphic symbol, since it is one of the visual vari-

*David J. Cuff, "Color on Temperature Maps," The Cartographic Journal, 10 (1), 17–21 (1973).

ables (Chapter 5). In that sense there are several classes of patterns.

1. *Line patterns* are composed of straight parallel lines. The lines may be of any width and spacing. When two sets of parallel lines are crossed, it is called crosshatching, and the sets of lines may intersect perpendicularly or at an angle (Fig. 14.15).
2. *Dot patterns* are those composed of round dots arranged either in a rectangular or triangular array (Fig. 14.15). An irregular arrangement of dots giving a more or less smooth appearance is called stippling.
3. *Miscellaneous patterns* are of great variety, ranging from the tufted grass symbols commonly used to show swamps and marshes to arrays of crosses or other distinctive marks (Fig. 14.15). Many such patterns have become conventional in specific subject matter fields, such as geology or soils.

Patterns alone may be either normal (dark marks on a white or light background) or reverse (white or light marks on a dark background). Sometimes these are referred to as being positive and negative, respectively.

Dimensions of Pattern

Very little study has been directed toward the understanding of patterns, their perceptual consequences, and the ability to discriminate among them. Some basic terminology is necessary in discussing them. Every pattern, whether composed of dots, lines, or a combination, has several dimensions (i.e., measurable attributes): (1) *texture* (i.e., the spacing of the marks), usually specified by the number of lines per inch; the greater the number the "finer" the texture; (2) *arrangement* (i.e., the relation of the positioning of the marks), such as parallel lines, crosshatching, square, or triangular arrangement of dots; (3) *orientation* (i.e., the positioning of the arrangement) relative to the viewer, such as vertical, horizontal, or diagonal; and (4) *value,* as measured on the gray scale.

These four basic attributes of pattern are perceived in integrated fashion in an impression that can be called pattern-value. It is impossible to describe this total impression by a single measure (just as it cannot be done for color); instead, we must rely on the listed dimensions, texture, arrangement, orientation, and value.[*]

The Employment of Pattern on Maps

As pointed out, the primary reason for employing pattern in symbology is to give areas distinctive qualities, either on a monochromatic map or to help differentiate among the several areas on a map employing color. For these purposes it is necessary to make the patterns distinctive. Often a simple addition of dark or light dots or lines is sufficient to enable the user to recognize easily the symbolized areas, as illustrated in Fig. 14.16.

When patterns alone are employed with a single color, these are often dot or line patterns. Because it is generally desirable to use fine-textured patterns instead of coarse patterns, it is sometimes a problem to obtain the desired contrast. To do so the cartographer may employ all the dimensions of pattern. Dot patterns having textures finer than 75 lines per inch will usually be perceived as gray value areas without any pattern, while those with textures coarser than 40 lines per inch will probably not convey much impression of value to most readers (see Fig. 14.17). As long as the marks making up the patterns are bold and the textures are relatively coarse, differences in both arrangement and orientation are quite noticeable. The most significant contrasts in these characteristics are the results of inducing visual lines or direction in the eyes of the viewer.

[*]Henry W. Castner and Arthur H. Robinson, Dot Area Symbols in Cartography: the influence of pattern on their perception, ACSM Monographs in Cartography, No. 1, American Congress on Surveying and Mapping, Washington, D.C., 1969.

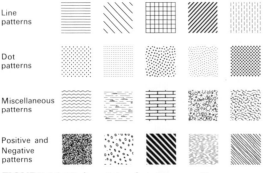

FIGURE 14.15 *A variety of common patterns.*

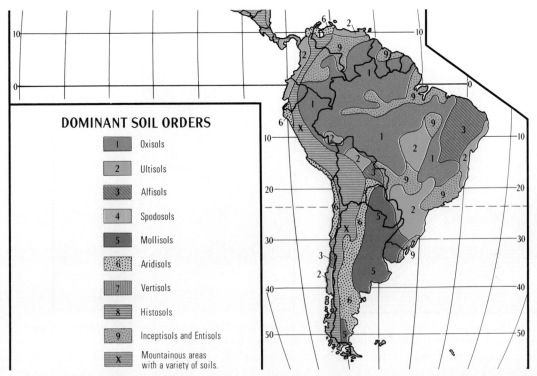

FIGURE 14.16 An example of a two-color map in which pattern has been employed to make areas appear more distinctive. (From Fundamentals of Physical Geography, 3d ed., by Trewartha, Robinson, Hammond and Horn, Copyright © 1977, McGraw-Hill Book Co., Inc.)

Contrasts in value among patterns are obtained by varying the percent area inked, but it is not possible to specify any simple lower threshold for discrimination. From the meager data now available it appears that a difference of 10 to 12% will usually be noticeable. The important point in obtaining contrasts among patterns is varying as many of the characteristics as possible instead of trying to hold some constant. The more characteristics that are held constant, the greater the difference must be in the quality that is varied.

Any line anywhere has, in the eye of the viewer, an orientation, and the viewer will tend to move his or her eyes in the direction of the line. If

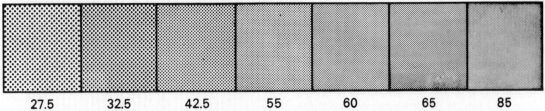

FIGURE 14.17 A series of dot area symbols (Artype) in which the dimensions, arrangement, orientation, and area inked are held essentially constant but texture is varied. The numbers below the boxes show the numbers of lines per inch in each dot screen. Note that textures coarser than about 40 are primarily seen as patterns without value, while a texture as fine as 85 appears primarily as a value area without visible pattern.

A

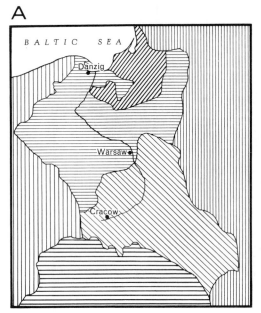

B

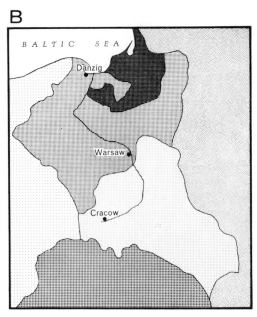

FIGURE 14.18 *A simple monochrome map contrasting the use of line and dot patterns. Line patterns are per-ceptually relatively unstable, and all but the finest-textured should be used with caution.*

irregular areas are differentiated by line patterns, as in Fig. 14.18A the reader's eyes will be forced to change direction frequently. Consequently, the reader will experience considerable difficulty in focusing on the boundaries. If the line patterns of Fig. 14.18A are replaced with dot patterns, as in Fig. 14.18B the map is seen to become much more stable, the eye no longer jumps, and the positions of the boundaries are much easier to distinguish. Lettering is also easier to read against a dot background than against a line background. If, however, the parallel lines are fine enough and closely enough spaced, the perceptual effect is largely one of value and it will have but a slight suggestion of direction.

Many parallel line patterns are definitely irritating to the eye. Figure 14.19 is an exaggerated example of the irritation that can occur from using parallel lines. It is probably because the eyes are unable quite to focus on one line. Whatever the reason, the effect is somewhat reduced if the lines, regardless of their width, are separated by white spaces greater than the thickness of the lines. Generally, a cartographer should be very wary

FIGURE 14.19 *Line patterns can irritate the eyes (and the map on which they are used!) if they are not chosen wisely.*

of using any but the finest-textured parallel line patterns.

SELECTED REFERENCES

Beck, J., "The Perception of Surface Color," *Scientific American, 233,* 62–75 (1975)

Birren, F., *The Story of Color,* Crimson Press, Westport, Conn, 1941.

Castner, H. W., and A. H. Robinson, *Dot Area Symbols in Cartography: the influence of pattern on their perception,* American Congress on Surveying and Mapping, Cartography Division, Technical Monograph No. CA-4, Washington, D.C., 1969.

Cuff, D. J., "Value Versus Chroma in Color Schemes on Quantitative Maps," *The Canadian Cartographer, 9,* 134–140 (1972).

Cuff, D. J., "Colour on Temperature Maps," *The Cartographic Journal, 10,* 17–21 (1973).

Evans, R. M., *An Introduction to Color,* John Wiley & Sons, New York, 1948.

Keates, J. S., "The Perception of Colour in Cartography," *Proceedings of the Cartographic Symposium,* University of Glasgow, Department of Geography, 1962, pp. 19–28.

Kimerling, A. J., "A Cartographic Study of Equal Gray Scales for Use with Screened Gray Areas," *The American Cartographer, 2,* 119–127 (1975).

Meyer, M. A., F. R. Broome and R. H. Schweitzer, Jr., "Color Statistical Mapping by the U.S. Bureau of the Census," *The American Cartographer, 2* 100–117 (1975).

Munsell, A. H., *A Color Notation,* 12th ed., Munsell Color, Baltimore, 1975.

Nimeroff, I., *Colorimetry,* National Bureau of Standards Monograph 104, U.S. Government Printing Office, Washington, D.C., 1968.

Olson, J. M., "The Organization of Color on Two-Variable Maps," Proceedings, International Symposium on Computer-Assisted Cartography, pp. 289–294 and color insert ff. p. 250, American Congress on Surveying and Mapping, Falls Church, VA, 1977.

Robinson, A. H., "Psychological Aspects of Color in Cartography," *International Yearbook of Cartography, 7,* 50–59 (1967).

Sargent, W., *The Enjoyment and Use of Color,* Dover Publications, Inc., New York, 1964.

Shellswell, M. A., "Towards Objectivity in the Use of Colour," *The Cartographic Journal, 13,* 72–84 (1976).

Standard Printing Color Identification System, Aeronautical Chart and Information Center (ACIC), Technical Report No. 72–2, St. Louis, MO, 1972.

The names on a map are vital in two respects. First, maps are to show where things are, and they must be identified to show this. Second, most words, when read or heard, are "transparent"; we pay little or no attention to their appearance or sounds. On the other hand, when used on a map, the names are a significant, "nontransparent" component of the visual display, and thus they are an important element in graphic design.

Perhaps no element of a map is so important in conveying an impression of quality as the styles and sizes of the type employed and the manner in which the map has been lettered. The graphic quality of a map is heavily dependent on its lettering; because of this we can often recognize at a glance the author or cartographic establishment from which a map comes.

The utility of a general map depends to a great extent on the characteristics of the type and its positioning. The recognition of the feature to which a name applies, the "search time" necessary to find names, and the ease with which the lettering can be read are all important to the function of the map. Like other aspects of cartographic design, we must approach the problem of cartographic typography and lettering by settling clearly on the objectives of the map and then fitting the selection of type and the positioning of the lettering to them.

The Function of Map Lettering

Like all other marks on a map, the type functions as a symbol, but it is considerably more complicated in its function this way than most symbols. Its most straightforward service is as a literal symbol: the individual letters of the alphabet, when arrayed, encode sounds that are the names of the features shown on the map. Generally, this is the most important role of the lettering in the communication system that is the map.

In several ways the type on a map is a graphic symbol. By its position within the structural framework, it helps to indicate the location of points (e.g., cities); it more directly shows by its spacing and array things such as linear or areal extent (e.g., mountain ranges and national areas); and by its arrangement with respect to the graticule, it can clearly indicate orientation. By systematically employing distinctions of style, form, and color, the type on a map may be used as a means of showing nominal classes to which the labeled feature belongs. For example, we can identify all hydrographic features in blue type; within that general class, we can further show open water by all capital letters, and running water by capitals and lowercase letter forms. By variations of size, it can portray the ordinal characteristics of geographical phenomena, ranking them in terms of relative area, importance, and so on.

CHAPTER 15
TYPOGRAPHY AND LETTERING

In a more subtle way, the type on a map serves as an indication of scale. Primarily by its size contrasts with respect to other factors such as line width and symbol size, we can often tell "by looking" that one map is a larger or smaller scale than another.

The History of Map Lettering

In the manuscript period, all map lettering was, of course, done freehand, and each copy of a map had to be laboriously hand lettered by the calligrapher. The styles were varied and ranged from cramped and severe to free flowing.

After the latter part of the fifteenth century, when maps began to be duplicated by woodcut printing and copper engraving, the lettering became the problem of the craftsman. Small lettering in woodcuts was difficult, and various ways of incorporating movable type were devised, such as inserting names in holes cut in the blocks or affixing stereotypes. The engraver cut letters with a burin or graver in reverse on the copper plate. The great Dutch atlas makers included many pictures of animals, ships, and wondrous other things for, as Hondius explained, "adornment and for entertainment," but their lettering was generally well planned in the classic style and well executed (see Fig. 15.1).

As might be expected, when hand lettering was done by those more interested in its execution than in its function, it tended to become excessively ornate. The trend toward poor lettering design seemed to accelerate after the development of lithography and continued well into the Victorian era. A process called cerography or wax engraving came into use in the 1840s and later made lettering a map quite easy. This led to considerable "overlettering."

FIGURE 15.1 The well-lettered Hondius map of America. This map was included in the Hondius-Mercator Atlas of 1606. Note also the ornate cartouches. The original is in the Newberry Library, Chicago. (Courtesy Rand McNally and Company.)

Lettering and type styles in general became so bad by the close of the nineteenth century that there was a general revolt against them, which caused a return to the classic styles and greater simplicity. Figure 15.2 is an example of ornate lettering in the title of a nineteenth-century geological map. The fancy lettering of that and earlier periods provide good examples of manual dexterity, since they are intricate and difficult to execute; but they are examples of poor communications design because they are difficult to read and call undue attention to themselves.

In the past half century, and especially during the past several decades, many changes have taken place in cartographic lettering practice. The introduction of the stickup process whereby type is printed on thin plastic, which is then wax-backed for adhesion to the map, has all but made freehand lettering a thing of the past as far as map construction is concerned, and it is safe to say that the cartographer now has an almost unlimited array of type styles and sizes from which to choose.

Although relatively few maps are hand lettered for reproduction in their construction stage, it is well to point out that even if the cartographer never executes freehand the ink lettering on a map, the ability to letter neatly is a most useful accomplishment. To be "sloppy" in almost anything usually requires more time in the long run than it does to be neat. In the cartographer's pencil compilations it is usually necessary to insert many names and to pay careful attention to positionings and spellings. If the cartographer hurries and is messy, mistakes of one kind or another will be made and these will cause trouble later. For this reason, if for no other, it is well for the student of cartography to study and practice freehand lettering; almost anyone can learn to do neat freehand lettering—especially with pencil—if he or she will but try.

Planning for Lettering

Lettering a map means the preparation of this aspect of the artwork, which includes all the names, numbers, and other typographic material. It involves the entire process from the beginning to the end and requires careful consideration of many factors.

The more elaborate and complex the map, of course, the more elements must be considered; but, in general, there are at least seven major headings to the planning "checklist." The complexities of the map and its purposes will add subheads to the following major elements.

1. The style of the type.

FIGURE 15.2 *Ornate lettering on a lithographed map of 1875. The name "Illinois" is more than 1 ft long on the original.*

2. The form of the type.
3. The size of the type.
4. The contrast between the lettering and its ground.
5. The method of lettering.
6. The positioning of the lettering.
7. The relation of the lettering to reproduction.

The style refers to the design character of the type; it includes elements such as thickness of line and serifs. The form refers to whether it is composed of capitals, lowercase, slant, upright, or combinations of these and other similar elements. The methods of lettering include the mechanical means whereby the type or lettering is affixed to the map. The positioning of the lettering involves the placement of the type. Also important is a consideration of when and where on the map, and in the construction schedule, the lettering is done.

Regardless of the kind of map, the lettering is there to be seen and read. Consequently, the elements of visibility and recognition are among the major yardsticks against which the choices and possibilities are to be measured.

ELEMENTS OF TYPOGRAPHIC DESIGN

Type Style

The cartographer is faced with an imposing array of possible choices when planning the lettering for a map. There is a surprising number of different alphabet designs from which to select, but the cartographer must also settle on the wanted combinations of capital letters, lowercase letters, small capitals, italic, slant, and upright forms of each alphabet. There is no other technique in cartography that provides such opportunity for individualistic treatment, and this is especially true with respect to the monochrome map. The cartographer who becomes well acquainted with type styles and their uses finds that every map or map series presents an interesting challenge.

Letter style exhibits a complex evolution since Roman times.* The immediate ancestors of

*A very well-illustrated, interesting treatment of the development of lettering and type styles is Alexander Nesbitt, The History and Technique of Lettering, *Dover Publications*, New York, 1957.

our present-day alphabets include grandparents such as the capital letters the Romans carved in stone and the manuscript writing of the long period prior to the discovery of printing. Subsequent to the development of printing, the types were copies of the manuscript writing, but it was not long until designers went to work to improve them. Using the classic Roman letters as models for capitals and the manuscript writing for the small letters, there evolved the alphabets of uppercase and lowercase letters that it is our custom to use today. There are three basic classes.

One class of designers kept much of the free-flowing, graceful appearance of some freehand calligraphy so that their letters carry something of the impression of having been formed with a brush. The distinction between thick and thin lines is not great, and the serifs with which the strokes are ended are smooth and easily attached. Such letters are known as Classic or Old Style. They appear "dignified" and have about them an air of quality and good taste that they tend to impart to the maps on which they are used. This style has a neat appearance but, at the same time, it lacks any pretense of the geometric. (see Fig. 15.3).

A radically different kind of face was devised later; for that reason it (unfortunately) is termed Modern. Actually, the Modern faces were tried out more than two centuries ago, although we think of

CHELTENHAM WIDE

Cheltenham Wide

CHELTENHAM WIDE ITALIC

Cheltenham Wide Italic

GOUDY BOLD

Goudy Bold

GOUDY BOLD ITALIC

Goudy Bold Italic

CASLON OPEN

FIGURE 15.3 *Some Classic or Old Style type faces. (Courtesy Monsen Typographers, Inc.)*

BODONI BOLD

Bodoni Bold

BODONI BOLD ITALIC

Bodoni Bold Italic

FIGURE 15.4 *Some Modern letter forms. (Courtesy Monsen Typographers, Inc.)*

CLOISTER BLACK

Cloister Black

STYMIE MEDIUM

Stymie Medium

FIGURE 15.6 *Examples of Text and Square Serif letter forms. (Courtesy Monsen Typographers, Inc.)*

them as coming into frequent use around 1800. These type faces look precise and geometric, as if they had been drawn with a straightedge and a compass, which they have. The difference between thick and thin lines is great and sometimes excessive in Modern Styles (see Fig. 15.4). Both Old Style and Modern types have serifs. Upright forms are sometimes loosely classed together as Roman.

A third style class includes some varieties that are definitely modern in time but not in name, as well as some of older origin. This class, more and more important in modern cartography, is called Sans Serif (without serifs), and has about it an up-to-date, clean-cut, new, and nontraditional appearance. There is nothing subtle about most Sans Serif forms. There are many variants in this class, some of which include variations in the thickness of the strokes (see Fig. 15.5). Sans Serif forms are sometimes called Gothic, as opposed to Roman.

MONSEN MEDIUM GOTHIC

Monsen Medium Gothic Italic

COPPERPLATE GOTHIC ITALIC

FUTURA MEDIUM

LYDIAN BOLD

DRAFTSMANS ITALIC

FIGURE 15.5 *Some Sans Serif letter forms. (Courtesy Monsen Typographers, Inc.)*

There are several other styles that are not common but that are occasionally used on maps, such as Text and Square Serif. Text, or black letter, is dark, heavy, and difficult to read. Square Serif is rarely seen any more, but it was popular during the last century (see Fig. 15.6).

This listing by no means exhausts the possibilities. There are literally hundreds of variations and modifications possible, such as the open letter, light or heavy face, expanded or condensed, and so on.

In the selection of type, the cartographer must be guided by certain general principles that have resulted from a considerable amount of research by the students of alphanumeric symbolism as well as from the evolved artistic principles of the typographer.

Recognition depends on the occurrence of familiar forms and on the distinctiveness of those forms from one another. For this reason, "fancy" lettering or ornate letter forms are hard to read, and text lettering is particularly difficult. Conversely, well-designed Classic, Modern, and Sans Serif forms stand at the top of the list, and apparently they rate about equally. Ease of recognition also depends, to some extent, on the thickness of the lines forming the lettering. The thinner the lines relative to the size of the lettering, the harder it is to read. The cartographer must, therefore, be careful in the selection of lettering since, although the bold lettering may be more easily seen, the thicker lines may overshadow or mask other equally important data. The type on a map is not always the most important element in the visual outline; instead, it may even be desirable that the type recede into the background. If so, light-line type may be an effective choice.

The problem of the positions of the lettering

in the visual outline is significant. For example, the title may be of great importance, while the balance of the type may be of value only as a secondary reference. Size is usually much more significant than style, in determining the relative prominence, but the general design of the type may also play an important part. For example, rounded lettering may be lost along a rounded, complex coastline, whereas in the same situation angular lettering of the same size may be sufficiently prominent.

It is the convention in cartography to utilize different styles of lettering for different nominal classes of features, but this may be easily overdone. Although there is a paucity of evidence, there does seem to be some indication that the average map reader is not nearly so discriminating in reactions to type differences as cartographers have hitherto thought. The use of many subtle distinctions in type form for fine classificational distinctions is probably a waste of effort.

As a general rule, the fewer the styles, the better harmony there will be. Most common type faces are available in several variants; it is better practice to utilize these as much as possible (see Fig. 15.7). If styles must be combined for emphasis or other reasons, good typographic practice allows Sans Serif to be used with either Classic or Modern. Classic and Modern are normally not combined. Differences in style are best utilized for nominal differentiation; size and boldness are more appropriate for ordinal, interval and ratio distinctions.

The cartographer is usually less concerned with what has been called the "congeniality" of type style, that is, the correspondence between the subjective impression one gains from a type design in relation to its use. There is no question that readers respond to various styles with reactions such as "authoritative," "masculine," "feminine," "arty," and "clean." Sans Serif seems to be assuming a dominant position in cartography, while the Old Style and Modern types are used mostly in literary composition. It is not uncommon to find a Sans Serif used in the body of a map with a contrasting style, usually Old Style, employed for the title and legend materials.

Type Form

Individual alphabets of any one style consist of two quite different letter forms called capitals and

Futura Light

Futura Light Italic

Futura Medium

Futura Medium Condensed

Futura Medium Italic

Futura Demibold

Futura Demibold Italic

Futura Bold

Futura Bold Italic

Futura Bold Condensed

Futura Bold Condensed Italic

FIGURE 15.7 *Variants of a single face. This face has a larger number of variants than is usual, but the list is representative except that expanded (opposite of condensed) is missing. (Courtesy Monsen Typographers, Inc.)*

lowercase letters. These two forms are used together in a systematic fashion in writing, but conventions as to their use are not so well established in cartography. In general, past practice has been to put more important names and titles in capitals and less important names and places in lowercase letters. Names requiring considerable separation of the letters are commonly placed in capitals.

Most tests have shown that capitals are more difficult to recognize or read than lowercase letters, since the latter contain more clues to letter form. A greater use of well-formed lowercase letters probably will improve the legibility of a map.

The tendency in cartography is for hydrography, landform, and other natural features to be labeled in slant or italic, and for cultural features (manmade) to be identified in upright forms. This can hardly be called a tradition, since departure

Kennerley—an upright Classic

Kennerley in the Italic form

Monsen Medium Gothic—upright

Monsen Medium Gothic Italic

FIGURE 15.8 *Differences between italic and slant forms. (Courtesy Monsen Typographers, Inc.)*

from it is frequent, except in the case of water features where this convention is very strong. The slant or italic form seems to suggest the fluidity of water.

There is a fundamental difference between slant and italic, although the terms are sometimes used synonymously (see Fig. 15.8). True italic in the Classic or Modern faces is a cursive form similar to script or handwriting. Gothic slant and Sans Serif italic are simply like the upright letters tilted forward. Roman italic forms are considerably harder to read than their upright counterparts; however, it is doubtful that there is much difference between the upright and slant letters of Gothic or Sans Serif insofar as recognition is concerned.

Type Size

The size of type is commonly designated by *points*, 1 point being nearly equal to 0.35 mm (¹/₇₂ in.). With the development of photo lettering machines, sizes may be expressed also in the height of letters. Lettering that is 6 mm (approximately ¹/₄ in.) high is roughly equal to 18-point type, although not exactly, since designation by points refers to the body height of type and not to the letter face on it (Fig. 15.9). Type (or the print from a photo matrix in a photo lettering machine) that is enlarged or reduced will be slightly different from the designed type at that size, since the line thickness of one font (one size of one style) is usually designed for that size only.

Perhaps the most common kinds of decisions concern the various sizes to be used for the great variety of items that must be named on maps. Traditionally, specifications for lettering are usually based on the magnitude of the object being named or the space to be filled, but the type must also be graded with respect to the total design

and intellectual content of the map. Much of the criticism, however, conscious or unconscious, that is leveled at map type is aimed specifically at size—or lack of it. There seems to be a widespread tendency among amateur (and even some professional) cartographers to overestimate the ability of the eye.

In Chapter 13 it was observed that, assuming no other complications (the assumption is a bit unreal), the eye reacts to size in relation to the angle the object subtends at the eye. Generally, with normal vision, an object that subtends an angle of 1 min. can just be recognized. Letter forms are complex, however, and it has been determined that about 3-point type is the smallest, just recognizable type at usual reading distance. Normal vision is, however, a misnomer; it certainly is not average vision. It is safer to generalize that probably 4-point or 5-point type comes closer to the lower limit of visibility for the average person.

A common employment of type size variations in cartography is for the purpose of ordinal, interval, or ratio ranking of places, usually settle-

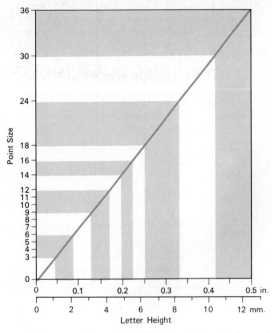

FIGURE 15.9 *The relation between letter height and nominal point size.*

ments, in terms of their relative magnitude as rated by their populations. The ability of map readers to recognize differences in size has not been extensively studied, but there is evidence that children generally require a height difference of at least 20%. This would be approximately 2 points in the range from 6 to 12 points. The average adult probably reacts in similar fashion. Furthermore, the number of categories of similar sizes that will normally be noted appears to be three, simply, small, medium, and large. For a greater number of categories we probably should introduce significant differences in form. In the interest of promoting ease of recognition ("findability" when looking for a name), the cartographer

should use the largest type sizes possible, consistent with good design.

Frequently, cartographers are called on to prepare presentations for groups. They must face questions such as, "What is the minimum type size that may be read, under normal conditions, on a wall map or chart from the back or middle of a 10-m (33-ft) room?" Figure 15.10 is a nomograph, the construction of which is based on the assumption that if a particular size, at normal reading distance from the eye (45 cm, approximately 18 in.), subtends a certain angle at the eye, then any size lettering, if viewed at such a distance that it subtends the same angle is, for practical purposes, the same size. Thus, 144-point type at

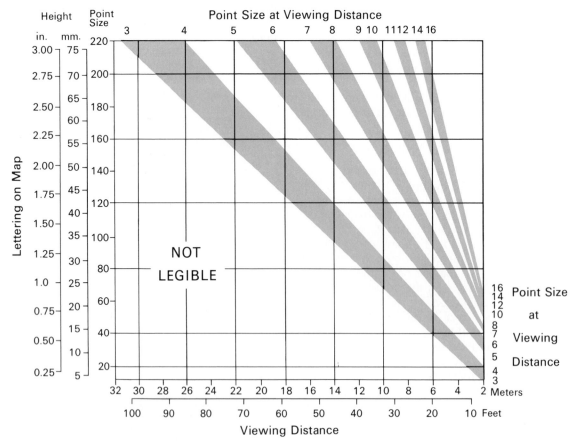

FIGURE 15.10 A nomograph showing the approximate apparent point size of lettering when viewed from various distances. Equivalent sizes less than 3 point are not legible.

about 10 m (33 ft) from the observer is the same as approximately 8-point type at normal reading distance, since each circumstance results in nearly the same angle subtended at the eye. It will be seen from the graph that legibility diminishes very rapidly with distance. For example, any type of 16-point size or smaller cannot be read even at 3 m (10 ft) from the chart or map.

The preparation of maps and diagrams to be projected by means of 35-mm slides or overhead transparencies involves several complicating factors that should be investigated with care.*

Type Color

In typography, "color" of type frequently refers to the relative overall tone of pages set with different faces. Here it is restricted to the actual hue (such as black, blue, gray, or white) of the letters and the relation between the hue and value of the type and that of the ground on which it appears. Commonly, lettering of equal intrinsic importance does not appear as equal in various parts of the map because of ground differences. Even when the same type face must be used in various graphic situations because of other design requirements, the cartographer should be aware of these possible effects. Perhaps, within the limits of the design, the cartographer may be able to correct or at least alleviate a graphically inequitable situation.

Stated in general terms, the perceptibility of lettering on a map (other effects being equal) depends on the amount of visual contrast between the type and its ground. Putting aside effects that result from texture of ground, type size, and so on, the basic variable is the degree of value contrast between the type and the ground on which it stands (Fig. 15.11). Thus, black type on a white ground would stand near the top of the scale and, as the tonal value of the lettering approached the tonal value of the ground, visibility would diminish. This is of concern either when large regional names are "spread" over a considerable area composed of units colored or shaded differently or when names of equal rank must be placed on areas of different values.

Commonly, the lettering on maps is either dark (e.g., black) on a light ground or light (e.g., white) on a dark ground. The latter is called *reverse* (or reversed) lettering. Also, type is often added to one of the color flaps of a multicolored map (e.g., blue for hydrography). Regardless of the color of the print and of the ground, if the value contrast is great, the lettering will be visible.

Screened lettering on a light or dark ground or reverse lettering on a relatively dark ground is an effective way to create contrast in the map lettering. The production of screened or reverse lettering is not difficult in the map construction process (see Chapter 17), and it provides a very effective way to categorize classes of lettering.

LETTERING THE MAP

Methods of Lettering

Until early in the present century the lettering for most maps was done freehand. To be sure, the cerographic technique (wax engraving) employed metal type pressed into a film of wax, which was then electroplated to produce a printing plate.*

---

*An excellent treatment of these problems with specific directions is given in H. W. Brockemuehl and Paul B. Wilson, "Minimum Letter Size for Visual Aids," The Professional Geographer, 28 (2), 185–189 (May 1976).

*Woodward, David, The All-American Map, Wax Engraving and Its Influence on Cartography, The University of Chicago Press, Chicago and London, 1977.

FIGURE 15.11 Perceptibility and legibility depend on lettering-background contrast.

Freehand lettering was slow and therefore costly; if it was to be done well, it required much more skill than most cartographers enjoyed. The desire for lower cost, speed, uniformity, and standardization led to the development of mechanical lettering devices. Today almost all mapmaking uses some method other than freehand. These methods may be grouped into two categories: (1) stickup or preprinted lettering, and (2) mechanical aids for ink lettering. Freehand lettering is now largely restricted to the occasional special map produced by free-lance cartographers or illustrators.

Stickup Lettering. Stickup lettering has a number of advantages over other methods of lettering, in addition to the fact that it is generally faster and requires less skill. Any of hundreds of type styles and sizes can be selected. The letters may be used as composed in a straight unit, or they may be cut apart and applied separately to fit curves. If the position first selected for the name turns out not to be suitable, the name may be relocated with no attendant problem of erasing.

The type used in this process is ordinarily printed on a thin plastic sheet, the back of which is then given a light coating of colorless wax. The words, numbers, or letters may then be cut from the sheet, positioned carefully, and then burnished onto the artwork by rubbing. Rubbing is usually done with a smooth tool, not directly on the film, but on a sheet laid over it to protect it. Light rubbing will hold it in place but makes it possible to reposition it. Final burnishing is postponed until all positioning decisions have been made.

It is possible, of course, to print or typewrite the letters on other materials and to affix them to the map in other ways. Gum-backed paper, thin tissue, and the like, have been employed, and liquid adhesives, self-adhesive materials, and so on have all been tried. The standard process, however, uses thin plastic film or cellophane with a wax backing.

There are three ways of producing such stickup materials: by photography, by special "typewriters," and by printing. There are a number of devices manufactured for the purpose, and each has its special characteristics. They will not be detailed here. Suffice it to say that photographic methods provide several advantages. These include the ability to vary the size of the let-

tering. This can be accomplished because in these devices the "type" appears as a photographic negative (matrix) and, by enlargement or reduction, the print can be scaled to the desired size. In others, a separate set of matrices or typewriter heads must be obtained for each size desired.

The cartographer must make a complete list of all names and other lettering to appear on the map or series of maps for which stickup is to be used. Spellings must, of course, be reviewed carefully. It is desirable to obtain several copies of the composition. Then, it is only necessary to repeat names, such as "river" or "lake," as many times as the total number of map occurrences of that size divided by the number of copies of the composition to be obtained. All lettering must be categorized in the listing according to type style, size, and capital and lowercase requirements. The names will be set as listed and, since possibilities of inadvertent omission are relatively high, it is good practice to include a request for alphabets of the styles and sizes being used. Prices for commercial composition vary considerably, depending on whether the type to be used must be hand set or machine set, so careful planning is desirable.

Generally, most type faces used for stickup were originally designed for printing books at size that is, without reduction. In many instances each such size of a type style has been designed separately for use at that size. In that case, when used on maps planned for reduction, there is a tendency for the lettering to appear somewhat light, since 18-point lettering reduced one-half is made of slightly thinner strokes than the 9-point size of the same style. It is often necessary, when using letterpress stickup lettering for maps to be reduced, to choose a slightly larger size than would be warranted ordinarily.

Another type of stickup lettering consists of alphabets (and symbols) on the "underside" of plastic. When burnished, the letters will come off the plastic and adhere to the map. It is primarily useful for titles and larger names.

Mechanical Lettering. Mechanical devices that enable someone unskilled in freehand lettering to produce acceptable ink lettering are available. With them we can obtain neat lettering, but it should be emphasized that the lettering produced with most of these aids appears rather mechanical and gives the impression of looking at a building blueprint. They are most used for just this purpose—engineering drawings, wherein variations in the character and orientation of letter forms are not particularly needed. The complexities of good quality map lettering require more versatility than can be easily obtained by such means. Nevertheless, mechanical lettering devices will continue to find a place in "do it yourself" cartography, especially for the production of graphs, charts, and diagrammatic maps.

The best-known devices require a special pen, feeding through a small tube, while the pen is guided either mechanically or by hand with the aid of a template.

Leroy is the patented name of a lettering system involving templates, a scriber, and special pens. A different template is necessary for each size lettering. The template is moved along the T-square or steel straightedge, and the scriber traces the depressed letters of the template and reproduces them with the pen beyond the template. A variety of letter weights and sizes in capitals and lowercase is possible by interchanging templates and pens.

Wrico is the patented name of a lettering system involving perforated templates or guides and special pens. The lettering guides are placed directly over the area to be lettered and are moved back and forth to form the various parts of a letter. The pen is held in the hand and is moved around the stencil cut in the guide. The guide rests on small blocks that hold it above the paper surface to prevent smearing. A considerable variety of letter forms are possible, including condensed and extended. A different guide is necessary for each size, although pens may be varied.

Varigraph is the patented name of a versatile lettering device that also involves a template with depressed letters and a stylus. The device is actually a sort of small, adjustable pantograph that fits over a template. The letters are traced from the template and are drawn by a pen at the other end of the pantographlike assembly. Adjustments may be made to make large or small, extended or condensed, lettering from a single template. Templates of a variety of letter styles are available.

Freehand Lettering. It is important that all cartographers become reasonably adept at freehand lettering with pencil so that the large amount of lettering necessary in compilation work will be neatly executed. Furthermore, the cartographer, in designing a graphic composition, will be called on to lay out titles, legends, and the like, and to work out the spaces and sizes needed, prior to obtaining stickup. Consequently, for executing this kind of lettering, the cartographer must gain facility in three basic aspects of freehand lettering: spacing, use of guidelines, and stroking. The term "freehand lettering" is a misnomer; such lettering well done is done with great care.

Spacing refers to the distance between letters. In letterpress type the spacing is mechanical and, as a consequence, words often appear ragged and uneven. Visual spacing in which the appropriate amount of space appears between each letter and its neighbor is far better (Fig. 15.12). For this reason, large names and especially titles done by stickup need to be positioned one letter at a time. The beginner will soon learn that there are different classes of letters according to their regular or irregular appearance and according to whether they are narrow, average, or wide. They must be separated differently, depending on the combination in the word. Mechanical spacing should always be avoided.

Guidelines are simply what their name implies, lines to serve as guides for the lettering, whether stickup, mechanical, or freehand. They may be drawn with a straightedge or a curve, but guidelines are more easily made with any of a number of patented plastic devices designed for this purpose. These devices have small holes in which a pencil point may be inserted. The device may then be moved along a straightedge or curve by moving the pencil. By placing the pencil in other holes, parallel lines may be drawn. Guidelines usually consist of three parallel lines, as shown in Fig. 15.13. The bottom two determine the height of the lowercase letters, and the upper

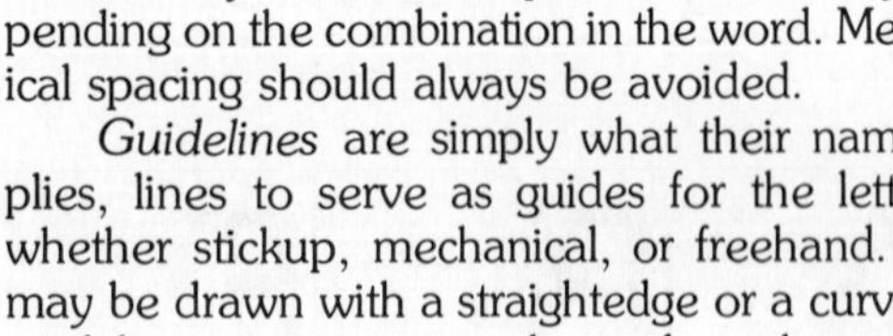

SPACING SPACING
MILES MILES

FIGURE 15.12 *Mechanical spacing on the right compared to the much better "visual" spacing on the left.*

FIGURE 15.13 Guidelines.

guideline indicates the height of the capitals and the ascenders of letters such as b, d, and f.

Stroking refers to the direction and order with which the various parts of letters are formed (Fig. 15.14). It is more or less important, depending on the kind of tools we are using. For example, we cannot make an upstroke with a sharp pen or

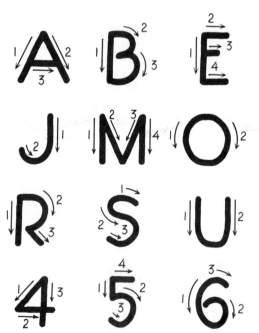

FIGURE 15.14 *Examples of proper letter stroking in some of the letters and numbers of a simple Gothic alphabet.*

pencil without digging into the surface. Once proper stroking becomes a habit, letters will be better formed, regardless of tools.

If the cartographer does any freehand lettering in ink, the efforts should be limited to simple designs. Such lettering is commonly confined to upright or slanted forms, often consisting of thick and thin strokes made with a square-tipped pen such as the Speedball Style C or the N or Z style of Graphos nibs. Simple lettering may be made with round or tubular nibs, so that whichever way the nib moves, the line thickness remains the same. Figure 15.15 shows an example of each style.

Positioning the Lettering

Map reading is affected greatly by the positioning of the names. When properly placed, the lettering clearly identifies the phenomenon to which it refers, without ambiguity. Equally important is the fact that the positioning of the type has as much effect on the graphic quality of the map as do the selections of type styles, forms, and sizes. Incongruous, sloppy positioning of lettering is as apparent to the reader as garish colors or poor line contrast.

Large mapmaking establishments tend to develop policies regarding the positioning of type partly to obtain uniformity, partly because it is cheaper, and partly because we can assign such activity to a machine. The excessive, systematic, overall result—*all* names parallel, for example—tends to set an unfortunate standard that should not be blindly followed. The cartographer, as in all other matters involving the structuring of graphic composition, must be guided by principles and precepts based on the functioning of the map as a medium of communication. Generally, the object to which a name applies should be easily recognized, the type should conflict with the other map material as little as possible, and the overall appearance should not be stiff and mechanical.

As previously observed, one of the important functions of type on a map is to serve as a locative device. The lettering can do this in three ways: (1) by referring to point locations, such as cities; (2) by indicating the orientation and length of linear phenomena, such as mountain ranges; and (3) by designating the form and extent of areas, such as regions or states. The first rule of positioning type

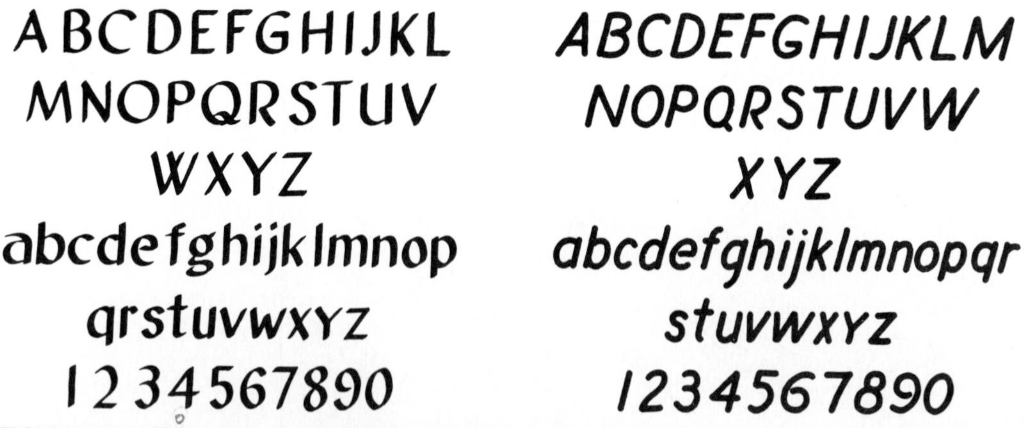

A B C D E F G H I J K L
M N O P Q R S T U V
W X Y Z
abcdefghijklmnop
qrstuvwxyz
1 2 3 4 5 6 7 8 9 0

ABCDEFGHIJKLM
NOPQRSTUVW
XYZ
abcdefghijklmnopqr
stuvwxyz
1234567890

FIGURE 15.15 Two simple freehand alphabets. The left alphabet is classed as upright (Roman); the right alphabet is classed as slanted or inclined.

is to position the lettering so that it enhances the locative function as much as possible.

It will be convenient in the listing of the major principles of type positioning to organize them as they refer to point, line, or area phenomena. There are a few general rules, however, that are independent of such a locative organization. These are as follows.

1. Names should be either entirely on the land or on the water.
2. Lettering should generally be oriented to match the orientation structure of the map. In large-scale maps this means parallel with the upper and lower edges; in small-scale maps, it means parallel with the parallels.
3. Type should not be curved (i.e., different from point 2 above) unless it is necessary to do so.
4. Disoriented lettering (point 2 above) should never be set in a straight line, but should always have a slight curve.
5. Names should be letter spaced as little as necessary.
6. Where the continuity of names and other map data, such as lines and tones, conflicts with the lettering, the data, not the names, should be interrupted.
7. Lettering should never be upside down.

It is not uncommon that conflicts in precepts will occur because of particular combinations of requirements. There is no general rule for deciding such issues; the cartographer must make a decision in light of all the special factors. Figure 15.16 has

been prepared to illustrate the foregoing principles of type positioning as well as the following precepts.*

Point Features. Generally, lettering that refers to point locations should be placed above or below the point in question, preferably above and to the right. The names of places located on one side of a river or boundary should be placed on that same side. Places on the shoreline of oceans and other large bodies of water should generally have their names entirely on the water. Where alternative names are shown (Köln, Cologne), the one should be symmetrically and indisputably arranged with the other so that no confusion can occur.

Linear Features. Lettering intended to identify linear features always should be placed alongside and "parallel" to the river, boundary, road, and so on, to which it refers, never separated from it by another symbol. Where there is curvature, the lettering should correspond. Ideally, we would place the type along an uncrowded extent where the lettering could be read horizontally, but this is usually not possible. Complicated curvatures should be avoided and, generally, the names

*The reader is specifically referred to the following article, which contains many illustrations of good and bad practice in positioning lettering. Eduard Imhof, "Positioning Names on Maps," The American Cartographer, 2 (2), 128–144 (1975).

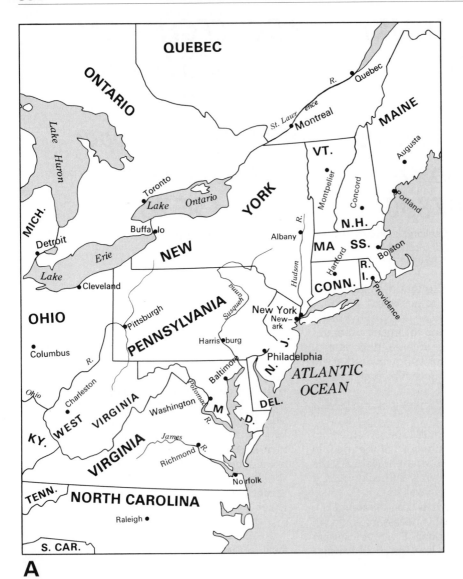

A

FIGURE 15.16 (A) *The general rules of type positioning have all been violated.* (B) *In contrast, this map attempts to set a better example.*

should be letter spaced somewhat, but not much. Such designations along rivers need to be repeated occasionally. If we must position lettering nearly vertically along a linear feature, if possible it should read upward on the left side of the map and downward on the right. Where it can be done, it is good practice to curve lettering so that the upper portions of the lowercase letters are closer together because there are more clues to letter form in the upper part. Names along linear items are better placed above than below the feature because there are fewer descenders than ascenders in lowercase lettering.

Areal Features. Normally, the lettering to identify areal phenomena will be placed within the

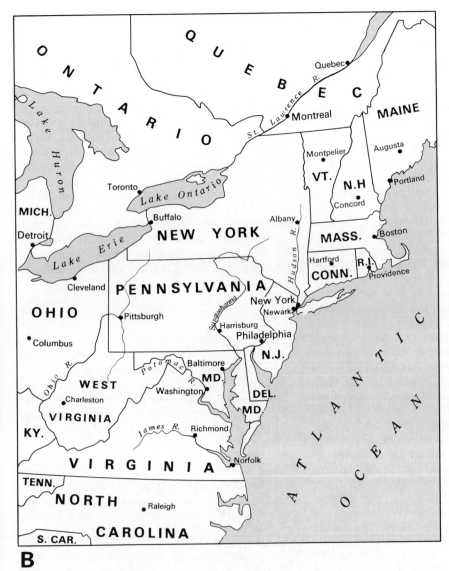

B

FIGURE 15.16 (Continued)

boundaries of the region. The name should be letter spaced to extend across the area, but it should not crowd against the boundaries. As observed in the general rules, where any tilting is necessary, there should be added a clearly noticeable bit of curvature so that the name will not look like a printed label simply cut out and pasted on the map. Curvatures should be very simple and constant.

Geographical Names

Difficult as the solutions may be to the various problems concerning the positioning of the lettering, often an even more difficult question is the proper or appropriate spelling of the names we wish to use. For example, do we name an important river in Europe *Donau* (German and Austrian), *Duna* (Hungarian), *Dunav* Yugoslavian and Bulgarian), *Dunarea* (Rumanian), or do we spell it

Danube, a form not used by any country through which it flows! Is it *Florence* or *Firenze, Rome* or *Roma, Wien* or *Vienna, Thessalonkiē, Thessoloniki, Salonika* or *Saloniki,* or any of a number of other variants? The problem is made even more difficult by the fact that names change because official languages are altered or because internal administrative changes occur. The problem of spelling is difficult indeed.

The difficulties are of sufficient moment that governments that produce many maps have established agencies whose sole job it is to formulate policy and to specify the spelling to be used for names on maps and in official documents. Examples are the British Permanent Committee on Geographical Names (PCGN) and the U.S. Board on Geographical Names (BGN) of the Defense Mapping Agency (DMA).* The majority of such governmental agencies concern themselves only with domestic problems, but the two named above include the spelling of all geographical names as part of their function.

One of the major tasks of such an agency (and of every cartographer) is the determination of how a name that exists in its original form in a non-Latin alphabet will be rendered in the Latin alphabet. Various systems of *transliteration* from one alphabet to another have been devised by experts, and the agencies have published the approved systems. The Board on Geographic Names has published numerous bulletins of place-name decisions and guides recommending treatment and sources of information for many foreign areas. These are available upon application. It is well for the cartographer to acquire, or at least to have available, such bulletins; frequently the cartographer is required to make decisions on matters of transliteration. Even more frequently, the cartographer will use map sources that contain other alphabets, characters, and ideographs.

*U.S. Board on Geographical Names, National Center, Stop 523, Reston, VA 22092.

The general rule is to use the conventional English form whenever such exists. Thus, *Finland* (instead of *Suomi*) and *Danube River* would be preferred. Names of places and features in countries using the Latin alphabet may, of course, be used in their local official form if that is desirable. The problem is very complex, and it is difficult to be consistent. It is easy for English speaking readers to accept *Napoli* and *Roma* for *Naples* and *Rome,* but it is more difficult for them to accept *Dilli, Mumbai,* and *Kalikata* for *Delhi, Bombay,* and *Calcutta.*

The problem is much too complex to be treated in any detail here, but it is important for the student of cartography to be aware of it. Above all, cartographers must not fall into toponymic blunders by placing on maps names such as *Rio Grande River, Lake Windermere,* or *Sierra Nevada Mountains.*

SELECTED REFERENCES

Bartz, B. S., "An Analysis of the Typographic Legibility Literature," *The Cartographic Journal, 7,* 10–16 (1970).

Brockemuehl, H. W., and P. B. Wilson, "Minimum Letter Size for Visual Aids," *The Professional Geographer, 28,* 185–189 (1976).

George, R. F., *Speedball Text Book,* 17th ed., Hunt Pen Co., Camden, NJ, 1948.

Imhof, E., "Positioning Names on Maps," *The American Cartographer, 2,* 128–144 (1975).

Keates, J. S., "Lettering," in *Cartographic Design and Production,* Longman Group Limited, London, 1973, pp. 201–211.

Nesbitt, A., *The History and Technique of Lettering,* Dover Publications, Inc., New York, 1957.

Updike, D. B., *Printing Types, Their History, Forms and Use; a Study in Survivals,* Harvard University Press, Cambridge, MA, 1922.

Yoeli, P., "The Logic of Automated Map Lettering," *The Cartographic Journal, 9,* 99–108 (1972).

MAP REPRODUCTION

When duplicate copies of maps were made by hand, before the advent of printing, the cartographer was free to prepare the map in any one of several methods. Even today, if a map is not to be reproduced, the cartographer is under few constraints as to design or the kinds of materials that may be used. On the other hand, since most of the maps are made to be duplicated, several factors, such as cost and requirements for artwork, place significant restrictions on the cartographer. Unfortunately, maps are sometimes prepared by people who are unfamiliar with reproduction; they give no thought to how the map is to be duplicated. In many cases it is discovered, too late, that there is no process that can adequately or economically reproduce the map.

It must be emphasized that the proper sequence in preparing artwork is first to choose the process by which it will be duplicated, and then to plan how the artwork can be prepared best to fit that method. In the selection of a process, we consider the kinds of copies we need, how many copies are required, the quality desired, and the size. To make intelligent choices and to be able to plan for the proper artwork, a knowledge of common duplicating processes is a necessity.

In this chapter we will acquaint students with several processes to enable them to choose the best one for a particular situation. An understanding of the possibilities and limitations of the processes will be needed as a background for Chapter 17, which deals with the preparation of artwork.

Because there are a number of terms used in the reproduction trades and in the cartographic construction process that are likely to be unfamiliar to the student, the following glossary is provided.

Blacklight. A slang term referring to electromagnetic energy rich in ultraviolet.

Continuous tone. Smooth and continuous transition of gray tones such as on an ordinary photograph.

Flap. Separate sheet on which a map or selected features of a map are prepared. (Other terms that are used to refer to these separate sheets are *overlay* and *separate*.) One map may require several flaps.

Mask. A film, usually considered to be in positive form, used to block light on particular parts of a sensitized material.

CHAPTER 16
MAP REPRODUCTION

Opaque. Impervious to the rays of light. Opaque also refers to any of a variety of substances, whether white, black, or red, that prevent transmission of light.

Opaque, actinically. Opaque to the wavelengths of light by which sensitized materials are affected.

Photoengraving. Photomechanical process for converting any object that can be photographed into an image in relief for letterpress printing.

Photolithography. Photomechanical process for converting an object that can be photographed into any of the various kinds of images on lithographic plates.

Pins, registry. Machined metal or plastic studs that are inserted into matched holes in flaps of a map to keep the flaps registered.

Proof. Any of various kinds of copies of a map that are used to check the accuracy or legibility of a map or to give an indication of its appearance in final printed form.

Registry. To fit precisely; when several pieces (flaps) of artwork are used to prepare a map, they must register.

Reverse. Changing the relationship of the transparent areas and opaque areas on photographic film. For example, if film is exposed to light passing through the transparent areas of a negative, those areas will be opaque on the new film and the remainder of the new film will be transparent.

Right reading. The image, such as type, is "readable" in the normal left to right situation. Whether the image-carrying side (emulsion) is up or down should be specified.

Scribecoat. An actinically opaque plastic coating (on a transparent base) into which lines and other symbols can be cut. The finished scribesheet can be used as a photographic negative in duplicating processes.

Screen, contact. A variable-opacity screen that can be used in a camera or a vacuum frame to convert continuous tone to varying sized dots. (Patterns other than dots are also available.)

Screen, halftone. A close network of perpen-

dicular lines etched into glass and filled with an opaque pigment. When used at the photographic stage, it converts continuous tone to varying sized dots.

Screen, tint. A film that is interposed between a negative and a sensitized surface to produce a pattern. Those with closely spaced small dots or narrow lines produce the illusion of solid gray.

Wrong reading. The image, such as type, is "unreadable" in that it is a mirror image and thus must be read from right to left. Whether the image-carrying side (emulsion) is up or down should be specified.

Printing Processes

Some of the earliest printing was done by the Chinese during the first millenium A.D. when they carved block characters in wood, inked the raised portion, and transferred the impression to paper. In the Western world the idea of a printing press developed later with the invention of movable type, in about 1450. Soon thereafter methods for printing maps were put into practice.* Letters of the alphabet were carved as single units that could then be assembled to produce text. By the process known as *letterpress,* a standard form of printing today, the paper receives the ink directly from surfaces standing in relief (Fig. 16.1). For maps and drawings, the raised lines could be left standing by cutting away the nonprinting areas of a smooth

*David Woodward (Editor), Five Centuries of Map Printing, *University of Chicago Press, Chicago, 1975.*

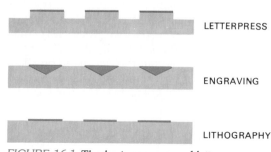

LETTERPRESS

ENGRAVING

LITHOGRAPHY

FIGURE 16.1 The basic processes of letterpress, engraving, and lithography operate in different ways to produce a surface from which ink may be transferred to paper.

woodblock to produce a printing surface called a *woodcut*. Woodcuts could be printed along with text for books in one operation, and were used well into the nineteenth century.

Another method of reproduction, *engraving,* also involved ink and an uneven surface. Grooves were cut in a flat metal plate, usually copper, and then filled with ink. The surface of the plate was then cleaned off and the plate, with its ink-filled grooves, was squeezed against a dampened sheet of paper. The paper "took hold" of the ink and, when removed from the metal plate, the pattern of grooves appeared as ink lines. In a sense, this process of printing from an *intaglio* surface is just the opposite of letterpress printing, since the inking area is "down" instead of "up." Until the late nineteenth century, most separate (not in a book) printed maps were reproduced by the engraving process.

Through the intervening years, until the development of photography, the printing plates from which copies were to be made were prepared by hand. To produce the printing elements for a map, for example, an engraver had to cut away or incise the relevant portions of the wood or metal to produce the printing areas for the lines and lettering. Since the impression was to be transferred directly to paper, all work was done backward, or wrong reading. Not only was this a laborious process, but the quality obviously depended greatly on the skill of the engraver and the interpretation of the cartographer's worksheet. Neither was it easy to produce the numerous fine lines necessary to give the illusion of changing values of tones; consequently, a very large proportion of printed matter consisted of text, with only relatively little illustrative material being used.

Progress in printing is probably most closely associated with advances in platemaking. An important development occurred in 1798 with the invention of lithography, a new printing process based on the incompatibility of grease and water. A drawing was made wrong reading directly on the smooth surface of a particular kind of limestone with greasy ink or crayon. The fats combined chemically with the elements in the stone to form a calcium oleate, which had the property of repelling water. The unmarked portion of the surface could then be dampened and, when a greasy printing ink, which was repelled by the water, was

rolled across the surface, the ink would adhere to the oily marked areas but not to the clean dampened ones. Paper pressed against the stone would pick up the ink. Because of the use of stone in the original form, this process was named *lithography.* Shading or tonal variations could be printed from the stone. Although thin metal plates have been substituted, it is still known as lithography. In contrast with printing from relief or intaglio surfaces, it is a planographic process, that is, the surface of the lithographic printing plate is a plane, and there is no significant difference in elevation between the inked and noninked areas.

All three methods of printing involved preparation of the plates by hand. There was no mechanical way to accomplish this; as a consequence, arts such as woodcutting, engraving with a burin or graver directly in the metal, or drawing directly on the lithographic stone were highly developed, and few possessed the necessary skills. The development of the specialized skills of the engraver or the lithographic artist had several effects on the development of cartography; the most important effect was that much of the detail of letter forms and other elements of the map design were delegated to the technician instead of to the cartographer.

Probably the most revolutionary development for the printing industry came during the nineteenth century when photography was incorporated into the letterpress and lithographic processes. This eliminated the necessity of having a technician copy a map onto a printing plate; the cartographer could now expect a facsimile. The amount of labor was reduced, and the amount of detail in the map had little effect on the cost of the plate.

In all the printing methods that have been discussed, it was not possible to print continuously changing tones of gray. The image, whether consisting of type or any other forms, were printed solid black, since only a full amount of ink was deposited. Closely spaced fine lines and other marks were used to convey an impression of tonal variations. *Continuous tone,* such as the changing tones of an ordinary photograph produced with a hand camera, could not be printed until the development of the *halftone screen* during the 1880s. To produce a halftone screen, fine, closely spaced parallel lines are engraved on the surface of two panes of glass and filled with an opaque pigment. The two pieces of glass, with the engraved surfaces in contact, are glued together with a transparent cement. Lines on one pane are positioned perpendicular to the lines on the other, forming a network of opaque lines and tiny transparent squares. When the screen is placed between the film and the lens of the camera, light reflected from different tones on the artwork passes through the openings in the screen and produces dots on the emulsion of the film that vary in size with the amount of light transmitted. Because all the dots are small and closely spaced, they blend when viewed from normal reading distance and create the illusion of continuously changing tone (Fig. 16.2).

Classification of Duplicating Methods

Each year new products are added to the growing list of materials and systems that can be used to duplicate maps. Some of the methods are simple and require little equipment; others may require a substantial investment in processing apparatus. Selection of a particular process may depend on considerations such as cost, physical nature of the copies, and availability of a given system. The cartographer should know enough about some of the more commonly used methods to be able to make intelligent choices and to prepare artwork in the most efficient manner.

FIGURE 16.2 How a halftone looks under magnification. The right-hand illustration shows clearly that the number of dots per unit area remains uniform; only their sizes vary. Compare with any photograph in this book (e.g., Fig. 16.7) under a strong magnifying glass. (Courtesy of the Chicago Lithographic Institute, Inc.)

It is difficult to classify reproduction methods in a satisfactory manner because many processes require more than one technique, and the intermediate techniques in one may be an end in themselves in another, or they may be intermediate in several different processes. For example, photography is a step in the printing process, but it can also be considered a separate reproduction process. Perhaps the most practical manner of classification is to group the reproduction methods on the basis of whether or not they involve a decreasing unit cost with increasing numbers of copies. It so happens that segregation on this basis also separates the common processes according to whether or not they require printing plates and printing ink, that is, whether they are printing or nonprinting processes.

In the following descriptions only the widely used and generally available processes are considered. It should, however, be pointed out that there are a number of other processes that produce excellent results in specific "requirement situations"; these processes range from stencil reproduction and silk screen to gravure and collotype. These and others of the same category (not widely used in cartography) are not considered here, but the interested student can find abundant information about them in the graphic arts literature.

Nonprinting processes are normally used when a relatively small number of copies is needed, since the unit cost does not decrease rapidly with increased numbers of copies. The cost of producing a second or third copy is as much or nearly as much as producing the first. The printing processes, on the other hand, provide low-cost, single or multicolor copies at a unit cost that decreases with the number of copies run. The initial preparation cost is high compared to the nonprinting methods. Some printing processes are, therefore, not appropriate for very small runs.

To decide whether a printing or a nonprinting method will produce copies most economically, we must consider the size of the map and the number of copies required. When the size is small and the number of copies required is large, the printing process tends to be more economical. Nonprinting methods tend to be more economical when the number of copies desired is small and the size is large.

The widely used nonprinting processes are: (1) direct contact positives or diazo, for example,

Ozalid; (2) xerography (*Xerox*), (3) photography, and (4) various others that fall into the category of proofing methods. This is an unsystematic listing in several respects; there are many similarities and differences among the processes that would enable one to group them differently. On the other hand, these are the categories usually used in "the trade," so it is convenient not to deviate. Today, the most widely used printing method by far is lithography, although letterpress is still employed to some extent. Copper engraving is no longer practiced.

A source of considerable confusion for the beginner is the terminology employed in the printing and duplicating businesses. Over the years, cartographers have adopted some of this terminology to describe materials that they furnish for duplication. Because of the kinds of materials currently being prepared by cartographers, many of the terms that have been used in the past no longer seem appropriate. For example, the word copy can mean either text material (words), artwork (drawings), or both. It also can be interpreted to mean one sheet of material, and it also seems to suggest that the material is in positive form. Referring to a duplicate of the copy as a copy or to duplicates as copies only adds to the confusion.

We will refer here to all map materials furnished for duplication as *artwork, art, original,* or *original drawing.* Negative materials will be referred to by the terms *negatives, negative scribecoat,* or *scribecoat,* and the term *negative* will be used to describe the film or other materials that can be used in a duplicating process that requires transmitted light to make copies. A *positive* is a print made from a negative, and a *film positive* is a print made on transparent or translucent film.

NONPRINTING PROCESSES

Nonprinting methods include some materials that the cartographer can process with rather inexpensive equipment. Other methods require certain kinds of graphic arts equipment such as process cameras, platemakers, and darkroom facilities. Ordinarily these processes are not used to produce great numbers of copies, and often only one color is possible. With some materials, however, color can be added to a copy for display, for teaching, or to facilitate research. Maps for which there is only an occasional need for a copy or maps that must be revised often can be reproduced most economically by these nonprinting methods.

In recent years, a number of new sensitized materials have been developed that can be used under ordinary lighting conditions. Some of these systems are invaluable for the cartographic process. Intermediate duplicating steps in the production of a map and the proofing process can be accomplished with a minimum of equipment. Use of the new materials and a trend toward having more duplicating apparatus in most cartographic establishments permits the mapmaker to be much more involved in the process of preparing the materials to produce a printing plate. It is not uncommon for the cartographer to furnish the printer with materials that have been proofed and are ready for platemaking.

Diazo

This process (e.g., *Ozalid*) relies on reaction of diazo compounds with other substances to produce colored dyes. Paper and other kinds of materials sensitized with diazo are exposed to an intense light while in contact with a positive opaque image on transparent or translucent material. Light (mainly ultraviolet) passing through the nonimage areas of the map decomposes the sensitized emulsion, while the opaque materials on the artwork protect the image areas by blocking the light. When the exposed material is developed in ammonia fumes, the unexposed diazo forms a colored dye.

It is not possible when making direct contact diazo prints to enlarge or reduce. Consequently, the artwork must be designed for "same-size" reproduction. Since the process depends on the translucence of the drawing surface, painting out imperfections with white is not possible because the paint is as opaque as the ink and would appear as a dark spot in the print. Creases on tracing paper or plastic and heavy erasures that affect the translucence are also frequently visible on the print. Prints are commonly obtained by feeding the original drawing into a machine against a roll of the diazo-sensitized paper; the exposure and dry developing take place rapidly within the machine. The drawing is returned and may be immediately reinserted for another exposure and copy. The image is not absolutely permanent; it fades especially rapidly when exposed to sunlight, and is not acceptable in circumstances where permanence is required.

The ultraviolet light source of diazo equipment furnishes a light of constant intensity. Exposure time is controlled by adjusting the speed with which the sensitized material passes through the machine. Original drawings on material of greater opacity require longer exposures in order to decompose the nonimage area so that it will appear colorless. Unless the drawing material is quite translucent and the image area is opaque, the resulting print is likely to suffer from lack of contrast.

Ordinary diazo paper has a poor drafting surface, so this process is not much used for obtaining copies of base maps on which other information is to be added. There are available, however, a great variety of sensitized diazo materials such as tracing cloth, cloth-backed paper, and various types of drafting films that do have adequate drafting surfaces. Cloth-backed paper, for example, is useful for the construction of wall maps. We can add with ink to the black image of the base map, and color can be applied by the use of ordinary artists' oils.*

Another useful diazo product is the intermediate paper or film that produces an actinically opaque (usually sepia) image. It is normally made wrong reading with the emulsion side up and can be used in place of the original artwork to provide more diazo copies. If a cartographer or researcher has available a base map that is to be used as a compilation base for a series of maps, but which should not be drawn on, the following procedure should be used. First, an intermediate may be prepared from the original; then data may be added to the back of the intermediate; and, finally, the corrected or modified intermediate can be used to produce subsequent direct contact positives.

Shading films, type stickup, and similar products on drawings produce best results in the direct contact process when they are placed on the back of the drawing to prevent "shadowing." To use stickup type in this fashion, the adhesive would have to be on the upper surface. An additional hazard with the standard forms of these

materials is present in the *Ozalid* diazo process because the drawing must be fed around a roller, and such materials tend to curl off the paper if it is heated and rolled. New types of preprinted sheets and stickup lettering with heat-resisting adhesive are available to obviate this difficulty.

Another way of circumventing this problem is to convert the drawing into a direct reading film positive. If the positive is prepared so the emulsion is on the bottom when it reads correctly, it can be fed onto the drum of an Ozalid machine with good results. It is not absolutely necessary to have the emulsion arranged as suggested, but the result will be more satisfactory.

Film Photograph Process

Although photography is an integral part of printing, it is also useful to cartographers as a nonprinting process. Copies can be had in the form of paper prints, film prints, or film negatives. Prints can be made on any of a variety of drafting films that provide good drafting surfaces as well as being dimensionally stable.*

Normally we consider photography to include those films and papers that are sensitized with silver-halide materials (Fig. 16.3). The base material for film can be acetate or some other plastic material such as polyester or polystyrene. Since acetate is not dimensionally stable, it is usually unsuitable for cartographic work. The base is coated with a light-sensitive emulsion composed of minute particles of silver salts suspended in a

---

*Cronaflex UC-4 Drafting Film, *E. I. Dupont de Nemours and Company, Photo Products Department;* Photact Contact Polyester Film, *Keuffel and Esser Company;* Kodagraph Ortho Matte Film, *Eastman Kodak Company.*

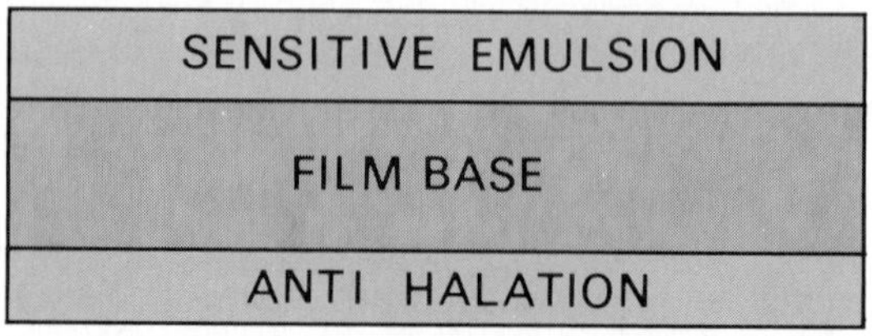

| SENSITIVE EMULSION |
| FILM BASE |
| ANTI HALATION |

FIGURE 16.3 *This exaggerated cross section of black and white film shows the position of the sensitized emulsion and the antihalation layer relative to the film base.*

---

*For details see Randall D. Sale, "A Technique for Producing Colored Wall Maps," The Professional Geographer, XIII (2), 19–21 (March 1961).

gelatin. The back of the film has an antihalation material that absorbs any light rays that penetrate the base during exposure. Also, this coating tends to minimize curling and compensates for distortion or dimensional changes caused by changes in the emulsion when it absorbs moisture.

Color Sensitivity. The color sensitivity of films is built in during manufacture and is indicated in film data books by wedge spectrograms (Fig. 16.4). The bands of blue, green, and red wavelengths are labeled. Yellow, an additive mixture of red and green, falls where those two colors merge. Wavelengths shorter than blue are called ultraviolet, and those longer than red are called infrared. The height and extent of the white area in each spectrogram indicates the sensitivity of a particular film to the various colors.

Blue-sensitive films record high negative densities for blue areas of an original. On a contact positive the tones are reversed, and the blue areas will be light while the greens, yellows, and reds will appear very dark. Since the film is blind to most of the spectrum, a bright yellow or red safelight can be used in the darkroom without fogging the film.

Orthochromatic emulsions are sensitive only to blue-green and yellow and are blind to red. This permits the use of a red safelight and also causes red to register the same as black on the film. Both red and black produce transparent areas on a negative; blue, green, and yellow cause the film to be opaque.

Panchromatic films are sensitive to all visible colors and render colored artwork in tones of gray. Because of the wide range of sensitivity, this film must be handled and processed in complete darkness.

Contrast. A characteristic curve for a particular film can be drawn by plotting the exposure time on a horizontal logarithmic scale and the corresponding densities produced on a vertical arithmetic scale (See Fig. 16.5). In the straight line portion of the curve there is a constant relationship between the resulting densities and the exposure lengths. The tangent of the angle between this part of the curve and the horizontal is a measure of the steepness of the curve and is referred to by the Greek letter gamma. The characteristic curve for high-contrast film has a steep straight line and is said to have high gamma. Small changes in exposure cause great increases in density, resulting in a clean break between image and nonimage areas. High-contrast films are not designed for copying continuous tone; however, some degree of success is possible if the developer is diluted to half strength and the developing time is increased by four or five times. Panchromatic films have low gamma and can faithfully copy continuous tones of gray or color to reproduce photographs or shaded drawings.

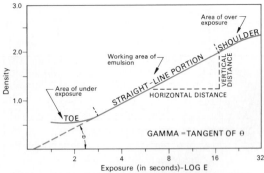

FIGURE 16.5 The straight-line portion of the characteristic curve shows a constant relationship between the resulting densities and the length of exposure. The tangent of the angle between the straight-line portion and the horizontal is a measure of the steepness of the curve and is referred to by the Greek letter gamma.

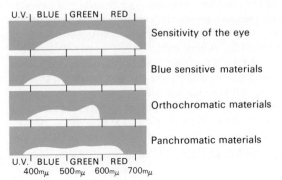

FIGURE 16.4 The sensitivity of each film to the various colors is indicated by the height and extent of the white area. (Yellow falls where red and green merge.)

Process Camera. Photography for cartographic purposes is usually accomplished with a camera that can make precise enlargements and reductions and can utilize large sheets of film. The camera consists of a movable copy frame that holds the artwork, a means for lighting the artwork, and a mount that permits positioning the lens at various distances from the film (Fig. 16.6). A vacuum is usually incorporated to hold the film flat, and either a vacuum or a glass cover is used to hold the artwork on the plane of the copy holder. Since both the copy frame and the lens are movable, the distance between the lens and the film and the lens and the copy holder can be varied to produce the desired reduction or enlargement.

Vacuum frame. Although a camera is necessary when a change in scale is desired, one-to-one reproduction is usually best accomplished by making contact copies. They can be made on either film or paper from negatives or positives or from translucent artwork. During exposure, it is necessary that the unexposed film be in tight contact with the piece being reproduced. This is usually accomplished in a vacuum frame that is equipped with a rubber pad, a glass lid, and a pump to remove air from inside the frame. As air is exhausted, atmospheric pressure holds the pieces of film snugly together and against the glass. The light source is usually regulated by a rheostat to vary the brightness, and the length of the exposure is controlled by an accurate timing device.

Films. Various types of film may be used in a camera or for contact work. For most cartographic work a film with high contrast is desirable, since most printing plates are exposed to negatives that have areas that are either transparent or completely opaque.

Often it is useful to be able to make duplicate copies of film, that is, make a negative from a negative or a positive from a positive. For this purpose a special film (duplicating film) has been developed. If a sheet of this film is developed with no exposure to light, it will be completely opaque. Exposure to white light, however, removes density, so that light passing through transparent areas of a negative will cause transparent areas on the duplicating film. To produce duplicate negatives or positives that have the same orientation (emulsion-image-reading relationship) as the original, the base (nonemulsion side) of the original should be placed in contact with the emulsion of the film. With a somewhat longer exposure, the same result can be achieved by placing the emulsion of the original in contact with the base of the duplicating film.

In special situations, an unusual film that can be exposed and then be "unexposed" at will is useful.* As with duplicating film, development without exposure results in solid black. If the film is exposed to yellow light before developing, this density will be removed. The density removed by the yellow light can be restored by reexposing the film to intense white light—and subsequently removed by yellow light.

Using this film, a same-size film positive of line copy on paper can be made without the use of a camera. Called *reflex* copying, it is accomplished by placing the film emulsion side down on the original and exposing to yellow light. Reflection from the paper removes the density in the nonimage areas, resulting in the positive copy on film.

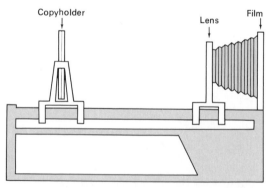

FIGURE 16.6 *The basic components of a camera are a holder, which keeps the film flat, and a lens and copyholder, which can be moved perpendicular to the plane of the film. The positions of the lens and copyholder control the amount of enlargement or reduction.*

*Three films that have this capability are Kodak Autopositive, DuPont Direct Positive, and Agfa-Gevaert Autoreversal.

When the artwork has continuous tone, panchromatic film, with less contrast, will record the variations in tones. A dark area on the artwork, which reflects little light, will appear more transparent than a lighter area that reflects more light. Making a print from the negative reverses the relationship and produces a replica of the original.

The continuous tone negative produced with panchromatic film is often not as useful in the cartographic process as is the halftone negative produced on orthochromatic film. Many of the sensitized materials used in the intermediate steps of map production and in proofing systems will not record tonal changes unless the tones have been changed to varying sized dots.

Corrections in the form of deletions are easily made on negatives by covering transparent areas with a water-soluble paint called *opaque,* or by covering the areas with red lithographer's tape. This is normally done on the base side of the film to prevent accidental damage to the emulsion. Additions are more difficult, since the rather brittle emulsion must be removed by cutting or scraping it away from the film base.

The photographic process can produce good quality copies at desired sizes on dimensionally stable materials; this makes it a useful process for compilation stages of map construction. Existing maps can be enlarged or reduced, and either negatives or positives can be used for tracing detail to a new map.

Xerography

Xerography (e.g., *Xerox*) occupies a unique position among photographic processes in that it is the only one that is completely dry; it uses no chemical solutions or chemical reactions. Producing the image depends on photoconductivity and surface electrification. A selenium-coated aluminum plate is given a positive charge that, upon exposure, disappears in those areas subjected to light. The remaining charged surface latent image is "developed" with a negatively charged powder that adheres to the image. Paper, which has been given a positive charge, is placed against the plate, receives the image by transfer of the powder and is fixed onto the paper with heat.

Because the *xerographic* process does not involve chemicals, time of immersion in solutions and subsequent drying are saved, and copies can be provided in just a few seconds. The image can be placed on a variety of kinds of paper or other surface materials, since the transfer is physical.

Xerography provides excellent copies at a reasonable cost. With the use of a lens arrangement, reductions or enlargements can be accomplished; however, most of the xerographic copying machines that are available produce copies the same size as the original.

In map construction, information from maps in books is sometimes difficult to trace because of difficulty in handling. Because of the thickness of books, copying from them with a process camera is not feasible. Xerography provides a handy solution by quickly providing copies (including transparencies) that can be copied by a camera or used flat on a drawing or light table.

Proofing Materials

Recently, a number of materials have been developed that are very useful in the cartographic process, especially for proofing. Since many of the materials are sensitive only to the violet and ultraviolet light, they can be handled under normal daylight conditions or under bright yellow light. A cartographer who does not have darkroom facilities can make full color proofs of maps and can even produce contact negatives and positives. The following is a brief description of some available materials; their specific uses will be discussed in Chapter 17.

3M Color-Key Proofing Film* is very thin clear acetate that has been coated with a colored, sensitized emulsion. A variety of colors are available, including the regular process colors: yellow, cyan, and magenta. The material can be used under daylight conditions, and the only equipment needed is a pressure or vacuum frame and a source of light, rich in ultraviolet.

A negative of a map or of one flap of a map is placed in contact with the proofing film (with the emulsion of the proofing film away from the source of light), and an exposure is made of a minute or so, depending on the intensity of the light. A special developer is then rubbed gently over the emulsion until the image is developed.

**This film is available from distributors for the Minnesota Mining and Manufacturing Company.*

Rinsing under a water tap and drying completes the operation, and the product is a positive print in the particular chosen color.

A series of positives in different colors, of separate flaps of a map, can be prepared and then superimposed to form a full color proof of the map. Tint screens and halftones are faithfully reproduced, and colors combine in the same manner as on a printing press.

The orange *3M Color-Key* film is actinically opaque, which means that it can be used to reverse negatives or positives. Consequently, one can convert translucent artwork to negatives without darkroom facilities. These, in turn, can be used with any of the other colors to produce a proof of a map.

3M Color-Key proofing sheets are also available for exposure with film positives. A special developer is used and a bleaching solution is applied after the film has been developed.

*General Color-Guide,** like the *3M* proofing system, is a negative-acting, ultraviolet sensitive film. It differs, however, in some important respects that are of special significance to cartographers. The film carrying the colored emulsion can easily be punched, whereas prepunched strips of some other material must be taped to the *3M* product for pin registration. *General Color-Guide* is developed by immersing it in a special salt bath and rinsing with warm water instead of by rubbing away the unexposed emulsion by hand. There is no shelf-life limitation and, if sheets are accidently exposed, they may be returned to the box and reused in 72 hr.

Colored bichromate emulsions are available under several trade names,† and can be applied to special plastics by whirling or by hand application (Fig. 16.7). Again, several colors are available; however, any desired hue can easily be obtained by mixing proper amounts of the primary colors.

Smoother application of color is possible

FIGURE 16.7 *Kwik-Proof emulsion is applied to the plastic sheet by using rather fast, gentle strokes with the applicator.*

using a whirler, but with practice we can apply the sensitizer evenly over fairly large areas by hand. A small amount of the color is poured onto a plastic sheet that has been taped to a flat smooth surface. A small block applicator covered with a soft pad is then used with soft even strokes to spread the color evenly. Exposure is made, again with light in the ultraviolet end of the spectrum. Developing consists merely of rinsing under a water tap, and then a thorough drying. Once completely dry, another color can be added over the first image and the process repeated. As many colors as are desired can be applied and the result is comparable to a map that has been printed in several colors by separate press runs.

Dylux 608 Registration Master Film is a DuPont product available on paper or dimensionally stable 0.007-in. Cronar polyester film. This produces an image immediately upon exposure to ultraviolet light through a negative or positive. The image requires no developing and self fixes under normal room lighting conditions. A pale yellow

---

*A product of General Photo Products Company Inc., 10 Paterson Avenue, Newton, NJ 07860. The material is available in various sheet sizes and in 42-in. rolls either 50 or 100 ft in length.

†Brite-Line, Camden Products Company, Box 2233 Gardner Station, St. Louis, MO; Kwik-Proof and Watercote, Direct Reproduction Corporation, 835 Union St., Brooklyn, NY 11215.

background disappears within an hour, but the process can be speeded by exposing the film to pulsed xenon or carbon arc lamps that are filtered with an ultraviolet blocking material.

The light source for the initial exposure must be exclusively ultraviolet to produce the maximum contrast. Best results are obtained with black light fluorescent lamps that have suitable phosphors, yet screen out nearly all visible light. The amount of density of the image is directly related to the length of exposure. Separation negatives for a map to be printed in more than one color can be given different exposures to produce images in different tones.

Since ultraviolet light forms an image and visible light deactivates the emulsion, positive images can be made from positive film. With the film in contact with the positive, an exposure is made with a bright light, which removes the sensitivity in the nonimage areas. After removing the positive, the sheet is exposed to ultraviolet to produce the image.

The *Cromalin** color proofing system provides proofs on single sheets in a wide range of colors. By successively laminating colorless film to a base sheet, exposing through a film positive, and applying color toners, a full color proof is produced.† The *Cromalin* film consists of a light-sensitive layer between two cover protective sheets. During lamination, the lower cover is discarded and the sensitized layer is made to adhere to a base sheet. After exposure to an ultraviolet source, the top cover sheet is stripped away, leaving the emulsion exposed. Finely powdered toner applied at this time clings to the sticky emulsion of the image area. After the toner is added, the base is again run through the laminator to sensitize it for the next color. The process is repeated until all the colors are in place.

Kodagraph Wash-Off Contact Film, produced by Kodak,‡ provides the cartographer who lacks darkroom facilities with a capability for contacting film positives or negatives. An added advantage is that the material has a good drafting surface enabling additions and corrections.

*A product of E. I. Dupont de Nemours and Company.
†Negative acting film is now in production. Modifications to the laminator are necessary to use this film.
‡Eastman Kodak Company, Rochester, NY 14650.

Again, only an ultraviolet source and a pressure or vacuum frame plus a flat tray for the developer are necessary. After a very short exposure, the film is developed for 1 min and then rinsed under warm water until the unwanted emulsion has been washed away.

The film also has the ability to accept bichromated emulsions, so colors can easily be added to a film positive of a base map.

PRINTING PROCESSES

Today most maps are printed by lithography, although if they are included in books, they might be done by letterpress. Until a decade or two ago letterpress was used for most books, partly because type was usually cast in metal and merely had to be assembled in wooden frames to be printed. With greater use of photographic means for producing type, lithography has replaced letterpress for most kinds of printing. Under ideal conditions each method can produce excellent results; however, under conditions less than ideal, the quality can vary to a degree that is of concern to the cartographer.

Steps in the Printing Process

The basic operations in the printing process, from the time the printer receives the artwork until the printed maps are delivered, are much the same, whether the process is letterpress or lithography, but the procedures and potentialities in each phase are different for the two kinds of printing. For a normal piece of art that involves no complications, the process consists mainly of the following operations.

1. Photographing the original drawing.
2. Processing the film.
3. Making the plate.
4. Presswork.

Photographing the original drawing or copy is an exacting process that requires the use of the copy camera. Relatively slow film is used in order to give the photographer greater control over the quality of the negative. For the printing processes (lithography and letterpress), the resulting negative must be composed of either opaque areas or transparent areas, and nothing intermediate. Grays are not permitted on the negative, as they are, of course, in ordinary photography. The printing plate, to be made subsequently from the negative, must be entirely divided into two kinds of surfaces, one that takes ink and one that does not.*

Any reduction that is to be made is done at the photographic stage. The photographer can adjust the camera precisely, either by calculating the ratio of reduction or by actually measuring the image in the camera before the exposure.

Since orthochromatic film is normally used, both red and black marks on the artwork will appear as clear, open areas on the negative, but light blue will not appear at all on the negative. For this reason, the cartographer should, as far as possible, draw any guidelines with a light blue pencil. Filters may be used in special circumstances to "drop" certain colors.

Another process, scribing, which will be discussed in Chapter 17, permits the cartographer to prepare the map in negative form and thereby eliminate the photographic step in the printing process. Lines and symbols are hand engraved in a rather soft coating on a sheet of clear plastic. Where the coating is removed, light can be transmitted; however, the remaining coating is opaque to the wavelengths of light by which light-sensitive emulsions are affected. The scribed negative can be used in the same manner as an ordinary film negative.

Processing the Negative

Processing the negative is one of the more important steps in the cartographic technique, since it is in this stage that some results can be obtained better than in the drafting stage. No matter what processing or modification is to take place, however, the negative must be brought to perfection by removing all "pinholes" and blemishes in the emulsion so that the image is left sharp and clear. This is done by placing the negative on a light table and opaquing with paint or scraping clean those spots and areas requiring repair or change (Fig. 16.8).

Depending on the circumstances, it is sometimes possible at this stage to add or subtract names and lines. If an important piece of informa-

*Experiments are being conducted in producing lithographic plates directly from continuous tone negatives. Short press runs and a lack of consistency from one plate to another are two problems being encountered.

FIGURE 16.8 Before being stripped into position for platemaking, pinholes or unwanted words or symbols are removed by applying a water-soluble opaque.

tion has been omitted or a word misspelled, a new piece of emulsion or emulsionlike material may be "stripped" into place on the negative. If material has been omitted, it can sometimes be added by "engraving" it in the emulsion. The emulsion, however, is quite brittle and tends to chip off, often resulting in ragged lines.

Screening of the negative involves interposing thin transparent sheets containing patterns of lines or dots (called screens) between the negative and the sensitized material. This accomplishes the same end result on the printing plate as applying *Zip-A-Tone* or similar shading film to, or drafting such shading on, the original drawing

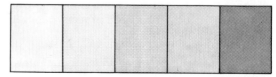

FIGURE 16.9 Preprinted dot patterns can be applied to the artwork to give an illusion of gray; however, the effect can be destroyed if too much reduction is necessary. When tint screens are introduced at the platemaking stage, the amount of reduction need not be a consideration. The preprinted patterns shown above were labeled as 10, 20, 30, 40, and 50%. It is evident that variations occur using these types of materials.

(Fig. 16.9). Better results are obtained with extra-fine patterns of lines or dots by applying them at this stage, after photography, than by making them "stand up" through the photographic process. Furthermore, opaquing can be done first and screening second; this is desirable because the negative is more difficult to opaque if the emulsion already contains a fine pattern of lines or dots in addition to other data.

After a negative has been processed, it is marked for position on the plate and in general made ready for platemaking.

Line and Halftone

All the material to be printed by conventional methods belongs to either of two classes, line or halftone. The distinguishing characteristic that places a drawing in one or the other class is whether or not it contains any color shading or gray tones (continuous tone). If it does, it must be processed using a glass or contact halftone screen.

The number of opaque lines per inch on glass halftone screens varies from less than 100 to more than 200. All other things being equal, the closer the lines, the smaller and closer together the dots will be. The closer together they are, the more difficult it is for the eye to see them individually, and the smoother and more natural the result will appear.

In recent years the contact halftone screen, which is cheaper and easier to use, has tended to replace the glass halftone screen. It produces the varying sized dots* through modulation of light by the optical action of a vignetted dot pattern of the screen acting on the film emulsion. The vignetted dots are produced by a dye that is thicker near the centers of the dots, and the size of the halftone dot depends on the amount of light that is able to pass through the dye.

The size of the printing dots relative to the white spaces between them is dependent on the darkness or lightness of the tones on the copy. It should be remembered, however, that unless special additional processing takes place, no part of a halftone will be without dots. All lines and lettering will therefore have fuzzy edges. Pure whites on the

*Contact halftone screens are available in patterns other than dots. Some, such as parallel lines and mezotints, are useful for cartographic presentations.

original drawing will, in ordinary halftoning, be printed with a covering of very small dots and therefore a light tone, whereas solid areas will reproduce with small white spaces instead of being completely solid. These effects can be removed by opaquing or scraping on the halftone negative, but this is difficult if the areas involved are complex.

Sometimes, for economic reasons, the cartographer is limited to line reproduction. However, in lieu of continuous tone, the cartographer may be able to achieve the desired effect by means of the following.

1. Shading either uniformly or for continuous-tone effect by hatching or stippling with pen and ink, "spatter painting" with an air brush, or shading on a coarse, rough surface.
2. Using preprinted symbols on the drawing.
3. Screening portions of the negative by covering them with shading film.
4. Using *Ross* or *Coquille* board for continuous-tone effect.

Presses

No advantage would be gained by attempting to describe various types of printing presses. There is, however, a basic difference between letterpress and lithographic printing that is of concern to the cartographer. Most lithographic presses incorporate an offset arrangement whereby the impression from the plate is transferred to another cylinder covered with a rubber mat, called a blanket, and the image is then transferred from the blanket to the paper (Fig. 16.10). Actually, the offset arrangement is so standard that the terms offset and lithography have become almost synonymous. In letterpress printing, on the other hand, the paper receives the ink directly from the metal relief surface of the printing plate.

Relief plates used on an offset press is not actually a form of lithography, since the image area is in high enough relief that a dampening solution is not required; this type of printing is often called "letterset," "dry offset," or low relief printing. The use of a wider range of inks is possible; however, there is a tendency for more image spread and less contrast than with traditional lithographic plates.

The soft rubber blanket that deposits the ink image on the paper in the offset process does not

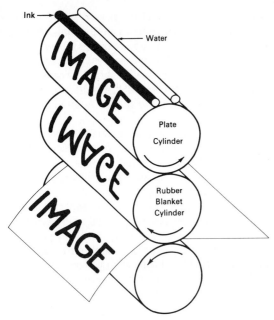

Ink →

← Water

Plate
Cylinder

Rubber
Blanket
Cylinder

FIGURE 16.10 Since the image is passed from the plate to a rubber blanket and from the blanket to the paper, the plate must be right-reading.

crush the paper fibers, and fine lines and dots can be faithfully reproduced, even on soft, absorbent paper. On the other hand, in the letterpress process, because pressure is necessary to cause the ink to adhere properly to the paper, the raised metal printing surface tends to squash the paper fiber. When soft unfinished paper is used in letterpress, fine lines and dots of halftones or screened areas tend to be enlarged and distorted. Very coarse halftone screens of 50 to 65 lines per inch must be used on newsprint paper, whereas 120 to 133 lines per inch can be used on finished paper.

If the cartographer can determine in advance the kind of printing and the grade of paper to be used, the map can be designed for that particular combination. If the cartographer is preparing a map for a book that is to be produced by letterpress on unfinished stock, the cartographer should not choose fine patterns or expect halftone or fine screens to turn out well.

Platemaking. Photolithography and photoengraving, although alike in principle, are each accomplished by ways whose differences are of sig-

nificance for the cartographer. Both the letterpress and lithographic plates must, of course, be divided into image and nonimage areas. On lithographic plates the image areas hold the greasy ink, while the nonimage areas attract water and repel the ink. In modern offset lithography the image on the plate must be *right reading,* whereas in early lithography a *wrong reading* or mirror image was necessary to produce the proper image on the paper.

Surface Plates. These plates have the image on the same plane as the surface of the plate; they may or may not be presensitized. Plates that have no sensitized coatings are known as direct-image plates or masters. The base material of the plate is often paper, but it can be plasticized paper, acetate, aluminum, or aluminum foil. Images can be placed on the plate by any of the following methods.

1. By use of a typewriter with a special ribbon or drawn by hand with pencil, ink, lithographic crayon, or ball-point pen. These materials must be slightly greasy to produce images that attract printing ink.
2. By printing using either letterpress or lithography.
3. By sensitizing plates with wipe-on solutions such as diazo. When coated, they are handled like presensitized plates and exposed through a negative.
4. By electrostatic methods.

Presensitized plates are surface coated with a light-sensitive emulsion during manufacture. Most sensitizers are of the nonphotographic variety; that is, the emulsions are slow and can be handled under normal room lighting conditions or under bright yellow lights.* Until quite recently most surface plates were ''manufactured'' in the printing shop. A sheet of metal, usually zinc, was first treated so it would accept and hold a thin film of water. Next it was coated with bichromated albumin and dried. After exposure through a negative, a developing ink was applied that produced the image. The unhardened, nonimage area was then washed away, and the plate was ready for the press.

Some plates are coated with silver halides and can be used in a camera to produce plates directly from artwork.

Today the most popular surface plate has a diazo emulsion applied by the manufacturer. To prepare this plate for presswork, it is only necessary to expose it to an ultraviolet light source through a negative and then process it for a few minutes to develop the image.

Overexposing the emulsion of a surface plate does not affect its printing qualities: therefore, a single plate can be successively exposed to several negatives. The process, called double burning or double printing, permits two or more separate flaps to be combined on one plate. Line and halftone or line and screening can easily be combined. Registry marks are not visible on the plate until it has been exposed separately to each negative and the composite image developed. To make certain that each negative is registered properly, a mechanical system, usually incorporating the use of registry pins, is used.

After proper exposures have been made, the plate is developed, placed on the press, and is ready to print. At this stage, most corrections on the plate are impractical if not impossible and are generally limited to deletions. If errors are found on the relatively inexpensive plate, it is usually discarded, the map artwork revised, a new negative made, and a new printing plate prepared.

Deep-etch plates are a form of lithographic printing in which the printing area is very slightly recessed into the plate. On a surface plate, the image tends to wear away with long press runs. Dots in a halftone or in a screened area become ragged and smaller. To compensate for this and to provide a generally sharper image, the deep-etch plate was developed.

A film positive instead of a negative is used to expose the plate. Areas under the image are protected from the light, whereas the emulsion on the rest of the plate is exposed and hardens. When the plate is developed, the emulsion in the image area is washed off, leaving the bare metal, which is subsequently etched to a depth of about 0.0002 to 0.0005 in.

Since positives and not negatives are used to expose the plate, successive burnings are not possible. As an alternative, one or more positives can be sandwiched together and the exposure of all made simultaneously.

Multimetal plates are multilevel in that the image area is either slightly raised or slightly etched below the surrounding surface. A metal that attracts ink is used for the printing surface, while the other level is a different metal that "likes" water. The upper layer of metal is electroplated to the base and is only 0.0002 to 0.0003 in. thick. The metal of the raised surface is normally one that wears better than usual plate materials. Multimetal plates have the advantages of more faithful reproduction and much longer press runs—as many as 2 million or more impressions.

An example of a multimetal plate is one with a steel or aluminum base coated with copper. The upper surface is coated with sensitized gum arabic and exposed to a negative. Where light passes through the negative (image area), the emulsion is hardened enough to resist the action of a developing solution. Where the emulsion has not been hardened by light through the negative, it is removed by the developer, exposing the bare copper. Next, the application of an etching solution removes the exposed copper from the nonimage area, while the hardened emulsion or stencil protects the image. Finally, the hardened emulsion is dissolved, leaving bare copper, which has great affinity for greasy ink, in the image area. The steel or aluminum of the nonimage portion of the plate readily accepts water which, in turn, repels ink.

Letterpress Plates. After exposure to a negative or to successive exposures to negatives, letterpress plates must be chemically etched to produce the raised or relief printing surface. Light transmitted through the transparent portion of the negative hardens the emulsion. This hardened emulsion protects the printing surfaces when the plate is exposed to an acid bath. The acid eats away the nonprinting areas, leaving a relief printing image on the surface of the plate.

The etching process is involved and requires several steps. Successive etchings (usually about four) are needed to reach the required depth for printing.* At this stage, a proof of a halftone from the plate would appear dull and lifeless. There would not be enough contrast, because the dots in the highlight areas tend to be too large, while

---

*With the use of zinc as the metal for the plate, the required depth can be reached with one etching.

those in the dark areas are likely to be too small. The engraver, therefore, must reduce chemically the size of the highlight dots and enlarge by polishing the dots in the darker areas. The engraver's work at this stage governs the quality of the halftone and, of course, adds to the cost of the plate.

In letterpress printing, elements of the composition (type and mounted engravings) are locked into a *form,* which is mounted in the press. Separate elements can be replaced quite easily, but errors on an engraving usually necessitate making a new plate. Because the letterpress plate is so expensive to produce, it is especially important for the cartographer to submit accurate artwork to the engraver.

Color Reproduction

The reproduction of maps in color does not differ from black reproduction except that different colored inks are used. Each separate ink requires, of course, a separate printing plate, and thus a complete duplication of the steps in the whole process. Thus, generally, the costs of color reproduction are many times that of the single color (usually black) reproduction. There are, however, two basically different color reproduction processes, and it is unwise to generalize further about relative costs. The two processes are called "flat color" and "process color." The major difference between them is that the color artwork for process color is prepared using color media in all hues on a single drawing, whereas that for flat color is prepared in black and white and requires at least one flap for each ink.

Flat color is the method most often used for maps and involves a straightforward procedure that varies little from the procedure described previously. For the flat color procedure the map is planned for a certain number of colored printing inks, and at least one separate drawing is prepared for each ink. Of course, many combinations of line and halftone effects are possible.

Process color or, more properly, four-color process, is the name applied to an essentially different procedure. This method is based on the fact that almost all color combinations can be obtained by varying mixtures of magenta, yellow, cyan, and black. The black flap usually contains the border (if any), lettering, and graticule outlines. On a second flap all color work is done by painting, airbrush, and similar methods.

The varying colors on the second flap are separated through the use of colored filters into negatives of the three component additive primary colors (Fig. C1.6). The three (positive) printing plates are halftones of varying amounts of the corresponding subtractive primaries (Fig. C1.5) With proper filters, four exposures are made on separate pieces of film. A halftone screen is rotated to the proper angle for each exposure, resulting in separation negatives that record colors of the original as black and white densities.

A filter transmits light of its color and absorbs the light rays of most other colors. For example, when white light strikes a blue filter, the blue wavelengths are allowed to pass through to expose the film, and the other wavelengths are blocked. The resulting negative is most dense in the blue areas and least dense in the green and red (yellow light) areas. The positive printing plate made from this negative will print most ink in the green and red areas and, therefore, it is inked with yellow. A green filter transmits green light and blocks blue and red (magenta light). Since most ink will be deposited in the blue and red areas, magenta ink is used. Finally, a red filter passes red light and absorbs green and blue (cyan light). The resulting plate deposits cyan ink in the areas of green and blue.

Yellow ink mixed with magenta results in reds, yellow combined with cyan forms the greens, and magenta with cyan produces the blues (Fig. C1.5). The halftone negative for the black plate can be made by using three partial exposures with each of the three filters. The black plate includes the line drawing and the halftone that is used as a toner to increase the shadow density and improve the overall contrast. When printed together again, the halftone dots and transparent inks perform their subtractive assignments, merge, and recreate the colors of the original drawing.

SELECTED REFERENCES

Basic Photography for the Graphic Arts, Eastman Kodak Company, Rochester, NY, 1963.

Carmichael, L. D., "The Relief of Map Making," *The Cartographic Journal, 6,* 18–20 (1969).

Halftone Methods for the Graphic Arts, Eastman Kodak Company, Rochester, NY, 1976.

Keates, J. S., *Cartographic Design and Production,* Longman Group Ltd., London, 1973.

Mertle, J. S., and Gordon L. Monsen, *Photomechanics and Printing,* Mertle Publishing Company, Chicago, 1957.

Moore, Lionel C., *Cartographic Scribing Materials, Instruments and Techniques,* 2nd ed., Technical Publication No. 3, American Congress on Surveying and Mapping, Cartography Division, Washington, D.C., 1968.

Ovington, J. J., "An Outline of Map Reproduction," *Cartography, 4,* 150–155, (1962).

Photography for the Printer, Eastman Kodak Company, Rochester, NY, 1962.

Pocket Pal: A Graphic Arts Production Handbook, International Paper Company, New York, 1974.

Schlemmer, Richard M., *Handbook of Advertising Art Production,* Prentice-Hall, Inc., Englewood Cliffs, NJ, 1966.

The Contact Screen Story, E. I. Du Pont de Nemours and Company, Inc., Wilmington, DE, 1972.

MAP CONSTRUCTION

Whether maps are printed or are duplicated by some other method, the quality of the copies is closely related to the nature of the original artwork. With the possible exception of reduction by photography, duplicating processes are not able to improve much the appearance of lines, type, or other map symbols. Flaws or irregularities in our work will usually be apparent on copies. The preparation of proper artwork involves choosing the most efficient and economical routine and using the materials and instruments best suited for a particular job.

The duplicating industries have made great strides during the past few years in perfecting new products and techniques. Some of these have solved problems that cartographers have lived with for years. New sensitized materials are available for intermediate steps in the cartographic process that permit more efficient map construction and facilitate the preparation of trial runs or proofing. Compiling can be accomplished in a greater variety of ways, and the compilations can easily be transferred from one surface to another.

The cartographer is no longer forced to rely on the printer to process artwork. Dimensionally stable materials are available that can be presented to the printer ready for platemaking. If desired, all the artwork necessary to produce multicolor maps can be produced by the cartographer without assistance from the printer.

To enable the cartographer to make use of new techniques and plan work properly, he or she must, of course, become acquainted with the new materials and processes. No one process is the best for all maps, and the production plan should be tailored for each situation. In a book of this size it is not possible to incorporate all the information available in the form of manuals, books, and catalogs. Also, because there are many combinations of materials and techniques that can be used, the treatment of the subject presents a problem in organization. This presentation is built on the background of the chapters on design, color, typography, and reproduction, and with the exercise of imagination, the student should be able to devise workable plans for the construction of maps that will result in well-designed maps that can be reproduced efficiently.

CHAPTER 17
MAP CONSTRUCTION

During the discussion of positive artwork, a variety of materials and equipment are mentioned along with information about possible uses and limitations. The list is not exhaustive but, on the other hand, those mentioned have been used and tested. The same holds true for the subsequent section, which deals with negative artwork. When reading both these parts, we should always keep in mind that there is a good deal of overlap, since negative and positive materials are often used in combination for one map. Finally, the section on complex artwork and proofing should provide the student with a basis for devising unique plans for solving complex construction and reproduction problems.

The rapid advances being made in computer-assisted cartography extend into map construction areas as well as data processing. In many respects the only significant difference is that mechanical or photographic plotters are doing the scribing or drawing instead of human hands. Computer output on microfilm can produce good quality negatives directly useable for printing color maps (Fig. C1.12).* Base maps, dot maps, terrain shading, inclined traces, isarithms, and the like can be drawn by machine, and the variety and sophistication of hard copy output, either directly by machine or indirectly by photographing cathode ray tube displays, will no doubt steadily increase. Most of the principles and processes dealt with in this chapter are applicable regardless of whether the map construction is by people or machines.

Map Artwork

Maps can be prepared in positive or negative form or in combinations of both. Preparing positive artwork involves placing symbols, lettering, and linework on a relatively white or translucent surface with ink or with a variety of preprinted materials. During the printing process, negatives of the artwork are prepared for use in making the printing plate, and the copies produced are then replicas of the original map.

Negative materials are prepared in an opposite manner, that is, producing the lines and symbols by cutting or scribing them from a coating that is actinically opaque.* The prepared negative can be used in the same manner as a film negative in duplicating processes.

Before beginning the map, the cartographer should decide which form of artwork would be most suitable, and select the one that will produce adequate copies at an economical cost. The cartographer must consider things such as the type of reproduction to be used, size of the copies, amount of detail, kinds of symbols, registration procedures, and the quality desired.

Whether artwork is prepared in negative or positive form or with a combination of the two, it often consists of more than one sheet of material. Each separate sheet is called a flap (Fig. 17.1). If a map is to be printed in one color, only one sheet of material may be necessary. More than one color or certain special effects such as areas of smooth gray tones usually require separate flaps that are combined at some stage in the reproduction process to produce the map on a single sheet.

POSITIVE ARTWORK

If artwork is prepared on translucent material and at the proposed reproduction size, negatives can be provided by contact methods in a vacuum frame. When opaque artwork is prepared, a process camera is usually required to produce the negatives. Use of a camera permits reduction; consequently, artwork can be prepared somewhat larger than the proposed reproduction size. Reduction sharpens the image somewhat, and slight irregularities tend to disappear. At the larger size, symbols are easier to construct and more detail can be placed on the map. There is a temptation, however, to add so much detail that it is either lost or difficult to read at the reduced size. With a great amount of reduction, it is not easy to anticipate how certain symbols or patterns will appear. For these reasons, artwork is usually prepared no more than twice the reproduction size. If, in unusual situations, it is necessary to make the map several times larger, a sample area should be

---

*M. A. Meyer, F. R. Broome, and R. H. Schweitzer, Jr., "Color Statistical Mapping by the U.S. Bureau of the Census," The American Cartographer, 2, 100–117 (1975).

*The negative need not be visually opaque, but opaque only to the wavelengths of light that affect light sensitive emulsions.

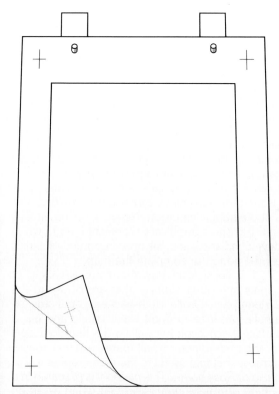

FIGURE 17.1 *Each separate sheet of material used in the preparation of artwork for a map can be referred to as a flap. The ¼-in. pins hold the flaps in register and the small crosses are used for positioning the printing plates on the press for successive colors.*

constructed according to the proposed design. A reduction of this sample will indicate whether or not the proposed design is acceptable.

Drawing Surfaces

The sheet materials to be used in the construction of the map may sometimes be selected on the basis of a particular attribute, but more often they are chosen on the basis of some overall qualities. Economy is sometimes a consideration, but the cost of materials is usually an insignificant part of the total cost of producing a map. The following brief listing includes the more important qualities for cartographic use.

1. *Dimensional Stability.* This refers to the degree of shrinkage or expansion in a material with changes in temperature and humidity. Probably the most stable materials used in cartography today are flexible polyester base films. Even these materials will undergo a relatively small change in dimension as temperature and humidity vary. Thermal and hygroscopic increases or decreases are additive in their effect on dimensional change. An increase in one and a decrease in the other, however, will have a canceling effect.

2. *Ink adherence.* A good drafting surface not only "holds on to" the ink well, but it accepts enough ink to provide an opaque image. Some special inks have been developed for special surfaces; however, most of them tend to clog many kinds of pens. The material that will accept standard drafting inks is usually most desirable.

3. *Translucence.* This refers to the ease with which it is possible to see through the material, and the ability of the material to transmit light for the purpose of exposing sensitized materials. Translucence is of special concern in cartographic drafting, not only because a considerable amount of tracing is usually done, but also because much drafting for reproduction is done on separate flaps, and it is desirable to be able to see what is where on other flaps when they are all in register. Translucent material makes it possible to produce contact negatives and enables the cartographer to make proofs or produce other materials for the cartographic process without the use of a camera.

4. *Erasing Quality.* The ability to remove ink without damaging the drafting surface is necessary both for making corrections and for revising artwork.

5. *Strength.* Some drawings must withstand repeated rolling and unrolling or even folding, and some receive wear from certain duplicating processes. For such drawings a durable material is required.

6. *Reaction to Wetting.* Many maps call for painting with various kinds of paints and inks. A material that curls excessively when wet is inappropriate for such a purpose.

There are many possible types of drawing surfaces, but only a few are useful for cartographic purposes. For maps that do not require great dimensional stability, a prepared tracing paper provides an excellent and economical drafting surface for ink.* Its translucence permits tracing even without the use of a light table, and good diazo copies can be made from it. It reacts to wetting, however, and it will curl if large areas are covered with ink or paint. Corrections can be made by gently scraping most of the ink away with a knife or razor blade and then rubbing gently with a medium-hard eraser. If we are careful and do not excessively loosen the fibers of the paper, further drafting can be done in the corrected area.

Tracing paper is not particularly stable and tends to shrink or expand with changes in temperature and humidity. For small drawings, a foot or so square, it can be used successfully, even if several flaps need to fit precisely. If, for some special reason, it is necessary to make separation drawings for large maps on tracing paper, we must take special precautions. Tracing paper does not change dimensions equally in both directions. The sheets, which should all be cut from the roll at one time, should have the grain of the paper running in the same direction. During the drafting, all the flaps should be kept at the same location. If some of the sheets are exposed to different temperature and humidity conditions, size changes will be unequal and registration will become a problem.

For preparing large maps where precise registry is required, a number of special plastic materials have been developed. Surfaces of these sheets have special coatings or have been roughened in some way to cause ink to adhere. Space does not permit a comparison of the various kinds; however, a material called *Cronaflex,* which has special and very desirable qualities, will be described.†

Cronaflex is available in rolls or in separate sheets and in varying thicknesses. Either side of the sheet provides an excellent drafting surface; however, lines produced with any given pen are slightly wider than when drawn on prepared trac-

ing paper. The surface can be cleaned of oil or grease with a soft cloth dampened with trichloroethylene. Such cleaning can be accomplished before drafting begins or at any time during production without affecting inked portions. Ink can be completely removed at any time by use of a cloth dampened with water without altering the drafting surface.

In addition to the ease of cleaning, correcting, and revising *Cronaflex,* its dimensional stability is a most important and useful quality. The material can be used in conjunction with very stable photographic films or scribe sheets and maintain precise registry. *Cronaflex* can also be provided with a photographic emulsion. Negatives of base maps can be photographically contacted and, when developed and dried, ink and preprinted materials can be added to complete the map.

For specialized jobs there are a variety of papers and other surfaces that are of use to the cartographer. *Ozalid* cloth-backed paper, for example, produces a good black image and has a surface that accepts ink well, and is therefore ideal for the preparation of wall maps. Another special purpose material useful to the cartographer is *Coquille* or *Ross* board. This material is prepared in a variety of rough surfaces so that when a carbon pencil or crayon is rubbed over it, the color remains only on the tops of the small bumps (Fig. 11.47). For this reason, varying shades of gray can be prepared with this material and can be reproduced without the necessity of being halftoned.

Drawing Pens

Although there are not as many kinds of pens as there are types of drawing surfaces, there are still a great number from which the cartographer may choose. Each pen has certain capabilities and limitations and, for an average job, the cartographer may use two or three different kinds. Common to all pens, however, is the need to keep them clean for proper operation. Ragged lines, gray instead of black lines, and lines of inconsistent width often result from dirty or clogged pens. We must clean our equipment even during use, since bits of lint or other debris are constantly being picked up from the drafting surface.

The standard ruling pen (Fig. 17.2) has probably been the most used of all drawing instru-

---

*Albanene Tracing Paper Number 107155 *can be purchased from Keuffel and Esser Company. A natural tracing paper is very thin tissue and is generally not suitable for cartographic work.*

†Cronflex UC-4 Drafting Film *is manufactured by E.I. Du Pont de Nemours and Company, Photo Products Department, Wilmington, DE.* Cartographic Drafting Film *developed by Keuffel and Esser Company has essentially the same qualities as* Cronaflex.

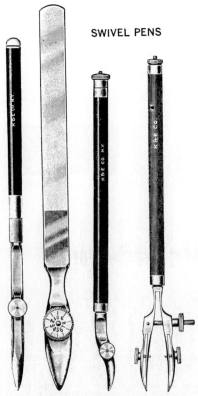

SWIVEL PENS

FIGURE 17.2 *Kinds of ruling pens. The standard pen on the left is most frequently used. (Courtesy Keuffel and Esser Company.)*

ments in the past. Although new pens have been developed that are easier to use and produce more precise results, we still find occasions when we must resort to the ruling pen. The spacing of the blades is adjustable with the small screw on the side, so that lines of different thickness may be made with the same pen. The pen is filled by placing ink with a dropper between the adjustable blades. Ink should not be left long in the pen because it will dry a bit and cause an unequal flow. To clean the pen, a cloth can be inserted between the blades to wipe them dry.

It is often difficult for the beginner to produce consistent widths with the ruling pen. Careful adjustment of the blades is necessary, and a trained eye is required to judge minute differences in line weights.

The *Pelican-Graphos* pen, which has come into wide use, can in most cases be used in lieu of the standard ruling pen (Fig. 17.3). A set of nibs, which have been machined to produce lines of specific widths, can quickly be interchanged on a fountain-type pen. With these nibs we no longer need estimate line widths, but must learn only to handle the pen in a consistent manner. Most other nibs available for the *Pelican-Graphos* pen are designed for free-hand work such as lettering or lines. Style "R" nibs are especially useful for producing round uniform dots.

Another pen that can produce excellent free-hand lines was designed for *Leroy* lettering templates (Fig. 17.4). The *Leroy* pen has a cylinder for a point, and ink is fed through a small hole in the cylinder. When used freehand in its special holder, it is useful for those lines that should maintain a constant width no matter which direction they follow. Varying pressure makes little difference in the width of the line.

Quill pens made of metal are needed to produce special kinds of lines and can be useful in drawing. A large variety is obtainable, and it is helpful to have a good selection on hand. Some are hard and stiff and make uniform lines; others are very flexible and are used for lines, such as rivers, that require a changing width on the drawing. A favorite is the one called a "crow quill," a relatively stiff pen, which requires a special holder. Quill pens of any type may be dipped in the ink bottle, but a better practice is to use the ink dropper to apply a drop to the underside of the pen. This procedure helps to produce a finer line and allows frequent cleaning without excessive waste of ink.

Compasses, which are used for drawing arcs or circles, usually have blades comparable to the standard ruling pen; however, special attachments also permit the use of *Pelican-Graphos* pens. Usually compasses are designed so a pen or pencil can be interchanged (Fig. 17.5). For circles or arcs with long radii, a special beam compass, which can be extended by the insertion of extra linkage, is necessary. The drop compass, on the other hand, is used for making small circles. The pen is loose on the pointed shaft; when the center has been located, the pen is dropped to the drafting surface and twirled. An alternative for constructing small circles is to use a *Leroy* pen inside the

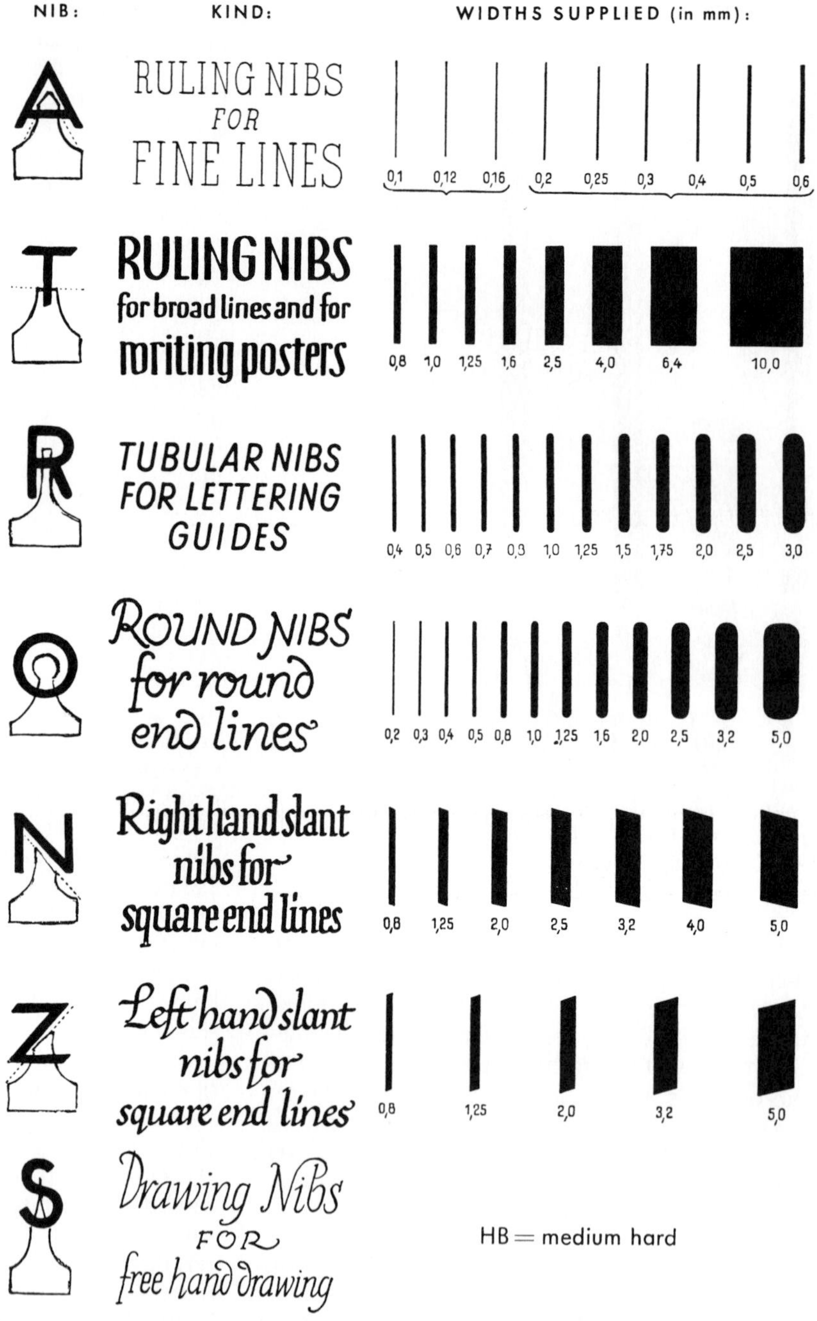

NIB:	KIND:	WIDTHS SUPPLIED (in mm):

RULING NIBS *FOR* FINE LINES

0,1 0,12 0,16 0,2 0,25 0,3 0,4 0,5 0,6

RULING NIBS for broad lines and for **writing posters**

0,8 1,0 1,25 1,6 2,5 4,0 6,4 10,0

TUBULAR NIBS FOR LETTERING GUIDES

0,4 0,5 0,6 0,7 0,9 1,0 1,25 1,5 1,75 2,0 2,5 3,0

ROUND NIBS for round end lines

0,2 0,3 0,4 0,5 0,8 1,0 1,25 1,6 2,0 2,5 3,2 5,0

Right hand slant nibs for square end lines

0,8 1,25 2,0 2,5 3,2 4,0 5,0

Left hand slant nibs for square end lines

0,8 1,25 2,0 3,2 5,0

Drawing Nibs FOR free hand drawing

HB = medium hard

FIGURE 17.3 *Illustration of the types of nibs and widths of lines that may be made by Pelican-Graphos nibs. All lines are made by individual nibs, that can be inserted in a single fountain-type pen. Special inserts are also available to provide the proper rate of flow from the reservoir, depending on the ink requirements of the nib. (Courtesy John Henschel and Company, New York.)*

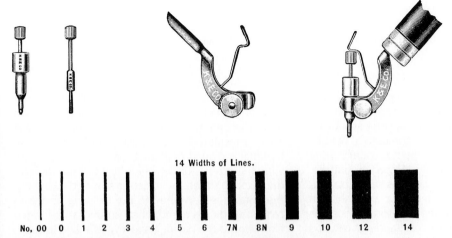

14 Widths of Lines.

No. 00   0   1   2   3   4   5   6   7N   8N   9   10   12   14

FIGURE 17.4 Leroy pens and penholder, and widths of lines made by various sizes of pens. (Courtesy Keuffel and Esser Company.)

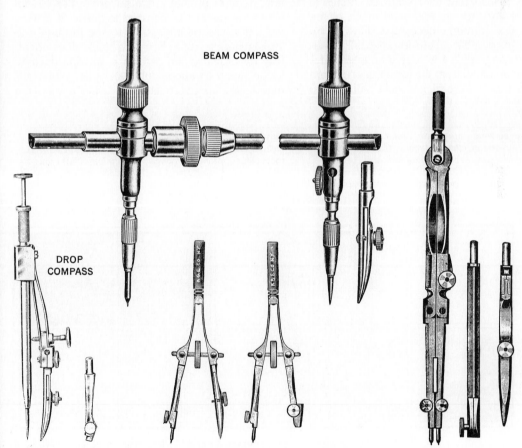

BEAM COMPASS

DROP COMPASS

FIGURE 17.5 Kinds of compasses. The beam of the beam compass may be several feet long. (Courtesy Keuffel and Esser Company.)

varying sized circles on any of several commercially available templates.

Positive Tints and Patterns

An indispensable part of many maps is the pattern of shading used to differentiate one area from another. In the past this was accomplished by drafting them laboriously, as in the case of parallel lines or dotted "stippling." Now it is usually done by applying a commercially prepared pattern. Continuous tone shades of gray, as might be prepared by shading with a pencil or a brush, can only be employed when the half-tone process is available. An appearance of shading can be created with *Coquille* or *Ross* board, but regular line work is relatively difficult on these surfaces.

Preprinted materials, such as lines and dots printed on transparent film with an adhesive backing, are easy to use and save much of the time that was formerly necessary (Fig. 17.6). There is a considerable variety commercially available from most graphic arts supply houses. Catalogs may be obtained from the local outlet or from the manufacturer.* In some instances patterns are

available in black, white, red, or other colors. The colors (except red) are not frequently used in cartography. The patterns are available on thin, transparent film coated with an adhesive, commonly wax, and protected by a translucent backing sheet. Recently, some manufacturers have discontinued use of the wax backing and have substituted heat-resisting adhesives. The material is placed over the area desired and is cut with a sharp needle or blade to fit. If cut without the backing sheet, the excess is stripped away and the pattern is burnished to the drawing. Extreme care must be exercised when cutting an adhesive film without the backing sheet over inked lines on a drawing. The stripping away of the excess will occasionally pull ink off the drawing.

For maps to be photographed for reproduction, patterns printed in translucent red are useful when the pattern to be used is dense; if it were black it would be impossible to see through it to know where to cut it. Red, like black, does not affect the ordinary film used in map reproduction. Large areas to appear ultimately as solid black may easily be constructed by using a solid red

---

*Para-Tone, Inc., 512 W. Burlington Avenue, La Grange, IL 60525, manufactures Zip-A-Tone; Craftint Manufacturing Company, 1615 Collamer Avenue, Cleveland, OH 44110, manufactures Craf-Tone; Chart-Pak, Inc., Leeds, MA 01053, manufactures Contak as well as an adhesive tape useful as a substitute for drafted lines; Artype, Inc.,

127 S. Northwest Highway, Barrington, IL 60010, manufactures dot and line screens with a regular progression of tonal values; Graphic Products Corporation, 3810 Industrial Avenue, Rolling Meadows, IL 60008, manufactures Formatt.

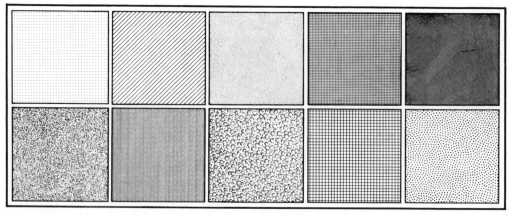

FIGURE 17.6 Examples of preprinted symbols, in this case Zip-A-Tone. Many of the maps and diagrams in this book have been prepared with the aid of these kinds of symbols.

film, especially on a paper that cannot be wetted. White patterns are useful for breaking up black areas or other patterns.

Knowledge of the relation of patterns to reduction and of the preparation of a graded series of values (darkness) requires considerable experimentation by the cartographer, as pointed out in Chapter 14.

Drafting the Map

Whether we make a map in positive or negative form, a worksheet of the compilation is prepared before any final drafting begins. The worksheet should be an accurate representation of everything to be shown on the map, and it is advantageous to prepare it at the scale planned for final drafting. Since it then can be merely traced, the cartographer can concentrate on drafting alone. When translucent material is used, tracing is easier and, if desired, the worksheet also can be used to make contact negatives or to contact images to other sensitized materials. For positive artwork the worksheet normally is done on one sheet of material. If categories of symbols are drawn with different colored pencils, this separation will assist at the drafting stage. One size pen, then, can be used to trace the features represented by one color. For separation drawings, in which categories of symbols are drawn on separate flaps, the colors assist in the same way.

When images are to be contacted to sensitized materials, the worksheet may take the form of more than one sheet of material. Separate categories of information can then be placed on separate sheets. Of course, whether one or several worksheets are prepared, it is important to use dimensionally stable material to minimize registry problems.

Registry. When the final art consists of more than one flap, some method for registering the flaps must be furnished to the printer. A common method is to draft small crosses in each of the four margins outside the map border (Fig. 17.1). During map preparation, these registry marks can be used to place visually each flap in the same position on the worksheet. The marks are retained when the negative is prepared and are transferred along with the map image to the printing plate. When the printer is satisfied that the plate is in the proper position on the press, the registry marks are removed with an abrasive. The crosses should be made with fine lines and should be placed far enough from the map proper to enable the printer to remove them easily without damaging the map. On the other hand, if they are placed far from the map, much more film is necessary to record their images.

Another, more precise method of registering the various flaps is to fasten them together with registry pins.* They consist of flat pieces of material to which machined $1/4$-in. studs have been attached. Studs of various lengths are available, some of which are short enough to allow their being used in vacuum frames. Precisely matched holes are punched into each proposed flap and into the worksheet before drafting begins. Special punches are manufactured for this purpose; however, they are quite expensive and, for maps of relatively small size, we can easily make our own. With an ordinary two- or three-hole adjustable office punch, on which the punches have been *welded into a fixed position,* we can produce perfectly matching holes on a series of sheets. For greater distances between holes, we can fasten separate punches to lengths of metal or wood. Two pins, however, may not provide the rigidity needed for large sheets of material. An extra pin may be incorporated at the bottom of the sheet or a pin may be used at the midpoint on each of the four sides.

Using three slots instead of $1/4$-in. circular holes results in more precise registry and is a more satisfactory system for large maps. These devices consist of a flat table surface with three movable punch assemblies that can be locked into position with the axes perpendicular. The rectangular pins used in this system are the same width as the slots but slightly shorter. This lengthwise clearance permits resetting the punch to conform with previously punched sheets by only approximate visual location of the sliding punches. The arrangement of the slots and the use of the rectangular pins keeps the center of the map in register and allocates proportionately on both sides of the center axis any misregister caused by contraction or expansion.

Available from the Chester F. Carlson Company, 2230 Edgewood Avenue, Minneapolis, MN 55426.

Another method of producing properly spaced holes is with the use of gummed tabs that have been prepunched.* Two tabs are placed on the worksheet and on the pins placed in the holes. Each flap is then successively placed on the worksheet and another set of tabs is pressed on after carefully positioning the holes over the pin.

When all flaps are properly pinned before drafting begins, each sheet will fit precisely throughout the entire drafting operation, and interchanging flaps is accomplished quickly and accurately. Registry marks should be used in conjunction with the pinning system, since the printer will need their images on the negatives and plates.

Preparation of Drawings. Final preparation of the map should begin with the line work and other symbols that require the application of ink. When preprinted materials are burnished onto the map, the adhesive tends to be squeezed out along the edges and can become distributed over the drafting surface, causing difficulty for further drafting. If ink must be added after this has happened, it is necessary to clean the area to be inked. Wax can be effectively removed by swabbing with trichloroethylene.†

Artwork furnished the printer is converted to negative form, either by use of a camera or by direct contact in a vacuum frame. In either case, good, sharp contrast between the symbols of the map and the drafting surface is important to produce quality negatives. The line artwork should be opaque black or red. Gray lines cause difficulty in determining exposure times, and the result is usually broken or weak line work on the printed copies. If the artwork is prepared specifically for direct contact negatives, white opaque cannot be used for deletions, because it registers on the film the same as opaque black.

When the artwork for a map consists partly of line and partly of continuous tone, they should be prepared on separate flaps. In the halftone process even solid black areas are converted to dots

| LINE | HALFTONE |

FIGURE 17.7 *Lettering used in conjunction with a halftone should be placed on a separate flap and burned separately into the plate. When halftoned, the lettering (or any line work) tends to be fuzzy.*

and will, therefore, appear as dark grays. Lettering and other line symbols have a fuzzy instead of a sharp black appearance (Fig. 17.7). When separate flaps are prepared, the line work is photographed separately to produce a regular line negative and a halftone screen is used only to produce the negative of the continuous tone flap. The two negatives are then combined by exposing the one printing plate to each negative separately. The line work will be solid and its edges will be less fuzzy.

Lettering Flap. Making a separate flap or flaps for lettering is standard procedure when a map involves the use of color. On single-color maps, however, the lettering is usually incorporated into a flap that contains other lines and symbols. In any case, there are usually situations when the lettering must fall over other map features. If colored areas or lines are not too dark, black lettering can usually be superimposed; however, dark or black lines should be interrupted to accommodate the names. A good practice is to leave spaces for the lettering when the line work is drafted.

NEGATIVE ARTWORK

Use of negative artwork usually reduces reproduction costs and generally improves cartographic quality. The capability for the cartographer to produce very precise lines in negative form came with the scribing process, which has come into wide use in the past few decades. Recently, many kinds of materials have appeared that can be used in conjunction with the scribecoat and permit the cartographer more easily to prepare artwork in negative form. The artwork is normally made at reproduction size, because the photographic step is not necessary and the printer can go directly to

---

Various kinds of tabs and pins are available from Berkey Technical Corporation, 25–15 50th St., Woodside, NY 11377.

†*Care must be exercised to prevent the trichloroethylene from coming in contact with preprinted materials, since damage may occur.*

the platemaking stage. Skipping the photographic stage means that the cartographer has control of the materials much further into the reproduction process.

Scribe Sheets and Coatings

Scribing is a technique almost the opposite of drawing with pen and ink. In drawing, the desired lines and marks are applied by the cartographer; in scribing, the desired marks are obtained by removing material. The cartographer starts with a sheet of plastic film to which a translucent coating has been applied. Then, working over a light table the coating is removed by cutting and scraping to produce the lines and symbols. Scribing can also be carried out on a flatbed plotter directed by a computer. When it is finished, the sheet has the same general appearance as a negative made photographically. The scribe coating is compounded so that, among other properties, it is visually translucent but actinically opaque, that is, opaque to those wavelengths of light by which light-sensitive emulsions are especially affected.

During the development stages of scribing, glass was used as the base material for the plastic scribe coating. Glass, of course, is breakable, heavy, and difficult to store; it has been replaced by various transparent plastics. The coating is available in several colors. The translucent quality permits tracing on a light table, although this method is seldom used; instead, an image of the worksheet to guide the cartographer is normally made to appear in the emulsion on the scribing surface. Most scribing is, however, done on a light table, and the color of the coating seems to have a noticeable effect on the degree of eyestrain. Experiments have indicated that orange or yellow causes the least discomfort.

The scribe sheet coating is thin and plastic enough to permit easy removal. The cutting tool must remove the coating without cutting into the base sheet, which causes variations in line width. Any fragments of the coating left in an open area will, of course, block light and cause broken lines on the printing plate. If such fragments remain, they can be eliminated by pressing a piece of drafting tape onto the area. When the tape is lifted, the unwanted bits of material adhere to it and are removed from the base sheet.

Corrections or alterations on the scribe coat are made by painting over with a specially prepared opaque. New lines can then be cut; however, scribing in areas covered with the hand-applied opaque is somewhat less satisfactory than removing the original coating.

Scribing Tools

There are a number of kinds and designs of scribing tools, and scribing points are made from a variety of materials (Fig. 17.8). There are three basic kinds of instruments: the pen type, the rigid graver, and the swivel graver. The pen-type is mainly for freehand work and is somewhat difficult to use in that it must be held at a correct and constant angle to produce consistent line widths. The rigid graver, on the other hand, with its tripod construction, assures a proper angle at all times. When points for different width lines are exchanged in the graver, adjustments must be made to produce the correct angle again. The most convenient and efficient method is to provide each of the several gravers with different size points; then, to change line widths, a different graver with the desired point is selected. The swivel graver is used for points that produce two or more parallel lines or for chisel-type points used in making wide lines.

Special kinds of gravers are designed for producing special symbols. A building graver, for example, can be adjusted so a chisel-type blade of a desired width is moved a specified distance. The adjustment, then, determines whether a square or a given size rectangle will be scribed. Another symbol that presents a special problem in scribing is the dot. The dot graver has a chisel point, which is spun into the coating and opens the dot. Dot gravers are available that are electrically operated.

The material from which the scribing points are made is of concern because it determines whether or not they will need sharpening. Because there is a certain amount of friction between the point and the base material, steel points tend to wear and, as a consequence, the line widths change. Jigs are commercially available with which they can be sharpened; however, this can be a tedious task. Jeweled points are advertised as never needing sharpening, and they do last for long periods of time. Although they are more expensive than steel points, time saved in sharpening may, in the long run, prove them to be more economical.

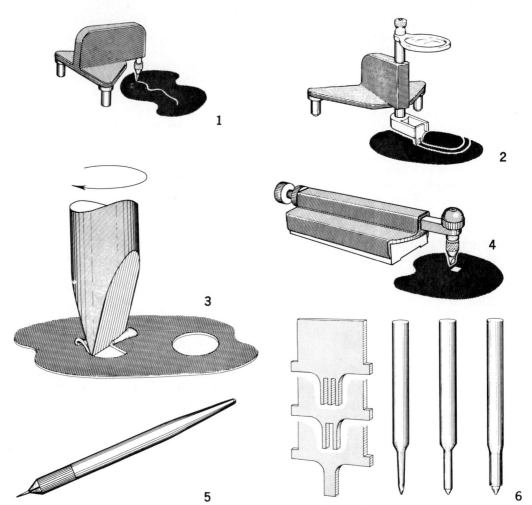

FIGURE 17.8 *Some of the kinds of scribing instruments used to remove the scribe coating. At the top are shown a rigid graver (1) and a swivel graver (2), both of which are used for line work. The dot graver (3) and the "building" graver (4) are used to make small circular of square openings in the coating, such as for dots or buildings on a topographic map. The pentype graver (5) is used for freehand work, and at the right (6) are shown some of the types of graver points, such as the single or multiple chisel-edged points and needle points. (Courtesy the U.S. Geological Survey.)*

Scribing the Map

As with positive artwork, a worksheet should be prepared from which the final artwork is prepared. In the scribing operation a worksheet is almost imperative, since scribing is most easily and accurately done if an image of the map is first placed on the scribecoat. Tracing an image through the scribecoat is difficult because of diffusion of the light.

The worksheet is usually produced in positive form on translucent material at the proposed scribing scale. This permits transfer of the image to the scribecoat base without the use of a camera or darkroom facilities. If, for some reason, it is necessary to prepare the worksheet at a larger scale, it can be reduced to the desired size by photography and then contacted to the scribe base.

Before beginning the worksheet, we should consider whether separation scribecoat negatives will be necessary. If line work is to appear on two or more scribecoats, it may be advisable to separate the information on separate flaps of the worksheet. For example, if drainage on the final map is to be printed in blue and the roads in red, each will be scribed on a separate sheet. If the information is separated on two flaps of a worksheet, a separate image can be placed on each scribesheet or both images can be contacted to the scribesheets in different colors. Actually, contacting all the flaps in different colors to each scribecoat can reveal discrepancies in the worksheet flaps. A road, for example, may have been plotted in error so that it crosses a small lake. The multiple image would show this, and the proper position of the road would need to be established.

For very detailed and complicated maps, a modification of the multiple printing of the worksheet flaps might be used. Assume that the hydrographic features for the map were compiled from very accurate large-scale topographic sheets, roads were selected from a somewhat smaller-scale, more generalized map, and lines such as soil boundaries had been originally plotted on aerial photographs. Since the hydrographic features are likely to be in the most accurate planimetric position, that image is transferred to the scribecoat and the lines scribed. The soil boundaries are related to the hydrographic features and, in some cases, might end abruptly at lakes or streams. The second scribecoat will include the image of the worksheet of the soil boundaries and the scribed image of the hydrographic features. The soil boundaries are then scribed with the hydrographic detail as a guide. On the last scribecoat, the already scribed images of the hydrography and soil boundaries are combined with the worksheet flap of the roads. The delineation of the roads can then be adjusted to fit the previously scribed features.

Scribecoat is available in a sensitized form for diazo duplication. To produce an image, we need only expose it to an ultraviolet source while it is in contact with a positive of the worksheet and develop the image in ammonia fumes. If we prefer we can purchase regular scribecoat and apply the diazo sensitizer ourselves.

Bichromate sensitizers are more versatile for contacting worksheet images, since successive applications of color enable the separation of categories of features. A disadvantage is that negatives and not positives are needed. We can, however, produce our own negatives, even without darkroom facilities, with either *Kodagraph Contact Film* or *3M* orange proofing foils.

Continuous tone images can be placed on *Stabilene Contone Film,** a white/rust scribecoat sensitized with a medium gamma orthochromatic emulsion on the scribe surface. Using a continuous-tone image such as an aerial photograph as the compilation, lines and other symbols can be scribed to produce a line negative for the map.

Letterpress—Lithography. The image placed on the scribecoat negative should be wrong reading if the lithographic process is to be used. In the lithographic process the plate is prepared by placing the emulsion of the film or the coating of the scribesheet face down in contact with the plate. If the emulsion side is "up," light tends to creep because of the thickness of the film, and the image on the plate will not be sharp.† A lithographic negative must, therefore, have a wrong reading image when viewed from the emulsion side, so that when it is placed emulsion side against the plate, the resulting image on the plate will be right reading.

On the other hand, in photoengraving a letterpress plate, a very thin film is used so it can be "flopped" to produce a wrong reading plate. For letterpress printing, then, the image should be scribed right reading on the scribesheet so that the scribecoat emulsion can be in contact with the plate during exposure.

Lettering. A separate positive flap is usually necessary when a map is scribed. If only a small number of names is required, it is possible to convert them to negative form, strip them to the scribe-

*Available from Keuffel and Esser Company, the material can be processed by standard darkroom methods to provide continuous-tone images.
†Using new improved films and a point source of light, the creeping of the light would be negligible; however, the best practice is to make exposures with the emulsion of the film in contact with the sensitized plate.

coat, and remove enough of the scribecoating to expose the name. This method is time consuming and is not often used.

To prepare the lettering flap, the scribecoat is usually flipped over on a light table to produce a positive or right reading image. A sheet of transparent or translucent material is placed over the scribecoat, and the scribed symbols are used as a guide for placing the lettering. Upon completion, the lettering flap is placed beneath the scribecoat, and the opaquing necessary to interrupt lines is done.

Very often, other symbols are placed on the lettering flap. Covering areas with prepared patterns can usually be done most conveniently on positive art, and some symbols that are difficult to scribe may be available in preprinted form. The lettering flap and the scribecoat are usually double burned at the platemaking stage; however, a negative of the lettering and the negative scribecoat can be combined in positive form on one piece of film. This film can then be treated as positive artwork for any of the duplicating processes.

Type can be applied directly to yellow or orange scribecoat and, by special processing with autopositive film, a positive including both the scribed lines and the type can be produced. First, an exposure in a vacuum frame is made using the scribecoat along with a yellow filter. Since yellow light removes density, the density remains only

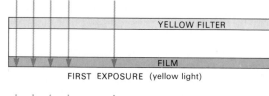

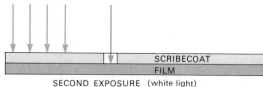

FIRST EXPOSURE (yellow light)

SECOND EXPOSURE (white light)

FIGURE 17.10 (Top) The yellow light removes density from the entire piece of film. (Bottom) With the scribecoat acting as a yellow filter, density is restored only where white light passes through scribed lines.

where type blocks all light. When subsequently exposed to white light (yellow filter removed), the scribecoat acts as the filter and only where light can pass through the scribed lines is the density restored (see Fig. 17.9).

By modifying the method it is possible to produce a positive of either the type or the scribed lines. To copy just the type, the film is developed after the first exposure in the above method. If the same system is used but without the scribecoat during the first exposure, only the scribed lines will be rendered (see Fig. 17.10).

COMPLEX ARTWORK AND PROOFING

The cartographer with a good knowledge of duplicating and printing methods can take advantage of their capabilities to produce special effects for improving design and legibility. To use the techniques of duplicating trades most efficiently, the artwork must be carefully planned and the use of particular materials is often necessary. The following section describes some of these materials and provides some examples of their use. It is hoped that students will be able to adapt and expand these techniques and procedures for solving their particular problems.

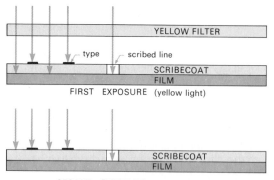

FIRST EXPOSURE (yellow light)

SECOND EXPOSURE (white light)

FIGURE 17.9 (Top) Yellow light removes all density except where type blocks the light rays. (Bottom) The scribecoat acts as the filter and density is restored when white light passes through the scribed lines.

Open Window Negatives

For laying tints in a given area, open window negatives are necessary. The negative must be actinically opaque except in that area in which the tint or color is desired. The open area, however, should fall exactly on a bounding line. Preparing such a negative completely by hand can be a tedious job, especially if the area boundary is complicated. It can be done by cutting a thin film and peeling it away from a clear plastic base.*

Open window negatives can also be prepared photomechanically, and the only hand operation is peeling away the proper areas and opaquing unwanted lines.†

The two kinds of material available are alike in appearance, since both have a thin red film adhered to a transparent sheet of plastic. Neither involves cutting, since the lines bounding areas can be photochemically etched from the red film. By lifting the edge of an area with a knife, the whole area can be lifted away from the base sheet, leaving the transparent opening. Both *Peelcoat* and *Striprite* may be processed under normal room lighting conditions and can, therefore, be used in cartographic drafting rooms that have limited equipment. Both materials are processed using negatives; however, *Striprite* is also available in a form that permits the use of positives. Both materials produce good-quality negatives that can be registered almost perfectly. A pin registry system is used throughout the process, of course, to insure registry of each open window negative with the lines on the scribecoat. Exposure is made with a source rich in ultraviolet light, after which the sheet is developed and etched.

Tint Screens

The rating of the tone produced by tint screens is designated in percentages of ink coverage; therefore, a 10% screen will produce a very light tone and an 80% screen will give a very dark tone. Tint screens are considered to be in negative form; therefore, an 80% screen actually has very small opaque dots, with most of the film transparent, whereas a 10% screen is mostly opaque with small transparent openings.‡

FIGURE 17.11 *Those with keen eyesight may be able to distinguish the dots on the 65-line tint screen (below) but are unable to do so on the 133-line screen (above).*

Screens produce dot patterns that are coarse or fine, depending on the number of dots per inch. A screen that has 65 dots per inch is very coarse and is called a 65 line screen. A 150 or 200 line screen, on the other hand, produces very fine dot patterns that can be detected only under magnification (Fig. 17.11). In letterpress printing, relatively coarse screens must be used except when printing on coated paper stock, whereas fine screens can be used in offset printing even on poor grades of paper.

An area on the map to be screened must be open on the negative. Positive artwork for this open window negative is prepared by covering the area on the drawing with black ink or with a red material such as *Zip-a-Tone*. This, of course, is done on a separate flap and is carefully registered to the line work. When the negatives are arranged for platemaking, they are carefully placed, using

---

*Rubilith *and* Amberlith *are manufactured by Ulano Graphic Arts Supplies, Inc., 610 Dean St., Brooklyn, NY 11238.*
†*Striprite *is produced by Direct Reproduction Corporation, 811–13 Union Street, Brooklyn, NY 11215. Peelcoat Film is available from the Keuffel and Esser Company.*
‡*Custom-made tint screens can be produced by the cartographer by using a contact halftone screen in a vacuum*

frame. Exposing orthochromatic film through the screen produces uniform dots, the size of which are determined by the length of exposure. Since contact screens are available in lines, dots, and other patterns, numerous kinds of tint screens can be produced.

the registry pins. If poor registry is discovered after the plate is made, adjustments cannot be made, and the plate must be discarded. This is especially significant in the case of expensive letterpress plates.

Tonal values can be added to maps by using a contact tint screen either at the platemaking stage or on film in a vacuum frame. It is necessary only to insert a screen between the negative and the sensitized material. These screens are composed of dots or closely spaced lines that block the light and produce the impression of gray on the printed copies. Since any reduction of the artwork has already been accomplished, there is not the danger of the dots disappearing as there is when preprinted dots are placed on the original artwork. When line work, type, or other symbols are screened, the dots can reduce the sharpness of the edges. The use of finer screens can improve the situation but, by using a method described below where we retain a fine line outline of the symbol, we not only retain the edge sharpness but also may preserve some of the detail that is normally lost with conventional screening.

Three separate exposures on an autopositive film are necessary to produce a negative on which the items are screened but retain a fine solid outline. First, an exposure made using a yellow filter and a positive of the desired tint produces a duplicate of the tint over the entire area of the film. The screen is then removed and replaced by the original drawing of the items to be screened. A second exposure to the yellow light removes the dots where the light is able to pass through the original. By overexposing at this point, some of the dot patterns along the edge of the image will be destroyed; the amount depends on the degree of overexposure. Leaving the film and original undisturbed, the yellow filter is removed and a normal exposure is made to white light. Density is not removed under the image of the original drawing; this results in a negative that shows the drawing as screened areas with a fine bounding line.

Up to three screens can be printed on one area and, in color printing, a fourth can be added. Extreme care must be exercised when positioning the screens, however, to prevent usually undesirable *moiré* effect (Fig. 17.12). Screens must be rotated so that the orientations of the lines of dots are separated by angles of 30 degrees. It is cus-

FIGURE 17.12 *Tint screens should be aligned with an angular separation of 30 degrees. Other angles can produce various moiré patterns such as the one illustrated.*

tomary to place one screen at 45 degrees, one at 75, and the third at 105. If a fourth screen must be used it should be assigned to the lightest color and rotated to an angle of 90 degrees.*

Because screens can be superimposed, it may not be necessary to prepare a separate flap for each tone. If, for example, we wanted values of 20, 40, and 60%, only two flaps would be required. On flap number one, the areas of 20 and 60% would be opaqued and a 20% screen specified. A second flap would include the areas of 40 and 60% and a 40% screen specified. Since the two flaps overlap, the 60% area would receive 20 plus 40% ink. The cost would be reduced because of one less negative and one less plate exposure.

Superimposing screens within a single color is a satisfactory method only if the proper screen angles are available. If more than one color is to be used to print a map, each color is normally assigned a screen angle 30 degrees from the others. Screens of different colors can then be superimposed, but screens at the same angle within a color cause a *moiré*.

Numerous graphic devices are used for positioning screens at the proper angle. Companies that manufacture tint screens usually have them available.

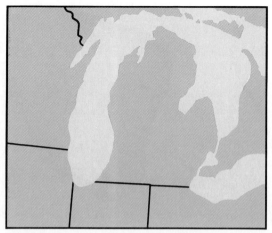 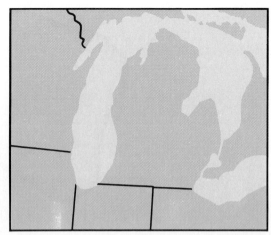

FIGURE 17.13   It is extremely difficult to make two tint screens match precisely (left). A tint superimposed on another (right), however, caused no difficulty.

Most often when gray tints are abutted, a black line is needed between them because making them join exactly is not usually possible. If, however, the tones are a result of the superposition of screens, the black line may not be necessary (Fig. 17.13). When a screened area is adjacent to a solid area, the artwork should always carry the tint into the solid. If we attempt to make them just join, white spaces are almost sure to occur.

A very useful technique is to screen the areas of preprinted patterns that have been applied to the artwork, which often tends to look rather harsh when printed in a solid color (Fig. 17.14). Also, lettering or symbols that might otherwise be lost can show clearly through the screened patterns. Again, a separate flap is prepared for screening. Selected patterns are substituted for the opaqued areas of regular screening. Unless very fine textured tint screens are to be used, coarse patterns should be chosen. It is also possible for a *moiré* to develop if fine dots or certain other preprinted patterns are used. As a precaution, we should test the orientation of patterns and screens by superimposing them on a light table. Lines, lettering, and symbols can also be screened, but very narrow lines and thin serifs on letters are likely to be lost. Lettering is probably most successfully screened with line screens, which do not produce the ragged edges caused by dot screens.

Bold lines and Sans-Serif lettering, when screened, can add much to the utility and overall appearance of a map.

Vignettes

With the use of a contact halftone screen, it is possible to produce dots of a diminishing size or a vignette (see Fig. 17.15). This can be a useful technique when, for example, we wish to indicate an indefinite zone instead of the sharp boundary produced by a line or the unaltered edge of a pattern. Quite commonly, a vignette is used along a shoreline or coastline instead of employing a constant tint over the entire water body.

FIGURE 17.14 The patterns in the lower row have been screened to 30% of black. The pattern appears less harsh, and lettering or symbols applied remain legible.

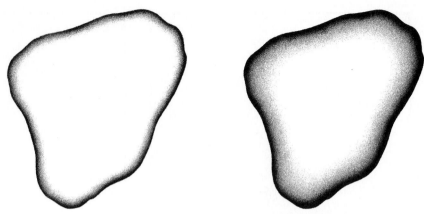

FIGURE 17.15 *An example of diminishing size dots or a vignette.*

To produce this effect on water bodies, we need an open window negative of the feature and a reverse or positive mask. In the contact printer we use a sandwich of film, negative, positive, and spacer, as shown in Fig. 17.16. Light is unable to reach the film where the positive mask, in contact with the screen, protects the land area. Spacers, consisting of clear sheets of film between the positive and the open window negative, permit the light to spread into the edge of the area of the water body. The amount of light reaching the film decreases with increasing distance from the edge, producing smaller and smaller dots. The size of the dots is related to the length of exposure, whereas the width of the vignette band is determined mainly by the amount of space between the negative and the positive and, to some extent, by the length of exposure.

A positive prepared as described above can also be used as a mask to vignette the edges of area patterns (Fig. 17.17). The positive blocks light

in decreasing amounts away from the area boundary while on open window negative is used to restrict the light to the image area. The area symbol must be in the form of a film negative.

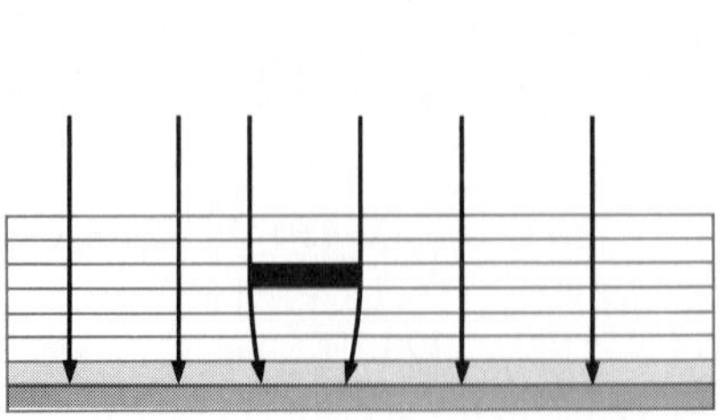

Reverse or Open Line Work

White line work on a background of black or color can be produced several ways, but usually the most satisfactory method is to reverse it at the platemaking stage. A photocopy changes black to white, and this can be used directly on the artwork; but such materials frequently do not have sharp images, and the results are often unsatisfactory. White lines and lettering are difficult to produce directly on the artwork, since opaque white will not flow properly in most pens and it tends to flake off when the artwork is handled. Reversing the line work, on the other hand, produces sharp images from artwork prepared with the usual black opaque materials.

The line work that is to be open is produced on a separate flap and prepared in the same way as any other flap. One of the other flaps will, of course, be an open window negative to produce either a tone or solid color for the background. Instructions to the printer merely indicate that a given flap is to be reversed from another. When the plate is exposed, a film positive of the flap to be reversed is placed, in registry, between the open window negative and the plate (Fig. 17.18). On the positive, the line work is opaque and blocks the light from those areas on the plate and

FIGURE 17.16 *The above arrangement of film, negative, spacers, and positive mask produces a vignette.*

Diffuser
Glass
Positive

Spacers

Contact Screen
Film

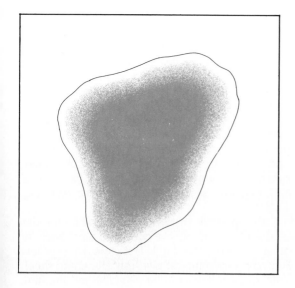

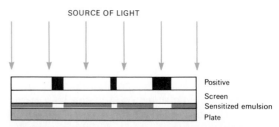

SOURCE OF LIGHT

Positive
Screen
Sensitized emulsion
Plate

FIGURE 17.18 White lettering, of course, is produced in the same manner by inserting a positive at the platemaking stage.

images. Suppose a shaded relief map is to be prepared for an area in which numerous lakes are to have a smooth gray tone. Shading the terrain can be done on the first flap without regard for the lake outlines and actually should overlap into the

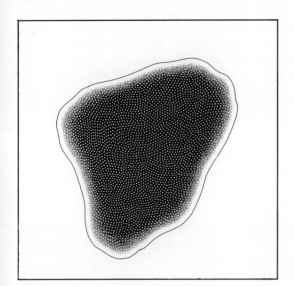

FIGURE 17.17 Vignette of a tint screen (above) and a dot pattern (below).

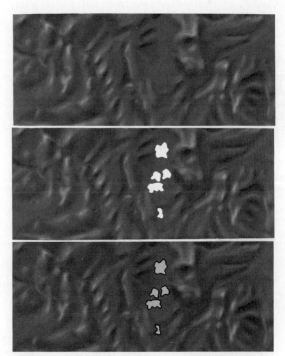

FIGURE 17.19 The shaded relief (above) has been accomplished without regard for the lakes. A film positive of the flap that is to lay the tint in the lake has been used to reverse the halftone dots out of the shaded relief (center). A negative of the lake flap then is screened and lays an even tint in the lakes (below).

produces a nonprinting area. The white paper then shows through the tint or solid area, and white line work is the result.

The positive used to reverse is sometimes referred to as a mask and is, indeed, useful in many cases as a mask to block out unwanted portions of lake area. A second flap, on which all the lakes are opaqued, will provide an open window negative to be screened to produce an even tint (Fig. 17.19). When making the plate, a film positive of the second flap is placed between the halftone negative made from the terrain drawing, which will leave the plate unexposed in the lake areas. The open window negative of the lake flap with an attached screen is then used to lay the even tint in the masked areas.

Screened areas provide the necessary background for open or reverse line work; at the same time they can permit overprinting of the solid color. If the screen is too light, white areas will not be distinguishable; if the screen is too dark, the solid color will not be discernible because the background will not provide enough contrast. A combination of a solid color and up to about 50 or 60% of the color is usually satisfactory for overprinting, while any tint down to about 30% is adequate for reversing (Fig. 17.20).

Chokes and Spreads

Sometimes it may be desirable to make lines wider or narrower instead of trying to maintain accurate widths. For example, we may wish to make lettering a bit thinner or a bit bolder. These effects, called chokes and spreads, are produced in a vacuum frame by separating the emulsions of a positive or negative and the film with spacers of clear sheets of acetate or film during exposure. A sheet of frosted material, such as drafting film, placed on top of the glass of the frame diffuses the light and allows it to undercut the image after passing through the spacers.*

With regular film, the use of a negative will result in a spread positive and the use of a positive will produce a choke negative. If duplicating film is used, a negative will produce a spread negative and a positive a choke positive.

An outline of type or other symbols can be made from solid symbols by contacting with a positive choke and a spread negative in register. Special effects, such as shadowing and highlighting, can be produced by shifting the positive and negative slightly out of register (see Fig. 17.21).

Autopositive or direct positive film can also be used to make outline effects on type, lines, or other symbols. The film is first exposed to yellow light to remove the density from the entire sheet. With spacers between the negative and the film, an exposure to white light restores density in the image area. The spacers allow the light to undercut and spread the image. A final exposure to yellow light through the negative with no spacers removes the density in the original image area and leaves only an outline.

Artwork for Color Printing

Colored maps are usually produced from artwork that is prepared in black and white. The color re-

*An alternative method is to place the light source off to the side at an angle of 20 to 45 degrees with the vertical and then spin the vacuum frame to undercut the image.

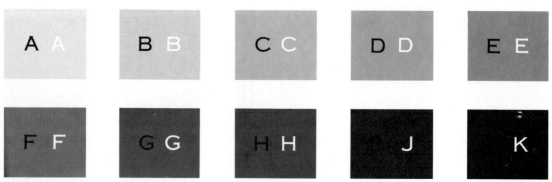

FIGURE 17.20 Lettering has been superimposed and reversed from tint screens from 10% in equal increments to solid black. We can determine from this illustration approximately the point where reversing or overprinting can be unsatisfactory.

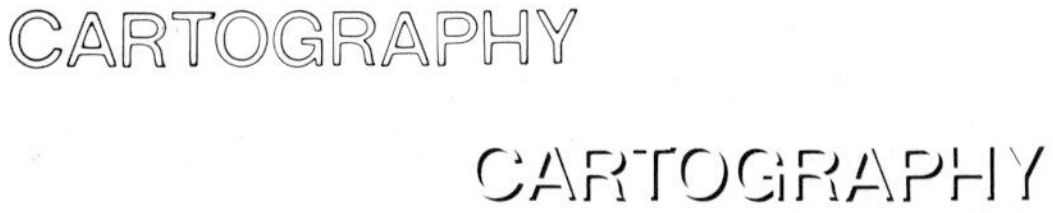

CARTOGRAPHY

CARTOGRAPHY

FIGURE 17.21 *A contact made with a spread negative and a choke positive slightly out of register can produce special effects.*

TABLE 17.1

Area	Colors	Result
1	Yellow solid, magenta 60%	Light orange
2	Yellow 60%, magenta 60%, cyan 10%	Reddish orange
3	Yellow 60%, magenta 10%	Medium yellow
4	Yellow solid, magenta 20%	Dark yellow
5	Yellow solid, magenta solid, cyan 10%	Dark orange
6	Yellow solid, magenta 10%, cyan 20%	Light green
7	Yellow 40%, magenta 40%, cyan 20%	Beige
8	Yellow 40%	Light yellow
9	Yellow 60%, magenta solid, cyan 10%	Red
10	Yellow solid, magenta 10%, cyan 80%	Dark green
Lakes	Cyan 20%	

sults from separate press runs using different colored inks. At least one flap and often more than one is necessary for each color. The cartographer furnishes a series of registered flaps to the printer with specific instructions for processing and printing.

A two-color map is usually one with black and one other color, although it can be two colors other than black. Three-color maps ordinarily would have black and any other two colors. A four-color map, however, most often uses the subtractive primary colors, magenta, yellow, and cyan, along with black. Theoretically, any other hue can be produced by combinations of varying amounts of the these primary colors. In color printing the combining is accomplished by successive press runs, and the amount of each color is controlled by tint screens. Charts showing the hues produced by superimposing different percentages of magenta, yellow, cyan, and black are prepared by most printing firms (Fig. CI.11). When choosing colors, the cartographer should use a chart prepared by the same printer who will reproduce the map. If another chart must be used, a copy of it should be furnished to the printer to indicate the precise color desired.

To demonstrate the procedure that we might follow in planning and preparing the artwork for a colored map, consider an actual case of a map that requires separate colors for 10 categories of information and light blue for numerous small lakes. From this listing we can prepare a chart that will reveal the number of flaps necessary to produce these colors and indicate which areas are to be opaqued on each flap (Fig. 17.22). Other flaps that are necessary will include a solid black flap with the base information, lettering, and area boundaries, a solid blue flap for the hydrographic features, and a 20% blue to lay a tint on the lakes. A separate flap for the lakes is also prepared to provide a film positive that can be used as a mask when exposing the plates for all other colors. This will prevent the other colors from falling within the lakes.

To produce the various flaps in positive form, it is customary to secure copies of the base map on a stable material with the image in light blue. Using the blue lines of the areas as a guide, black opaque is applied to the appropriate areas on each flap. When the open window negatives are prepared, the light blue images will disappear. A specified tint screen is then attached to the negatives and the plates are prepared.

To produce a map on which line work is to be

	YELLOW	RED	BLUE
Solid	1 4 5 6 10	5 9	
80 %			10
60 %	2 3 9	1 2	
40 %	7 8	7	
20 %		4	6 7 Lakes
10 %		3 6 10	2 5 9

FIGURE 17.22 *When complicated color combinations are necessary, a chart of this sort can be a helpful guide for the cartographer.*

in different colors, another method can be used in which all line work is placed on one flap. When the art is photographed, multiple negatives are made, one for each proposed color. Then each negative is opaqued to leave only that part of the image that is to be printed in a particular color. Since most film has relatively good dimensional stability, excellent registry can be maintained.

Multiple Use of Negatives and Positives

With the capabilities of the printer to produce tints and reverse line work, the cartographer has many opportunities to enhance the map by various combinations of white, grays, and black. The decisions concerning the use of such combinations are made at the design stage, so that before drafting is begun, the artwork has been planned in a fashion that will produce the best results with a minimum of effort. A good knowledge of the possibilities and limitations of the printing processes is necessary for us to visualize how the various pieces of film are to be used; for example, we must keep in mind that a printing plate cannot be exposed to two negatives at the same time; that is, negatives require successive exposures. On the other hand, more than one film positive, or film positives and a negative, can be *sandwiched* together and used for a single exposure. On one printing plate, then, there can be an image that resulted from a series of exposures, any of which can incorporate more than one piece of film.

The following example will demonstrate that cartographers are not as limited in one-color printing as one might believe.

Suppose a design program specifies that the hydrographic features and section lines of a map are to appear white, while the swamp areas, county boundaries, and township lines are to appear solid black on a background flat tone of 40% black (Fig. 17.23). Hydrographic features must be reversed from the background tint and from the swamp areas, but the black boundaries are to print over the white.

Four flaps are used to produce the map.
Flap A All lowland areas.
Flap B Entire area of map opaque to provide open window negative for overall tint screen.
Flap C Hydrographic features and section lines
Flap D County boundaries and township lines.

All the flaps can be prepared in positive form with black ink and red *Zip-A-Tone.*

The following are the instructions given to the printer.
1. Lay a 60% tint screen on a negative of flap B, and, using a positive of flap A as a mask, contact to a new film.
2. Double print on the plate:

   a. The new film with a positive of flap C as a mask.
   b. The negative of flap D.

When a 60% screen is used with the negative of flap B and contacted to the new film with a positive of flap A, the screen is converted to 40%. It will not cover the low wet areas, however, since the mask holds the light back and transparent areas result. If the plate were to be exposed to the new film, a 40% tint would result over the whole map except in the wet areas, which would be solid black. But, by combining a positive of flap C with the new film, the hydrographic features and the section lines block the light and produce a nonprinting area in the background tint and in the low wet areas.

Proofing

When artwork is completed, it must be carefully and thoroughly checked before being sent for duplication. When facilities and materials are available, it is wise to proof a map if it consists of more than one flap. Even one-color maps can consist of as many as 10 flaps or more, greatly increasing the chances for error and making editing difficult. A properly prepared proof will closely resemble the copies that will be printed from the artwork, and can, therefore, evaluate the success of the design as well as reveal errors.

Artwork prepared on translucent material can be used to prepare proofs in the drafting room, even though darkroom facilities are not available, as described in Chapter 16. The various proofing processes, such as *General Color Guide Film, 3M* proofing materials, and the use of bichromate sensitizers, require the use of negatives for exposure. These direct contact negatives can be made using *Kodagraph Contact Film* or with *Orange 3M Color Key,* both of which are actinically opaque; either of these negatives can be used to make the

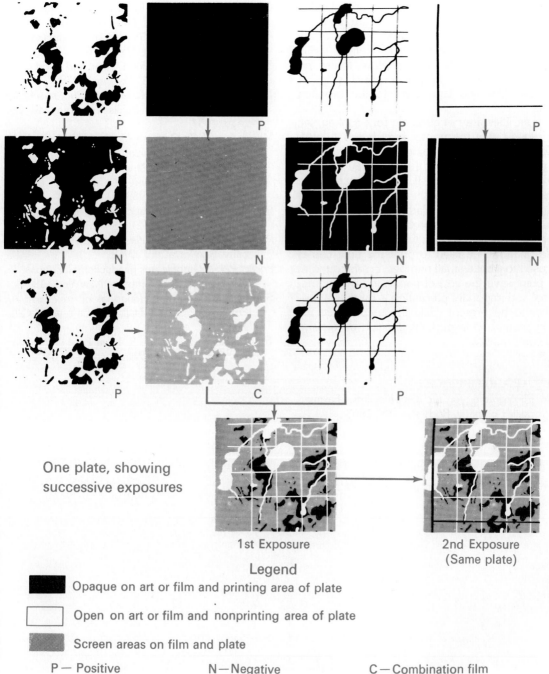

One plate, showing
successive exposures

1st Exposure

2nd Exposure
(Same plate)

Legend

Opaque on art or film and printing area of plate

Open on art or film and nonprinting area of plate

Screen areas on film and plate

P — Positive   N—Negative   C—Combination film

FIGURE 17.23 *Possibilities for use of black and white are unlimited. The student only need use imagination.*

proofs with any of the proofing systems. During each step of the proofing operation, the pin registration system must be employed. Because of the nature of the 3M material, it cannot be easily punched and it is necessary, therefore, to use the gummed prepunched tabs or to punch strips of discarded film and tape them to the proofing sheets.

If facilities are not available for proofing, we can request that the printer furnish proofs after making the negatives but before making the printing plates. Errors found at this stage can often be corrected on the negative. If the artwork must be revised, we have lost only the cost of the negatives and not the cost of the plates, paper, and press time. For complicated colored maps we may wish to request press proofs, in which case the actual plates that will be used to print the final copies are used to print a small number. Errors found at this point save the cost of paper and press time. A final and important precaution is for the cartographer to be present while the map is actually being printed to inspect things such as registry and color quality.

SELECTED REFERENCES

Basic Photography for the Graphic Arts, Eastman Kodak Company, Rochester, NY, 1963.

Carmichael, L. D., "The Relief of Map Making," *The Cartographic Journal, 6,* 18–20 (1969).

Halftone Methods for the Graphic Arts, Eastman Kodak Company, Rochester, NY, 1976.

Keates, J. S., *Cartographic Design and Production,* Longman Group Ltd., London, 1973.

Mertle, J. S. and Gordon L. Monsen, *Photomechanics and Printing,* Mertle Publishing Company, Chicago, 1957.

Moore, Lionel C., *Cartographic Scribing Materials, Instruments and Techniques,* 2nd ed., Technical Publication No. 3, American Congress on Surveying and Mapping, Cartography Division, Washington, D.C., 1968.

Ovington, J. J., "An Outline of Map Reproduction," *Cartography, 4,* (1962), 150–155.

Photography for the Printer, Eastman Kodak Company, Rochester, NY, 1962.

Pocket Pal: A Graphic Arts Production Handbook, International Paper Company, New York, 1974.

Schlemmer, Richard M., *Handbook of Advertizing Art Production,* Prentice-Hall, Inc., Englewood Cliffs, NJ, 1966.

The Contact Screen Story, E. I. Du Pont de Nemours and Company, Inc., Wilmington, Del, 1972.

DIMENSIONS, CONSTANTS, AND FORMULAS

Equatorial circumference of earth	40,075.4 km	24,901.7 mi
Area of earth (approximate)	510,900,000 km²	197,260,000 mi²
Radius of sphere of equal area	6,370.9 km	3,958.7 mi
Circumference to the diameter of a circle (π)	3.141 593	
Area of a circle	πr^2	
Area of a sphere	$4\pi r^2$	

CONVERSIONS

Note. SI refers to the International System of Units (metric), USCS to United States Customary System. The values in the table are based on the generally employed conversion ratio that 1.0 inch of the International Foot equals 25.4 millimeters exactly. Note, however, that in the United States, for surveying and large-scale mapping and charting purposes, the slightly different ratio (established in 1866) that 1.0 inch of the "U.S. Survey Foot" equals 25.40005 millimeters (1.0 meter = 39.37 inches exactly) has been retained. The difference is about 2 parts per million.

SI A	USCS B	To convert from A to B, multiply A by	To convert from B to A, multiply B by
Meters	Feet	3.280 840	0.304 8*
Meters	Yards	1.093 613	0.914 4*
Centimeters	Inches	0.393 700 8	2.54*
Centimeters	Feet	0.032 808 40	30.48*
Kilometers	Miles (U.S. statute)	0.621 371 1	1.609 344*
Kilometers	Miles (international nautical)	0.539 956 8	1.852*
Square meters	Square feet	10.763 91	0.092 903 04*
Square meters	Square yards	1.195 990	0.836 127 36*
Square centimeters	Square inches	0.155 000 3	6.451 6*
Square centimeters	Square feet	0.001 076 39	929.030 4*
Square kilometers	Square miles (U.S. statute)	0.386 102 1	2.589 988
Hectares	Acres	2.471 054	0.404 685 6
Hectares	Square miles	0.003 861 02	258.998 8

From Units of Weights and Measures, National Bureau of Standards Miscellaneous Publication 286, Supt. of Documents, U.S. Government Printing Office, Washington, D.C., 1967.
* Figures identified with an asterisk are exact. The others are generally given to seven significant figures.

APPENDIX A
USEFUL DIMENSIONS, CONSTANTS, FORMULAS, AND CONVERSIONS

Logarithms are used for the multiplication and division of complex numbers when a calculator is not available and to determine powers and roots of numbers.

The common logarithm of a number is the power to which 10 must be raised to equal that number. For example, the number $356 = 10^{2.55145}$; accordingly, the logarithm of 356 is 2.55145. The integer to the left of the decimal point in a logarithm is called the *characteristic*. The decimal fractional part of the logarithm (the numbers to the right of the decimal point) is called the *mantissa*. The logarithm of any number in which the same integers are in the same order, for example, 35,600, or 35.60, or 0.03560, has the same mantissa but a different characteristic. Characteristics are easy to determine; hence, tables of logarithms show only the mantissas.

When any number is equal to or greater than 1, then the characteristic of its logarithm is positive, and it is numerically *one less* than the number of places to the left of the decimal point in the original number. When any number is less than one, then the characteristic of its logarithm is negative, and it is numerically *one more* than the number of zeros immediately to the right of the decimal point. For example:

Number	Logarithm
Etc.	
35,600.0	4.55145
3,560.0	3.55145
356.0	2.55145
35.60	1.55145
3.560	0.55145
0.3560	− 1.55145 or 9.55145 − 10
0.03560	− 2.55145 or 8.55145 − 10
0.003560	− 3.55145 or 7.55145 − 10
Etc.	

When we add or subtract logarithms that have negative characteristics, it is usually convenient to use the method of notation shown at the far right in the preceding table.

Multiplication is easily accomplished since when the logarithm of any number is *added* to the logarithm of any other number (e.g., $\log x + \log y$), the sum of the two logarithms is the logarithm of the *product* of the two numbers, that is, $\log x + \log y = \log$ of xy. The numerical value of the product of xy is obtained by finding the mantissa of the sum in the table, noting the number (N) to which it refers (called the *antilogarithm*), and then placing the decimal point according to the value of the characteristic. Division is equally simple, since $\log x - \log y = \log$ of $x \div y$. Consequently, multiplication and division of large or complex numbers are enormously simplified by using logarithms.

Powers and roots of numbers may be easily obtained with the aid of logarithms as illustrated below. The logarithm of any number (N), when *directly multiplied* by a number, provides the logarithm of that power of the original number. For example, $356 \times 356 = 356^2 = 126{,}736$; the logarithm of 356 is 2.55145; $2.55145 \times 2 = 5.10290$, which is the logarithm of 126,736. Note that the logarithm is directly multiplied by the numerical value of the power, 2 in this example. Similarly, any root of any number (N) may be obtained by *directly dividing* the logarithm of the number by the numerical value of the root desired. For example, the logarithm 2.55145 divided by $2 = 1.275725$, which is the logarithm of 18.86, which is the square root of 356.

INTERPOLATION

The mantissas in Table B.1 refer to values of N to four places. To find the logarithm of a number to five places, interpolation between successive mantissas is necessary. Likewise, when a mantissa value lies between two given in the table, interpo-

APPENDIX B
COMMON LOGARITHMS

lation is necessary to obtain the antilogarithm to five places. Interpolation is facilitated by the tables of proportional parts shown alongside the mantissas, for example.

18	
1	1.8
2	3.6
3	5.4
4	7.2
5	9.0
6	10.8
7	12.6
8	14.4
9	16.2

The heading of the above table, 18, is the difference between the successive mantissas in a portion of the table near where it appears. The numbers on the left show the fifth place in N, and the numbers opposite show the corresponding decimal divisions of the mantissa interval.

TABLE B.1
Five-Place Logarithms: 100–150

N	0	1	2	3	4	5	6	7	8	9
100	00 000	043	087	130	173	217	260	303	346	389
01	432	475	518	561	604	647	689	732	775	817
02	00 860	903	945	988	*030	*072	*115	*157	*199	*242
03	01 284	326	368	410	452	494	536	578	620	662
04	01 703	745	787	828	870	912	953	995	*036	*078
05	02 119	160	202	243	284	325	366	407	449	490
06	531	572	612	653	694	735	776	816	857	898
07	02 938	979	*019	*060	*100	*141	*181	*222	*262	*302
08	03 342	383	423	463	503	543	583	623	663	703
09	03 743	782	822	862	902	941	981	*021	*060	*100
110	04 139	179	218	258	297	336	376	415	454	493
11	532	571	610	650	689	727	766	805	844	883
12	04 922	961	999	*038	*077	*115	*154	*192	*231	*269
13	05 308	346	385	423	461	500	538	576	614	652
14	05 690	729	767	805	843	881	918	956	994	*032
15	06 070	108	145	183	221	258	296	333	371	408
16	446	483	521	558	595	633	670	707	744	781
17	06 819	856	893	930	967	*004	*041	*078	*115	*151
18	07 188	225	262	298	335	372	408	445	482	518
19	555	591	628	664	700	737	773	809	846	882
120	07 918	954	990	*027	*063	*099	*135	*171	*207	*243
21	08 279	314	350	386	422	458	493	529	565	600
22	636	672	707	743	778	814	849	884	920	955
23	08 991	*026	*061	*096	*132	*167	*202	*237	*272	*307
24	09 342	377	412	447	482	517	552	587	621	656
25	09 691	726	760	795	830	864	899	934	968	*003
26	10 037	072	106	140	175	209	243	278	312	346
27	380	415	449	483	517	551	585	619	653	687
28	10 721	755	789	823	857	890	924	958	992	*025
29	11 059	093	126	160	193	227	261	294	327	361
130	394	428	461	494	528	561	594	628	661	694
31	11 727	760	793	826	860	893	926	959	992	*024
32	12 057	090	123	156	189	222	254	287	320	352
33	385	418	450	483	516	548	581	613	646	678
34	12 710	743	775	808	840	872	905	937	969	*001
35	13 033	066	098	130	162	194	226	258	290	322
36	354	386	418	450	481	513	545	577	609	640
37	672	704	735	767	799	830	862	893	925	956
38	13 988	*019	*051	*082	*114	*145	*176	*208	*239	*270
39	14 301	333	364	395	426	457	489	520	551	582
140	613	644	675	706	737	768	799	829	860	891
41	14 922	953	983	*014	*045	*076	*106	*137	*168	*198
42	15 229	259	290	320	351	381	412	442	473	503
43	534	564	594	625	655	685	715	746	776	806
44	15 836	866	897	927	957	987	*017	*047	*077	*107
45	16 137	167	197	227	256	286	316	346	376	406
46	435	465	495	524	554	584	613	643	673	702
47	16 732	761	791	820	850	879	909	938	967	997
48	17 026	056	085	114	143	173	202	231	260	289
49	319	348	377	406	435	464	493	522	551	580
150	17 609	638	667	696	725	754	782	811	840	869
N	0	1	2	3	4	5	6	7	8	9

Prop. Parts

	44	43	42
1	4.4	4.3	4.2
2	8.8	8.6	8.4
3	13.2	12.9	12.6
4	17.6	17.2	16.8
5	22.0	21.5	21.0
6	26.4	25.8	25.2
7	30.8	30.1	29.4
8	35.2	34.4	33.6
9	39.6	38.7	37.8

	41	40	39
1	4.1	4	3.9
2	8.2	8	7.8
3	12.3	12	11.7
4	16.4	16	15.6
5	20.5	20	19.5
6	24.6	24	23.4
7	28.7	28	27.3
8	32.8	32	31.2
9	36.9	36	35.1

	38	37	36
1	3.8	3.7	3.6
2	7.6	7.4	7.2
3	11.4	11.1	10.8
4	15.2	14.8	14.4
5	19.0	18.5	18.0
6	22.8	22.2	21.6
7	26.6	25.9	25.2
8	30.4	29.6	28.8
9	34.2	33.3	32.4

	35	34	33
1	3.5	3.4	3.3
2	7.0	6.8	6.6
3	10.5	10.2	9.9
4	14.0	13.6	13.2
5	17.5	17.0	16.5
6	21.0	20.4	19.8
7	24.5	23.8	23.1
8	28.0	27.2	26.4
9	31.5	30.6	29.7

	32	31	30
1	3.2	3.1	3
2	6.4	6.2	6
3	9.6	9.3	9
4	12.8	12.4	12
5	16.0	15.5	15
6	19.2	18.6	18
7	22.4	21.7	21
8	25.6	24.8	24
9	28.8	27.9	27

TABLE B.1

Five-Place Logarithms: 150–200

Prop. Parts	N	0	1	2	3	4	5	6	7	8	9
	150	17 609	638	667	696	725	754	782	811	840	869
	51	17 898	926	955	984	*013	*041	*070	*099	*127	*156
	52	18 184	213	241	270	298	327	355	384	412	441
	53	469	498	526	554	583	611	639	667	696	724
	54	18 752	780	808	837	865	893	921	949	977	*005
	55	19 033	061	089	117	145	173	201	229	257	285
	56	312	340	368	396	424	451	479	507	535	562
	57	590	618	645	673	700	728	756	783	811	838
	58	19 866	893	921	948	976	*003	*030	*058	*085	*112
	59	20 140	167	194	222	249	276	303	330	358	385
	160	412	439	466	493	520	548	575	602	629	656
	61	683	710	737	763	790	817	844	871	898	925
	62	20 952	978	*005	*032	*059	*085	*112	*139	*165	*192
	63	21 219	245	272	299	325	352	378	405	431	458
	64	484	511	537	564	590	617	643	669	696	722
	65	21 748	775	801	827	854	880	906	932	958	985
	66	22 011	037	063	089	115	141	167	194	220	246
	67	272	298	324	350	376	401	427	453	479	505
	68	531	557	583	608	634	660	686	712	737	763
	69	22 789	814	840	866	891	917	943	968	994	*019
	170	23 045	070	096	121	147	172	198	223	249	274
	71	300	325	350	376	401	426	452	477	502	528
	72	553	578	603	629	654	679	704	729	754	779
	73	23 805	830	855	880	905	930	955	980	*005	*030
	74	24 055	080	105	130	155	180	204	229	254	279
	75	304	329	353	378	403	428	452	477	502	527
	76	551	576	601	625	650	674	699	724	748	773
	77	24 797	822	846	871	895	920	944	969	993	*018
	78	25 042	066	091	115	139	164	188	212	237	261
	79	285	310	334	358	382	406	431	455	479	503
	180	527	551	575	600	624	648	672	696	720	744
	81	25 768	792	816	840	864	888	912	935	959	983
	82	26 007	031	055	079	102	126	150	174	198	221
	83	245	269	293	316	340	364	387	411	435	458
	84	482	505	529	553	576	600	623	647	670	694
	85	717	741	764	788	811	834	858	881	905	928
	86	26 951	975	998	*021	*045	*068	*091	*114	*138	*161
	87	27 184	207	231	254	277	300	323	346	370	393
	88	416	439	462	485	508	531	554	577	600	623
	89	646	669	692	715	738	761	784	807	830	852
	190	27 875	898	921	944	967	989	*012	*035	*058	*081
	91	28 103	126	149	171	194	217	240	262	285	307
	92	330	353	375	398	421	443	466	488	511	533
	93	556	578	601	623	646	668	691	713	735	758
	94	28 780	803	825	847	870	892	914	937	959	981
	95	29 003	026	048	070	092	115	137	159	181	203
	96	226	248	270	292	314	336	358	380	403	425
	97	447	469	491	513	535	557	579	601	623	645
	98	667	688	710	732	754	776	798	820	842	863
	99	29 885	907	929	951	973	994	*016	*038	*060	*081
	200	30 103	125	146	168	190	211	233	255	276	298
Prop. Parts	N	0	1	2	3	4	5	6	7	8	9

Proportional Parts

	29	28
1	2.9	2.8
2	5.8	5.6
3	8.7	8.4
4	11.6	11.2
5	14.5	14.0
6	17.4	16.8
7	20.3	19.6
8	23.2	22.4
9	26.1	25.2

	27	26
1	2.7	2.6
2	5.4	5.2
3	8.1	7.8
4	10.8	10.4
5	13.5	13.0
6	16.2	15.6
7	18.9	18.2
8	21.6	20.8
9	24.3	23.4

	25
1	2.5
2	5.0
3	7.5
4	10.0
5	12.5
6	15.0
7	17.5
8	20.0
9	22.5

	24	23
1	2.4	2.3
2	4.8	4.6
3	7.2	6.9
4	9.6	9.2
5	12.0	11.5
6	14.4	13.8
7	16.8	16.1
8	19.2	18.4
9	21.6	20.7

	22	21
1	2.2	2.1
2	4.4	4.2
3	6.6	6.3
4	8.8	8.4
5	11.0	10.5
6	13.2	12.6
7	15.4	14.7
8	17.6	16.8
9	19.8	18.9

TABLE B.1
Five-Place Logarithms: 200–250

N	0	1	2	3	4	5	6	7	8	9	Prop. Parts
200	30 103	125	146	168	190	211	233	255	276	298	
01	320	341	363	384	406	428	449	471	492	514	
02	535	557	578	600	621	643	664	685	707	728	
03	750	771	792	814	835	856	878	899	920	942	
04	30 963	984	*006	*027	*048	*069	*091	*112	*133	*154	
05	31 175	197	218	239	260	281	302	323	345	366	
06	387	408	429	450	471	492	513	534	555	576	
07	597	618	639	660	681	702	723	744	765	785	
08	31 806	827	848	869	890	911	931	952	973	994	
09	32 015	035	056	077	098	118	139	160	181	201	
210	222	243	263	284	305	325	346	366	387	408	
11	428	449	469	490	510	531	552	572	593	613	
12	634	654	675	695	715	736	756	777	797	818	
13	32 838	858	879	899	919	940	960	980	*001	*021	
14	33 041	062	082	102	122	143	163	183	203	224	
15	244	264	284	304	325	345	365	385	405	425	
16	445	465	486	506	526	546	566	586	606	626	
17	646	666	686	706	726	746	766	786	806	826	
18	33 846	866	885	905	925	945	965	985	*005	*025	
19	34 044	064	084	104	124	143	163	183	203	223	
220	242	262	282	301	321	341	361	380	400	420	
21	439	459	479	498	518	537	557	577	596	616	
22	635	655	674	694	713	733	753	772	792	811	
23	34 830	850	869	889	908	928	947	967	986	*005	
24	35 025	044	064	083	102	122	141	160	180	199	
25	218	238	257	276	295	315	334	353	372	392	
26	411	430	449	468	488	507	526	545	564	583	
27	603	622	641	660	679	698	717	736	755	774	
28	793	813	832	851	870	889	908	927	946	965	
29	35 984	*003	*021	*040	*059	*078	*097	*116	*135	*154	
230	36 173	192	211	229	248	267	286	305	324	342	
31	361	380	399	418	436	455	474	493	511	530	
32	549	568	586	605	624	642	661	680	698	717	
33	736	754	773	791	810	829	847	866	884	903	
34	36 922	940	959	977	996	*014	*033	*051	*070	*088	
35	37 107	125	144	162	181	199	218	236	254	273	
36	291	310	328	346	365	383	401	420	438	457	
37	475	493	511	530	548	566	585	603	621	639	
38	658	676	694	712	731	749	767	785	803	822	
39	37 840	858	876	894	912	931	949	967	985	*003	
240	38 021	039	057	075	093	112	130	148	166	184	
41	202	220	238	256	274	292	310	328	346	364	
42	382	399	417	435	453	471	489	507	525	543	
43	561	578	596	614	632	650	668	686	703	721	
44	739	757	775	792	810	828	846	863	881	899	
45	38 917	934	952	970	987	*005	*023	*041	*058	*076	
46	39 094	111	129	146	164	182	199	217	235	252	
47	270	287	305	322	340	358	375	393	410	428	
48	445	463	480	498	515	533	550	568	585	602	
49	620	637	655	672	690	707	724	742	759	777	
250	39 794	811	829	846	863	881	898	915	933	950	
N	0	1	2	3	4	5	6	7	8	9	Prop. Parts

Prop. Parts

	22	21
1	2.2	2.1
2	4.4	4.2
3	6.6	6.3
4	8.8	8.4
5	11.0	10.5
6	13.2	12.6
7	15.4	14.7
8	17.6	16.8
9	19.8	18.9

	20
1	2
2	4
3	6
4	8
5	10
6	12
7	14
8	16
9	18

	19
1	1.9
2	3.8
3	5.7
4	7.6
5	9.5
6	11.4
7	13.3
8	15.2
9	17.1

	18
1	1.8
2	3.6
3	5.4
4	7.2
5	9.0
6	10.8
7	12.6
8	14.4
9	16.2

	17
1	1.7
2	3.4
3	5.1
4	6.8
5	8.5
6	10.2
7	11.9
8	13.6
9	15.3

TABLE B.1

Five-Place Logarithms: 250–300

Prop. Parts	N	0	1	2	3	4	5	6	7	8	9
	250	39 794	811	829	846	863	881	898	915	933	950
	51	39 967	985	*002	*019	*037	*054	*071	*088	*106	*123
18	52	40 140	157	175	192	209	226	243	261	278	295
1 1.8	53	312	329	346	364	381	398	415	432	449	466
2 3.6											
3 5.4	54	483	500	518	535	552	569	586	603	620	637
4 7.2	55	654	671	688	705	722	739	756	773	790	807
5 9.0	56	824	841	858	875	892	909	926	943	960	976
6 10.8											
7 12.6	57	40 993	*010	*027	*044	*061	*078	*095	*111	*128	*145
8 14.4	58	41 162	179	196	212	229	246	263	280	296	313
9 16.2	59	330	347	363	380	397	414	430	447	464	481
	260	497	514	531	547	564	581	597	614	631	647
	61	664	681	697	714	731	747	764	780	797	814
17	62	830	847	863	880	896	913	929	946	963	979
1 1.7	63	41 996	*012	*029	*045	*062	*078	*095	*111	*127	*144
2 3.4											
3 5.1	64	42 160	177	193	210	226	243	259	275	292	308
4 6.8	65	325	341	357	374	390	406	423	439	455	472
5 8.5	66	488	504	521	537	553	570	586	602	619	635
6 10.2											
7 11.9	67	651	667	684	700	716	732	749	765	781	797
8 13.6	68	813	830	846	862	878	894	911	927	943	959
9 15.3	69	42 975	991	*008	*024	*040	*056	*072	*088	*104	*120
	270	43 136	152	169	185	201	217	233	249	265	281
	71	297	313	329	345	361	377	393	409	425	441
16	72	457	473	489	505	521	537	553	569	584	600
1 1.6	73	616	632	648	664	680	696	712	727	743	759
2 3.2											
3 4.8	74	775	791	807	823	838	854	870	886	902	917
4 6.4	75	43 933	949	965	981	996	*012	*028	*044	*059	*075
5 8.0	76	44 091	107	122	138	154	170	185	201	217	232
6 9.6											
7 11.2	77	248	264	279	295	311	326	342	358	373	389
8 12.8	78	404	420	436	451	467	483	498	514	529	545
9 14.4	79	560	576	592	607	623	638	654	669	685	700
	280	716	731	747	762	778	793	809	824	840	855
	81	44 871	886	902	917	932	948	963	979	994	*010
15	82	45 025	040	056	071	086	102	117	133	148	163
1 1.5	83	179	194	209	225	240	255	271	286	301	317
2 3.0											
3 4.5	84	332	347	362	378	393	408	423	439	454	469
4 6.0	85	484	500	515	530	545	561	576	591	606	621
5 7.5	86	637	652	667	682	697	712	728	743	758	773
6 9.0											
7 10.5	87	788	803	818	834	849	864	879	894	909	924
8 12.0	88	45 939	954	969	984	*000	*015	*030	*045	*060	*075
9 13.5	89	46 090	105	120	135	150	165	180	195	210	225
	290	240	255	270	285	300	315	330	345	359	374
	91	389	404	419	434	449	464	479	494	509	523
14	92	538	553	568	583	598	613	627	642	657	672
1 1.4	93	687	702	716	731	746	761	776	790	805	820
2 2.8											
3 4.2	94	835	850	864	879	894	909	923	938	953	967
4 5.6	95	46 982	997	*012	*026	*041	*056	*070	*085	*100	*114
5 7.0	96	47 129	144	159	173	188	202	217	232	246	261
6 8.4											
7 9.8	97	276	290	305	319	334	349	363	378	392	407
8 11.2	98	422	436	451	465	480	494	509	524	538	553
9 12.6	99	567	582	596	611	625	640	654	669	683	698
	300	47 712	727	741	756	770	784	799	813	828	842
Prop. Parts	**N**	**0**	**1**	**2**	**3**	**4**	**5**	**6**	**7**	**8**	**9**

TABLE B.1

Five-Place Logarithms: 300–350

N	0	1	2	3	4	5	6	7	8	9
300	47 712	727	741	756	770	784	799	813	828	842
01	47 857	871	885	900	914	929	943	958	972	986
02	48 001	015	029	044	058	073	087	101	116	130
03	144	159	173	187	202	216	230	244	259	273
04	287	302	316	330	344	359	373	387	401	416
05	430	444	458	473	487	510	515	530	544	558
06	572	586	601	615	629	643	657	671	686	700
07	714	728	742	756	770	785	799	813	827	841
08	855	869	883	897	911	926	940	954	968	982
09	48 996	*010	*024	*038	*052	*066	*080	*094	*108	*122
310	49 136	150	164	178	192	206	220	234	248	262
11	276	290	304	318	332	346	360	374	388	402
12	415	429	443	457	471	485	499	513	527	541
13	554	568	582	596	610	624	638	651	665	679
14	693	707	721	734	748	762	776	790	803	817
j15	831	845	859	872	886	900	914	927	941	955
16	49 969	982	996	*010	*024	*037	*051	*065	*079	*092
17	50 106	120	133	147	161	174	188	202	215	229
18	243	256	270	284	297	311	325	338	352	365
19	379	393	406	420	433	447	461	474	488	501
320	515	529	542	556	569	583	596	610	623	637
21	651	664	678	691	705	718	732	745	759	772
22	786	799	813	826	840	853	866	880	893	907
23	50 920	934	947	961	974	987	*001	*014	*028	*041
24	51 055	068	081	095	108	121	135	148	162	175
25	188	202	215	228	242	255	268	282	295	308
26	322	335	348	362	375	388	402	415	428	441
27	455	468	481	495	508	521	534	548	561	574
28	587	601	614	627	640	654	667	680	693	706
29	720	733	746	759	772	786	799	812	825	838
330	851	865	878	891	904	917	930	943	957	970
31	51 983	996	*009	*022	*035	*048	*061	*075	*088	*101
32	52 114	127	140	153	166	179	192	205	218	231
33	244	257	270	284	297	310	323	336	349	362
34	375	388	401	414	427	440	453	466	479	492
35	504	517	530	543	556	569	582	595	608	621
36	634	647	660	673	686	699	711	724	737	750
37	763	776	789	802	815	827	840	853	866	879
38	52 892	905	917	930	943	956	969	982	994	*007
39	53 020	033	046	058	071	084	097	110	122	135
340	148	161	173	186	199	212	224	237	250	263
41	275	288	301	314	326	339	352	364	377	390
42	403	415	428	441	453	466	479	491	504	517
43	529	542	555	567	580	593	605	618	631	643
44	656	668	681	694	706	719	732	744	757	769
45	782	794	807	820	832	845	857	870	882	895
46	53 908	920	933	945	958	970	983	995	*008	*020
47	54 033	045	058	070	083	095	108	120	133	145
48	158	170	183	195	208	220	233	245	258	270
49	283	295	307	320	332	345	357	370	382	394
350	54 407	419	432	444	456	469	481	494	506	518
N	0	1	2	3	4	5	6	7	8	9

Prop. Parts

	15
1	1.5
2	3.0
3	4.5
4	6.0
5	7.5
6	9.0
7	10.5
8	12.0
9	13.5

	14
1	1.4
2	2.8
3	4.2
4	5.6
5	7.0
6	8.4
7	9.8
8	11.2
9	12.6

	13
1	1.3
2	2.6
3	3.9
4	5.2
5	6.5
6	7.8
7	9.1
8	10.4
9	11.7

	12
1	1.2
2	2.4
3	3.6
4	4.8
5	6.0
6	7.2
7	8.4
8	9.6
9	10.8

TABLE B.1

Five-Place Logarithms: 350–400

N	0	1	2	3	4	5	6	7	8	9
350	54 407	419	432	444	456	469	481	494	506	518
51	531	543	555	568	580	593	605	617	630	642
52	654	667	679	691	704	716	728	741	753	765
53	777	790	802	814	827	839	851	864	876	888
54	54 900	913	925	937	949	962	974	986	998	*011
55	55 023	035	047	060	072	084	096	108	121	133
56	145	157	169	182	194	206	218	230	242	255
57	267	279	291	303	315	328	340	352	364	376
58	388	400	413	425	437	449	461	473	485	497
59	509	522	534	546	558	570	582	594	606	618
360	630	642	654	666	678	691	703	715	727	739
61	751	763	775	787	799	811	823	835	847	859
62	871	883	895	907	919	931	943	955	967	979
63	55 991	*003	*015	*027	*038	*050	*062	*074	*086	*098
64	56 110	122	134	146	158	170	182	194	205	217
65	229	241	253	265	277	289	301	312	324	336
66	348	360	372	384	396	407	419	431	443	455
67	467	478	490	502	514	526	538	549	561	573
68	585	597	608	620	632	644	656	667	679	691
69	703	714	726	738	750	761	773	785	797	808
370	820	832	844	855	867	879	891	902	914	926
71	56 937	949	961	972	984	996	*008	*019	*031	*043
72	57 054	066	078	089	101	113	124	136	148	159
73	171	183	194	206	217	229	241	252	264	276
74	287	299	310	322	334	345	357	368	380	392
75	403	415	426	438	449	461	473	484	496	507
76	519	530	542	553	565	576	588	600	611	623
77	634	646	657	669	680	692	703	715	726	738
78	749	761	772	784	795	807	818	830	841	852
79	864	875	887	898	910	921	933	944	955	967
380	57 978	990	*001	*013	*024	*035	*047	*058	*070	*081
81	58 092	104	115	127	138	149	161	172	184	195
82	206	218	229	240	252	263	274	286	297	309
83	320	331	343	354	365	377	388	399	410	422
84	433	444	456	467	478	490	501	512	524	535
85	546	557	569	580	591	602	614	625	636	647
86	659	670	681	692	704	715	726	737	749	760
87	771	782	794	805	816	827	838	850	861	872
88	883	894	906	917	928	939	950	961	973	984
89	58 995	*006	*017	*028	*040	*051	*062	*073	*084	*095
390	59 106	118	129	140	151	162	173	184	195	207
91	218	229	240	251	262	273	284	295	306	318
92	329	340	351	362	373	384	395	406	417	428
93	439	450	461	472	483	494	506	517	528	539
94	550	561	572	583	594	605	616	627	638	649
95	660	671	682	693	704	715	726	737	748	759
96	770	780	791	802	813	824	835	846	857	868
97	879	890	901	912	923	934	945	956	966	977
98	59 988	999	*010	*021	*032	*043	*054	*065	*076	*086
99	60 097	108	119	130	141	152	163	173	184	195
400	60 206	217	228	239	249	260	271	282	293	304

Prop. Parts

13		12		11		10	
1	1.3	1	1.2	1	1.1	1	1.0
2	2.6	2	2.4	2	2.2	2	2.0
3	3.9	3	3.6	3	3.3	3	3.0
4	5.2	4	4.8	4	4.4	4	4.0
5	6.5	5	6.0	5	5.5	5	5.0
6	7.8	6	7.2	6	6.6	6	6.0
7	9.1	7	8.4	7	7.7	7	7.0
8	10.4	8	9.6	8	8.8	8	8.0
9	11.7	9	10.8	9	9.9	9	9.0

TABLE B.1

Five-Place Logarithms: 400–450

N	0	1	2	3	4	5	6	7	8	9	Prop. Parts
400	60 206	217	228	239	249	260	271	282	293	304	
01	314	325	336	347	358	369	379	390	401	412	
02	423	433	444	455	466	477	487	498	509	520	
03	531	541	552	563	574	584	595	606	617	627	
04	638	649	660	670	681	692	703	713	724	735	
05	746	756	767	778	788	799	810	821	831	842	
06	853	863	874	885	895	906	917	927	938	949	
07	60 959	970	981	991	*002	*013	*023	*034	*045	*055	
08	61 066	077	087	098	109	119	130	140	151	162	
09	172	183	194	204	215	225	236	247	257	268	
410	278	289	300	310	321	331	342	352	363	374	
11	384	395	405	416	426	437	448	458	469	479	
12	490	500	511	521	532	542	553	563	574	584	
13	595	606	616	627	637	648	658	669	679	690	
14	700	711	721	731	742	752	763	773	784	794	
15	805	815	826	836	847	857	868	878	888	899	
16	61 909	920	930	941	951	962	972	982	993	*003	
17	62 014	024	034	045	055	066	076	086	097	107	
18	118	128	138	149	159	170	180	190	201	211	
19	221	232	242	252	263	273	284	294	304	315	
420	325	335	346	356	366	377	387	397	408	418	
21	428	439	449	459	469	480	490	500	511	521	
22	531	542	552	562	572	583	593	603	613	624	
23	634	644	655	665	675	685	696	706	716	726	
24	737	747	757	767	778	788	798	808	818	829	
25	839	849	859	870	880	890	900	910	921	931	
26	62 941	951	961	972	982	992	*002	*012	*022	*033	
27	63 043	053	063	073	083	094	104	114	124	134	
28	144	155	165	175	185	195	205	215	225	236	
29	246	256	266	276	286	296	306	317	327	337	
430	347	357	367	377	387	397	407	417	428	438	
31	448	458	468	478	488	498	508	518	528	538	
32	548	558	568	579	589	599	609	619	629	639	
33	649	659	669	679	689	699	709	719	729	739	
34	749	759	769	779	789	799	809	819	829	839	
35	849	859	869	879	889	899	909	919	929	939	
36	63 949	959	969	979	988	998	*008	*018	*028	*038	
37	64 048	058	068	078	088	098	108	118	128	137	
38	147	157	167	177	187	197	207	217	227	237	
39	246	256	266	276	286	296	306	316	326	335	
440	345	355	365	375	385	395	404	414	424	434	
41	444	454	464	473	483	493	503	513	523	532	
42	542	552	562	572	582	591	601	611	621	631	
43	640	650	660	670	680	689	699	709	719	729	
44	738	748	758	768	777	787	797	807	816	826	
45	836	846	856	865	875	885	895	904	914	924	
46	64 933	943	953	963	972	982	992	*002	*011	*021	
47	65 031	040	050	060	070	079	089	099	108	118	
48	128	137	147	157	167	176	186	196	205	215	
49	225	234	244	254	263	273	283	292	302	312	
450	65 321	331	341	350	360	369	379	389	398	408	
N	0	1	2	3	4	5	6	7	8	9	Prop. Parts

Prop. Parts

	11
1	1.1
2	2.2
3	3.3
4	4.4
5	5.5
6	6.6
7	7.7
8	8.8
9	9.9

	10
1	1.0
2	2.0
3	3.0
4	4.0
5	5.0
6	6.0
7	7.0
8	8.0
9	9.0

	9
1	0.9
2	1.8
3	2.7
4	3.6
5	4.5
6	5.4
7	6.3
8	7.2
9	8.1

TABLE B.1

Five-Place Logarithms: 450–500

Prop. Parts	N	0	1	2	3	4	5	6	7	8	9
	450	65 321	331	341	350	360	369	379	389	398	408
	51	418	427	437	447	456	466	475	485	495	504
	52	514	523	533	543	552	562	571	581	591	600
	53	610	619	629	639	648	658	667	677	686	696
	54	706	715	725	734	744	753	763	772	782	792
	55	801	811	820	830	839	849	858	868	877	887
	56	896	906	916	925	935	944	954	963	973	982
	57	65 992	*001	*011	*020	*030	*039	*049	*058	*068	*077
	58	66 087	096	106	115	124	134	143	153	162	172
	59	181	191	200	210	219	229	238	247	257	266
	460	276	285	295	304	314	323	332	342	351	361
	61	370	380	389	398	408	417	427	436	445	455
	62	464	474	483	492	502	511	521	530	539	549
	63	558	567	577	586	596	605	614	624	633	642
	64	652	661	671	680	689	699	708	717	727	736
	65	745	755	764	773	783	792	801	811	820	829
	66	839	848	857	867	876	885	894	904	913	922
	67	66 932	941	950	960	969	978	987	997	*006	*015
	68	67 025	034	043	052	062	071	080	089	099	108
	69	117	127	136	145	154	164	173	182	191	201
	470	210	219	228	237	247	256	265	274	284	293
	71	302	311	321	330	339	348	357	367	376	385
	72	394	403	413	422	431	440	449	459	468	477
	73	486	495	504	514	523	532	541	550	560	569
	74	578	587	596	605	614	624	633	642	651	660
	75	669	679	688	697	706	715	724	733	742	752
	76	761	770	779	788	797	806	815	825	834	843
	77	852	861	870	879	888	897	906	916	925	934
	78	67 943	952	961	970	979	988	997	*006	*015	*024
	79	68 034	043	052	061	070	079	088	097	106	115
	480	124	133	142	151	160	169	178	187	196	205
	81	215	224	233	242	251	260	269	278	287	296
	82	305	314	323	332	341	350	359	368	377	386
	83	395	404	413	422	431	440	449	458	467	476
	84	485	494	502	511	520	529	538	547	556	565
	85	574	583	592	601	610	619	628	637	646	655
	86	664	673	681	690	699	708	717	726	735	744
	87	753	762	771	780	789	797	806	815	824	833
	88	842	851	860	869	878	886	895	904	913	922
	89	68 931	940	949	958	966	975	984	993	*002	*011
	490	69 020	028	037	046	055	064	073	082	090	099
	91	108	117	126	135	144	152	161	170	179	188
	92	197	205	214	223	232	241	249	258	267	276
	93	285	294	302	311	320	329	338	346	355	364
	94	373	381	390	399	408	417	425	434	443	452
	95	461	469	478	487	496	504	513	522	531	539
	96	548	557	566	574	583	592	601	609	618	627
	97	636	644	653	662	671	679	688	697	705	714
	98	723	732	740	749	758	767	775	784	793	801
	99	810	819	827	836	845	854	862	871	880	888
	500	69 897	906	914	923	932	940	949	958	966	975
Prop. Parts	**N**	**0**	**1**	**2**	**3**	**4**	**5**	**6**	**7**	**8**	**9**

Proportional Parts

10		9		8	
1	1.0	1	0.9	1	0.8
2	2.0	2	1.8	2	1.6
3	3.0	3	2.7	3	2.4
4	4.0	4	3.6	4	3.2
5	5.0	5	4.5	5	4.0
6	6.0	6	5.4	6	4.8
7	7.0	7	6.3	7	5.6
8	8.0	8	7.2	8	6.4
9	9.0	9	8.1	9	7.2

TABLE B.1

Five-Place Logarithms: 500–550

N	0	1	2	3	4	5	6	7	8	9
500	69 897	906	914	923	932	940	949	958	966	975
01	69 984	992	*001	*010	*018	*027	*036	*044	*053	*062
02	70 070	079	088	096	105	114	122	131	140	148
03	157	165	174	183	191	200	209	217	226	234
04	243	252	260	269	278	286	295	303	312	321
05	329	338	346	355	364	372	381	389	398	406
06	415	424	432	441	449	458	467	475	484	492
07	501	509	518	526	535	544	552	561	569	578
08	586	595	603	612	621	629	638	646	655	663
09	672	680	689	697	706	714	723	731	740	749
510	757	766	774	783	791	800	808	817	825	834
11	842	851	859	868	876	885	893	902	910	919
12	70 927	935	944	952	961	969	978	986	995	*003
13	71 012	020	029	037	046	054	063	071	079	088
14	096	105	113	122	130	139	147	155	164	172
15	181	189	198	206	214	223	231	240	248	257
16	265	273	282	290	299	307	315	324	332	341
17	349	357	366	374	383	391	399	408	416	425
18	433	441	450	458	466	475	483	492	500	508
19	517	525	533	542	550	559	567	575	584	592
520	600	609	617	625	634	642	650	659	667	675
21	684	692	700	709	717	725	734	742	750	759
22	767	775	784	792	800	809	817	825	834	842
23	850	858	867	875	883	892	900	908	917	925
24	71 933	941	950	958	966	975	983	991	999	*008
25	72 016	024	032	041	049	057	066	074	082	090
26	099	107	115	123	132	140	148	156	165	173
27	181	189	198	206	214	222	230	239	247	255
28	263	272	280	288	296	304	313	321	329	337
29	346	354	362	370	378	387	395	403	411	419
530	428	436	444	452	460	469	477	485	493	501
31	509	518	526	534	542	550	558	567	575	583
32	591	599	607	616	624	632	640	648	656	665
33	673	681	689	697	705	713	722	730	738	746
34	754	762	770	779	787	795	803	811	819	827
35	835	843	852	860	868	876	884	892	900	908
36	916	925	933	941	949	957	965	973	981	989
37	72 997	*006	*014	*022	*030	*038	*046	*054	*062	*070
38	73 078	086	094	102	111	119	127	135	143	151
39	159	167	175	183	191	199	207	215	223	231
540	239	247	255	263	272	280	288	296	304	312
41	320	328	336	344	352	360	368	376	384	392
42	400	408	416	424	432	440	448	456	464	472
43	480	488	496	504	512	520	528	536	544	552
44	560	568	576	584	592	600	608	616	624	632
45	640	648	656	664	672	679	687	695	703	711
46	719	727	735	743	751	759	767	775	783	791
47	799	807	815	823	830	838	846	854	862	870
48	878	886	894	902	910	918	926	933	941	949
49	73 957	965	973	981	989	997	*005	*013	*020	*028
550	74 036	044	052	060	068	076	084	092	099	107
N	0	1	2	3	4	5	6	7	8	9

Prop. Parts

	9		8		7
1	0.9	1	0.8	1	0.7
2	1.8	2	1.6	2	1.4
3	2.7	3	2.4	3	2.1
4	3.6	4	3.2	4	2.8
5	4.5	5	4.0	5	3.5
6	5.4	6	4.8	6	4.2
7	6.3	7	5.6	7	4.9
8	7.2	8	6.4	8	5.6
9	8.1	9	7.2	9	6.3

TABLE B.1

Five-Place Logarithms: 550–600

Prop. Parts	N	0	1	2	3	4	5	6	7	8	9
	550	74 036	044	052	060	068	076	084	092	099	107
	51	115	123	131	139	147	155	162	170	178	186
	52	194	202	210	218	225	233	241	249	257	265
	53	273	280	288	296	304	312	320	327	335	343
	54	351	359	367	374	382	390	398	406	414	421
	55	429	437	445	453	461	468	476	484	492	500
	56	507	515	523	531	539	547	554	562	570	578
	57	586	593	601	609	617	624	632	640	648	656
	58	663	671	679	687	695	702	710	718	726	733
	59	741	749	757	764	772	780	788	796	803	811
	560	819	827	834	842	850	858	865	873	881	889
	61	896	904	912	920	927	935	943	950	958	966
	62	74 974	981	989	997	*005	*012	*020	*028	*035	*043
	63	75 051	059	066	074	082	089	097	105	113	120
	64	128	136	143	151	159	166	174	182	189	197
	65	205	213	220	228	236	243	251	259	266	274
	66	282	289	297	305	312	320	328	335	343	351
	67	358	366	374	381	389	397	404	412	420	427
	68	435	442	450	458	465	473	481	488	496	504
	69	511	519	526	534	542	549	557	565	572	580
	570	587	595	603	610	618	626	633	641	648	656
	71	664	671	679	686	694	702	709	717	724	732
	72	740	747	755	762	770	778	785	793	800	808
	73	815	823	831	838	846	853	861	868	876	884
	74	891	899	906	914	921	929	937	944	952	959
	75	75 967	974	982	989	997	*005	*012	*020	*027	*035
	76	76 042	050	057	065	072	080	087	095	103	110
	77	118	125	133	140	148	155	163	170	178	185
	78	193	200	208	215	223	230	238	245	253	260
	79	268	275	283	290	298	305	313	320	328	335
	580	343	350	358	365	373	380	388	395	403	410
	81	418	425	433	440	448	455	462	470	477	485
	82	492	500	507	515	522	530	537	545	552	559
	83	567	574	582	589	597	604	612	619	626	634
	84	641	649	656	664	671	678	686	693	701	708
	85	716	723	730	738	745	753	760	768	775	782
	86	790	797	805	812	819	827	834	842	849	856
	87	864	871	879	886	893	901	908	916	923	930
	88	76 938	945	953	960	967	975	982	989	997	*004
	89	77 012	019	026	034	041	048	056	063	070	078
	590	085	093	100	107	115	122	129	137	144	151
	91	159	166	173	181	188	195	203	210	217	225
	92	232	240	247	254	262	269	276	283	291	298
	93	305	313	320	327	335	342	349	357	364	371
	94	379	386	393	401	408	415	422	430	437	444
	95	452	459	466	474	481	488	495	503	510	517
	96	525	532	539	546	554	561	568	576	583	590
	97	597	605	612	619	627	634	641	648	656	663
	98	670	677	685	692	699	706	714	721	728	735
	99	743	750	757	764	772	779	786	793	801	808
	600	77 815	822	830	837	844	851	859	866	873	880
Prop. Parts	**N**	**0**	**1**	**2**	**3**	**4**	**5**	**6**	**7**	**8**	**9**

Prop. Parts (against rows 61–69):

	8
1	0.8
2	1.6
3	2.4
4	3.2
5	4.0
6	4.8
7	5.6
8	6.4
9	7.2

Prop. Parts (against rows 81–89):

	7
1	0.7
2	1.4
3	2.1
4	2.8
5	3.5
6	4.2
7	4.9
8	5.6
9	6.3

TABLE B.1
Five-Place Logarithms: 600–650

N	0	1	2	3	4	5	6	7	8	9	Prop. Parts
600	77 815	822	830	837	844	851	859	866	873	880	
01	887	895	902	909	916	924	931	938	945	952	
02	77 960	967	974	981	988	996	*003	*010	*017	*025	
03	78 032	039	046	053	061	068	075	082	089	097	
04	104	111	118	125	132	140	147	154	161	168	
05	176	183	190	197	204	211	219	226	233	240	
06	247	254	262	269	276	283	290	297	305	312	
07	319	326	333	340	347	355	362	369	376	383	
08	390	398	405	412	419	426	433	440	447	455	
09	462	469	476	483	490	497	504	512	519	526	
610	533	540	547	554	561	569	576	583	590	597	
11	604	611	618	625	633	640	647	654	661	668	
12	675	682	689	696	704	711	718	725	732	739	
13	746	753	760	767	774	781	789	796	803	810	
14	817	824	831	838	845	852	859	866	873	880	
15	888	895	902	909	916	923	930	937	944	951	
16	78 958	965	972	979	986	993	*000	*007	*014	*021	
17	79 029	036	043	050	057	064	071	078	085	092	
18	099	106	113	120	127	134	141	148	155	162	
19	169	176	183	190	197	204	211	218	225	232	
620	239	246	253	260	267	274	281	288	295	302	
21	309	316	323	330	337	344	351	358	365	372	
22	379	386	393	400	407	414	421	428	435	442	
23	449	456	463	470	477	484	491	498	505	511	
24	518	525	532	539	546	553	560	567	574	581	
25	588	595	602	609	616	623	630	637	644	650	
26	657	664	671	678	685	692	699	706	713	720	
27	727	734	741	748	754	761	768	775	782	789	
28	796	803	810	817	824	831	837	844	851	858	
29	865	872	879	886	893	900	906	913	920	927	
630	79 934	941	948	955	962	969	975	982	989	996	
31	80 003	010	017	024	030	037	044	051	058	065	
32	072	079	085	092	099	106	113	120	127	134	
33	140	147	154	161	168	175	182	188	195	202	
34	209	216	223	229	236	243	250	257	264	271	
35	277	284	291	298	305	312	318	325	332	339	
36	346	353	359	366	373	380	387	393	400	407	
37	414	421	428	434	441	448	455	462	468	475	
38	482	489	496	502	509	516	523	530	536	543	
39	550	557	564	570	577	584	591	598	604	611	
640	618	625	632	638	645	652	659	665	672	679	
41	686	693	699	706	713	720	726	733	740	747	
42	754	760	767	774	781	787	794	801	808	814	
43	821	828	835	841	848	855	862	868	875	882	
44	889	895	902	909	916	922	929	936	943	949	
45	80 956	963	969	976	983	990	996	*003	*010	*017	
46	81 023	030	037	043	050	057	064	070	077	084	
47	090	097	104	111	117	124	131	137	144	151	
48	158	164	171	178	184	191	198	204	211	218	
49	224	231	238	245	251	258	265	271	278	285	
650	81 291	298	305	311	318	325	331	338	345	351	
N	0	1	2	3	4	5	6	7	8	9	Prop. Parts

Prop. Parts

8	
1	0.8
2	1.6
3	2.4
4	3.2
5	4.0
6	4.8
7	5.6
8	6.4
9	7.2

7	
1	0.7
2	1.4
3	2.1
4	2.8
5	3.5
6	4.2
7	4.9
8	5.6
9	6.3

6	
1	0.6
2	1.2
3	1.8
4	2.4
5	3.0
6	3.6
7	4.2
8	4.8
9	5.4

TABLE B.1

Five-Place Logarithms: 650–700

Prop. Parts	N	0	1	2	3	4	5	6	7	8	9
	650	81 291	298	305	311	318	325	331	338	345	351
	51	358	365	371	378	385	391	398	405	411	418
	52	425	431	438	445	451	458	465	471	478	485
	53	491	498	505	511	518	525	531	538	544	551
	54	558	564	571	578	584	591	598	604	611	617
	55	624	631	637	644	651	657	664	671	677	684
	56	690	697	704	710	717	723	730	737	743	750
	57	757	763	770	776	783	790	796	803	809	816
	58	823	829	836	842	849	856	862	869	875	882
	59	889	895	902	908	915	921	928	935	941	948
	660	81 954	961	968	974	981	987	994	*000	*007	*014
	61	82 020	027	033	040	046	053	060	066	073	079
	62	086	092	099	105	112	119	125	132	138	145
	63	151	158	164	171	178	184	191	197	204	210
	64	217	223	230	236	243	249	256	263	269	276
	65	282	289	295	302	308	315	321	328	334	341
	66	347	354	360	367	373	380	387	393	400	406
	67	413	419	426	432	439	445	452	458	465	471
	68	478	484	491	497	504	510	517	523	530	536
	69	543	549	556	562	569	575	582	588	595	601
	670	607	614	620	627	633	640	646	653	659	666
	71	672	679	685	692	698	705	711	718	724	730
	72	737	743	750	756	763	769	776	782	789	795
	73	802	808	814	821	827	834	840	847	853	860
	74	866	872	879	885	892	898	905	911	918	924
	75	930	937	943	950	956	963	969	975	982	988
	76	82 995	*001	*008	*014	*020	*027	*033	*040	*046	*052
	77	83 059	065	072	078	085	091	097	104	110	117
	78	123	129	136	142	149	155	161	168	174	181
	79	187	193	200	206	213	219	225	232	238	245
	680	251	257	264	270	276	283	289	296	302	308
	81	315	321	327	334	340	347	353	359	366	372
	82	378	385	391	398	404	410	417	423	429	436
	83	442	448	455	461	467	474	480	487	493	499
	84	506	512	518	525	531	537	544	550	556	563
	85	569	575	582	588	594	601	607	613	620	626
	86	632	639	645	651	658	664	670	677	683	689
	87	696	702	708	715	721	727	734	740	746	753
	88	759	765	771	778	784	790	797	803	809	816
	89	822	828	835	841	847	853	860	866	872	879
	690	885	891	897	904	910	916	923	929	935	942
	91	83 948	954	960	967	973	979	985	992	998	*004
	92	84 011	017	023	029	036	042	048	055	061	067
	93	073	080	086	092	098	105	111	117	123	130
	94	136	142	148	155	161	167	173	180	186	192
	95	198	205	211	217	223	230	236	242	248	255
	96	261	267	273	280	286	292	298	305	311	317
	97	323	330	336	342	348	354	361	367	373	379
	98	386	392	398	404	410	417	423	429	435	442
	99	448	454	460	466	473	479	485	491	497	504
	700	84 510	516	522	528	535	541	547	553	559	566

| Prop. Parts | N | 0 | 1 | 2 | 3 | 4 | 5 | 6 | 7 | 8 | 9 |

Proportional Parts:

7
1 | 0.7
2 | 1.4
3 | 2.1
4 | 2.8
5 | 3.5
6 | 4.2
7 | 4.9
8 | 5.6
9 | 6.3

6
1 | 0.6
2 | 1.2
3 | 1.8
4 | 2.4
5 | 3.0
6 | 3.6
7 | 4.2
8 | 4.8
9 | 5.4

TABLE B.1

Five-Place Logarithms: 700–750

N	0	1	2	3	4	5	6	7	8	9	Prop. Parts
700	84 510	516	522	528	535	541	547	553	559	566	
01	572	578	584	590	597	603	609	615	621	628	
02	634	640	646	652	658	665	671	677	683	689	
03	696	702	708	714	720	726	733	739	745	751	
04	757	763	770	776	782	788	794	800	807	813	
05	819	825	831	837	844	850	856	862	868	874	
06	880	887	893	899	905	911	917	924	930	936	
07	84 942	948	954	960	967	973	979	985	991	997	
08	85 003	009	016	022	028	034	040	046	052	058	
09	065	071	077	083	089	095	101	107	114	120	
710	126	132	138	144	150	156	163	169	175	181	
11	187	193	199	205	211	217	224	230	236	242	
12	248	254	260	266	272	278	285	291	297	303	
13	309	315	321	327	333	339	345	352	358	364	
14	370	376	382	388	394	400	406	412	418	425	
15	431	437	443	449	455	461	467	473	479	485	
16	491	497	503	509	516	522	528	534	540	546	
17	552	558	564	570	576	582	588	594	600	606	
18	612	618	625	631	637	643	649	655	661	667	
19	673	679	685	691	697	703	709	715	721	727	
720	733	739	745	751	757	763	769	775	781	788	
21	794	800	806	812	818	824	830	836	842	848	
22	854	860	866	872	878	884	890	896	902	908	
23	914	920	926	932	938	944	950	956	962	968	
24	85 974	980	986	992	998	*004	*010	*016	*022	*028	
25	86 034	040	046	052	058	064	070	076	082	088	
26	094	100	106	112	118	124	130	136	141	147	
27	153	159	165	171	177	183	189	195	201	207	
28	213	219	225	231	237	243	249	255	261	267	
29	273	279	285	291	297	303	308	314	320	326	
730	332	338	344	350	356	362	368	374	380	386	
31	392	398	404	410	415	421	427	433	439	445	
32	451	457	463	469	475	481	487	493	499	504	
33	510	516	522	528	534	540	546	552	558	564	
34	570	576	581	587	593	599	605	611	617	623	
35	629	635	641	646	652	658	664	670	676	682	
36	688	694	700	705	711	717	723	729	735	741	
37	747	753	759	764	770	776	782	788	794	800	
38	806	812	817	823	829	835	841	847	853	859	
39	864	870	876	882	888	894	900	906	911	917	
740	923	929	935	941	947	953	958	964	970	976	
41	86 982	988	994	999	*005	*011	*017	*023	*029	*035	
42	87 040	046	052	058	064	070	075	081	087	093	
43	099	105	111	116	122	128	134	140	146	151	
44	157	163	169	175	181	186	192	198	204	210	
45	216	221	227	233	239	245	251	256	262	268	
46	274	280	286	291	297	303	309	315	320	326	
47	332	338	344	349	355	361	367	373	379	384	
48	390	396	402	408	413	419	425	431	437	442	
49	448	454	460	466	471	477	483	489	495	500	
750	87 506	512	518	523	529	535	541	547	552	558	
N	0	1	2	3	4	5	6	7	8	9	Prop. Parts

Prop. Parts:

	7
1	0.7
2	1.4
3	2.1
4	2.8
5	3.5
6	4.2
7	4.9
8	5.6
9	6.3

	6
1	0.6
2	1.2
3	1.8
4	2.4
5	3.0
6	3.6
7	4.2
8	4.8
9	5.4

	5
1	0.5
2	1.0
3	1.5
4	2.0
5	2.5
6	3.0
7	3.5
8	4.0
9	4.5

TABLE B.1

Five-Place Logarithms: 750–800

Prop. Parts	N	0	1	2	3	4	5	6	7	8	9
	750	87 506	512	518	523	529	535	541	547	552	558
	51	564	570	576	581	587	593	599	604	610	616
	52	622	628	633	639	645	651	656	662	668	674
	53	679	685	691	697	703	708	714	720	726	731
	54	737	743	749	754	760	766	772	777	783	789
	55	795	800	806	812	818	823	829	835	841	846
	56	852	858	864	869	875	881	887	892	898	904
	57	910	915	921	927	933	938	944	950	955	961
	58	87 967	973	978	984	990	996	*001	*007	*013	*018
	59	88 024	030	036	041	047	053	058	064	070	076
	760	081	087	093	098	104	110	116	121	127	133
	61	138	144	150	156	161	167	173	178	184	190
	62	195	201	207	213	218	224	230	235	241	247
	63	252	258	264	270	275	281	287	292	298	304
	64	309	315	321	326	332	338	343	349	355	360
	65	366	372	377	383	389	395	400	406	412	417
	66	423	429	434	440	446	451	457	463	468	474
	67	480	485	491	497	502	508	513	519	525	530
	68	536	542	547	553	559	564	570	576	581	587
	69	593	598	604	610	615	621	627	632	638	643
	770	649	655	660	666	672	677	683	689	694	700
	71	705	711	717	722	728	734	739	745	750	756
	72	762	767	773	779	784	790	795	801	807	812
	73	818	824	829	835	840	846	852	857	863	868
	74	874	880	885	891	897	902	908	913	919	925
	75	930	936	941	947	953	958	964	969	975	981
	76	88 986	992	997	*003	*009	*014	*020	*025	*031	*037
	77	89 042	048	053	059	064	070	076	081	087	092
	78	098	104	109	115	120	126	131	137	143	148
	79	154	159	165	170	176	182	187	193	198	204
	780	209	215	221	226	232	237	243	248	254	260
	81	265	271	276	282	287	293	298	304	310	315
	82	321	326	332	337	343	348	354	360	365	371
	83	376	382	387	393	398	404	409	415	421	426
	84	432	437	443	448	454	459	465	470	476	481
	85	487	492	498	504	509	515	520	526	531	537
	86	542	548	553	559	564	570	575	581	586	592
	87	597	603	609	614	620	625	631	636	642	647
	88	653	658	664	669	675	680	686	691	697	702
	89	708	713	719	724	730	735	741	746	752	757
	790	763	768	774	779	785	790	796	801	807	812
	91	818	823	829	834	840	845	851	856	862	867
	92	873	878	883	889	894	900	905	911	916	922
	93	927	933	938	944	949	955	960	966	971	977
	94	89 982	988	993	998	*004	*009	*015	*020	*026	*031
	95	90 037	042	048	053	059	064	069	075	080	086
	96	091	097	102	108	113	119	124	129	135	140
	97	146	151	157	162	168	173	179	184	189	195
	98	200	206	211	217	222	227	233	238	244	249
	99	255	260	266	271	276	282	287	293	298	304
	800	90 309	314	320	325	331	336	342	347	352	358
Prop. Parts	N	0	1	2	3	4	5	6	7	8	9

Prop. Parts (alongside rows 61–69):

	6
1	0.6
2	1.2
3	1.8
4	2.4
5	3.0
6	3.6
7	4.2
8	4.8
9	5.4

Prop. Parts (alongside rows 81–89):

	5
1	0.5
2	1.0
3	1.5
4	2.0
5	2.5
6	3.0
7	3.5
8	4.0
9	4.5

TABLE B.1

.ve-Place Logarithms: 800–850

N	0	1	2	3	4	5	6	7	8	9	Prop. Parts
800	90 309	314	320	325	331	336	342	347	352	358	
01	363	369	374	380	385	390	396	401	407	412	
02	417	423	428	434	439	445	450	455	461	466	
03	472	477	482	488	493	499	504	509	515	520	
04	526	531	536	542	547	553	558	563	569	574	
05	580	585	590	596	601	607	612	617	623	628	
06	634	639	644	650	655	660	666	671	677	682	
07	687	693	698	703	709	714	720	725	730	736	
08	741	747	752	757	763	768	773	779	784	789	
09	795	800	806	811	816	822	827	832	838	843	
810	849	854	859	865	870	875	881	886	891	897	
11	902	907	913	918	924	929	934	940	945	950	
12	90 956	961	966	972	977	982	988	993	998	*004	
13	91 009	014	020	025	030	036	041	046	052	057	
14	062	068	073	078	084	089	094	100	105	110	
15	116	121	126	132	137	142	148	153	158	164	
16	169	174	180	185	190	196	201	206	212	217	
17	222	228	233	238	243	249	254	259	265	270	
18	275	281	286	291	297	302	307	312	318	323	
19	328	334	339	344	350	355	360	365	371	376	
820	381	387	392	397	403	408	413	418	424	429	
21	434	440	445	450	455	461	466	471	477	482	
22	487	492	498	503	508	514	519	524	529	535	
23	540	545	551	556	561	566	572	577	582	587	
24	593	598	603	609	614	619	624	630	635	640	
25	645	651	656	661	666	672	677	682	687	693	
26	698	703	709	714	719	724	730	735	740	745	
27	751	756	761	766	772	777	782	787	793	798	
28	803	808	814	819	824	829	834	840	845	850	
29	855	861	866	871	876	882	887	892	897	903	
830	908	913	918	924	929	934	939	944	950	955	
31	91 960	965	971	976	981	986	991	997	*002	*007	
32	92 012	018	023	028	033	038	044	049	054	059	
33	065	070	075	080	085	091	096	101	106	111	
34	117	122	127	132	137	143	148	153	158	163	
35	169	174	179	184	189	195	200	205	210	215	
36	221	226	231	236	241	247	252	257	262	267	
37	273	278	283	288	293	298	304	309	314	319	
38	324	330	335	340	345	350	355	361	366	371	
39	376	381	387	392	397	402	407	412	418	423	
840	428	433	438	443	449	454	459	464	469	474	
41	480	485	490	495	500	505	511	516	521	526	
42	531	536	542	547	552	557	562	567	572	578	
43	583	588	593	598	603	609	614	619	624	629	
44	634	639	645	650	655	660	665	670	675	681	
45	686	691	696	701	706	711	716	722	727	732	
46	737	742	747	752	758	763	768	773	778	783	
47	788	793	799	804	809	814	819	824	829	834	
48	840	845	850	855	860	865	870	875	881	886	
49	891	896	901	906	911	916	921	927	932	937	
850	92 942	947	952	957	962	967	973	978	983	988	
N	**0**	**1**	**2**	**3**	**4**	**5**	**6**	**7**	**8**	**9**	**Prop. Parts**

Prop. Parts — 6

1	0.6
2	1.2
3	1.8
4	2.4
5	3.0
6	3.6
7	4.2
8	4.8
9	5.4

Prop. Parts — 5

1	0.5
2	1.0
3	1.5
4	2.0
5	2.5
6	3.0
7	3.5
8	4.0
9	4.5

TABLE B.1

Five-Place Logarithms: 850–900

Prop. Parts	N	0	1	2	3	4	5	6	7	8	9
	850	92 942	947	952	957	962	967	973	978	983	988
	51	92 993	998	*003	*008	*013	*018	*024	*029	*034	*039
	52	93 044	049	054	059	064	069	075	080	085	090
	53	095	100	105	110	115	120	125	131	136	141
	54	146	151	156	161	166	171	176	181	186	192
	55	197	202	207	212	217	222	227	232	237	242
	56	247	252	258	263	268	273	278	283	288	293
	57	298	303	308	313	318	323	328	334	339	344
	58	349	354	359	364	369	374	379	384	389	394
	59	399	404	409	414	420	425	430	435	440	445
	860	450	455	460	465	470	475	480	485	490	495
	61	500	505	510	515	520	526	531	536	541	546
	62	551	556	561	566	571	576	581	586	591	596
	63	601	606	611	616	621	626	631	636	641	646
	64	651	656	661	666	671	676	682	687	692	697
	65	702	707	712	717	722	727	732	737	742	747
	66	752	757	762	767	772	777	782	787	792	797
	67	802	807	812	817	822	827	832	837	842	847
	68	852	857	862	867	872	877	882	887	892	897
	69	902	907	912	917	922	927	932	937	942	947
	870	93 952	957	962	967	972	977	982	987	992	997
	71	94 002	007	012	017	022	027	032	037	042	047
	72	052	057	062	067	072	077	082	086	091	096
	73	101	106	111	116	121	126	131	136	141	146
	74	151	156	161	166	171	176	181	186	191	196
	75	201	206	211	216	221	226	231	236	240	245
	76	250	255	260	265	270	275	280	285	290	295
	77	300	305	310	315	320	325	330	335	340	345
	78	349	354	359	364	369	374	379	384	389	394
	79	399	404	409	414	419	424	429	433	438	443
	880	448	453	458	463	468	473	478	483	488	493
	81	498	503	507	512	517	522	527	532	537	542
	82	547	552	557	562	567	571	576	581	586	591
	83	596	601	606	611	616	621	626	630	635	640
	84	645	650	655	660	665	670	675	680	685	689
	85	694	699	704	709	714	719	724	729	734	738
	86	743	748	753	758	763	768	773	778	783	787
	87	792	797	802	807	812	817	822	827	832	836
	88	841	846	851	856	861	866	871	876	880	885
	89	890	895	900	905	910	915	919	924	929	934
	890	939	944	949	954	959	963	968	973	978	983
	91	94 988	993	998	*002	*007	*012	*017	*022	*027	*032
	92	95 036	041	046	051	056	061	066	071	075	080
	93	085	090	095	100	105	109	114	119	124	129
	94	134	139	143	148	153	158	163	168	173	177
	95	182	187	192	197	202	207	211	216	221	226
	96	231	236	240	245	250	255	260	265	270	274
	97	279	284	289	294	299	303	308	313	318	323
	98	328	332	337	342	347	352	357	361	366	371
	99	376	381	386	390	395	400	405	410	415	419
	900	95 424	429	434	439	444	448	453	458	463	468
Prop. Parts	**N**	**0**	**1**	**2**	**3**	**4**	**5**	**6**	**7**	**8**	**9**

Prop. Parts (left margin):

6
1	0.6
2	1.2
3	1.8
4	2.4
5	3.0
6	3.6
7	4.2
8	4.8
9	5.4

5
1	0.5
2	1.0
3	1.5
4	2.0
5	2.5
6	3.0
7	3.5
8	4.0
9	4.5

4
1	0.4
2	0.8
3	1.2
4	1.6
5	2.0
6	2.4
7	2.8
8	3.2
9	3.6

TABLE B.1

Five-Place Logarithms: 900–950

N	0	1	2	3	4	5	6	7	8	9	Prop. Parts
900	95 424	429	434	439	444	448	453	458	463	468	
01	472	477	482	487	492	497	501	506	511	516	
02	521	525	530	535	540	545	550	554	559	564	
03	569	574	578	583	588	593	598	602	607	612	
04	617	622	626	631	636	641	646	650	655	660	
05	665	670	674	679	684	689	694	698	703	708	
06	713	718	722	727	732	737	742	746	751	756	
07	761	766	770	775	780	785	789	794	799	804	
08	809	813	818	823	828	832	837	842	847	852	
09	856	861	866	871	875	880	885	890	895	899	
910	904	909	914	918	923	928	933	938	942	947	
11	952	957	961	966	971	976	980	985	990	995	
12	95 999	*004	*009	*014	*019	*023	*028	*033	*038	*042	
13	96 047	052	057	061	066	071	076	080	085	090	
14	095	099	104	109	114	118	123	128	133	137	
15	142	147	152	156	161	166	171	175	180	185	
16	190	194	199	204	209	213	218	223	227	232	
17	237	242	246	251	256	261	265	270	275	280	
18	284	289	294	298	303	308	313	317	322	327	
19	332	336	341	346	350	355	360	365	369	374	
920	379	384	388	393	398	402	407	412	417	421	
21	426	431	435	440	445	450	454	459	464	468	
22	473	478	483	487	492	497	501	506	511	515	
23	520	525	530	534	539	544	548	553	558	562	
24	567	572	577	581	586	591	595	600	605	609	
25	614	619	624	628	633	638	642	647	652	656	
26	661	666	670	675	680	685	689	694	699	703	
27	708	713	717	722	727	731	736	741	745	750	
28	755	759	764	769	774	778	783	788	792	797	
29	802	806	811	816	820	825	830	834	839	844	
930	848	853	858	862	867	872	876	881	886	890	
31	895	900	904	909	914	918	923	928	932	937	
32	942	946	951	956	960	965	970	974	979	984	
33	96 988	993	997	*002	*007	*011	*016	*021	*025	*030	
34	97 035	039	044	049	053	058	063	067	072	077	
35	081	086	090	095	100	104	109	114	118	123	
36	128	132	137	142	146	151	155	160	165	169	
37	174	179	183	188	192	197	202	206	211	216	
38	220	225	230	234	239	243	248	253	257	262	
39	267	271	276	280	285	290	294	299	304	308	
940	313	317	322	327	331	336	340	345	350	354	
41	359	364	368	373	377	382	387	391	396	400	
42	405	410	414	419	424	428	433	437	442	447	
43	451	456	460	465	470	474	479	483	488	493	
44	497	502	506	511	516	520	525	529	534	539	
45	543	548	552	557	562	566	571	575	580	585	
46	589	594	598	603	607	612	617	621	626	630	
47	635	640	644	649	653	658	663	667	672	676	
48	681	685	690	695	699	704	708	713	717	722	
49	727	731	736	740	745	749	754	759	763	768	
950	97 772	777	782	786	791	795	800	804	809	813	
N	0	1	2	3	4	5	6	7	8	9	Prop. Parts

Proportional parts (upper box):

	5
1	0.5
2	1.0
3	1.5
4	2.0
5	2.5
6	3.0
7	3.5
8	4.0
9	4.5

Proportional parts (lower box):

	4
1	0.4
2	0.8
3	1.2
4	1.6
5	2.0
6	2.4
7	2.8
8	3.2
9	3.6

TABLE B.1
Five-Place Logarithms: 950–1000

Prop. Parts	N	0	1	2	3	4	5	6	7	8	9
	950	97 772	777	782	786	791	795	800	804	809	813
	51	818	823	827	832	836	841	845	850	855	859
	52	864	868	873	877	882	886	891	896	900	905
	53	909	914	918	923	928	932	937	941	946	950
	54	97 955	959	964	968	973	978	982	987	991	996
	55	98 000	005	009	014	019	023	028	032	037	041
	56	046	050	055	059	064	068	073	078	082	087
	57	091	096	100	105	109	114	118	123	127	132
	58	137	141	146	150	155	159	164	168	173	177
	59	182	186	191	195	200	204	209	214	218	223
	960	227	232	236	241	245	250	254	259	263	268
	61	272	277	281	286	290	295	299	304	308	313
	62	318	322	327	331	336	340	345	349	354	358
5	63	363	367	372	376	381	385	390	394	399	403
1 | 0.5	64	408	412	417	421	426	430	435	439	444	448
2 | 1.0	65	453	457	462	466	471	475	480	484	489	493
3 | 1.5	66	498	502	507	511	516	520	525	529	534	538
4 | 2.0	67	543	547	552	556	561	565	570	574	579	583
5 | 2.5	68	588	592	597	601	605	610	614	619	623	628
6 | 3.0	69	632	637	641	646	650	655	659	664	668	673
7 | 3.5	970	677	682	686	691	695	700	704	709	713	717
8 | 4.0	71	722	726	731	735	740	744	749	753	758	762
9 | 4.5	72	767	771	776	780	784	789	793	798	802	807
	73	811	816	820	825	829	834	838	843	847	851
	74	856	860	865	869	874	878	883	887	892	896
	75	900	905	909	914	918	923	927	932	936	941
	76	945	949	954	958	963	967	972	976	981	985
	77	98 989	994	998	*003	*007	*012	*016	*021	*025	*029
	78	99 034	038	043	047	052	056	061	065	069	074
	79	078	083	087	092	096	100	105	109	114	118
	980	123	127	131	136	140	145	149	154	158	162
4	81	167	171	176	180	185	189	193	198	202	207
1 | 0.4	82	211	216	220	224	229	233	238	242	247	251
2 | 0.8	83	255	260	264	269	273	277	282	286	291	295
3 | 1.2	84	300	304	308	313	317	322	326	330	335	339
4 | 1.6	85	344	348	352	357	361	366	370	374	379	383
5 | 2.0	86	388	392	396	401	405	410	414	419	423	427
6 | 2.4	87	432	436	441	445	449	454	458	463	467	471
7 | 2.8	88	476	480	484	489	493	498	502	506	511	515
8 | 3.2	89	520	524	528	533	537	542	546	550	555	559
9 | 3.6	990	564	568	572	577	581	585	590	594	599	603
	91	607	612	616	621	625	629	634	638	642	647
	92	651	656	660	664	669	673	677	682	686	691
	93	695	699	704	708	712	717	721	726	730	734
	94	739	743	747	752	756	760	765	769	774	778
	95	782	787	791	795	800	804	808	813	817	822
	96	826	830	835	839	843	848	852	856	861	965
	97	870	874	878	883	887	891	896	900	904	909
	98	913	917	922	926	930	935	939	944	948	952
	99	99 957	961	965	970	974	978	983	987	991	996
	1000	00 000	004	009	013	017	022	026	030	035	039
Prop. Parts	N	0	1	2	3	4	5	6	7	8	9

TABLE C.1

Length of Degrees of the Parallel

Lat.	Meters	Statute miles	Lat.	Meters	Statute miles	Lat.	Meters	Statute miles
° ′			° ′			° ′		
0 00	111 321	69.172	30 00	96 488	59.956	60 00	55 802	34.674
1 00	111 304	69.162	31 00	95 506	59.345	61 00	54 110	33.623
2 00	111 253	69.130	32 00	94 495	58.716	62 00	52 400	32.560
3 00	111 169	69.078	33 00	93 455	58.071	63 00	50 675	31.488
4 00	111 051	69.005	34 00	92 387	57.407	64 00	48 934	30.406
5 00	110 900	68.911	35 00	91 290	56.725	65 00	47 177	29.315
6 00	110 715	68.795	36 00	90 166	56.027	66 00	45 407	28.215
7 00	110 497	68.660	37 00	89 014	55.311	67 00	43 622	27.106
8 00	110 245	68.504	38 00	87 835	54.579	68 00	41 823	25.988
9 00	109 959	68.326	39 00	86 629	53.829	69 00	40 012	24.862
10 00	109 641	68.129	40 00	85 396	53.063	70 00	38 188	23.729
11 00	109 289	67.910	41 00	84 137	52.281	71 00	36 353	22.589
12 00	108 904	67.670	42 00	82 853	51.483	72 00	34 506	21.441
13 00	108 486	67.410	43 00	81 543	50.669	73 00	32 648	20.287
14 00	108 036	67.131	44 00	80 208	49.840	74 00	30 781	19.127
15 00	107 553	66.830	45 00	78 849	48.995	75 00	28 903	17.960
16 00	107 036	66.510	46 00	77 466	48.136	76 00	27 017	16.788
17 00	106 487	66.169	47 00	76 058	47.261	77 00	25 123	15.611
18 00	105 906	65.808	48 00	74 628	46.372	78 00	23 220	14.428
19 00	105 294	65.427	49 00	73 174	45.469	79 00	21 311	13.242
20 00	104 649	65.026	50 00	71 698	44.552	80 00	19 394	12.051
21 00	103 972	64.606	51 00	70 200	43.621	81 00	17 472	10.857
22 00	103 264	64.166	52 00	68 680	42.676	82 00	15 545	9.659
23 00	102 524	63.706	53 00	67 140	41.719	83 00	13 612	8.458
24 00	101 754	63.228	54 00	65 578	40.749	84 00	11 675	7.255
25 00	100 952	62.729	55 00	63 996	39.766	85 00	9 735	6.049
26 00	100 119	62.212	56 00	62 395	38.771	86 00	7 792	4.842
27 00	99 257	61.676	57 00	60 774	37.764	87 00	5 846	3.632
28 00	98 364	61.122	58 00	59 135	36.745	88 00	3 898	2.422
29 00	97 441	60.548	59 00	57 478	35.716	89 00	1 949	1.211
						90 00	0	0

Tables C.1 and C.2 are from U.S. Coast and Geodetic Survey; Table C.3 is from *Smithsonian Geographical Tables*.

APPENDIX C
GEOGRAPHIC TABLES

TABLE C.2

Lengths of Degrees of the Meridian

Lat.	Meters	Statute miles	Lat.	Meters	Statute miles	Lat.	Meters	Statute miles
°			°			°		
0–1	110 567.3	68.703	30–31	110 857.0	68.883	60–61	111 423.1	69.235
1–2	110 568.0	68.704	31–32	110 874.4	68,894	61–62	111 439.9	69.246
2–3	110 569.4	68.705	32–33	110 892.1	68.905	62–63	111 456.4	69.256
3–4	110 571.4	68.706	33–34	110 910.1	68.916	63–64	111 472.4	69.266
4–5	110 574.1	68.707	34–35	110 928.3	68.928	64–65	111 488.1	69.275
5–6	110 577.6	68.710	35–36	110 946.9	68.939	65–66	111 503.3	69.285
6–7	110 581.6	68.712	36–37	110 965.6	68.951	66–67	111 518.0	69.294
7–8	110 586.4	68.715	37–38	110 984.5	68.962	67–68	111 532.3	69.303
8–9	110 591.8	68.718	38–39	111 003.7	68.974	68–69	111 546.2	69.311
9–10	110 597.8	68.722	39–40	111 023.0	68.986	69–70	111 559.5	69.320
10–11	110 604.5	68.726	40–41	111 042.4	68.998	70–71	111 572.2	69.328
11–12	110 611.9	68.731	41–42	111 061.9	69.011	71–72	111 584.5	69.335
12–13	110 619.8	68.736	42–43	111 081.6	69.023	72–73	111 596.2	69.343
13–14	110 628.4	68.741	43–44	111 101.3	69.035	73–74	111 607.3	69.349
14–15	110 637.6	68.747	44–45	111 121.0	69.047	74–75	111 617.9	69.356
15–16	110 647.5	68.753	45–46	111 140.8	69.060	75–76	111 627.8	69.362
16–17	110 657.8	68.759	46–47	111 160.5	69.072	76–77	111 637.1	69.368
17–18	110 668.8	68.766	47–48	111 180.2	69.084	77–78	111 645.9	69.373
18–19	110 680.4	68.773	48–49	111 199.9	69.096	78–79	111 653.9	69.378
19–20	110 692.4	68.781	49–50	111 219.5	69.108	79–80	111 661.4	69.383
20–21	110 705.1	68.789	50–51	111 239.0	69.121	80–81	111 668.2	69.387
21–22	110 718.2	68.797	51–52	111 258.3	69.133	81–82	111 674.4	69.391
22–23	110 731.8	68.805	52–53	111 277.6	69.145	82–83	111 679.9	69.395
23–24	110 746.0	68,814	53–54	111 296.6	69.156	83–84	111 684.7	69.398
24–25	110 760.6	68.823	54–55	111 315.4	69.168	84–85	111 688.9	69.400
25–26	110 775.6	68.833	55–56	111 334.0	69.180	85–86	111 692.3	69.402
26–27	110 791.1	68.842	56–57	111 352.4	69.191	86–87	111 695.1	69.404
27–28	110 807.0	68.852	57–58	111 370.5	69.202	87–88	111 697.2	69.405
28–29	110 823.3	68.862	58–59	111 388.4	69.213	88–89	111 698.6	69.406
29–30	110 840.0	68.873	59–60	111 405.9	69.224	89–90	111 699.3	69.407

TABLE C.3

Areas of Quadrilaterals of Earth's Surface of 1° Extent in Latitude and Longitude

Lower latitude of quadrilateral °	Area in square miles	Lower latitude of quadrilateral °	Area in square miles
0	4 752.16	45	3 354.01
1	4 750.75	46	3 294.71
2	4 747.93	47	3 234.39
3	4 743.71	48	3 173.04
4	4 738.08	49	3 110.69
5	4 731.04		
6	4 722.61	50	3 047.37
7	4 712.76	51	2 983.08
8	4 701.52	52	2 917.85
9	4 688.89	53	2 851.68
		54	2 784.62
10	4 674.86	55	2 716.67
11	4 659.43	56	2 647.85
12	4 642.63	57	2 578.19
13	4 624.44	58	2 507.70
14	4 604.87	59	2 436.42
15	4 583.92		
16	4 561.61	60	2 364.34
17	4 537.93	61	2 291.51
18	4 512.90	62	2 217.94
19	4 486.51	63	2 143.66
		64	2 068.68
20	4 458.78	65	1 993.04
21	4 429.71	66	1 916.75
22	4 399.30	67	1 839.84
23	4 367.57	68	1 762.33
24	4 334.52	69	1 684.24
25	4 300.17		
26	4 264.51	70	1 605.62
27	4 227.56	71	1 526.46
28	4 189.33	72	1 446.81
29	4 149.83	73	1 366.69
		74	1 286.12
30	4 109.06	75	1 205.13
31	4 067.05	76	1 123.75
32	4 023.79	77	1 041.99
33	3 979.30	78	959.90
34	3 933.59	79	877.49
35	3 886.67		
36	3 838.56	80	794.79
37	3 789.26	81	711.83
38	3 738.80	82	628.64
39	3 687.18	83	545.24
		84	461.66
40	3 634.42	85	377.93
41	3 580.54	86	294.08
42	3 525.54	87	210.12
43	3 469.44	88	126.10
44	3 412.26	89	42.04

Table D.1 gives the values of the *sine, cosine, tangent,* and *cotangent* of degrees from 0 to 90 degrees: For degrees at the left use the column headings at the top; for degrees at the right use column headings at the bottom.

The values of the *secant* and *cosecant* may be derived as follows: secant = 1 ÷ cos; cosecant = 1 ÷ sin.

Trigonometric Functions of an Acute Angle in a Right Triangle

$$sine = \frac{opposite\ side}{hypotenuse}$$

$$cosine = \frac{adjacent\ side}{hypotenuse}$$

$$tangent = \frac{opposite\ side}{adjacent\ side}$$

$$cotangent = \frac{adjacent\ side}{opposite\ side}$$

$$secant = \frac{hypotenuse}{adjacent\ side}$$

$$cosecant = \frac{hypotenuse}{opposite\ side}$$

APPENDIX D
NATURAL TRIGONOMETRIC FUNCTIONS

TABLE D.1

Natural Trigonometric Functions

X		sin x	cos x	tan x	cot x		
0°	0′	.0000	1.0000	.0000	∞	90°	0′
	10′	.0029	1.0000	.0029	343.774		50′
	20′	.0058	1.0000	.0058	171.885		40′
	30′	.0087	1.0000	.0087	114.589		30′
	40′	.0116	.9999	.0116	85.9398		20′
	50′	.0145	.9999	.0145	68.7501		10′
1°	0′	.0175	.9998	.0175	57.2900	89°	0′
	10′	.0204	.9998	.0204	49.1039		50′
	20′	.0233	.9997	.0233	42.9641		40′
	30′	.0262	.9997	.0262	38.1885		30′
	40′	.0291	.9996	.0291	34.3678		20′
	50′	.0320	.9995	.0320	31.2416		10′
2°	0′	.0349	.9994	.0349	28.6363	88°	0′
	10′	.0378	.9993	.0378	26.4316		50′
	20′	.0407	.9992	.0407	24.5418		40′
	30′	.0436	.9990	.0437	22.9038		30′
	40′	.0465	.9989	.0466	21.4704		20′
	50′	.0494	.9988	.0495	20.2056		10′
3°	0′	.0523	.9986	.0524	19.0811	87°	0′
	10′	.0552	.9985	.0553	18.0750		50′
	20′	.0581	.9983	.0582	17.1693		40′
	30′	.0610	.9981	.0612	16.3499		30′
	40′	.0640	.9980	.0641	15.6048		20′
	50′	.0669	.9978	.0670	14.9244		10′
4°	0′	.0698	.9976	.0699	14.3007	86°	0′
	10′	.0727	.9974	.0729	13.7267		50′
	20′	.0756	.9971	.0758	13.1969		40′
	30′	.0785	.9969	.0787	12.7062		30′
	40′	.0814	.9967	.0816	12.2505		20′
	50′	.0843	.9964	.0846	11.8262		10′
5°	0′	.0872	.9962	.0875	11.4301	85°	0′
	10′	.0901	.9959	.0904	11.0594		50′
	20′	.0929	.9957	.0934	10.7119		40′
	30′	.0958	.9954	.0963	10.3854		30′
	40′	.0987	.9951	.0992	10.0780		20′
	50′	.1016	.9948	.1022	9.7882		10′
6°	0′	.1045	.9945	.1051	9.5144	84°	0′
	10′	.1074	.9942	.1080	9.2553		50′
	20′	.1103	.9939	.1110	9.0098		40′
	30′	.1132	.9936	.1139	8.7769		30′
	40′	.1161	.9932	.1169	8.5555		20′
	50′	.1190	.9929	.1198	8.3450		10′
7°	0′	.1219	.9925	.1228	8.1443	83°	0′
	10′	.1248	.9922	.1257	7.9530		50′
	20′	.1276	.9918	.1287	7.7704		40′
	30′	.1305	.9914	.1317	7.5958		30′
		cos x	sin x	cot x	tan x	x	

TABLE D.1 (continued)

X		sin x	cos x	tan x	cot x		
	30′	.1305	.9914	.1317	7.5958		30′
	40′	.1334	.9911	.1346	7.4287		20′
	50′	.1363	.9907	.1376	7.2687		10′
8°	0′	.1392	.9903	.1405	7.1154	82°	0′
	10′	.1421	.9899	.1435	6.9682		50′
	20′	.1449	.9894	.1465	6.8269		40′
	30′	.1478	.9890	.1495	6.6912		30′
	40′	.1507	.9886	.1524	6.5606		20′
	50′	.1536	.9881	.1554	6.4348		10′
9°	0′	.1564	.9877	.1584	6.3138	81°	0′
	10′	.1593	.9872	.1614	6.1970		50′
	20′	.1622	.9868	.1644	6.0844		40′
	30′	.1650	.9863	.1673	5.9758		30′
	40′	.1679	.9858	.1703	5.8708		20′
	50′	.1708	.9853	.1733	5.7694		10′
10°	0′	.1736	.9848	.1763	5.6713	80°	0′
	10′	.1765	.9843	.1793	5.5764		50′
	20′	.1794	.9838	.1823	5.4845		40′
	30′	.1822	.9833	.1853	5.3955		30′
	40′	.1851	.9827	.1883	5.3093		20′
	50′	.1880	.9822	.1914	5.2257		10′
11°	0′	.1908	.9816	.1944	5.1446	79°	0′
	10′	.1937	.9811	.1974	5.0658		50′
	20′	.1965	.9805	.2004	4.9894		40′
	30′	.1994	.9799	.2035	4.9152		30′
	40′	.2022	.9793	.2065	4.8430		20′
	50′	.2051	.9787	.2095	4.7729		10′
12°	0′	.2079	.9781	.2126	4.7046	78°	0′
	10′	.2108	.9775	.2156	4.6382		50′
	20′	.2136	.9769	.2186	4.5736		40′
	30′	.2164	.9763	.2217	4.5107		30′
	40′	.2193	.9757	.2247	4.4494		20′
	50′	.2221	.9750	.2278	4.3897		10′
13°	0′	.2250	.9744	.2309	4.3315	77°	0′
	10′	.2278	.9737	.2339	4.2747		50′
	20′	.2306	.9730	.2370	4.2193		40′
	30′	.2334	.9724	.2401	4.1653		30′
	40′	.2363	.9717	.2432	4.1126		20′
	50′	.2391	.9710	.2462	4.0611		10′
14°	0′	.2419	.9703	.2493	4.0108	76°	0′
	10′	.2447	.9696	.2524	3.9617		50′
	20′	.2476	.9689	.2555	3.9136		40′
	30′	.2504	.9681	.2586	3.8667		30′
	40′	.2532	.9674	.2617	3.8208		20′
	50′	.2560	.9667	.2648	3.7760		10′
15°	0′	.2588	.9659	.2679	3.7321	75°	0′
		cos x	sin x	cot x	tan x	x	

TABLE D.1 (continued)

X		sin x	cos x	tan x	cot x		
15°	0'	.2588	.9659	.2679	3.7321	75°	0'
	10'	.2616	.9652	.2711	3.6891		50'
	20'	.2644	.9644	.2742	3.6470		40'
	30'	.2672	.9636	.2773	3.6059		30'
	40'	.2700	.9628	.2805	3.5656		20'
	50'	.2728	.9621	.2836	3.5261		10'
16°	0'	.2756	.9613	.2867	3.4874	74°	0'
	10'	.2784	.9605	.2899	3.4495		50'
	20'	.2812	.9596	.2931	3.4124		40'
	30'	.2840	.9588	.2962	3.3759		30'
	40'	.2868	.9580	.2994	3.3402		20'
	50'	.2896	.9572	.3026	3.3052		10'
17°	0'	.2924	.9563	.3057	3.2709	73°	0'
	10'	.2952	.9555	.3089	3.2371		50'
	20'	.2979	.9546	.3121	3.2041		40'
	30'	.3007	.9537	.3153	3.1716		30'
	40'	.3035	.9528	.3185	3.1397		20'
	50'	.3062	.9520	.3217	3.1084		10'
18°	0'	.3090	.9511	.3249	3.0777	72°	0'
	10'	.3118	.9502	.3281	3.0475		50'
	20'	.3145	.9492	.3314	3.0178		40'
	30'	.3173	.9483	.3346	2.9887		30'
	40'	.3201	.9474	.3378	2.9600		20'
	50'	.3228	.9465	.3411	2.9319		10'
19°	0'	.3256	.9455	.3443	2.9042	71°	0'
	10'	.3283	.9446	.3476	2.8770		50'
	20'	.3311	.9436	.3508	2.8502		40'
	30'	.3338	.9426	.3541	2.8239		30'
	40'	.3365	.9417	.3574	2.7980		20'
	50'	.3393	.9407	.3607	2.7725		10'
20°	0'	.3420	.9397	.3640	2.7475	70°	0'
	10'	.3448	.9387	.3673	2.7228		50'
	20'	.3475	.9377	.3706	2.6985		40'
	30'	.3502	.9367	.3739	2.6746		30'
	40'	.3529	.9356	.3772	2.6511		20'
	50'	.3557	.9346	.3805	2.6279		10'
21°	0'	.3584	.9336	.3839	2.6051	69°	0'
	10'	.3611	.9325	.3872	2.5826		50'
	20'	.3638	.9315	.3906	2.5605		40'
	30'	.3665	.9304	.3939	2.5386		30'
	40'	.3692	.9293	.3973	2.5172		20'
	50'	.3719	.9283	.4006	2.4960		10'
22°	0'	.3746	.9272	.4040	2.4751	68°	0'
	10'	.3773	.9261	.4074	2.4545		50'
	20'	.3800	.9250	.4108	2.4342		40'
	30'	.3827	.9239	.4142	2.4142		30'
		cos x	sin x	cot x	tan x		x

TABLE D.1 (continued)

X		sin x	cos x	tan x	cot x		
	30′	.3827	.9239	.4142	2.4142		30′
	40′	.3854	.9228	.4176	2.3945		20′
	50′	.3881	.9216	.4210	2.3750		10′
23°	0′	.3907	.9205	.4245	2.3559	67°	0′
	10′	.3934	.9194	.4279	2.3369		50′
	20′	.3961	.9182	.4314	2.3183		40′
	30′	.3987	.9171	.4348	2.2998		30′
	40′	.4014	.9159	.4383	2.2817		20′
	50′	.4041	.9147	.4417	2.2637		10′
24°	0′	.4067	.9135	.4452	2.2460	66°	0′
	10′	.4094	.9124	.4487	2.2286		50′
	20′	.4120	.9112	.4522	2.2113		40′
	30′	.4147	.9100	.4557	2.1943		30′
	40′	.4173	.9088	.4592	2.1775		20′
	50′	.4200	.9075	.4628	2.1609		10′
25°	0′	.4226	.9063	.4663	2.1445	65°	0′
	10′	.4253	.9051	.4699	2.1283		50′
	20′	.4279	.9038	.4734	2.1123		40′
	30′	.4305	.9026	.4770	2.0965		30′
	40′	.4331	.9013	.4806	2.0809		20′
	50′	.4358	.9001	.4841	2.0655		10′
26°	0′	.4384	.8988	.4877	2.0503	64°	0′
	10′	.4410	.8975	.4913	2.0353		50′
	20′	.4436	.8962	.4950	2.0204		40′
	30′	.4462	.8949	.4986	2.0057		30′
	40′	.4488	.8936	.5022	1.9912		20′
	50′	.4514	.8923	.5059	1.9768		10′
27°	0′	.4540	.8910	.5095	1.9626	63°	0′
	10′	.4566	.8897	.5132	1.9486		50′
	20′	.4592	.8884	.5169	1.9347		40′
	30′	.4617	.8870	.5206	1.9210		30′
	40′	.4643	.8857	.5243	1.9074		20′
	50′	.4669	.8843	.5280	1.8940		10′
28°	0′	.4695	.8829	.5317	1.8807	62°	0′
	10′	.4720	.8816	.5354	1.8676		50′
	20′	.4746	.8802	.5392	1.8546		40′
	30′	.4772	.8788	.5430	1.8418		30′
	40′	.4797	.8774	.5467	1.8291		20′
	50′	.4823	.8760	.5505	1.8165		10′
29°	0′	.4848	.8746	.5543	1.8040	61°	0′
	10′	.4874	.8732	.5581	1.7917		50′
	20′	.4899	.8718	.5619	1.7796		40′
	30′	.4924	.8704	.5658	1.7675		30′
	40′	.4950	.8689	.5696	1.7556		20′
	50′	.4975	.8675	.5735	1.7437		10′
30°	0′	.5000	.8660	.5774	1.7321	60°	0′
		cos x	sin x	cot x	tan x	x	

TABLE D.1 (continued)

X		sin x	cos x	tan x	cot x		
30°	0'	.5000	.8660	.5774	1.7321	60°	0'
	10'	.5025	.8646	.5812	1.7205		50'
	20'	.5050	.8631	.5851	1.7090		40'
	30'	.5075	.8616	.5890	1.6977		30'
	40'	.5100	.8601	.5930	1.6864		20'
	50'	.5125	.8587	.5969	1.6753		10'
31°	0'	.5150	.8572	.6009	1.6643	59°	0'
	10'	.5175	.8557	.6048	1.6534		50'
	20'	.5200	.8542	.6088	1.6426		40'
	30'	.5225	.8526	.6128	1.6319		30'
	40'	.5250	.8511	.6168	1.6212		20'
	50'	.5275	.8496	.6208	1.6107		10'
32°	0'	.5299	.8480	.6249	1.6003	58°	0'
	10'	.5324	.8465	.6289	1.5900		50'
	20'	.5348	.8450	.6330	1.5798		40'
	30'	.5373	.8434	.6371	1.5697		30'
	40'	.5398	.8418	.6412	1.5597		20'
	50'	.5422	.8403	.6453	1.5497		10'
33°	0'	.5446	.8387	.6494	1.5399	57°	0'
	10'	.5471	.8371	.6536	1.5301		50'
	20'	.5495	.8355	.6577	1.5204		40'
	30'	.5519	.8339	.6619	1.5108		30'
	40'	.5544	.8323	.6661	1.5013		20'
	50'	.5568	.8307	.6703	1.4919		10'
34°	0'	.5592	.8290	.6745	1.4826	56°	0'
	10'	.5616	.8274	.6787	1.4733		50'
	20'	.5640	.8258	.6830	1.4641		40'
	30'	.5664	.8241	.6873	1.4550		30'
	40'	.5688	.8225	.6916	1.4460		20'
	50'	.5712	.8208	.6959	1.4370		10'
35°	0'	.5736	.8192	.7002	1.4281	55°	0'
	10'	.5760	.8175	.7046	1.4193		50'
	20'	.5783	.8158	.7089	1.4106		40'
	30'	.5807	.8141	.7133	1.4019		30'
	40'	.5831	.8124	.7177	1.3934		20'
	50'	.5854	.8107	.7221	1.3848		10'
36°	0'	.5878	.8090	.7265	1.3764	54°	0'
	10'	.5901	.8073	.7310	1.3680		50'
	20'	.5925	.8056	.7355	1.3597		40'
	30'	.5948	.8039	.7400	1.3514		30'
	40'	.5972	.8021	.7445	1.3432		20'
	50'	.5995	.8004	.7490	1.3351		10'
37°	0'	.6018	.7986	.7536	1.3270	53°	0'
	10'	.6041	.7969	.7581	1.3190		50'
	20'	.6065	.7951	.7627	1.3111		40'
	30'	.6088	.7934	.7673	1.3032		30'
		cos x	sin x	cot x	tan x	x	

TABLE D.1 (continued)

X		sin x	cos x	tan x	cot x		
	30′	.6088	.7934	.7673	1.3032		30′
	40′	.6111	.7916	.7720	1.2954		20′
	50′	.6134	.7898	.7766	1.2876		10′
38°	0′	.6157	.7880	.7813	1.2799	52°	0′
	10′	.6180	.7862	.7860	1.2723		50′
	20′	.6202	.7844	.7907	1.2647		40′
	30′	.6225	.7826	.7954	1.2572		30′
	40′	.6248	.7808	.8002	1.2497		20′
	50′	.6271	.7790	.8050	1.2423		10′
39°	0′	.6293	.7771	.8098	1.2349	51°	0′
	10′	.6316	.7753	.8146	1.2276		50′
	20′	.6338	.7735	.8195	1.2203		40′
	30′	.6361	.7716	.8243	1.2131		30′
	40′	.6383	.7698	.8292	1.2059		20′
	50′	.6406	.7679	.8342	1.1988		10′
40°	0′	.6428	.7660	.8391	1.1918	50°	0′
	10′	.6450	.7642	.8441	1.1847		50′
	20′	.6472	.7623	.8491	1.1778		40′
	30′	.6494	.7604	.8541	1.1708		30′
	40′	.6517	.7585	.8591	1.1640		20′
	50′	.6539	.7566	.8642	1.1571		10′
41°	0′	.6561	.7547	.8693	1.1504	49°	0′
	10′	.6583	.7528	.8744	1.1436		50′
	20′	.6604	.7509	.8796	1.1369		40′
	30′	.6626	.7490	.8847	1.1303		30′
	40′	.6648	.7470	.8899	1.1237		20′
	50′	.6670	.7451	.8952	1.1171		10′
42°	0′	.6691	.7431	.9004	1.1106	48°	0′
	10′	.6713	.7412	.9057	1.1041		50′
	20′	.6734	.7392	.9110	1.0977		40′
	30′	.6756	.7373	.9163	1.0913		30′
	40′	.6777	.7353	.9217	1.0850		20′
	50′	.6799	.7333	.9271	1.0786		10′
43°	0′	.6820	.7314	.9325	1.0724	47°	0′
	10′	.6841	.7294	.9380	1.0661		50′
	20′	.6862	.7274	.9435	1.0599		40′
	30′	.6884	.7254	.9490	1.0538		30′
	40′	.6905	.7234	.9545	1.0477		20′
	50′	.6926	.7214	.9601	1.0416		10′
44°	0′	.6947	.7193	.9657	1.0355	46°	0′
	10′	.6967	.7173	.9713	1.0295		50′
	20′	.6988	.7153	.9770	1.0235		40′
	30′	.7009	.7133	.9827	1.0176		30′
	40′	.7030	.7112	.9884	1.0117		20′
	50′	.7050	.7092	.9942	1.0058		10′
45°	0′	.7071	.7071	1.0000	1.0000	45°	0′
		cos x	sin x	cot x	tan x	x	

The general equation to produce class limits by either increasing or decreasing (monotonic) arithmetic or geometric series is:

$$L + B_1X + B_2X + \ldots + B_nX = H$$

where L = lowest value
H = highest value
B_n = the value of the nth term in the progression

The equation is then solved for X for any given values of L and H to establish the class limits.

A. For arithmetic progressions the quantity

$$B_n = a + [(n - 1)d]$$

where a = value of the first term in the progression
n = number of the term being determined in the progression (first, second, etc.)
d = the stated difference

B. For geometric progressions the quantity

$$B_n = gr^{n-1}$$

where g = value of the first nonzero term
n = number of the term being determined in the progression, (first, second, etc.)
r = the stated ratio

Six forms of each type of progression are possible:

1. Increasing at a constant rate or ratio.
2. Increasing at an increasing rate or ratio.
3. Increasing at a decreasing rate or ratio.
4. Decreasing at a constant rate or ratio.
5. Decreasing at an increasing rate or ratio.
6. Decreasing at a decreasing rate or ratio.

The following examples use the data in Table 7.2. The four class limits in each example are derived by using the six forms of each increasing or decreasing arithmetic and geometric progressions.

Note. In calculating many such series, fractions are common. In the following, all such fractions have been rounded to the most convenient whole number.

A. Arithmetic Progression Examples (data from Table 7.2)

1. Increasing at a constant rate

$$a = 1$$
$$d = 1$$
$$n = 1, 2, 3, 4$$
$$B_1 = 1 + (0)1 = 1$$
$$B_2 = 1 + (1)1 = 2$$
$$B_3 = 1 + (2)1 = 3$$
$$B_4 = 1 + (3)1 = 4$$

$$L + \sum_i B_iX = H$$

$$0 + 10X = 100$$
$$X = 10$$

Upper class limits are   0 to 10
                          10 to 30
                          30 to 60
                          60 to 100

APPENDIX E

CALCULATION OF ARITHMETIC AND GEOMETRIC CLASS INTERVALS

2. Increasing at an increasing rate

$a = 1$

$d = n - 1$

$n = 1, 2, 3, 4$

$B_1 = 1 + (0)(0) = 1$

$B_2 = 1 + (1)(1) = 2$

$B_3 = 1 + (2)(2) = 5$

$B_4 = 1 + (3)(3) = 10$

$$L + \sum_i B_i X = H$$

$0 + 18X = 100$

$X = 5.556$

Upper class limits are   0 to 6

6 to 17

17 to 44

44 to 100

3. Increasing at a decreasing rate

$a = 1$

$d = 1/n$

$n = 1, 2, 3, 4$

$B_1 = 1 + (0)(1) = 1$

$B_2 = 1 + (1)(\frac{1}{2}) = 1.5$

$B_3 = 1 + (2)(\frac{1}{3}) = 1.67$

$B_4 = 1 + (3)(\frac{1}{4}) = 1.75$

$$L + \sum_i B_i X = H$$

$0 + 5.92X = 100$

$X = 16.89$

Upper class limits are   0 to 17

17 to 42

42 to 70

70 to 100

4. Decreasing at a constant rate

$a = 4$

$d = 1$

$n = 1, 2, 3, 4$

$B_1 = 4 + (0)(-1) = 4$

$B_2 = 4 + (1)(-1) = 3$

$B_3 = 4 + (2)(-1) = 2$

$B_4 = 4 + (3)(-1) = 1$

$$L + \sum_i B_i X = H$$

$0 + 10X = 100$

$X = 10$

Upper class limits are   0 to 40

40 to 70

70 to 90

90 to 100

5. Decreasing at an increasing rate

$a = 10$

$d = -(n - 1)$

$n = 1, 2, 3, 4$

$B_1 = 10 + (0)(-0) = 10$

$B_2 = 10 + (1)(-1) = 9$

$B_3 = 10 + (2)(-2) = 6$

$B_4 = 10 + (3)(-3) = 1$

$$L + \sum_i B_i X = H$$

$0 + 26X = 100$

$X = 3.846$

Upper class limits are   0 to 38

38 to 73

73 to 96

96 to 100

6. Decreasing at a decreasing rate

$a = 1$

$d = -1/n$

$n = 1, 2, 3, 4$

$B_1 = 1 + (0)(-1) = 1 = 1.0$

$B_2 = 1 + (1)(-\frac{1}{2}) = \frac{1}{2} = 0.5$

$B_3 = 1 + (2)(-\frac{1}{3}) = \frac{1}{3} = 0.333$

$B_4 = 1 + (3)(-\frac{1}{4}) = \frac{1}{4} = 0.25$

$$L + \sum_i B_i X = H$$

$0 + 2.083X = 100$

$X = 48$

Upper class limits are   0 to 48

48 to 72

72 to 88

88 to 100

B. Geometric Progression Examples (Data from Table 7.2)

1. Increasing at a constant ratio

$a = 1$

$r = 2$

$n = 1, 2, 3, 4$

$B_1 = (1)(2)^{(0)} = 1$

$B_2 = (1)(2)^{(1)} = 2$

$B_3 = (1)(2)^{(2)} = 4$

$B_4 = (1)(2)^{(3)} = 8$

$$L + \sum_i B_i X = H$$

$$0 + 15X = 100$$
$$X = 6.6667$$

Upper class limits are   0 to 7
7 to 21
21 to 47
47 to 100

2. Increasing at an increasing ratio
$a = 1$
$r = n$
$n = 1, 2, 3, 4$
$B_1 = (1)(1)^{(0)} = 1$
$B_2 = (1)(2)^{(1)} = 2$
$B_3 = (1)(3)^{(2)} = 9$
$B_4 = (1)(4)^{(3)} = 64$

$$L + \sum_i B_i X = H$$

$$0 + 76X = 100$$
$$X = 1.316$$
Upper class limits are   0 to 1
1 to 4
4 to 16
16 to 100

3. Increasing at a decreasing ratio
$a = 1$
$r = n/n - 1$
$n = 1, 2, 3, 4$
$B_1 = (1)(0)^0 = 1$
$B_2 = (1)(2)^1 = 2$
$B_3 = (1)(^3/_2)^2 = 2.25$
$B_4 = (1)(^4/_3)^3 = 2.37$

$$L + \sum_i B_i X = H$$

$$0 + 7.62X = 100$$
$$X = 13.1233$$

Upper class limits are   0 to 13
13 to 39
39 to 69
69 to 100

4. Decreasing at a constant ratio
$a = 1$
$r = \frac{1}{2}$
$n = 1, 2, 3, 4$
$B_1 = (1)(\frac{1}{2})^{(0)} = 1 = 1.0$
$B_2 = (1)(\frac{1}{2})^{(1)} = \frac{1}{2} = 0.5$

$B_3 = (1)(\frac{1}{2})^{(2)} = \frac{1}{4} = 0.25$
$B_4 = (1)(\frac{1}{2})^{(3)} = \frac{1}{8} = 0.125$

$$L + \sum_i B_i X = H$$

$$0 + 1.875X = 100$$
$$X = 53.3333$$

Upper class limits are   0 to 53
53 to 80
80 to 93
93 to 100

5. Decreasing at an increasing ratio
$a = 1$
$r = 1/n$
$n = 1, 2, 3, 4$
$B_1 = (1)(1)^{(0)} = 1 = 1.0$
$B_2 = (1)(\frac{1}{2})^{(1)} = \frac{1}{2} = 0.50$
$B_3 = (1)(\frac{1}{3})^{(2)} = \frac{1}{9} = 0.111111$
$B_4 = (1)(\frac{1}{4})^{(3)} = \frac{1}{64} = 0.015625$

$$L + \sum_i B_i X = H$$

$$0 + 1.627X = 100$$
$$X = 61.463$$

Upper class limits are   0 to 61
61 to 92
92 to 99
99 to 100

6. Decreasing at a decreasing ratio
$a = 1$

$$r = \frac{n - 1}{n}$$

$n = 1, 2, 3, 4$
$B_1 = (1)(0)^{(0)} = 1.0$
$B_2 = (1)(\frac{1}{2})^{(1)} = 0.5$
$B_3 = (1)(\frac{2}{3})^{(2)} = 0.44444$
$B_4 = (1)(\frac{3}{4})^{(3)} = 0.421875$

$$L + \sum_i B_i X = H$$

$$0 + 2.3663X = 100$$
$$X = 42.26$$

Upper class limits are   0 to 42
42 to 63
63 to 82
82 to 100

Table F.1, devised by John K. Wright, enables us to solve without multiplication or division the fundamental equation required to estimate densities of fractional areas as diagrammed in Fig. F.1.*

The basic equation is

$$D_n = \frac{D}{1 - a_m} - \frac{D_m a_m}{1 - a_m}$$

in which D_n = the density in area n (see Fig. F.1)

D = average density of the area as a whole (number of units ÷ area)

D_m = estimated density in part m of area

a_m = the fraction (0.1 to 0.9) of the total area comprised in m

$1 - a_m$ = the fraction (0.1 to 0.9) of the total area comprised in n

Values of $D/(1 - a_m)$ and $D_m a_m /(1 - a_m)$ may be extracted from the table as follows. The table is entered at the top with a_m and entered at the side with D or D_m as arguments. When D is the argument, the right-hand column under the particular value of a_m gives values of $D_m a_m /(1 - a_m)$.

In order to obtain D_n, subtract the value obtained from entering the table with D_m as argument from the value obtained from entering the table with D as argument. (If this value is a minus quantity, then the value of D_m is too large to be consistent with the values of D and of a_m.)

For example: $D = 90$, $D_m = 15$, $a_m = 0.7$.

*The explanation and tables are here presented by permission of Geographical Review, published by the American Geographical Society of New York.

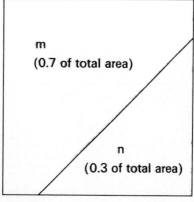

D = 90
Dm = 15
am = 0.7

FIGURE F.1 *A hypothetical enumeration district with a density (D) of 90. The assumed density of area m (Dm) is 15, and the area of m (am) is estimated to be 0.7 of the total area of the district.*

From row 90 and the left-hand column under 0.7, extract value 300; from row 15 and the right-hand column under 0.7, extract value 35; $300 - 35 = 265 = D_n$.

APPENDIX F

TABLE FOR ESTIMATING VALUES OF FRACTIONAL AREAS

TABLE F.1

Tabular Aid to Consistency in Estimating Densities of Parts

The paired columns under each value of a_m are headed $\dfrac{D}{1-a_m}$ and $\dfrac{D_m a_m}{1-a_m}$.

D or D_m	$a_m=0.1$ $\frac{D}{1-a_m}$	$\frac{D_m a_m}{1-a_m}$	$a_m=0.2$ $\frac{D}{1-a_m}$	$\frac{D_m a_m}{1-a_m}$	$a_m=0.3$ $\frac{D}{1-a_m}$	$\frac{D_m a_m}{1-a_m}$	$a_m=0.4$ $\frac{D}{1-a_m}$	$\frac{D_m a_m}{1-a_m}$	$a_m=0.5$ $\frac{D}{1-a_m}$	$\frac{D_m a_m}{1-a_m}$	$a_m=0.6$ $\frac{D}{1-a_m}$	$\frac{D_m a_m}{1-a_m}$	$a_m=0.7$ $\frac{D}{1-a_m}$	$\frac{D_m a_m}{1-a_m}$	$a_m=0.8$ $\frac{D}{1-a_m}$	$\frac{D_m a_m}{1-a_m}$	$a_m=0.9$ $\frac{D}{1-a_m}$	$\frac{D_m a_m}{1-a_m}$	D or D_m
1	1	0.1	1	0.3	1	0.4	2	0.7	2	1	3	1.5	3	2.3	5	4	10	9	1
2	2	0.2	3	0.5	3	0.9	3	1.3	4	2	5	3.0	7	4.7	10	8	20	18	2
3	3	0.3	4	0.8	4	1.3	5	2.0	6	3	8	4.5	10	7.0	15	12	30	27	3
4	4	0.4	5	1.0	6	1.7	7	2.7	8	4	10	6.0	13	9.3	20	16	40	36	4
5	6	0.6	6	1.3	7	2.1	8	3.3	10	5	13	7.5	17	11.7	25	20	50	45	5
6	7	0.7	8	1.5	9	2.6	10	4.0	12	6	15	9.0	20	14.0	30	24	60	54	6
7	8	0.8	9	1.8	10	3.0	12	4.7	14	7	18	10.5	23	16.3	35	28	70	63	7
8	9	0.9	10	2.0	11	3.4	13	5.3	16	8	20	12.0	27	18.7	40	32	80	72	8
9	10	1.0	11	2.3	13	3.9	15	6.0	18	9	23	13.5	30	21.0	45	36	90	81	9
10	11	1.1	13	2.5	14	4.3	17	6.7	20	10	25	15.0	33	23.3	50	40	100	90	10
11	12	1.2	14	2.8	16	4.7	18	7.3	22	11	28	16.5	37	25.7	55	44	110	99	11
12	13	1.3	15	3.0	17	5.1	20	8.0	24	12	30	18.0	40	28.0	60	48	120	108	12
13	14	1.4	16	3.3	19	5.6	22	8.7	26	13	33	19.5	43	30.3	65	52	130	117	13
14	16	1.6	18	3.5	20	6.0	23	9.3	28	14	35	21.0	47	32.7	70	56	140	126	14
15	17	1.7	19	3.8	21	6.4	25	10.0	30	15	38	22.5	50	35.0	75	60	150	135	15
16	18	1.8	20	4.0	23	6.9	27	10.7	32	16	40	24.0	53	37.3	80	64	160	144	16
17	19	1.9	21	4.3	24	7.3	28	11.3	34	17	43	25.5	57	39.7	85	68	170	153	17
18	20	2.0	23	4.5	26	7.7	30	12.0	36	18	45	27.0	60	42.0	90	72	180	162	18
19	21	2.1	24	4.8	27	8.1	32	12.7	38	19	48	28.5	63	44.3	95	76	190	171	19
20	22	2.2	25	5.0	29	8.6	33	13.3	40	20	50	30.0	67	46.7	100	80	200	180	20
21	23	2.3	26	5.3	30	9.0	35	14.0	42	21	53	31.5	70	49.0	105	84	210	189	21
22	24	2.4	28	5.5	31	9.4	37	14.7	44	22	55	33.0	73	51.3	110	88	220	198	22
23	26	2.6	29	5.8	33	9.9	38	15.3	46	23	58	34.5	77	53.7	115	92	230	207	23
24	27	2.7	30	6.0	34	10.3	40	16.0	48	24	60	36.0	80	56.0	120	96	240	216	24
25	28	2.8	31	6.3	36	10.7	42	16.7	50	25	63	37.5	83	58.3	125	100	250	225	25
26	29	2.9	33	6.5	37	11.1	43	17.3	52	26	65	39.0	87	60.7	130	104	260	234	26
27	30	3.0	34	6.8	39	11.6	45	18.0	54	27	68	40.5	90	63.0	135	108	270	243	27
28	31	3.1	35	7.0	40	12.0	47	18.7	56	28	70	42.0	93	65.3	140	112	280	252	28
29	32	3.2	36	7.3	41	12.4	48	19.3	58	29	73	43.5	97	67.7	145	116	290	261	29
30	33	3.3	38	7.5	43	12.9	50	20.0	60	30	75	45.0	100	70.0	150	120	300	270	30
31	34	3.4	39	7.8	44	13.3	52	20.7	62	31	78	46.5	103	72.3	155	124	310	279	31
32	36	3.6	40	8.0	46	13.7	53	21.3	64	32	80	48.0	107	74.7	160	128	320	288	32
33	37	3.7	41	8.3	47	14.1	55	22.0	66	33	83	49.5	110	77.0	165	132	330	297	33
34	38	3.8	43	8.5	49	14.6	57	22.7	68	34	85	51.0	113	79.3	170	136	340	306	34
35	39	3.9	44	8.8	50	15.0	58	23.3	70	35	88	52.5	117	81.7	175	140	350	315	35
36	40	4.0	45	9.0	51	15.4	60	24.0	72	36	90	54.0	120	84.0	180	144	360	324	36
37	41	4.1	46	9.3	53	15.9	62	24.7	74	37	93	55.5	123	86.3	185	148	370	333	37
38	42	4.2	48	9.5	54	16.3	63	25.3	76	38	95	57.0	127	88.7	190	152	380	342	38
39	43	4.3	49	9.8	56	16.7	65	26.0	78	39	98	58.5	130	91.0	195	156	390	351	39
40	44	4.4	50	10.0	57	17.1	67	26.7	80	40	100	60.0	133	93.3	200	160	400	360	40
41	46	4.6	51	10.3	59	17.6	68	27.3	82	41	103	61.5	137	95.7	205	164	410	369	41
42	47	4.7	53	10.5	60	18.0	70	28.0	84	42	105	63.0	140	98.0	210	168	420	378	42
43	48	4.8	54	10.8	61	18.4	72	28.7	86	43	108	64.5	143	100.3	215	172	430	387	43
44	49	4.9	55	11.0	63	18.9	73	29.3	88	44	110	66.0	147	102.7	220	176	440	396	44
45	50	5.0	56	11.3	64	19.3	75	30.0	90	45	113	67.5	150	105.0	225	180	450	405	45
46	51	5.1	58	11.5	66	19.7	77	30.7	92	46	115	69.0	153	107.3	230	184	460	414	46
47	52	5.2	59	11.8	67	20.1	78	31.3	94	47	118	70.5	157	109.7	235	188	470	423	47
48	53	5.3	60	12.0	69	20.6	80	32.0	96	48	120	72.0	160	112.0	240	192	480	432	48
49	54	5.4	61	12.3	70	21.0	82	32.7	98	49	123	73.5	163	114.3	245	196	490	441	49

50	450	500	200	250	116.5	167	75.0	125	50	100	50	33.3	83	22.2	72	12.5	63	5.6	56	50
50	450	500	200	250	116.5	167	75.0	125	50	100	50	33.3	83	22.2	72	12.5	63	5.6	56	50
51	459	510	204	255	118.8	170	76.5	127	51	102	51	34.0	85	22.6	73	12.8	64	5.7	57	51
52	468	520	208	260	121.2	173	78.0	130	52	104	52	34.6	86	23.1	75	13.0	65	5.8	58	52
53	477	530	212	265	123.5	177	79.5	132	53	106	53	35.3	88	23.5	76	13.3	66	5.9	59	53
54	486	540	216	270	125.8	180	81.0	135	54	108	54	36.0	90	24.0	78	13.5	68	6.0	60	54
55	495	550	220	275	128.2	183	82.5	137	55	110	55	36.6	91	24.4	79	13.8	69	6.1	61	55
56	504	560	224	280	130.5	187	84.0	140	56	112	56	37.3	93	24.9	81	14.0	70	6.2	62	56
57	513	570	228	285	132.8	190	85.5	142	57	114	57	38.0	95	25.3	82	14.3	71	6.3	63	57
58	522	580	232	290	135.1	193	87.0	145	58	116	58	38.6	96	25.8	84	14.5	73	6.4	64	58
59	531	590	236	295	137.5	197	88.5	147	59	118	59	39.3	98	26.2	85	14.8	74	6.5	65	59
60	540	600	240	300	139.8	200	90.0	150	60	120	60	40.0	100	26.6	86	15.0	75	6.7	67	60
61	549	610	244	305	142.1	203	91.5	152	61	122	61	40.6	101	27.1	88	15.3	76	6.8	68	61
62	558	620	248	310	144.5	207	93.0	155	62	124	62	41.3	103	27.5	89	15.5	78	6.9	69	62
63	567	630	252	315	146.8	210	94.5	157	63	126	63	42.0	105	28.0	91	15.8	79	7.0	70	63
64	576	640	256	320	149.1	213	96.0	160	64	128	64	42.6	106	28.4	92	16.0	80	7.1	71	64
65	585	650	260	325	151.5	217	97.5	162	65	130	65	43.3	108	28.9	94	16.3	81	7.2	72	65
66	594	660	264	330	153.8	220	99.0	165	66	132	66	44.0	110	29.3	95	16.5	83	7.3	73	66
67	603	670	268	335	156.1	223	100.5	167	67	134	67	44.6	111	29.7	96	16.8	84	7.4	74	67
68	612	680	272	340	158.4	227	102.0	170	68	136	68	45.3	113	30.2	98	17.0	85	7.5	75	68
69	621	690	276	345	160.8	230	103.5	172	69	138	69	46.0	115	30.6	99	17.3	86	7.7	77	69
70	630	700	280	350	163.1	233	105.0	175	70	140	70	46.6	116	31.1	101	17.5	88	7.8	78	70
71	639	710	284	355	165.4	237	106.5	177	71	142	71	47.3	118	31.5	102	17.8	89	7.9	79	71
72	648	720	288	360	167.8	240	108.0	180	72	144	72	48.0	120	32.0	104	18.0	90	8.0	80	72
73	657	730	292	365	170.1	243	109.5	182	73	146	73	48.6	121	32.4	105	18.3	91	8.1	81	73
74	666	740	296	370	172.4	247	111.0	185	74	148	74	49.3	123	32.9	107	18.5	93	8.2	82	74
75	675	750	300	375	174.8	250	112.5	187	75	150	75	50.0	125	33.3	108	18.8	94	8.3	83	75
76	684	760	304	380	177.1	253	114.0	190	76	152	76	50.6	126	33.7	109	19.0	95	8.4	84	76
77	693	770	308	385	179.4	257	115.5	192	77	154	77	51.3	128	34.2	111	19.3	96	8.5	85	77
78	702	780	312	390	181.7	260	117.0	195	78	156	78	51.9	129	34.6	112	19.5	98	8.7	87	78
79	711	790	316	395	184.1	263	118.5	197	79	158	79	52.6	131	35.1	114	19.8	99	8.8	88	79
80	720	800	320	400	186.4	267	120.0	200	80	160	80	53.3	133	35.5	115	20.0	100	8.9	89	80
81	729	810	324	405	188.7	270	121.5	202	81	162	81	53.9	134	36.0	117	20.3	101	9.0	90	81
82	738	820	328	410	191.1	273	123.0	205	82	164	82	54.6	136	36.4	118	20.5	103	9.1	91	82
83	747	830	332	415	193.4	277	124.5	207	83	166	83	55.3	138	36.9	120	20.8	104	9.2	92	83
84	756	840	336	420	195.7	280	126.0	210	84	168	84	55.9	139	37.3	121	21.0	105	9.3	93	84
85	765	850	340	425	198.1	283	127.5	212	85	170	85	56.6	141	37.7	122	21.3	106	9.4	94	85
86	774	860	344	430	200.4	287	129.0	215	86	172	86	57.3	143	38.2	124	21.5	108	9.5	95	86
87	783	870	348	435	202.7	290	130.5	217	87	174	87	57.9	144	38.6	125	21.8	109	9.7	97	87
88	792	880	352	440	205.0	293	132.0	220	88	176	88	58.6	146	39.1	127	22.0	110	9.8	98	88
89	801	890	356	445	207.4	297	133.5	222	89	178	89	59.3	148	39.5	128	22.3	111	9.9	99	89
90	810	900	360	450	209.7	300	135.0	225	90	180	90	59.9	149	40.0	130	22.5	113	10.0	100	90
91	819	910	364	455	212.0	303	136.5	227	91	182	91	60.6	151	40.4	131	22.8	114	10.1	101	91
92	828	920	368	460	214.4	307	138.0	230	92	184	92	61.3	153	40.8	132	23.0	115	10.2	102	92
93	837	930	372	465	216.7	310	139.5	232	93	186	93	61.9	154	41.3	134	23.3	116	10.4	103	93
94	846	940	376	470	219.0	313	141.0	235	94	188	94	62.6	156	41.7	135	23.5	118	10.5	104	94
95	855	950	380	475	221.4	317	142.5	237	95	190	95	63.3	158	42.2	137	23.8	119	10.7	105	95
96	864	960	384	480	223.7	320	144.0	240	96	192	96	63.9	159	42.6	138	24.0	120	10.8	107	96
97	873	970	388	485	226.0	323	145.5	242	97	194	97	64.6	161	43.1	140	24.3	121	10.9	108	97
98	882	980	392	490	228.3	327	147.0	245	98	196	98	65.3	163	43.5	141	24.5	123	11.0	109	98
99	891	990	396	495	230.7	330	148.5	247	99	198	99	65.9	164	44.0	143	24.8	124	11.0	110	99
100	900	1000	400	500	233.0	333	150.0	250	100	200	100	66.6	166	44.4	144	25.0	125	11.1	111	100

The values in the body of Table G.1 provide directly the radius indices for scaling graduated circles. The values obtained from the table need only be multiplied by the chosen unit radius value to be used for scaling the circles. The basic procedure is described in Chapter 10.

The values in the table are the antilogarithms of the logarithms of $N \times 0.5718$. First suggested by John L. Thompson, they were machine computed according to a plan devised by George F. McCleary, Jr., Joel L. Morrison, and Morton W. Scripter. The constant 0.5718 is derived from the work of James J. Flannery.

5. Choose an appropriate radius for one of the values. Ordinarily this should be the smallest in order to ensure that the smallest symbol will be of sufficient size. Then find the *unit radius,* that is, the radius that would represent the value 1.0. The unit radius is the chosen radius divided by the index value it is to represent. For example, if city B is to be represented by a circle 8 mm in radius, then the unit radius is $8 \div 10.69 = 0.748$ mm. Each index value is then multiplied by 0.748 to find the appropriate radius for that value.

City	Index Value		Unit Radius	Plotting Value
A	32.26	x	0.748	24.1 mm
B	10.69	x	0.748	8.0 mm
C	15.96	x	0.748	11.9 mm

EXAMPLE 1.
Suppose we wanted to draw graduated circles representing the populations of three cities with the following populations:

A—435,210
B— 62,647
C—126,538

The procedure would be as follows.
1. By rounding, drop the last three digits.
2. For A enter the table horizontally with $N = 430$ and vertically in the column headed 5; find the index value 32.26.
3. For B enter the table with 60 and 3 as arguments; find the index value 10.69.
4. For C enter the table with 120 and 7 as arguments; find the index value 15.96.

EXAMPLE 2.
To find the radius index for N greater than 999, the largest argument in the table, for example, $N = 1,622$.
1. Divide N into two whole numbers whose product equals 1622, for example, 811 and 2.
2. Enter the table with these as arguments and obtain, respectively, 46.07 and 1.49.
3. Multiply these two together to obtain the required radius index value, 68.64, $(46.07 \times 1.49 = 68.64)$.
4. Proceed as in 5. in Example 1.

APPENDIX G
RADIUS INDEX VALUES FOR GRADUATED CIRCLES

N	0	1	2	3	4	5	6	7	8	9
0	0	1.00	1.49	1.87	2.21	2.51	2.79	3.04	3.28	3.51
10	3.73	3.94	4.14	4.33	4.52	4.70	4.88	5.05	5.22	5.39
20	5.55	5.70	5.86	6.01	6.15	6.30	6.44	6.58	6.72	6.86
30	6.99	7.12	7.26	7.38	7.51	7.64	7.76	7.88	8.00	8.12
40	8.24	8.36	8.48	8.59	8.70	8.82	8.93	9.04	9.15	9.26
50	9.36	9.47	9.58	9.68	9.79	9.89	9.99	10.09	10.19	10.29
60	10.39	10.49	10.59	10.69	10.78	10.88	10.98	11.07	11.16	11.26
70	11.35	11.44	11.54	11.63	11.72	11.81	11.90	11.99	12.08	12.16
80	12.25	12.34	12.43	12.51	12.60	12.68	12.77	12.85	12.94	13.02
90	13.10	13.19	13.27	13.35	13.43	13.52	13.60	13.68	13.76	13.84
100	13.92	14.00	14.08	14.16	14.23	14.31	14.39	14.47	14.54	14.62
110	14.70	14.77	14.85	14.93	15.00	15.08	15.15	15.23	15.30	15.37
120	15.45	15.52	15.59	15.67	15.74	15.81	15.89	15.96	16.03	16.10
130	16.17	16.24	16.31	16.38	16.45	16.52	16.59	16.66	16.73	16.80
140	16.87	16.94	17.01	17.08	17.15	17.21	17.28	17.35	17.42	17.48
150	17.55	17.62	17.68	17.75	17.82	17.88	17.95	18.01	18.08	18.15
160	18.21	18.28	18.34	18.40	18.47	18.53	18.60	18.66	18.73	18.79
170	18.85	18.92	18.98	19.04	19.10	19.17	19.23	19.29	19.35	19.42
180	19.48	19.54	19.60	19.66	19.73	19.79	19.85	19.91	19.97	20.03
190	20.09	20.15	20.21	20.27	20.33	20.39	20.45	20.51	20.57	20.63
200	20.69	20.75	20.81	20.87	20.92	20.98	21.04	21.10	21.16	21.22
210	21.27	21.33	21.39	21.45	21.50	21.56	21.62	21.68	21.73	21.79
220	21.85	21.90	21.96	22.02	22.07	22.13	22.19	22.24	22.30	22.35
230	22.41	22.47	22.52	22.58	22.63	22.69	22.74	22.80	22.85	22.91
240	22.96	23.02	23.07	23.13	23.18	23.23	23.29	23.34	23.40	23.45
250	23.50	23.56	23.61	23.66	23.72	23.77	23.82	23.88	23.93	23.98
260	24.04	24.09	24.14	24.20	24.25	24.30	24.35	24.41	24.46	24.51
270	24.56	24.61	24.67	24.72	24.77	24.82	24.87	24.92	24.97	25.03
280	25.08	25.13	25.18	25.23	25.28	25.33	25.38	25.43	25.48	25.54
290	25.59	25.64	25.69	25.74	25.79	25.84	25.89	25.94	25.99	26.04
300	26.09	26.14	26.19	26.24	26.28	26.33	26.38	26.43	26.48	26.53
310	26.58	26.63	26.68	26.73	26.78	26.82	26.87	26.92	26.97	27.02
320	27.07	27.12	27.16	27.21	27.26	27.31	27.36	27.40	27.45	27.50
330	27.55	27.60	27.64	27.69	27.74	27.79	27.83	27.88	27.93	27.97
340	28.02	28.07	28.12	28.16	28.21	28.26	28.30	28.35	28.40	28.44
350	28.49	28.54	28.58	28.63	28.68	28.72	28.77	28.81	28.86	28.91
360	28.95	29.00	29.04	29.09	29.14	29.18	29.23	29.27	29.32	29.36
370	29.41	29.46	29.50	29.55	29.59	29.64	29.68	29.73	29.77	29.82
380	29.86	29.91	29.95	30.00	30.04	30.09	30.13	30.18	30.22	30.26
390	30.31	30.35	30.40	30.44	30.49	30.53	30.57	30.62	30.66	30.71
400	30.75	30.79	30.84	30.88	30.93	30.97	31.01	31.06	31.10	31.14
410	31.19	31.23	31.28	31.32	31.36	31.41	31.45	31.49	31.53	31.58
420	31.62	31.66	31.71	31.75	31.79	31.84	31.88	31.92	31.96	32.01
430	32.05	32.09	32.13	32.18	32.22	32.26	32.30	32.35	32.39	32.43
440	32.47	32.51	32.56	32.60	32.64	32.68	32.73	32.77	32.81	32.85
450	32.89	32.94	32.98	33.02	33.06	33.10	33.14	33.18	33.23	33.27
460	33.31	33.35	33.39	33.43	33.47	33.52	33.56	33.60	33.64	33.68
470	33.72	33.76	33.80	33.84	33.89	33.93	33.97	34.01	34.05	34.09
480	34.13	34.17	34.21	34.25	34.29	34.33	34.37	34.41	34.45	34.49
490	34.53	34.57	34.62	34.66	34.70	34.74	34.78	34.82	34.86	34.90

N	0	1	2	3	4	5	6	7	8	9
500	34.94	34.98	35.02	35.06	35.10	35.14	35.17	35.21	35.25	35.29
510	35.33	35.37	35.41	35.45	35.49	35.53	35.57	35.61	35.65	35.69
520	35.73	35.77	35.81	35.85	35.89	35.92	35.96	36.00	36.04	36.08
530	36.12	36.16	36.20	36.24	36.28	36.31	36.35	36.39	36.43	36.47
540	36.51	36.55	36.58	36.62	36.66	36.70	36.74	36.78	36.82	36.85
550	36.89	36.93	36.97	37.01	37.05	37.08	37.12	37.16	37.20	37.24
560	37.27	37.31	37.35	37.39	37.43	37.46	37.50	37.54	37.58	37.62
570	37.65	37.69	37.73	37.77	37.80	37.84	37.88	37.92	37.96	37.99
580	38.03	38.07	38.11	38.14	38.18	38.22	38.25	38.29	38.33	38.37
590	38.40	38.44	38.48	38.52	38.55	38.59	38.63	38.66	38.70	38.74
600	38.77	38.81	38.85	38.89	38.92	38.96	39.00	39.03	39.07	39.11
610	39.14	39.18	39.22	39.25	39.29	39.33	39.36	39.40	39.44	39.47
620	39.51	39.54	39.58	39.62	39.65	39.69	39.73	39.76	39.80	39.84
630	39.87	39.91	39.94	39.98	40.02	40.05	40.09	40.12	40.16	40.20
640	40.23	40.27	40.30	40.34	40.38	40.41	40.45	40.48	40.52	40.55
650	40.59	40.63	40.66	40.70	40.73	40.77	40.80	40.84	40.88	40.91
660	40.95	40.98	41.02	41.05	41.09	41.12	41.16	41.19	41.23	41.26
670	41.30	41.34	41.37	41.41	41.44	41.48	41.51	41.55	41.58	41.62
680	41.65	41.69	41.72	41.76	41.79	41.83	41.86	41.90	41.93	41.97
690	42.00	42.04	42.07	42.10	42.14	42.17	42.21	42.24	42.28	42.31
700	42.35	42.38	42.42	42.45	42.49	42.52	42.55	42.59	42.62	42.66
710	42.69	42.73	42.76	42.80	42.83	42.86	42.90	42.93	42.97	43.00
720	43.04	43.07	43.10	43.14	43.17	43.21	43.24	43.27	43.31	43.34
730	43.38	43.41	43.44	43.48	43.51	43.55	43.58	43.61	43.65	43.68
740	43.71	43.75	43.78	43.82	43.85	43.88	43.92	43.95	43.98	44.02
750	44.05	44.09	44.12	44.15	44.19	44.22	44.25	44.29	44.32	44.35
760	44.39	44.42	44.45	44.49	44.52	44.55	44.59	44.62	44.65	44.69
770	44.72	44.75	44.79	44.82	44.85	44.89	44.92	44.95	44.98	45.02
780	45.05	45.08	45.12	45.15	45.18	45.22	45.25	45.28	45.31	45.35
790	45.38	45.41	45.45	45.48	45.51	45.54	45.58	45.61	45.64	45.67
800	45.71	45.74	45.77	45.81	45.84	45.87	45.90	45.94	45.97	46.00
810	46.03	46.07	46.10	46.13	46.16	46.20	46.23	46.26	46.29	46.33
820	46.36	46.39	46.42	46.45	46.49	46.52	46.55	46.58	46.62	46.65
830	46.68	46.71	46.74	46.78	46.81	46.84	46.87	46.90	46.94	46.97
840	47.00	47.03	47.06	47.10	47.13	47.16	47.19	47.22	47.26	47.29
850	47.32	47.35	47.38	47.42	47.45	47.48	47.51	47.54	47.57	47.61
860	47.64	47.67	47.70	47.73	47.76	47.80	47.83	47.86	47.89	47.92
870	47.95	47.98	48.02	48.05	48.08	48.11	48.14	48.17	48.20	48.24
880	48.27	48.30	48.33	48.36	48.39	48.42	48.46	48.49	48.52	48.55
890	48.58	48.61	48.64	48.67	48.71	48.74	48.77	48.80	48.83	48.86
900	48.89	48.92	48.95	48.98	49.02	49.05	49.08	49.11	49.14	49.17
910	49.20	49.23	49.26	49.29	49.33	49.36	49.39	49.42	49.45	49.48
920	49.51	49.54	49.57	49.60	49.63	49.66	49.69	49.73	49.76	49.79
930	49.82	49.85	49.88	49.91	49.94	49.97	50.00	50.03	50.06	50.09
940	50.12	50.15	50.18	50.21	50.24	50.27	50.31	50.34	50.37	50.40
950	50.43	50.46	50.49	50.52	50.55	50.58	50.61	50.64	50.67	50.70
960	50.73	50.76	50.79	50.82	50.85	50.88	50.91	50.94	50.97	51.00
970	51.03	51.06	51.09	51.12	51.15	51.18	51.21	51.24	51.27	51.30
980	51.33	51.36	51.39	51.42	51.45	51.48	51.51	51.54	51.57	51.60
990	51.63	51.66	51.69	51.72	51.75	51.78	51.81	51.84	51.87	51.90

The most efficient way of obtaining graticules (and base data) on desired projection systems is to have them computer plotted. If such facilities or programs are unavailable, the graticule of any uncopyrighted projection may be copied. Only occasionally may a cartographer find it necessary to construct a projection, but more often it may be necessary to modify, extend, recenter, or fit some ready-made projection. A knowledge of how projections can be constructed manually is a considerable advantage in such circumstances.

However a projection is to be obtained, it must be done to a chosen scale. This is done by constructing the projection so that the scale along a standard line bears the proper relationship to that line on the sphere, or equals its length on the nominal globe. The length of the line segment on the projection or globe in relation to the length of the same line on a spherical earth is the principal scale. To determine this is very easy, of course; if D represents the diameter of the sphere, then (1) the length of a great circle (the equator or a pair of meridians) on the sphere is πD, and (2) the length of a parallel is the cosine of the latitude (ϕ) times the circumference of the sphere ($\cos\phi \times \pi D$). These may be determined either by (1) first calculating their true lengths on the earth and then reducing the values by the desired ratio or, (2) by first reducing the earth (radius or diameter) by the chosen ratio to the nominal globe and then calculating the map lengths desired.

The difference between a sphere and the spheroid need be taken into account only for maps with scales larger than 1:5,000,000. At that scale, some two-thirds of all points will lie within 1 mm of their spheroidal position if mapped on a sphere (Tobler).

The directions and tabular data in this appendix are arranged in four sections dealing with the construction of (1) cylindrical projections, (2) conic projections, (3) pseudocylindrical projections, and (4) azimuthal projections. Only some of the projection systems that are commonly employed for smaller-scale maps are included here. Directions for the construction of others may be found in technical treatises and manuals that will also usually contain the equations necessary for their programming for computer-plotter construction. Several such treatises are listed in the Selected References for Chapter 4.

Directions for the construction of the following projections are given as numbered.
A. Cylindrical Projections.
   1. Plane chart.
   2. Cylindrical equal-area.
B. Conic Projections.
   3. Lambert's conformal.
   4. Albers' (equal-area).
C. Pseudocylindrical Projections.
   5. Sinusoidal (equal-area).
   6. Mollweide's (equal-area).
   7. Flat polar quartic (equal-area).
D. Azimuthal Projections.
   8. Orthographic.
   9. Lambert's equal-area.
   10. Azimuthal equidistant.

A. CYLINDRICAL PROJECTIONS

In the conventional form of these projections all meridians are the same length and all parallels are the same length; merely the spacing varies. In practice, the length of the equator or standard parallel and the length of a meridian are determined. These are drawn at right angles to one another. The standard parallel is then subdivided for the

APPENDIX H
THE CONSTRUCTION OF MAP PROJECTIONS

longitudinal interval desired. The meridians are drawn through these points as parallel lines. The spacing of the parallels is then plotted along a meridian, and the parallels are drawn.

1. Plane Chart

The conventional form of this projection is more precisely called the equirectangular projection, the name plane chart being reserved for the phase wherein the equator is standard. In any case, its construction is relatively simple.

A standard parallel is chosen, and its length is calculated or determined from tables. The length of any parallel (ϕ) may be calculated by multiplying the circumference of the earth by the cosine of the latitude. It may also be obtained by referring to the table of lengths of the degrees of the parallel (Table C.1, Appendix C) and multiplying the value given for the latitude by the number of degrees the map is to extend. It must then, of course, be reduced to scale. The results, however determined, are then marked on a straight horizontal line. This is the standard parallel. Vertical lines are drawn through the points to establish the meridians. The other parallels are determined by pointing off the actual distance between the parallels (reduced to scale) as determined from the table of meridional parts (Table C.2, Appendix C).

If the equator is made the standard parallel, the projection will be made up of squares; any other standard parallels will make rectangles whose north-south dimension is the long one.

2. Cylindrical (equal-area)

The cylindrical equal-area projection is, like all cylindrical projections, relatively easy to construct, requiring only a straightedge, dividers, scale, and a triangle. In its conventional form the equator alone or any pair of parallels spaced equally from the equator may be chosen as standard. Since the angular deformation is zero along the great circle or small circles chosen as standard, these may be selected so that they pass through the areas of significance. If the projection is desired for only a portion of the earth rather than the whole, the projection is designed merely as a part of a larger, incomplete world projection.

This projection requires some calculation, but the formulas are elementary and are accomplished merely through the use of arithmetic. The general formulas for any form of the projection are as follows:

1. Length of all parallels is $2R\pi \times \cos\theta$.
2. Length of all meridians is $2R \div \cos\theta$.
3. Distance of each parallel from equator is $R \sin\phi \div \cos\theta$.

R is the radius of the generating globe of chosen *area scale;* θ is the standard parallel; ϕ is the latitude.

The procedure for construction is similar to that for other rectangular projections. Perpendicular lines are drawn to represent the equator and a meridian (Fig. H.1). The length of the parallel chosen as standard (*ab*) is marked off on the equator. The length of a meridian (*cd*) is determined. These dimensions define a rectangle forming the poles and the bounding meridian of the projection. The distances of the parallels from the equator are then laid off on a meridian, and the parallels are drawn parallel to the equator, as in quadrant *C*, Fig. H.1. The parallels are, of course, equally subdivided by the meridians as in quadrant *D*.

B. CONIC PROJECTIONS

Conic projections may be constructed either from tables that provide *X* and *Y* plane coordinate values with which to locate the intersections of the graticule, or by the use of a straightedge and a beam compass capable of drawing large arcs. In the latter procedure it is necessary to determine the radii of the arcs representing the parallels and the spacing of the meridians on the parallels. In most conventional conic projections the meridians are equally spaced along each parallel, and the parallels are arcs of circles that may or may not be concentric; the meridians are usually straight lines.

In conventional form all conic projections are symmetrical around the central meridian. Thus, it is only necessary to draw one side of the projection; the other side may be copied by either folding the paper or copying it onto another paper with the aid of a light table. Another easy procedure is to place a sheet of paper over the projection and prick the intersections of the graticule with a needle. The paper may then be "flopped over" and the points pricked through in their proper place on the other side of the central meridian.

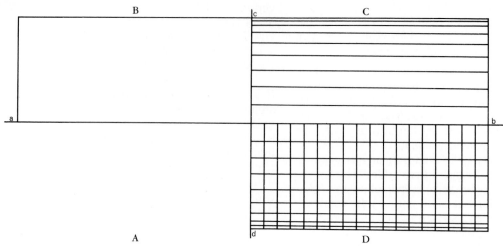

FIGURE H.1 *Construction of a cylindrical equal-area projection.*

3. Lambert's Conformal Conic

This projection is similar in appearance to Albers' and the simple conic. It, too, has straight-line meridians that meet at a common center: the parallels are arcs of circles, two of which are standard; and the parallels and meridians meet at right angles. The only difference is in the spacings of the parallels and meridians. In Lambert's conic they are so spaced as to satisfy the condition of conformality, that is, $a = b$ at every point. It is advisable, if a satisfactory distribution of scale error is desired, to space the standard parallels so that they include between them about two-thirds of the meridional section to be mapped.

The calculation of this projection requires considerable mathematical computations. Because of its relatively wide use for air-navigation maps, many tables for its construction with various standard parallels have been published. For example, a table for the construction of a map of the United States with standard parallels at 29 and 45 degrees is given in the U.S. Coast and Geodetic Survey, *Special Publication No. 52.*

Lambert's conic with standard parallels at 36 and 54 degrees is useful for middle-latitude areas. Table H.1 gives the radii of the parallels in meters for a map at a scale of 1:1. It is, of course, necessary to reduce each value to the scale desired.

To construct the projection draw a line, *cd* in Fig. H.2, which will be the central meridian. The line must be sufficiently long so that it will include

TABLE H.1

Table for the Construction of Lambert's Conformal Conic Projection. Standard Parallels 36° and 54°

Latitude	Radii, Meters	Latitude	Radii, Meters
75°	2 787 926	40°	6 833 183
70°	3 430 294	35°	7 386 250
65°	4 035 253	30°	7 946 911
60°	4 615 579	25°	8 519 065
55°	5 179 774	20°	9 106 796
50°	5 734 157	15°	9 714 516
45°	6 283 826		

From Deetz and Adams, *Elements of Map Projection.*

the center of the arcs of latitude. With a beam compass, describe arcs with radii, reduced to scale, taken from Table H.1.

To determine the meridians, it is necessary to calculate the chord distance on a lower parallel from its intersection with the central meridian (0 degree) to its intersection with an outer meridian. This is done by the following formula.

$$\text{Chord} = 2r \sin \frac{n\lambda}{2}$$

where $n = 0.7101$; $\lambda = $ longitude out from central meridian; and $r = $ radius of the parallel in question.

This procedure is similar to the general process for obtaining the length of a chord that subtends any central angle. The formula intro-

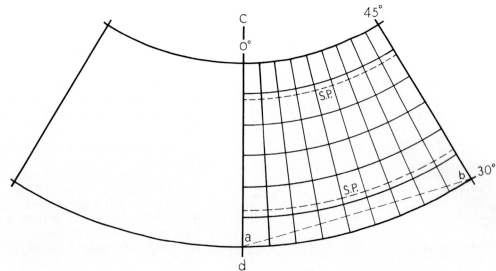

FIGURE H.2 *Construction of Lambert's conformal conic projection.*

duces a factor n to take into account the special meridian spacing of the projection.

EXAMPLE.

On parallel 30 degrees the chord of 45 degrees (see Fig. H.2) out from the central meridian =
1. $0.7101 \times 45° = 31° \, 57' \, 14'' = n\lambda$.
2. $n\lambda/2 = 15° \, 58' \, 37''$.
3. $\text{Sin } 15° \, 58' \, 37'' = 0.27534$.
4. $2r = 15{,}893{,}822$ m.
5. $15{,}893{,}822 \times 0.27534 = 4{,}376{,}200$ m.

The value thus determined, ab in Fig. H.2, is reduced by the desired scale ratio and measured out from the intersection of the parallel and the central meridian to the bounding meridian. If point b, thus located, is connected by means of a straightedge with the same center used in describing the parallels, this will determine the outer meridian. If a long straightedge is not available, the same procedure may be followed (determining the chord) for another parallel in the upper part of the map, and the two points, thus determined, may be joined by a straight line. This will produce the same result. Since the parallels are equally subdivided by the meridians, the other meridians may be easily located.

4. Albers' (equal-area)

The construction procedure for this projection is essentially the same as for the preceding projec-

tion, Lambert's conic. Like Lambert's, Albers' conic, in conventional orientation, is suited to representation of an area predominantly east-west in extent in the middle latitudes.

Table H.2 gives the radii of the parallels and the lengths of chords on two parallels for a map of the United States with standard parallels 29° 30' and 45° 30'. The scale of the table is 1:1, and the values are in meters. As in the preceding projection, reduction to the desired scale is necessary.

C. PSEUDOCYLINDRICAL PROJECTIONS

Most useful projections of this class are equivalent, and they are constructed to an area scale. The linear dimensions, for construction purposes, of a pseudocylindrical projection depend on the shape of the bounding line (usually a meridian) that encloses the projection. It is obvious that the axes of two dissimilar shapes would be different if both shapes enclose the same area, that is, if they were the same area scale. Most such projections have a vertical axis half the length of the horizontal axis; in the conventional "equatorial" form, this is the relation to be expected between a meridian and the equator. The relationship between the scale of the nominal globe and the particular projection being constructed is merely one that states the length of the central meridian on the projection

TABLE H.2

Table for Construction of Albers'
Equal-Area Projection with Standard
Parallels 29°30′ and 45°30′

Latitude	Radius of Parallel, Meters
20°	10 253 177
21°	10 145 579
22°	10 037 540
23°	9 929 080
24°	9 820 218
25°	9 710 969
26°	9 601 361
27°	9 491 409
28°	9 381 139
29°	9 270 576
29° 30′	9 215 188
30°	9 159 738
31°	9 048 648
32°	8 937 337
33°	8 825 827
34°	8 714 150
35°	8 602 328
36°	8 490 392
37°	8 378 377
38°	8 266 312
39°	8 154 228
40°	8 042 163
41°	7 930 152
42°	7 818 231
43°	7 706 444
44°	7 594 828
45°	7 483 426
45° 30′	7 427 822
46°	7 372 288
47°	7 261 459
48°	7 150 987
49°	7 040 925
50°	6 931 333
51°	6 822 264
52°	6 713 780

Longitude from Central Meridian	Chord Distances in Meters On Latitude 25°	On Latitude 45°
1°	102 185	78 745
5°	510 867	393 682
25°	2 547 270	1 962 966
30°		2 352 568

From Deetz and Adams, *Elements of Map Projection.*

compared to the radius (R) of the nominal globe of the same area scale. Since the equator is twice the length of the central meridian, no further calculation is necessary.

In the majority of the pseudocylindrical projections, the meridians are equally spaced along the parallels. The spacing of the parallels along the central meridian varies from projection to projection. These values are available in tabular form.

5. Sinusoidal

The sinusoidal projection is particularly simple to construct, since the spacings of the parallels on the central meridian and meridians on the parallels are the same (to scale) as they are on the earth. The length of the central meridian is 3.1416 times the radius (R) of a nominal globe of equal area. The equator is twice the length of the central meridian.

To construct the projection a horizontal line representing the equator, ab in Fig. H.3 is drawn twice the length of the central meridian. The equator is bisected, and at point o a perpendicular central meridian (cd) is constructed. The positions of the parallels on cd are determined by spacing them as they are on the globe. For a small-scale projection this means spacing them equally. Through the points thus established the parallels are drawn parallel to the equator as in quadrant B. The lengths of the various parallels are their true lengths (to scale) as they are on the earth and may be determined by multiplying the length of the equator (ab) by the cosine of the latitude. One-half this value is plotted on each side of the central meridian. The meridians are drawn by subdividing, as in quadrant C, each parallel equally, with dividers, and drawing smooth curves (with a French curve) through homologous points as in quadrant D.

6. Mollweide's

Mollweide's projection does not have the simple relationship to the sphere that characterizes the sinusoidal. The meridians of the sinusoidal are sine curves, which produce a pointed appearance, whereas in Mollweide's the meridians are ellipses, which provide a projection shape that is somewhat less of a radical departure from the globe impression. The length of the central meridian is 2.8284 times the radius (R) of the nominal globe of equal area. The equator is twice the length of the central meridian.

To construct the projection, a horizontal line representing the equator, ab in Fig. H.4, is drawn twice the length of the central meridian. The equator is bisected, and at point o a perpendicular central meridian (cd) is constructed. A circle whose radius is oc is constructed around point o. This contains a hemisphere. The spacing of the parallels on the central meridian is given in Table H.3, in which oc equals 1.

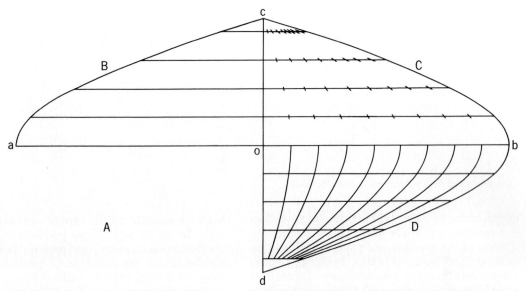

FIGURE H.3 Construction of a sinusoidal projection.

These positions are plotted on the central meridian, and the parallels are drawn through the points parallel to the equator. Each parallel is extended outside the hemisphere circle a distance equal to its length inside the circle. Thus, in quadrant *B* of Fig. H.4, *ef = fg, hi = ij, kl = lm,* and so on. Each parallel is subdivided equally with dividers, as in quadrant *C*, to establish the posi-

tion of the meridians. The meridians are drawn through homologous points, as in quadrant *D*, with the aid of a curve.

7. Flat Polar Quartic
This projection is representative of a group in which a line represents the pole in conventional form, but in which that line is less than half the

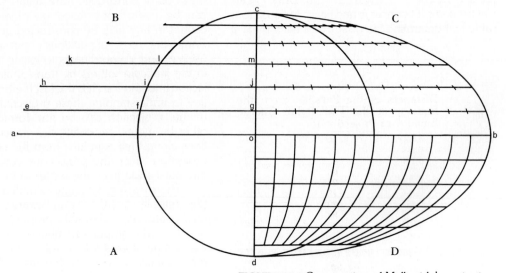

FIGURE H.4 Construction of Mollweide's projection.

TABLE H.3

Distances of the Parallels from the Equator in Mollweide's Projection (*oc* = 1)

0°	0.000	50°	0.651
5°	0.069	55°	0.708
10°	0.137	60°	0.762
15°	0.205	65°	0.814
20°	0.272	70°	0.862
25°	0.339	75°	0.906
30°	0.404	80°	0.945
35°	0.468	85°	0.978
40°	0.531	90°	1.000
45°	0.592		

From Deetz and Adams, *Elements of Map Projection*

TABLE H.5

Lengths of the Parallels in the Flat Polar Quartic Equal-Area Projection (*ob* = 1)

0°	1.000	50°	0.752
5°	0.998	55°	0.700
10°	0.990	60°	0.643
15°	0.979	65°	0.581
20°	0.961	70°	0.517
25°	0.939	75°	0.453
30°	0.911	80°	0.394
35°	0.879	85°	0.350
40°	0.842	90°	0.333
45°	0.800		

length of the equator and the bounding meridian is a complex curve. On this account, neither the spacing of the parallels nor their lengths can be easily derived by construction. Tables of parallel lengths and spacings are necessary. Tables H.4 and H.5 list the necessary information to construct the projection with a 5 degree grid interval. In the flat polar quartic equal-area projection, the length of the central meridian is 2.6513 times the radius (*R*) of a nominal globe of equal area. The equator is 2.2214 times the length of the central meridian, and the length of the line representing the pole is one-third the length of the equator.

To construct the projection, draw a vertical line, *cd* in Fig. H.5, representing the central meridian to scale. Draw a perpendicular *ab* through *o*, midway between *c* and *d*. Lay off distance *ao* equal to 1.1107 *cd*. Length *ob* = *ao*. Draw *fg* parallel to *ab* at *c*. Distances *fc* and *cg* = 1/3 *ao*. From Table H.4 determine the distance of each parallel

TABLE H.4

Distances of the Parallels from the Equator in the Flat Polar Quartic Equal-Area Projection (*oc* = 1)

0°	0.000	50°	0.668
5°	0.070	55°	0.727
10°	0.140	60°	0.784
15°	0.209	65°	0.837
20°	0.278	70°	0.886
25°	0.346	75°	0.930
30°	0.413	80°	0.966
35°	0.479	85°	0.991
40°	0.544	90°	1.000
45°	0.607		

From Coast and Geodetic Survey, *Special Publication* No. 245.

from the equator on the central meridian. In the table, length *oc* = 1. These positions are plotted on the central meridian, and through these points parallels to *ab* are drawn as in quadrant *B*. The length of each parallel is obtained from Table H.5 in which it is shown as a proportion of length *ob*. Subdivide each parallel equally with dividers as in quadrant *C* of Fig. H.5 to establish the positions of the meridians. The meridians are drawn through homologous points, as in quadrant *D*, with the aid of a curve.

D. AZIMUTHAL PROJECTIONS

Although azimuthal projections seem to have more in common with each other than with any other class of projections, the uses and common methods of construction are quite varied. Some can be easily constructed geometrically; some cannot in any way be constructed geometrically. Some are most expeditiously put together by using *X* and *Y* coordinates to locate intersections in the graticule; others by transforming one projection into another. It is this last method that is the key to understanding these projections. Any azimuthal projection can be transformed into any other by merely relocating the graticule intersections along their azimuths from the center of the projection, since the projections vary only as to the radial scale from the center of the projection.

There are a great many azimuthal projections (theoretically, an infinite number are possible), but only a few have desirable properties. Of these, only the orthographic, Lambert's equal-area and the azimuthal equidistant will be treated here as representative of construction methods.

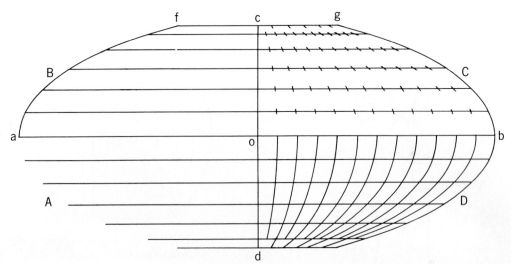

FIGURE H.5 Construction of a flat polar quartic equal-area projection.

8. Orthographic

The orthographic projection is rather like an architect's elevation. The principle of its construction can be seen in Fig. H.6 (showing the projection centered on the pole), where the latitudinal spacing on the globe is projected by parallel lines to the central meridian of the projection. Being an azimuthal projection, all great circles through the center are straight lines and azimuths from the center are correct. Since in this form the pole is the center, all great circles that pass through it are meridians; hence, all meridians on the projection are straight lines and are correctly arranged around the pole.

The construction of the projection centered on the equator is no more involved. The procedure is illustrated in Fig. H.7. The parallel spacing on the central meridian is the same as in the polar case, but the parallels are horizontal lines. The positions of the meridians on the parallels are carried over from the polar case as illustrated. Since all four quadrants are images of one another, only one need be drawn. The others may be traced.

The orthographic projection centered on the pole or the equator is seldom used. The projection is more often centered on some point of interest between the pole and the equator. An appropriate method is to trace a photograph of the globe cen-

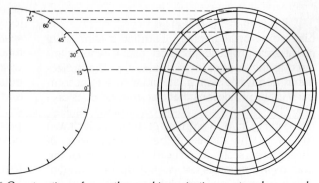

FIGURE H.6 Construction of an orthographic projection centered on a pole.

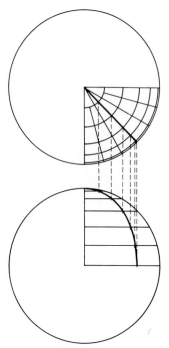

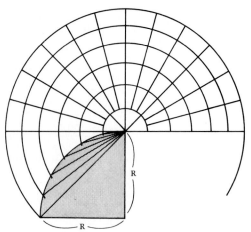

FIGURE H.8 *Construction of Lambert's equal-area projection centered on a pole.*

FIGURE H.7 *Construction of an orthographic projection centered on the equator.*

tered at the desired spot. A photograph of the globe is not a true orthographic, since some perspective convergence is bound to occur in the photographing, even when the camera is at a considerable distance. Nevertheless, the result is a true azimuthal projection and is very nearly the same as the orthographic. Since the only useful precise property is its azimuthality, nothing is lost by using a photograph.

9. Lambert's (equal-area)

Like the orthographic, Lambert's azimuthal equal-area projection is most useful when centered in the area of interest, although the projection is frequently seen in the polar case to accompany other projections that distort areas considerably. The polar case is easily constructed, as illustrated in Fig. H.8. A segment of the globe is drawn with R as the radius of the nominal globe of chosen area scale. The chord distances from the pole to the parallels are carried up to a tangent with a compass and establish the positions of the

parallels on the projection. The meridians, as in other azimuthal projections, are straight lines through the poles.

As with most of the azimuthal projections, the oblique case centered on some area of interest is the most useful. Coordinates are given in Table H.6 for a graticule centered at latitude 40 degrees, which is an appropriate place for maps of the United States or North America, among others, to be centered. When using a table, such as Table H.6, a convenient procedure is to construct first a grid for the xy coordinates at the desired scale. This makes much calculation unnecessary.

10. Azimuthal Equidistant

It is not difficult to construct the azimuthal equidistant projection centered on any spot. The polar case is constructed by first drawing an appropriate set of meridians and then constructing equally spaced circles concentric around the pole. This may be extended to include the whole earth, in which case the bounding circle is the opposite pole, as in Fig. H.9. As may be expected, the area and linear distortions become large as the periphery is approached. A world map centered on a pole is not a very appropriate use of the azimuthal equidistant projection.

For an oblique case the simplest procedure is to prepare first a stereographic projection centered at the desired latitude. This may then be trans-

TABLE H.6

Table for the Construction of Lambert's Equal-Area Projection Centered at Latitude 40°. Coordinates in Units of the Earth's Radius ($R = 1$)

Lat.	Long. 0°		Long. 10°		Long. 20°		Long. 30°		Long. 40°	
	x	y	x	y	x	y	x	y	x	y
90°	0	+0.845	0.000	+0.845	0.000	+0.845	0.000	+0.845	0.000	+0.845
80°	0	+0.684	0.032	+0.686	0.063	+0.692	0.093	+0.704	0.120	+0.718
70°	0	+0.518	0.062	+0.521	0.121	+0.533	0.175	+0.553	0.231	+0.580
60°	0	+0.347	0.088	+0.352	0.174	+0.369	0.257	+0.396	0.334	+0.434
50°	0	+0.174	0.112	+0.181	0.222	+0.200	0.328	+0.234	0.427	+0.280
40°	0	0.000	0.133	+0.007	0.264	+0.029	0.391	+0.067	0.510	+0.120
30°	0	−0.174	0.151	−0.166	0.300	−0.166	0.445	−0.102	0.582	−0.045
20°	0	−0.347	0.166	−0.338	0.331	−0.314	0.489	−0.272	0.642	−0.214
10°	0	−0.518	0.177	−0.509	0.353	−0.484	0.524	−0.442	0.689	−0.383
0°	0	−0.684	0.185	−0.675	0.369	−0.651	0.548	−0.610	0.722	−0.553
−10°	0	−0.845	—	—	—	—	—	—	—	—

Lat.	Long. 50°		Long. 60°		Long. 70°		Long. 80°		Long. 90°		Long. 100°	
	x	y	x	y	x	y	x	y	x	y	x	y
90°	0.000	+0.845	0.000	+0.845	0.000	+0.845	0.000	+0.845	0.000	+0.845	0.000	+0.845
80°	0.143	+0.736	0.163	+0.758	0.178	+0.782	0.188	+0.808	0.192	+0.834	—	—
70°	0.278	+0.613	0.318	+0.646	0.349	+0.701	0.371	+0.751	0.382	+0.804	—	—
60°	0.403	+0.481	0.463	+0.538	0.511	+0.602	0.547	+0.674	0.567	+0.752	0.570	+0.833
50°	0.518	+0.338	0.583	+0.408	0.663	+0.489	0.713	+0.580	0.744	+0.679	0.755	+0.785
40°	0.620	+0.186	0.702	+0.267	0.801	+0.361	—	—	—	—	—	—
30°	0.710	+0.027	0.825	+0.115	—	—	—	—	—	—	—	—
20°	0.785	−0.138	—	—	—	—	—	—	—	—	—	—
10°	0.844	−0.307	—	—	—	—	—	—	—	—	—	—
0°	0.887	−0.478	—	—	—	—	—	—	—	—	—	—

Adapted from Deetz and Adams.

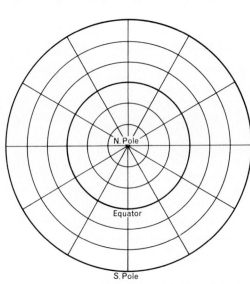

FIGURE H.9 *An azimuthal equidistant projection centered on a pole.*

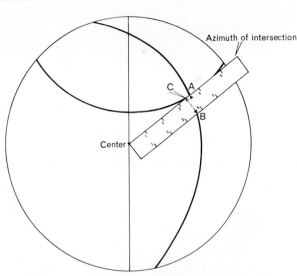

FIGURE H.10 *The transformation of one azimuthal to another. A is the distance from the center to a point according to the radial scale of the stereographic. B is the distance of the same point according to the radial scale of the equidistant. C is the location of the point along the same aximuth on the equidistant.*

formed (to any other azimuthal projection) by merely relocating the positions of the intersections of the new graticule along their azimuths from the center. This is accomplished by marking off on one edge of a strip of paper or plastic the radial scale of the stereographic and, on the other edge, the radial scale of the equidistant. The strip is then placed on the stereographic, and the distance of an intersection in the graticule from the center is noted. This distance is transferred to the radial scale of the equidistant and the position plotted *along the same azimuth.* Figure H.10 illustrates this procedure. Of course, the scale of the new projection may be made different, if desired, at the same time.

Positions outside the inner hemisphere of the stereographic may be located in the following manner. When the equidistant hemisphere has been completed, the stereographic is no longer needed. The position of every point on the earth is obviously 180 degrees from its antipode; every point and its antipode lie on a great circle through the center of the projection; the diameter of the hemisphere is 180 degrees; and the scale is uniform from the center. Thus, all that is necessary is to mark on a straightedge the diameter of the hemisphere and, keeping the edge on the center of the projection, locate all the outer intersections of the graticule from their antipodes in the inner hemisphere.

The law of deformation was developed by M. A. Tissot and appears in full in his *Mémoire sur la représentation des surfaces et les projections des cartes géographiques,* Paris, 1881, which includes 60 pages of deformation tables for various projections. The late Oscar S. Adams, a noted authority on the mathematics of map projections, included an account of it in his "General Theory of Polyconic Projections," U.S. Coast and Geodetic Survey *Special Publication 57,* Washington, D.C. 1934, pp. 153–163.

The following demonstration of the proof of the law of deformation consists of slightly reworded extracts taken from the above source with the permission of the Director, U.S. Coast and Geodetic Survey. The modifications are those of the authors.

To represent one surface on another, it is necessary to imagine that each surface is composed of two systems of lines that divide them into infinitesimal parallelograms. To each line of the first surface is made to correspond one of the lines of the second. The intersection of two lines of the different systems on the one surface and the intersection of the two corresponding lines on the other, therefore, determine two corresponding points. The totality of the points of the second surface, which correspond to the points of the first, forms the representation or the projection of the first surface. Different methods of representation are obtained by varying the two series of lines that form the graticule on one of the surfaces.

If two surfaces are not applicable to each other, that is, cannot be transformed without compression or expansion, it is impossible to choose a method of projection where there is similarity between every figure traced on the first and the corresponding figure on the second. On the other hand, whatever the two surfaces may be, there exists an infinity of systems of projection that preserve the angles; consequently, each *infinitely small* figure and its representation are similar to each other. There is also an infinity of systems that preserve the areas. These two classes of projections are exceptions, however. In a method of projection being taken by chance, the angles will be generally changed, except possibly at particular points, and the corresponding areas will not have a constant ratio to each other. The lengths on the one surface will thus be altered on the second.

Consider two curves, one on one surface and one on the other, which correspond to each other. In Fig. I.1 let O and M be two points of the one, O' and M' the corresponding points of the other, and let OT be the tangent at O to the first curve. If the point M lies infinitely close to the point O, then point M' will approach infinitely close to the point O'. The ratio of the length of the arc $O'M'$ to that of the arc OM will therefore tend toward a certain limit; this limit is what is called the ratio of lengths at the point O on the curve OM or in the direction OT. In a system of projection preserving the angles, the ratio of lengths thus defined has the same value for all directions at a given point; but it varies with the position of this point, unless the two surfaces are applicable to each other. When the representation does not preserve the angles except at particular points, the ratio of lengths at all other points changes with the direction.

The deformation produced around each point is subjected to a law that depends neither on the nature of the surfaces nor on the method of projection.

APPENDIX I
PROOF OF TISSOT'S LAW OF DEFORMATION

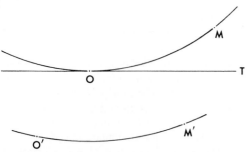

FIGURE I.1 *A curve (above) and its projection (below).*

Every representation of one surface on another can be replaced by an infinity of orthogonal projections, each made at a suitable scale.

It is noted, first, that there always exists at every point of the first surface two tangents perpendicular to each other such that the directions that correspond to them on the second surface also intersect at right angles. In Fig. I.2 let *CE* and *OD* be two tangents perpendicular to each other at the point *O* on the first surface; let *C'E'* and *O'D'* be the corresponding tangents on the second surface. Suppose, further, that of the two angles *C'O'D'* and *D'O'E'*, the first is acute, and imagine that a right angle having its vertex at *O* turns from left to right around this point in the plane *CDE* starting from the position *COD* and

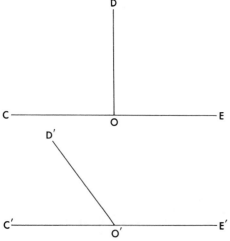

FIGURE I.2 *Two tangents at right angles and their projections.*

arriving at the position *DOE*. The corresponding angle in the plane tangent at *O* to the second surface will first coincide with *C'O'D'* and will be acute; in its final position it will coincide with *D'O'E'* and will be obtuse; within the interval it will have passed through a right angle. Therefore, there exists a system of two tangents satisfying the condition stated, except at certain singular points. From this property it may be concluded that in every system of representation there is on the first of the two surfaces a system of two series of orthogonal curves whose projections on the second surface are also orthogonal. The two surfaces are thus divided into infinitesimal rectangles that correspond the one to the other.

With this fact established, let *M* be a point in Fig. I.3 infinitely near to *O* on the first surface and let *OPMQ* be that one of the infinitesimal rectangles that has just been described that has *OM* as a diagonal. Then let *OP'M'Q'* be the rectangle on the second surface that corresponds to *OPMQ* on the first. Move the second surface and place it in such a position that *O* on each surface coincides, and that the sides *OP'* and *OQ'* on the second surface fall on the sides *OP* and *OQ* on the first surface. Designate as *N* the point of intersection of the lines *OM'* and *PM*. This point can be considered as the orthogonal projection of the point *M* if the plane of the rectangle *OPMQ* were turned through a suitable angle with *OP* as an axis. But this angle, which depends only on the ratio of the two lines *NP* and *MP*, is the same whatever point *M* may be; for denoting, respectively, by *a* and *b* the ratios of the lengths in the directions *OP* and *OQ*—that is, on setting

$$\frac{OP'}{OP} = a \text{ and } \frac{OQ'}{OQ} = b$$

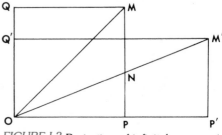

FIGURE I.3 *Projection of infinitely near points.*

then

$$\frac{NP}{M'P'} = \frac{OP}{OP'} = \frac{1}{a}, \text{ and } \frac{MP}{M'P'} = \frac{OQ}{OQ'} = \frac{1}{b}$$

and, consequently,

$$\frac{NP}{MP} = \frac{b}{a}$$

Thus if M moves on an infinitesimal curve traced around O, the locus described by N is obtained by turning this curve through a certain angle around OP as an axis and by then projecting orthogonally on the plane tangent at O. On the other hand,

$$\frac{OM'}{ON} = \frac{OP'}{OP} = a$$

so that the locus of the points M' is homothetic to that of the points N; the center of similitude is O, and the ratio of similitude has the value a. The representation of the infinitesimal figure described by the point M is then in reality an orthogonal projection of this figure made on a suitable scale, or the figure formed by the points N and that formed by the points M' are formed by parallel sections of the same cone. Any map projection can therefore be considered as produced by the juxtaposition of orthogonal projections of the surface elements of the earth sphere, provided that, from one element to the other, both the scale of the reduction and the position of the element with respect to the plane of the map are varied.

Of all the right angles that are formed by the tangents at the point O, those of the lines OP and OQ and their prolongations are the only ones one side of which remains parallel to the tangent plane after the rotation that was just described. These are, therefore, the only right-angled tangents that are projected into right angles. An addition to the proposition that has just been proved can now be stated, and the whole can be expressed in the following form: at every point of the spherical surface that is to be represented, there are two perpendicular tangents and, if the angles are not preserved, there are only two, such that those that correspond to them on the other surface (the map plane) also intersect at right angles. Thus, on each of the two surfaces there is a system of orthogonal trajectories and, if the method of representation does not preserve the angles, there is only one of them, the projections of which on the other surface are also orthogonal.

The two perpendicular tangents, the angle between which is not altered by the projection, are designated as first and second principal tangents. The ratio of lengths in the directions of these two tangents is designated respectively by a and b. Ratio a is assumed to be greater than b.

If the infinitesimal curve drawn around the point O is a circumference of which O is the center, the representation of this curve will be an ellipse, the axes of which will fall on the principal tangents, and these will have the values $2a$ and $2b$ when the radius of the circle is taken as unity. This ellipse constitutes at each point a sort of indicatrix of the system of projection.

In place of (1) projecting orthogonally the circumference (the locus of the points M in Fig. I.3), which gives the ellipse (the locus of the points N), and (2) then increasing this in the ratio of a to unity (which gives the locus of the points M'), one may perform these two operations in the inverse order. Then, in Fig. I.4 the point M' of the elliptic indicatrix can be obtained that corresponds to a given point M of the circle. This is done by prolonging the radius OM until it meets at R the circumference described on the major axis as diameter, and then by dropping a perpendicular from R on OA, the semimajor axis and, finally, by reducing this perpendicular RS, starting from its foot S in the ratio of b to a. The point M' thus determined is the required point.

If in Fig. I.4 OM' is drawn, then the angles AOM and AOM' that correspond on the two surfaces may be designated respectively as U and U'.

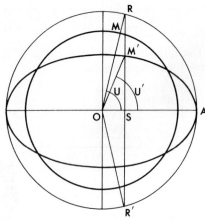

FIGURE I.4 Tissot's indicatrix.

Inasmuch as the second is the smaller of the two, it may be seen that the representation diminishes all the acute angles, one side of which coincides with the first principal tangent. Between U and U' there is, moreover, the relation

$$\tan U' = \frac{b}{a} \tan U$$

since

$$\tan U = \frac{RS}{OS}$$

$$\tan U' = \frac{M'S}{OS}$$

and, consequently,

$$\tan U' = \frac{M'S}{RS} \tan U = \frac{b}{a} \tan U$$

Prolong the line RS to R' and then join O and R'. The two triangles ORM' and $OR'M'$ give

$$\sin(U - U') = \frac{a - b}{a + b} \sin(U + U')$$

which is obtained by equating two expressions for the ratio of the areas of the triangles. The same relation follows at once analytically from the tangent relation first given. The angle U increasing from zero to $\pi/2$, its alteration $U - U'$ increases from zero up to a certain value ω, then decreases to zero. The maximum is produced at the moment when the sum $U + U'$ becomes equal to $\pi/2$. From the tangent formula the following are their values.

$$\tan U = \frac{\sqrt{a}}{\sqrt{b}} \quad \text{and} \quad \tan U' = \frac{\sqrt{b}}{\sqrt{a}}$$

The quantity ω can be computed by any one of the following formulas.

$$\sin \omega = \frac{a - b}{a + b}$$

$$\cos \omega = \frac{2\sqrt{ab}}{a + b}$$

$$\tan \omega = \frac{a - b}{2\sqrt{ab}}$$

$$\tan \frac{\omega}{2} = \frac{\sqrt{a} - \sqrt{b}}{\sqrt{a} + \sqrt{b}}$$

$$\tan\left(\frac{\pi}{4} + \frac{\omega}{2}\right) = \frac{\sqrt{a}}{\sqrt{b}} \quad \text{and} \quad \tan\left(\frac{\pi}{4} - \frac{\omega}{2}\right) = \frac{\sqrt{b}}{\sqrt{a}}$$

If one wishes to calculate directly the alteration that any given angle U is subject to, he may use one of the two formulas

$$\tan(U - U') = \frac{(a - b) \tan U}{a + b \tan {}^2U'}$$

or

$$\tan(U - U') = \frac{(a - b) \sin 2U}{a + b + (a - b) \cos 2U}$$

which follow immediately from the previous formulas.

Consider now an angle MON in Figs. I.5 and I.6, which has for sides neither one nor the other of the principal tangents OA and OB. Assume the two directions OM and ON to be to the right of OB and the one of them OM above OA. According to whether the other ON will be above OA (Fig. I.5) or below OA (Fig. I.6), the corresponding angle $M'ON'$ can be calculated by taking the difference or the sum of the angles AOM' and AON', which would be given by the formula stated above. The alteration $MON - M'ON'$ would also in the first case be the difference and, in the second case, would be the sum of the alterations of the angles AOM and AON. When the angle AON (Fig. I.5) is equal to the angle BOM', its alteration is the same as that of the angle AOM, so that the angle MON will then be reproduced in its true magnitude by the angle $M'ON'$. Thus to every given direction another can be joined and only one other, such that their angle is preserved in the projection.

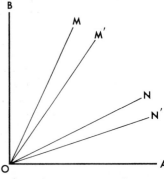

FIGURE 1.5 Angular change in projections, first case.

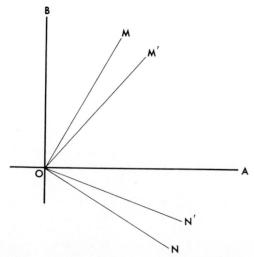

FIGURE I.6 *Angular change in projections, second case.*

However, the second direction will coincide with the first when it makes with *OA* the angle that has been denoted by *U*.

The angle the most altered is that which this direction forms with the point symmetric to it with respect to *OA*; it is represented on the projection by its supplement. The maximum alteration thus produced is equal to 2ω. This can never be found applicable to two directions that are perpendicular to each other.

The length *OM* in Fig. I.4 having been taken as unity, the ratio of lengths in the direction *OM* is measured by *OM'*. If this ratio is designated as *r*, it may be calculated by one of the formulas

$$r \cos U' = a \cos U$$
$$r \sin U' = b \sin U$$

or

$$r^2 = a^2 \cos^2 U + b^2 \sin^2 U.$$

There is also among *r*, *U*, and the alteration $U - U'$ of the angle *U* the relation

$$2r \sin(U - U') = (a - b) \sin 2U$$

which expresses that, in the triangle *ORM'*, the sines of two of the angles are to each other as the sides opposite.

The maximum and the minimum of *r* correspond to the principal tangents and are, respectively, *a* and *b*.

If *r* and r_1 are the ratios of lengths in two directions at right angles to each other and if θ is the alteration that the right angle formed by these two directions is subjected to then, from the properties of conjugate diameters in the ellipse,

$$r^2 + r_1^2 = a^2 + b^2$$
$$rr_1 \cos \theta = ab$$

In terms of the scale along a parallel (*h*) and along a meridian (*k*) at a point on a projection, the semiaxes are given by the equations

$$a^2 + b^2 = h^2 + k^2$$
$$ab = hk \cos \theta$$

INDEX